ENERGY CONSERVATION IN HEATING, COOLING, AND VENTILATING BUILDINGS

SERIES IN THERMAL AND FLUIDS ENGINEERING

JAMES P. HARTNETT and THOMAS F. IRVINE, JR., Editors
JACK P. HOLMAN, Senior Consulting Editor

Cebeci and Bradshaw	• Momentum Transfer in Boundary Layers
Chang	• Control of Flow Separation: Energy Conservation, Operational Efficiency, and Safety
Chi	• Heat Pipe Theory and Practice: A Sourcebook
Eckert and Goldstein	• Measurements in Heat Transfer, 2d edition
Edwards, Denny, and Mills	• Transfer Processes: An Introduction to Diffusion, Convection, and Radiation
Fitch and Surjaatmadja	• Introduction to Fluid Logic
Ginoux	• Two-Phase Flows and Heat Transfer with Application to Nuclear Reactor Design Problems
Hsu and Graham	• Transport Processes in Boiling and Two-Phase Systems, Including Near-Critical Fluids
Kestin	• A Course in Thermodynamics, revised printing
Kreith and Kreider	• Principles of Solar Engineering
Lu	• Introduction to the Mechanics of Viscous Fluids
Moore and Sieverding	• Two-Phase Steam Flow in Turbines and Separators: Theory, Instrumentation, Engineering
Nogotov	• Applications of Numerical Heat Transfer
Richards	• Measurement of Unsteady Fluid Dynamic Phenomena
Sparrow and Cess	• Radiation Heat Transfer, augmented edition
Tien and Lienhard	• Statistical Thermodynamics, revised printing
Wirz and Smolderen	• Numerical Methods in Fluid Dynamics

PROCEEDINGS

Hoogendoorn and Afgan	• Energy Conservation in Heating, Cooling, and Ventilating Buildings: Heat and Mass Transfer Techniques and Alternatives
Keairns	• Fluidization Technology
Spalding and Afgan	• Heat Transfer and Turbulent Buoyant Convection: Studies and Applications for Natural Environment, Buildings, Engineering Systems
Zarić	• Thermal Effluent Disposal from Power Generation

A publication of the International Centre for Heat and Mass Transfer
Belgrade

ENERGY CONSERVATION IN HEATING, COOLING, AND VENTILATING BUILDINGS

Heat and Mass Transfer Techniques and Alternatives

Volume 1

Edited by

C. J. Hoogendoorn

Applied Physics Department
Delft University of Technology
Delft, The Netherlands

and

N. H. Afgan

University of Belgrade
Belgrade, Yugoslavia

HEMISPHERE PUBLISHING CORPORATION

Washington   London

**ENERGY CONSERVATION IN HEATING, COOLING,
AND VENTILATING BUILDINGS**

Copyright © 1978 by Hemisphere Publishing Corporation. All rights reserved.
Printed in the United States of America. No part of this publication may be
reproduced, stored in a retrieval system, or transmitted, in any form or by
any means, electronic, mechanical, photocopying, recording, or otherwise,
without the prior written permission of the publisher.

1 2 3 4 5 6 7 8 9 0   D O D O   7 8 3 2 1 0 9 8

Library of Congress Cataloging in Publication Data

Main entry under title:

Energy conservation in heating, cooling, and ventilating
  buildings.

  Lectures and papers presented at a seminar sponsored
by the International Centre for Heat and Mass Transfer,
held at Dubrovnik, Yugoslavia, Aug. 29–Sept. 2, 1977.
  Includes indexes.
  1. Buildings—Energy conservation—Congresses.
2. Heat—Transmission—Congresses. 3. Mass transfer—
Congresses. I. Hoogendoorn, C. J. II. Afgan, Naim.
III. International Center for Heat and Mass Transfer.
TJ163.5.B84E528   697   78-1108
ISBN 0-89116-094-9 (vol. 1)

697
856
v. 1

CONTENTS

v

VENTILATION AND AIR MOVEMENT INSIDE BUILDINGS

MATHEMATICAL MODELLING AND EVALUATION OF ENERGY REQUIREMENTS IN HEATING AND COOLING OF BUILDINGS

PREFACE

The International Centre for Heat and Mass Transfer (ICHMT) organizes an annual seminar on a subject related to transport phenomena. The one on "Heat and Mass Transfer in Buildings," held at Dubrovnik, Yugoslavia, August 29–September 2, 1977, was the forum for presentation of the lectures and papers contained in these volumes. Dealing with the energy requirements for heating, cooling, and ventilating buildings, these papers are from heat and mass transfer scientists, from engineers working in the field of heating and ventilation, and from building engineers.

The recent awareness of the need to conserve our primary energy resources has emphasized the necessity of energy conservation strategies in industrialized societies. Many studies on the energy problem have been made. The 1975 ICHMT seminar, "Future Energy Production—Heat and Mass Transfer Problems,"* followed the energy crisis of 1973 and stressed the need to find new sources of energy. A recent report from the Workshop on Alternative Energy Strategies (WAES) stresses that in the next 25 years decline in the supply of primary energy will cause serious problems. Various energy scenarios have been made for the period 1985–2000. All show a decline in oil and natural gas production as well as the need to change over to drastically different energy policies. As these changes often require lead times of five to ten years, timely decisions are needed to avoid catastrophic consequences of present policies. This is, in fact, the greatest challenge our technological societies have to face in the next 25 years.

As use of primary energy for heating and cooling buildings is about one-third of total consumption for many countries, we have here an important field for changes in energy policies. Building practice was until recently based on a large and relatively cheap energy supply. In the future we will still have sufficient energy, certainly if we include that derived from solar and fusion sources, but it will not be cheap or easily available. If we are to introduce new energy-conserving techniques on a large scale in building practice, we must make timely decisions as necessary. As lifetimes of buildings are from 25 to over 100 years, decisions taken now will have to be based on future energy requirements. This is certainly very important, because our well being and the full development of our communities strongly depend on the environmental control of our living and working areas. Not only technical, but also economic and human factors are involved.

This volume contains papers on the many factors that the energy requirements of buildings involve. These range from human comfort conditions and economic aspects to heat and moisture transfer in building materials, ventilation of buildings, mathematical modelling, and new techniques like heat pumps and solar energy.

We would like to acknowledge the seminar committee members who cooperated in

---

*Published as J. C. Denton and N. H. Afgan (eds.), *Future Energy Production Systems*, vols. 1 and 2, Hemisphere, Washington, 1976.

setting up the programme, selecting the papers, and organizing the various sessions of the meeting: V. P. Motulevich, F. Kreith, A. F. E. Wise, K. Gertis, and B. Givoni. They all contributed review lectures in their special fields, which are included in this volume.

C. J. Hoogendoorn
Chairman, Seminar Committee

N. H. Afgan
Scientific Secretary, ICHMT

ENERGY CONSERVATION
IN HEATING AND
COOLING OF BUILDINGS

C. J. HOOGENDOORN

Department of Applied Physics
Technical University of Delft
Delft, The Netherlands

ABSTRACT

The need of conserving our primary energy resources has led to an increased awereness of the need of energy savings in heating of buildings.
Effects should be made to minimise the use of primary energy, within the
constraints of human comfort, economics and architectural design.
Many factors play a role in the energy requirements for heating and
cooling of buildings. Thermal factors influence the transmission and ventilation losses. These losses will depend on the micro-meteorological conditions outside the building. Temperature control in rooms together with the
thermal transient behaviour of a building can be quite important. Different
technical factors can lead to energy savings, of course better insulation,
but also the use of district heating and heat pumps. Solar loads are important both for cooling and heating of buildings. The application of solar
heating to save on primary energy resources finds widespread interest nowa-
days.
To evaluate the effects of the many factors that play a role mathematical modelling of the thermal behaviour of buildings is a very fruitful
tool. Coupled to hourly micro-meteorological data optimisation studies can
easily be made. From a case study in the Netherlands the accuracy of these
models are discussed. From the same study it can also be concluded that
energy savings of 33 % could easily be obtained by better insulation
techniques.
Finally it is stressed that in an early stage of studies on new urban
developments a "thermal" urban planning should be done. These studies should
include all new energy saving techniques together with a thermal evaluation
of new architectural designs.
Realisation of the goal of energy conservation will require cooperation of
architects, engineers and scientists.

INTRODUCTION

World-wide concern about depletion of our energy resources has stressed
the need of energy conservation strategies. One of the main fields of primary
energy consumption is heating and cooling of buildings. This can be shown from
data of a study [1] in the Netherlands on the consumption of primary energy
for different purposes in 1972 (see table 1). It reveals that a total of 33.6%
of the primary energy is required for residental heating, including hot
water supply, and heating of commercial buildings and greenhouses. Also for
other countries, such as the U.S.A. and Germany it has been reported that
about 1/3 of the primary energy is needed for heating of buildings.

1

Table 1. Primary Energy Consumption in the Netherlands, 1972

INDUSTRY	32 %
ELECTRICITY	20 %
HEATING: RESIDENTIAL AND COMMERC.	30 %
HORTICULTURE	3.6%
TRANSPORTATION	12 %
OTHERS	2.4%

It is clear that there is a great need to save on the utilization of such large quantities of primary energy, which in general is used only in the form of low-grade thermal energy at temperature levels between 50 and 120°C.

This use of thermal energy is primarily associated with the maintenance of comfortable ambient conditions for the occupants of the buildings. These buildings cover a wide range, such as single-family homes, high-rise apartment buildings, offices, workshops, hospitals, schools, stores, warehouses, greenhouses and stables. In general, they include all man-made enclosures containing living beings.

The goal of energy conservation in heating and cooling of buildings is to minimise the use of primary energy resources by: minimising the net losses from the building and by maximising the use of non-depletable energy sources (like solar and wind energy), and by using more efficient techniques (like district heating) within the constraints of human comfort and economics.

This goal could also comprise the energy content of building materials and the energy needed in the construction phase. However, in a recent study at Delft University we found that on the basis of a 30-year lifetime of buildings this can be neglected, because it is only 10% of the total energy required over the 30-year period. This holds at least within normal economic constraints. The total energy required might be reduced, for example, by making extremely thick walls, but this is unrealistic.

For the Netherlands the energy requirement per unit time for heating of buildings was for the year 1972 on an average about 1.7 kW per capita. This value is much higher than 25 years earlier. This can be attributed to two causes. One is that the increase in economic prosperity increased the housing space available from about 30 m³ to 60 m³ per capita. Secondly, over the same period there was an increase in indoor temperature for living areas in winter time from 18 to 22°C, at an average outside temperature of 6°C. Here one can speak of thermal prosperity, also reflecting changes in clothing habits.

Finally, in that period there was an increase in relative window area in the outside wall. For economic reasons single-pane windows are in general use in the Netherlands.

The general awareness in recent years that energy conservation is important has also affected the heating and cooling of buildings. Architects, building engineers, heating and ventilation engineers together with scientists working in the field of heat and mass transfer will contribute to achieving the goal of energy conservation.

In general it is expected that primary energy savings of 25 up to 40% are easily possible in this area [1,2,8]. Various countries have started national programmes on research and development to realise the incentives. A typical example is the Swedish programme on energy conservation in buildings [3].

THERMAL FACTORS

Let us consider how the energy requirement for heating of living areas is determind. It is based on the human comfort condition, which are related to the heat and mass transfer from the human body to its surroundings. Depending on human activity and clothing this will set requirements on the air and radiant temperatures of the enclosure, on the air humidity and on the air velocities (draught).

The net heat losses from the enclosure will depend on its temperature and are due to the three main forms of heat transfer.

(a). Transmission losses. Owing to conduction through the walls, roof and floor, heat is lost to a cooler environment.
The thermal transport in structural materials as well as the thermal capacity of the walls play an important role.
In general these transmission losses take place through complicated structures, composite materials and walls with internal air gaps. Owing to connections of floors and walls, cantilevered constructions and colomns in façades, thermal bridges may locally give rise to high heat losses in outside walls. In a good design these thermal bridges should be minimised.

(b). Convection and ventilation losses. Air movements inside and outside the building will have two main effects. On the one hand they will affect the human comfort conditions inside the enclosure. On the other and they will influence the ventilation losses of a building. Less important is their effect on the convection from air to a well-insulated wall. However, for single-pane windows the natural convection along a cold window will be important.

(c). Radiation. Solar radiation may reach the enclosure through windows. In winter this can be a positive gain in the energy required for heating. In summer it may require additional cooling, depending on the thermal transient characteristics of the building. View factors and shading are important in calculating the solar load. Transmission and reflection from special kinds of glass can influence the thermal comfort in summer and winter time. Furthermore, thermal radiation exchanges between surfaces in the enclosures can be significant. The radiation temperature has a direct effect in human comfort conditions.

HEAT LOSS MEASUREMENTS AND CALCULATIONS

A study on the relative effects of the various factors influencing the heat losses of houses has recently been made by the Institute of Applied Physics, TNO. in the Netherlands [4].

This study has been sponsored by a "Foundation for Building Research". In this organisation the Dutch building industries cooperate to stimulate research activities. Energy conservation in buildings is an important item in these activities.

A number of newly built single-family homes in the town of Oss were equiped with different heating systems, including a solar heater, and were differently insulated. These homes were fully instrumented with thermocouples, heat-flux meters and flow meters.

During a 3-month winter period (January - April 1976) continuous measurements
were done using a data-logger. The data were evaluated by a computer programme
making heat balances and calculating the k-values. Fig. 7 at the end of this
paper shows the front face of the houses with solar collectors. The total
building volume of each house is about 350 m^3.

The reduction in measured heat losses for the well-insulated houses
compared with the poorly-insulated houses was about 33%. These reductions
were obtained by simple and standard methods of insulation, Table 2 gives
some k-values for these houses.

Table 2. Insulation values k of houses tested.

	Poorly-insulated house	Well-insulated house
WALLS	1.7	0.55
ROOF	1.4	0.58
FLOOR	2.1	0.62
SINGLE P. WINDOW	5.8	–
DOUBLE P. WINDOW	–	3.3

From the data a comparison could also be made between measured k-values and
k-values on the basis of material properties. It was found that the k-values
measured at different points showed a great spread. Fig. 1 gives some data
averaged over one week and measured at points in the roofs of the four houses.
However, the mean value for k-roof = 0.58 W/m^2°C compares well with the
calculated k_r = 0.6 W/m^2°C.

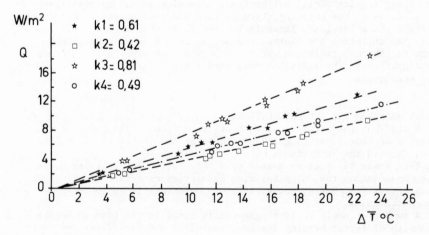

Fig. 1   Heat losses through roofs of 4 identical, well insulated houses.
           Data averaged over periods of 1 week.

Fig. 2 gives the mean of the measured heat losses for the well-insulated
walls. The k-values thus found agreed reasonably well with the expected ones.
The effects of non-stationary heating and cooling of the walls on the measured
heat fluxes were probably less important when averaged over periods of 1 week.

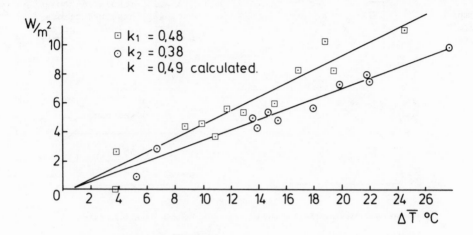

Fig. 2   Heat losses through walls of 2 well insulated houses. Data
            averaged over periods of 1 week.

The total heat losses through walls, floor, roof and by ventilation amounted
to 0.36 kW/°C for the poorly and 0.24 kW/°C for the well insulated houses.
It is of interest to indicate the distribution of the energy losses and
inputs. Figures 3 b and d show the relative heat losses through walls, windows,
roofs, floor and due to ventilation In figures 3 a and c the energy input is
also given for the test period. It turns out that for the well-insulated
houses about 15% of the heat requirements comes from the gain of solar energy
through the windows in the south wall. This balances the heat loss through
these double-glass windows, which is about 14%. For the houses with single-
glass windows the gain is 10% and the loss 13%.
     Moreover, the internal heat loads – cooking, lighting and man's own heat
production – amount to 20% (well-insulated) and 15% (poorly-insulated) of the
heat requirement for the test period. The solar collector on some of these
houses still gave operating problems during the measuring period and contributed only about 5% of the energy. It can be calculated that when the solar
heater, 21 m2 collector area per house, is operating correctly, it could
contribute about 22% of the heat supply.
     Besides, it was found that the consumption of primary energy (natural
gas) was about 20% higher in air-heated houses than in the ones using a hot
water radiator system, the cause being that in air-heated houses the inhabi-
tants set higher mean temperatures in the various rooms than in the case of
water heating systems.

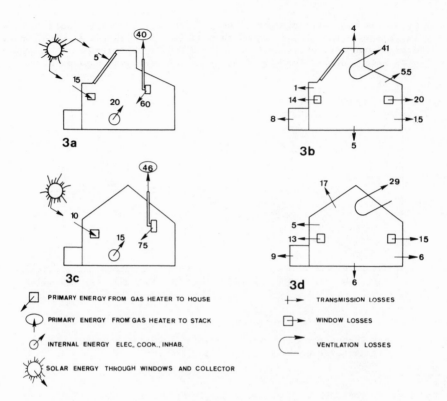

Fig. 3   Energy inputs to well-(a) and poorly-(c) insulated houses in %
           of total inputs.

           Energy losses of well-(b) and poorly-(d) insulated houses in %
           of total loss.

Not included in the figures for the heat losses are the stack losses. In the
figures 3 a and c they are given as a fraction of the imput of energy that
goes to the house. The efficiency of the gas heater used is not very good
(about 61% on average), which gives a considerable increase in primary
energy requirements. In the well-insulated house the absolute value of the
stack loss is about half of that in the other case. However it still offsets
the heat gains by the solar and internal loads.
        To compare different ways of obtaining energy savings in buildings
in general, energy requirement calculations have been made. In Germany the
VDI has given rules for these calculations (VDI-Richtlinien 2067 [5]) based
on a DIN-4701 code.

For the period of test the average weekly consumption of primary energy for heating as measured was compared with the calculated values on the basis of VDI-2067. Fig. 4 shows reasonable agreement, but there seems to be a constant shift in the calculated values, which are somewhat too high. For the well-insulated houses this shift is larger (750 MJ/week) than for the poorly-insulated houses (600MJ/week). These values approximate the internal load of 460 MJ/week, which was measured during the full test period. Moreover there is also a solar load through the windows of about 310 MJ/week which is also neglected in VDI-2067. The conclusion is that on the basis of VDI-2067 accurate calculations on different energy saving systems can be made provided the internal loads are taken into account.

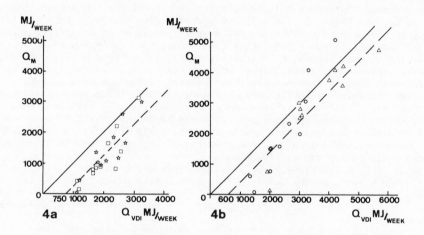

Fig. 4   Comparison between measured consumption of primary energy and calculated values on basis of VDI-2067. 4a Well-insulated, 4b poorly-insulated houses.

HUMAN FACTORS
        Studies on the heat requirements of buildings have revealed a large spread in energy consumptions in identical dwellings. This is due to human factors. The comfort conditions are not completely identical for all persons; moreover, they depend on activity level and clothing, which may differ in different households. In addition, the manual control of the thermostat setting may be uneconomic. Opening of windows when overheating has occurred will increase the losses considerably.
        In general there is not a seperate control of the temperatures in the various rooms of a house. This may lead to appreciable differences between different households. In the preciously discussed measurements in the Netherlands it caused differences between air and water heating. People with water heating tend to close off radiators in sleeping rooms, except on cold days. However in the air heated houses these rooms were heated for the full period. Separate control for each room would be more favourable.

From the above mentioned and similar Dutch studies it has been found
that differences between thermostat settings and window opening for various
households gave a spread in heat requirements, resulting in a standard
deviation of 25%. Neighbouring houses show heat exchanges of up to 20% of the
total requirements.

ENERGY-SAVING TECHNIQUES
        Apart from better insulation of buildings, there are various other
options for savings of primary energy. An important one is the use of district heating. In a city or part of it buildings can be heated by using the
waste heat of a specially designed power plant. This requires studies on the
long-distance transport of hot water. Problems are heat losses in buried hot
water lines and the distribution of heat over many different buildings.
In several countries this method is being used or is under consideration [6].
        Another option is the use of solar energy for heating and cooling-
of buildings. In Australia, Israel, and various other countries this has
already been done on a small scale for over 15 years. Other countries, often
with less favourable solar insolation conditions, are developing solar heating
systems with higher thermal efficiencies. This requires fundamental work on
the heat transfer characteristics of solar collectors. Radiation properties of
spectral selective surface materials and natural convection data in such
collectors determine the efficiency. New developments like honeycomb and vacuum
collectors are of interest in this field. The coupling of a collector to a
heating system requires optimization. This can be done by using mathematical
models as discussed next. These models are especially fruitful in such studies.
Using the meteorological data of the country or city under consideration the
required optimization can easily be made.
In general, a heat storage will be required. This may be a simple water or
rock storage vessel, or the building structure itself can be used for this
purpose. For storage over longer periods the ground or dry wells can be used,
giving rise ot a number of interesting heat transport problems for non-stationary conditions.
In this connection also heat pumps can be considered as an energy saving
option. Using heat from ground water, a heat pump with a good efficiency can
also be attractive to save on primary energy resources.

MATHEMATICAL MODELLING
        To evaluate the various options on energy savings mathematical mod-
eling is an important tool. By making a full thermal model of a building and
the heating-or cooling-system, calculations on the energy requirements can
be made. Transient conditions can be taken into account by using the appropriate response functions. Coupling this thermal model to micro-meteorological
data makes accurate predicitons of seasonal energy requirements possible. The
data on the micro-climate, like air temperature, wind velocity and solar
radiation should be available for various years on computer tape. Hourly or
half-hourly data are needed for these calculations. In various countries a
"standard" meteorological year is under consideration consisting op typical
average months from a 10- or 20-year period.
        Models including the full thermal dynamic behaviour of buildings
have originally been developed for cooling load calculations of air-conditioned
buildings. The same type of models are now frequently used in energy requirement studies for the heating of buildings. They make it possible to evaluate
the various possibilities on energy conservation measures. Together with the
economic and architecturical constraints these models give the possibility of

designing buildings with minimised energy requirements for heating or cooling
[7] .
    The different energy saving techniques can be optimized. Especially
for the development of solar heating systems these studies are very fruitful.
This because it can insure that for each particular set of micro-meteorological data a good choice of the system to be applied can be made. Especially
for those regions with a not to abundant solar insolation it can prevent
installation of less efficient systems, which would lead to disappointment
in the application of solar energy.

PLANNING
    To realise the many options on energy savings for heating an inte-
grated study of a new project is needed before full architectural planning.
Already in an early stage of planning new communities, towns or parts of
cities, the factor of energy savings for heating should be taken into account.
District heating, solar heating and long-time heat storage should be included
in "thermal" urban planning studies. Different architectural designs of a
new building should be compared on the basis of thermal energy requirements.
Façade design with factors like height-to-length-to-depth ratios,window areas
should be considered. The goal should be the design of energy-conserving
structures. Many crucial decisions should be made very early in the design
process.
    A very simple picture on how the architectural design of dwellings
influences the energy requirements for heating has recently been shown in a
British report on energy conservation [8] . Figure 5 shows the heat requirements
of dwellings, having the same volume and same fraction of window areas, but
of different designs. Differences of a factor of 3 are shown. Of course one
should on basis of this not start building only high-rise dwellings with the
lowest energy requirements. Still where possible it should be given attention. Moreover it shows that better insulation is most important for the
high energy requirement dwellings.
    Using mathematical modelling techniques it is possible to indicate
optimum conditions and to make an unbiased choice between the alternatives.
Also the use of different heating systems and energy-saving techniques can
be evaluated at this stage. Generalised rules in this respect would be very
valuable for the architect.

Fig. 5   Energy requirements for heating of dwellings with various
          architectural designs, all dwellings have same volume.

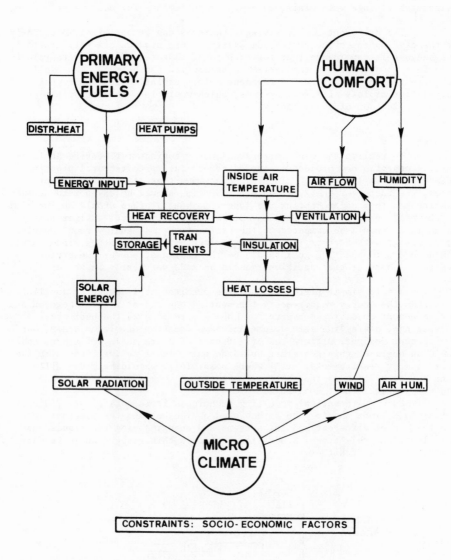

Fig. 6   Matrix of factors influencing the energy requirements for
heating of buildings.

Fig. 7   Front face of Dutch test houses with solar collectors.

CONCLUSIONS
        Energy conservation measures offer good prospects in the field of
heating of buildings. Savings of primary energy up to 15% of the total
national requirements can be expected for countries with moderate and nordic
climates. Many aspects play a role, which are summarised in Fig. 6 in a matrix
of factors. They are of a technical, meteorological and human nature with
the socio-economic factors as a constraint. Realization of this goal will
require co-operation of architects, engineers and scientists.
        The use of mathemathematical modelling for heat requirement calcula-
tions may help to evaluate the different options that are available.
Solar heating, heat pumps and district heating techniques offer good prospects
for future application. Research in the field of Heat and Mass Transfer will
help in further developments in energy-saving techniques.

REFERENCES:

1   Over, J.A. and Sjoerdsma, A.C. Energy conservation:
      ways and means. Future shape of techn. publ. 19, 1974
      Princessegracht 23. The Hague, the Netherlands.

2   Heat losses from dwellings. Building Research Establishment
      Digest 190, June 1976. p.1.-4 Dept. Environment, Great Britain.

3   Energy Conservation in Buildings R & D. Swedish Council for
      Building Research, 1976.

4   Ouden, C. den. and Brethouwer, D.E.
      Study on the effect of energy conservation measures in a series
      of dwellings in the city of Oss. Institute of Applied Physics-
      TPD-TNO report no. 403.246, Delft, 1977.

5   VDI-Richtlinien 2067, January 1974
      "Wirtschaftlichkeits berechnungen von Wärmeverbrauchsanlagen-
      Raumheizungsanlagen".

6   Proceedings Congress on: District Heating.
      Congrès des Union Int. des Distr. de Chaleur, Paris, 1975.

7   Bonsteel, D.L. and Kippenham, C. Collaborative utilization of
      computing techniques to achieve energy-effective design of
      buildings. ASME paper no 76-WAIHT - 23, Department of
      Architecture and Mech. Eng. Un. of Washington, Seattle 1 (1976).

8   Energy conservation: a study of energy consumption in buildings
      and possible means of saving energy in housing. Building
      Res. Establ. report CP 56/75, Watford, 1975.

HEAT AND MOISTURE TRANSPORT IN STRUCTURAL MATERIALS

ECONOMICALLY OPTIMAL
HEAT PROTECTION
IN BUILDINGS

K. GERTIS

Fraunhofer–Gesellschaft zur Förderung der Angewandten Forschung E.V.
Institut für Bauphysik
Stuttgart, Federal Republic of Germany

Using energy economically!

The general economic recession, a reduction in spending on the
part of the consumer caused by the "oil crisis shock", and two
relatively mild winters have almost led to an energy surplus on
the market during the last few years. These symptoms however are
deceiving. During the coming decades the energy supply will be
tight and an economical use of energy will be the order of the
day. This will be especially true in building construction,
since approximately 1/3 of the total energy needed in the Federal
Republic of Germany is consumed via the exterior walls. Appropriate energy saving measures in the building sector would therefore contribute significantly towards easing our country's
dependency on energy imports.

The total cost as the decisive factor!

Energy saving measures in building construction cost money.
Since the difficulties of financing building projects are already almost intolerable, these energy preserving measures
should also result in a favourable cost benefit relationship.
In building construction this means minimizing construction cost
including necessary subsequent operating expenses.

As demonstrated by figure 1, the opposite tendencies of heating
cost (installation and operating cost) and the cost of the
transmission element (such as exterior walls, basement ceiling,
roof slab) result in a total cost minimum based on a specific
thermal resistance of the transmission element, which is called
the "optimal thermal resistance". Thermally independent cost
(such as interior work, cost of building site, other construction
related cost) do not influence the optimal thermal resistance
but rather the total cost. The total cost can be calculated
on the basis of DM/m² living area or DM/m² a (so-called total
annual cost).

15

Figure 1. Schematic diagram
showing optimization of thermal resistance of the building elements.
The total cost K results from
the sum of specific cost items
such as thermally independent
costs, a (interior work),
heating cost b comprising
installation and operation,
cost of transmission element
c (such as exterior walls,
basement ceiling and roof slab).

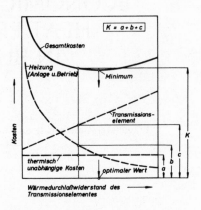

Establishing a cost minimum

Establishing a cost minimum is not a particularly easy task,
since a multitude of factors must be considered. Apart from
meteorological conditions at the site, thermal-constructive and
building-geometrical data, essential factors are economic-finan-
cial data concerning capital borrowed, conditions of loans and
mortgages as well as the utilization period of the building.
Also the calculation method itself is an important consideration.
The substitution principle, often used in business for investment
calculations, by which specific costs are exchangeable, is frequently difficult to employ in building construction, since a
number of substitutions have no easily recognizable functional
relationship with their subsequent cost. Furthermore the substitution of the construction costs for running costs is sometimes
used incorrectly. For these reasons it seems advisable to present in more detail some of the investment calculation methods
normally used.

Figure 2. Schematic
presentation of methods
commonly used in invest-
ment calculations.

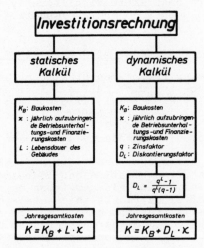

The schematic presentation in figure 2 demonstrates these methods in more detail.

Static calculation method

By means of investment calculations one can either obtain information about probable profits from prospective investments (yield), or attempt to minimize cost of construction including subsequent costs as in the case of public investment in building construction or in private housing. Here the difficulty arises of relating initial building costs (investment) to running costs (operating and financing costs), in order to find a comprehensible means of measuring the total cost. The following equation may serve to demonstrate, especially to the layman, the concept of "total cost".

$$K = K_B + L \cdot x \tag{1}$$

K_B indicates the construction costs arising during the building period, x symbolizes the annual operating and financing costs, and L refers to the utilization period (life span) of the building. Thus K represents the total cost for construction and use of the building during its life span. Since the time element with regard to the interest on payments, due at different times, is not considered, equation 1 is called a "static calculation method". The annual costs are added to the initial construction costs as linear costs. Due to its inherent inadequacy, the static method should therefore only be used for estimates.

Dynamic calculation method

To calculate the feasibility of an investment more closely, it is impossible to avoid using the dynamic calculation method, in other words, to take into account the influence of time on an investment. Assuming this means specifically considering income and expenditure due to interest on capital, the calculations can be based quite simply on the rules of compound interest accounting. If costs, payable at different times, such as construction costs and subsequent related expenses are to be compared, a "common point of convergence" must be established at a certain date. This means determining mathematically an "equivalent present value" K_0 of a capital K_n, existing in the year n [1]. K_0 is specified as "capital value" of K_n and is calculated with the following equation:

$$K_0 = K_n / q^n \tag{2}$$

in which

$$q = 1 + \frac{p}{100} \tag{3}$$

represents the interest term and p the per cent interest rate. If this equation 2 is solved for K_n

$$K_n = K_0 \cdot q^n \tag{2a}$$

the generally known equation is obtained for calculating the "equivalent future value" K_n of a starting capital K_0 after n years, including compound interest.

If x is invested not only once but annually in the same amount, the capital increases in n years as follows ([2] p. 97):

$$K_n = x \cdot \frac{q^n - 1}{q - 1} \tag{4}$$

The corresponding capital value of K_n is obtained using equation 4 and 2:

$$K_0 = x \cdot \frac{q^n - 1}{q^n (q - 1)} \tag{5}$$

In this way it is possible to compare annual payments such as operating expenses and financing costs with initial costs (construction cost). The annual costs, occuring throughout the life span of the building, are converged at the date of construction, which determines the capital value of the annual payment according to equation 5 (so called capital value method). Operating, maintenance and financing costs, through computation of their capital value, have thus become comparable to the initial construction cost. By using equation 1 they can consequently be added up in the following way:

$$K = K_B + D_L \cdot x \tag{6}$$

in which

$$\mathcal{D}_L = \frac{q^L - 1}{q^L \cdot (q - 1)}$$ (7)

is called the "interest term".

The difference between the static and the dynamic calculation methods becomes apparent, when comparing equation 1 and 6. Here it is evident that the so called "capital value method" (equation 6) results in lower subsequent costs the higher the assumed interest and the longer the building's life span.

What do optimization calculations produce?

The result of optimizations may be looked at from different perspectives. As a rule the investor will be guided by other interests than those of the user/renter. The investor will naturally resist higher initial investments, whereas the renter is faced with resulting higher heating costs for several decades. Below the case of a single family house ([3]) will demonstrate some results of optimization calculations, as seen with regard to likely increased energy costs in the future.

Figure 3 shows - from the investor's point of view - the dependence of various costs upon the energy price. In one case the single family house conforms to standard accepted practise (standard case, index O), whereas in the other case a superior insulation is chosen (optimal case, index O).

Investor (Bauherr)

Figure 3. The influence of the energy price on 1m² (m³) of a single family house per year.

S : standard case

O : optimal case (optimal case value of thermal resistance for exterior walls, roof and basement ceiling).

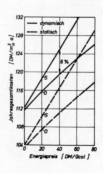 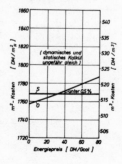

In both calculation methods the total cost (figure 3, left) rises considerably with an increased energy price. The favourable influence of optimizations is especially noticeable with in-

creased energy costs. In the standard case the investment costs
(per m²) are independent of the energy price, since in this case
the building elements are dimensioned under different criteria.
In the optimal case the investment costs increase with the
rising energy price, since the thermal quality of the enclosure
must be improved with rising energy costs, in order to maintain
an economical energy consumption. The fact that in the optimal
case investment costs are even lower than in the standard case
- provided the energy price is below 30 DM/Gcal - is due to
lower heating installation costs, and thereby reduced costs,
if the heat insulation is improved. Based on a 1974/75 energy
price of 40 DM/Gcal, it is possible to achieve by optimization
a 6 % reduction of total costs - compared to the standard case
which corresponds to less than an 0,5 % increase in investment
costs.

Nutzer (Mieter)

Figure 4. The influence of
the energy price "rent incl.
heat", "rent not incl. heat"
and heating costs (calculated
dynamically).

S : standard case

O : optimal case (optimal value
    of thermal resistance for
    enclosures).

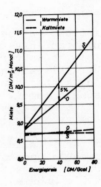
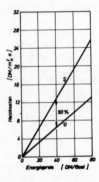

The user/renter is interested in factors such as rent and
heating expenses. Therefore figure 4 shows the influence of the
energy price on the "rent incl. heat", and the "rent not incl.
heat" (left diagram) and heating expenses (right diagram). It
becomes evident that, contrary to the "rent not incl. heat",
the "rent incl. heat"* depends greatly on increased energy
prices. It is therefore in the interest of the user to live in
a house, which is well insulated. In the optimal case he can

---
*

The "rent incl. heat" comprises the "rent not incl. heat", the
heating costs and the costs of the hot water and the cold water
consumption. The term "rent incl. heat" is still in discussion
for reasons, which are not to be treated here.

save, by calculation, almost 50 % of his heating expenses[*], as
compared to the standard case. Based on today's energy price
this amounts to a 5 % reduction of "rent incl. heat".

The building physicist, architect or structural engineer is mainly interested in 2 figures, namely a building's median coefficient
of thermal transmittance (see the newly defined DIN 4108 Beiblatt),
and the optimal thermal resistance of exterior walls, roof and
basement ceiling. The left diagram in figure 5 shows the median
coefficient of thermal transmittance in the standard and the
optimal case, compared to the max. permissible k_m value recommended in the DIN 4108 Beiblatt, based on the area/volume relationship (A/V = 1,02 m^{-1}) of the single family house considered.

Figure 5. The influence of the
energy price on the median
coefficient of thermal transmittance and on the optimal
thermal resistance of the transmission elements (calculated
dynamically).

Bauphysiker

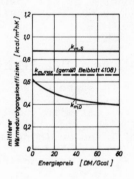
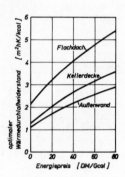

The left diagram demonstrates
the maximal median coefficient
of thermal transmittance, as
defined in the DIN 4108 Beiblatt,
based on the area/volume relationship of the single family house
(A/V = 1,02 m^{-1}).

S : standard case

O : optimal case (optimal value
of thermal resistance for
exterior walls, roof and
basement ceiling).

The decline of the k_m value in the optimal case, as can be seen
in the right diagram, is due to the increased optimal thermal resistance of the transmission elements. It follows from the left
diagram in figure 5 that, based on the 1974/1975 energy price of
40 DM/Gcal, the DIN 4108 Beiblatt assumes a position between the
standard and the optimal case.

---

[*] Due to the imponderables, which always exist in praxis, especially due to the ventilating customs of the user, which cannot be
calculated in advance, the de-facto-savings have therefore to be
fixed below the calculated value.

This realization is of the greatest importance when assuming a
position in the frequently subjective discussion of the Beiblatt
by the public. As far as the construction of single family houses
is concerned, the DIN 4108 Beiblatt has gone practically half way
towards economically optimal heat protection.

Looking at the right diagram, figure 5, it becomes abvious that
the economically optimal heat insulation of the enclosures lies
considerably above the recommendations of DIN 4108, which are
based on the standard case, and that it increases further with
rising energy prices. It is of the utmost importance to insulate
the roof sufficiently, since this can be done relatively inex-
pensively in a single family house. Even a layer of mineral wool
or foam, 12 cm thick, can result in a thermal resistance of
approx. 3,0 m²hK/kcal, a value which should not be lower in a
modern single family house. Insulation values of basement ceilings
should be raised to about 2,5 m²hK/kcal thermal resistance.
Exterior walls should have a thermal resistance above 1,5 m²hK/kcal.
Because of today's energy prices thermal resistance values of at
least 2 m²hK/kcal would be economically optimal for single family houses.

Concerning the practical application of thermal insulation

Careful consideration must be given to the practical realization
of thermal insulation values of the above mentioned scale. Here
we are concerned above all with the translation of the increased
thermal insulation values, resulting from economic necessity,
into building material and building methods, at the same time
preventing building damage. Figure 6 shows the insulation values
of different building materials and building parts.

Praktisch erreichbare Dämmwerte

Figure 6. Showing thermal
resistance and thermal
conductance to be obtained
by various building
materials.

Baustoff bzw. Bauteil	Wärmedurch-laßwider-stand [m²h K/kcal]	Wärmedurch-laßkoef-fizient [kcal/m²hK]
Mauerwerk	0,55 - 1,2	1,4 - 0,7
Leichtbeton (haufwerkporig)	0,55 - 0,8	1,4 - 1,0
Gasbeton	1,0 - 1,8	0,8 - 0,5
Wärmedämmstoffe	0,6 - 3,0	1,3 - 0,3
Fenster (zwei-und dreifach)	-	3,0 - 1,4

It is obvious that in the case of traditional monolithic masonry
construction only a max. thermal resistance of 1,2 m²hK/kcal can
be obtained, or in the case of aerated concrete blocks a max. of
1,8 m²hK/kcal. Higher insulation values require the installation
of layered thermal insulation. This generally means "layered
construction".

The transition from monolithic masonry construction to layered constructions will be difficult in practise, since especially in the field of housing construction, traditional monolithic building methods have been used successfully for a long time. In view of future energy prices however, it will be unavoidable to "make the leap" to layered construction methods.

How much has to be invested?

In a building recession which is not yet over, it is of parti- cular interest for anyone involved in building construction, to know how much additional investment is needed for energy saving measures. Research has shown that the necessary invest- ment can vary greatly depending on the quantity of heating energy to be saved, and the type of building. Figure 7 points out a relationship between increased construction costs and the resulting real saving of energy for objects constructed according to traditional building technology and furnished with traditional mechanical equipment.

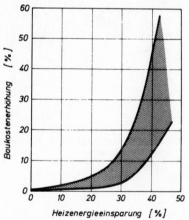

Figure 7. Relationship between increased construction costs for heat protection measures and thereby resulting, real energy saving. The per cent values are based on building standards existing before the energy crisis. The shaded area indicates considerable individual differences in cost.

The per cent values, for both increased building construction costs and energy savings, are based on building standards before the energy crisis. It is evident that - compared to the status quo ante - for example energy savings of 20 % require additional investment, varying between approx. 1 % (lower limit) and approx. 5 % (upper limit). It follows that, depending on the building project, the increase in construction costs can vary greatly. As demonstrated by the shaded area in the graph, this is especially the case, if larger savings are to be effected. Generally speaking it is relatively inexpensive to achieve real savings up to approx. 20 %. Higher savings, however, may in some cases be very expensive.

Bibliography

[1]   Küsgen, H.: Planungsökonomie. Was kosten Planungsentschei-
       dungen? Karl-Krämer-Verlag, Stuttgart (1970).

[2]   Brzoska, F. und Bartsch, W.: Mathematische Formelsammlung.
       Fachbuchverlag Leipzig (1957).

[3]   Werner, H. und Gertis, K.: Wirtschaftlich optimaler Wärme-
       schutz bei Einfamilienhäusern. Kritische Gedanken zu Opti-
       mierungsrechnungen. Ges.-Ing. 97 (1976), H. 1/2, S. 27-31;
       H. 5, S. 97 - 103.

FINITE ELEMENT ANALYSIS
OF HEAT AND MASS TRANSFER
IN BUILDINGS

G. COMINI

Università di Trieste
Istituto di Fisica Tecnica
Via A. Valerio 10
Trieste, Italy 34127

S. DEL GIUDICE

Università di Padova
Istituto di Fisica Tecnica
Via Marzolo 9
Padova, Italy 35100

ABSTRACT

The finite element method is used for the solution of two-dimensional heat and mass transfer problems in buildings. The versatility of this technique results in a viable method of solution for practical applications of the Luikov system of equations.

Examples presented in the paper include heat and mass transfer in the foundation basement of a cold store, moisture movement in porous walls and a direct solution of quasi-stationary heat conduction problems in building components.

NOMENCLATURE

c_m moisture capacity ($kg_{moisture}/kg_{dry\,body} \cdot °M$)

c_q heat capacity ($J/kg \cdot K$)

I imaginary part of complex amplitude τ (K)

j_m specific mass flux ($kg_{moisture}/m^2 \cdot s$)

j_q specific heat flux (W/m^2)

k_m moisture conductivity ($kg_{moisture}/m \cdot s \cdot °M$)

k_q thermal conductivity ($W/m \cdot K$)

ℓ reference length (m)

R real part of complex amplitude τ (K)

t temperature ($°C$)

u mass transfer potential ($°M$)

x,y Cartesian coordinates (m)

α_m convective mass transfer coefficient ($kg_{moisture}/m^2 \cdot s \cdot °M$)

α_q convective heat transfer coefficient ($W/m^2 \cdot K$)

γ direction cosines of the outward normal

Γ boundary surface (m^2)

25

δ thermo-gradient coefficient ($^{\circ}M/K$)

ε ratio of the vapour diffusion coefficient to the coefficient of the total diffusion of moisture

φ phase angle of temperature oscillations (rad)

ϑ time (s)

τ complex amplitude of temperature oscillations (K)

λ heat of phase change (J/kg)

ω circular frequency (rad/s)

Ω domain of definition (m^3)

ρ dry body density ($kg_{dry\ body}/m^3$)

Subscripts and superscripts

a   ambient

e   element

h   time level

i   initial

m   mass

q   heat

x,y   in direction of x,y

w   surface

δ thermo-diffusion

ε heat sink due to internal evaporation

o   reference

*   equivalent

INTRODUCTION

        The interrelation between heat and mass transfer in porous bodies was first established by Luikov [1,2,3] who proposed a two term relationship for non isothermal mass diffusion and also determined experimentally the coefficients of diffusion and thermo-diffusion for a number of moist materials. Later, via the use of thermodynamics of irreversible processes, he defined a coupled system of partial differential equations for heat and mass transfer potential distributions in porous bodies [4].
        These basic relationships are fully appropriate to those systems, consisting of various construction materials containing pores, voids and cracks of different shapes, to which buildings pertain.
        Analytical solution of heat and mass transfer problems pres-

ents great mathematical difficulties and, consequently, solutions are given for only the simplest of geometrical configurations and boundary conditions [5]. In any realistic problem, where geometrically complicate building components are considered, resort must be made to numerical techniques. These have usually been based on the finite difference method [6] but an alternative technique, which utilizes the finite element method, has been recently applied to situations of great engineering significance [7,8].

In this paper a finite element analysis of coupled heat and mass transfer in buildings is presented. The accuracy and the versatility of this procedure are then illustrated by several examples of application.

TRANSFER EQUATIONS AND FINITE ELEMENT FORMULATION

If the total pressure is assumed constant throughout the moist body and a zonal system of calculation is used [5,9], the heat and mass exchange in porous materials can be described in every zone of the entire domain of definition by the following equations:

$$\rho c_q \frac{\partial t}{\partial \vartheta} = k_q \, \nabla^2 t + \varepsilon \lambda \rho \, c_m \frac{\partial u}{\partial \vartheta}$$

$$\rho c_m \frac{\partial u}{\partial \vartheta} = k_m \delta \, \nabla^2 t + k_m \, \nabla^2 u \tag{1}$$

where t and u are the heat and mass transfer potentials and the coefficients are taken as constant and equal to their respective mean values in each zone.

A general set of boundary conditions for the system of equations (1) is given by:

$$t = t_w \tag{2}$$

on Γ_1, i.e. the portion of the boundary with a known temperature

$$k_q \, \nabla t \, \underset{\sim}{n} + j_q + \alpha_q \, (t - t_a) + (1-\varepsilon) \, \lambda \, \alpha_m \, (u - u_a) = 0 \tag{3}$$

on Γ_2, being that part of the boundary subjected to heat flux conditions.

Also, for the mass transfer we have:

$$u = u_w \tag{4}$$

on Γ_3, i.e. the portion of the boundary with a known moisture potential and

$$k_m \, \nabla u \, \underset{\sim}{n} + j_m + k_m \, \delta \, \nabla t \, \underset{\sim}{n} + \alpha_m \, (u - u_a) = 0 \tag{5}$$

on Γ_4, which is that portion of the boundary subjected to moisture flux conditions.

The problem defined by equations (1)-(5) can be rewritten in a generalized two-dimensional form:

$$C_q \frac{\partial T}{\partial \theta} = K_q \left(\frac{\partial^2 T}{\partial x^2} + \frac{\partial^2 T}{\partial y^2} \right) + K_\varepsilon \left(\frac{\partial^2 U}{\partial x^2} + \frac{\partial^2 U}{\partial y^2} \right) \tag{6}$$

$$C_m \frac{\partial U}{\partial \theta} = K_\delta \left(\frac{\partial^2 T}{\partial X^2} + \frac{\partial^2 T}{\partial Y^2} \right) + K_m \left(\frac{\partial^2 U}{\partial X^2} + \frac{\partial^2 U}{\partial Y^2} \right)$$

with boundary conditions:

$$T = T_w \qquad \text{on } \Gamma_1, \tag{7}$$

$$K_q \left(\frac{\partial T}{\partial X} \gamma_x + \frac{\partial T}{\partial Y} \gamma_y \right) + J_q^* = 0 \quad \text{on } \Gamma_2, \tag{8}$$

$$U = U_w \qquad \text{on } \Gamma_3, \tag{9}$$

and

$$K_m \left(\frac{\partial U}{\partial X} \gamma_x + \frac{\partial U}{\partial Y} \gamma_y \right) + J_m^* = 0 \quad \text{on } \Gamma_4. \tag{10}$$

In the above, dimensionless variables: $T=t/t_o$, $U=u/u_o$, $\theta = \vartheta/\vartheta_o$, $X=x/\ell$ and $Y=y/\ell$ are utilized, with $t_o (=t_a)$, $u_o (=u_a)$, and ϑ_o taken as reference values.

Generalized capacities C's, generalized transfer coefficients K's and generalized equivalent fluxes J^*'s defined from generalized fluxes J's and generalized convective coefficients A's

$$J_q^* = A_q (T - T_a) + A_\varepsilon (U - U_a) + J_q$$
$$J_m^* = A_\delta (T - T_a) + A_m (U - U_a) + J_m \tag{11}$$

are also referred to. Boundary conditions (8), (10) are formulated in such a manner as to retain the symmetry of the problem. Also, with a suitable definition of the generalized coefficients, K_ε can be made equal to K_δ thus making the system of equations (6) symmetric [3,7].

In the finite element formulation, the variable potentials T and U are approximated, throughout the solution domain Ω, by the relationships

$$T \cong \sum_{n=1}^{n} N_r (X,Y) \; T_r (\theta) = N \, T$$
$$U \cong \sum_{n=1}^{n} N_r (X,Y) \; U_r (\theta) = N \, U \tag{12}$$

where T_r and U_r are the nodal values and N_r are the usual shape functions which represent the potential distributions and are defined piecewise element by element [10].

If the approximations given by equation (12) are substituted into the governing differential equations (6) a residual is obtained which can be minimized using Galerkin's approach. This requires that the weighted error over the domain must be zero, with the shape functions being utilized as the weighting functions [3,7].

Applying Green's theorem then results in the following system of differential equations [3,7]

$$K \, \phi + \zeta \, \dot{\phi} + J = 0 \tag{13}$$

where K, ζ are 2nx2n symmetric matrices:

$$K = \begin{bmatrix} K_q & K_\varepsilon \\ K_\delta & K_m \end{bmatrix} \; ; \qquad \zeta = \begin{bmatrix} \zeta_q & 0 \\ 0 & \zeta_m \end{bmatrix} . \tag{14}$$

The dot indicates differentiation with respect to time, and the vectors ϕ, J are defined by:

$$\phi = \left[\underset{\sim}{T}, \underset{\sim}{U} \right]^{T} \; ; \qquad\qquad \underset{\sim}{J} = \left[\underset{\sim}{J}_{q}, \underset{\sim}{J}_{m} \right]^{T} . \tag{15}$$

Typical matrix elements are [3,7]

$$K_{r,s} = \sum \int_{\Omega^{e}} K \left(\frac{\partial N_{r}}{\partial x} \frac{\partial N_{s}}{\partial x} + \frac{\partial N_{r}}{\partial y} \frac{\partial N_{s}}{\partial y} \right) d\Omega \tag{16}$$

$$C_{r,s} = \sum \int_{\Omega^{e}} C \, N_{r} \, N_{s} \, d\Omega \tag{17}$$

and

$$\left(J_{r} \right)_{q} = \sum \int_{\Gamma^{e}} \left(J_{q}^{*} + J_{m}^{*} \, K_{\varepsilon} / K_{m} \right) N_{r} \, d\Gamma \tag{18}$$

$$\left(J_{r} \right)_{m} = \sum \int_{\Gamma^{e}} \left(J_{m}^{*} + J_{q}^{*} \, K_{\delta} / K_{q} \right) N_{r} \, d\Gamma \tag{19}$$

where (r,s=1,n).

The preceding summations are taken over the contributions of each element and the boundary conditions are applied only on the appropriate surfaces.

The system of equations (13) is linear in $\underset{\sim}{K}$ and $\underset{\sim}{C}$ but nonlinear in $\underset{\sim}{J}$ as the fluxes are functions of the external node potentials. Values at three consecutive time steps are then used to march in time, which results in the following recurrence scheme:

$$\phi^{h+1} = - \left[\underset{\sim}{K}^{h} / 3 + \underset{\sim}{C}^{h} / (2 \Delta\theta) \right]^{-1}$$
$$\times \left[\underset{\sim}{K}^{h} \phi^{h} / 3 + \underset{\sim}{K}^{h} \phi^{h-1} / 3 - \underset{\sim}{C}^{h} \phi^{h-1} / (2 \Delta\theta) + \underset{\sim}{J}^{h} \right] \tag{20}$$

where the superscript h refers to the time level and $\Delta\theta$ is the time step.

It can be seen that central values of the matrices are used in equation (20) which circumvents the necessity for iterating on $\underset{\sim}{J}$. The scheme requires two starting values of ϕ for initiation, but this presents no difficulty as known stationary values can be easily assumed.

SOME PARTICULAR CASES OF PRACTICAL INTEREST

Sometimes, as a first approximation, the influence of thermal fields on moisture movement can be neglected ($K_{\varepsilon} = K_{\delta} \cong 0$). This is the case, for example, with capillarity being the main mechanism for mass transfer. Moisture migration occurs mostly in liquid form and equations (1) are uncoupled but for the boundary conditions ($A_{\delta} \cong 0$; A_{q}, A_{m} and $A_{\varepsilon} \neq 0$).

Another situation of practical interest arises when moisture movement does not influence the transfer of heat and the boundary conditions on temperature are periodic functions of time of the form

$$t_{w} = \tau_{w} \, \exp\left(i \, \omega \, \vartheta \right) \tag{21}$$

$$t_{a} = \tau_{a} \, \exp\left(i \, \omega \, \vartheta \right) \tag{22}$$

with $i = \sqrt{-1}$ being the imaginary unit.
Then the quasi-stationary temperature field can be described by the expression

$$t = \tau \, exp \, (i \, \omega \, \vartheta).$$

(23)

In equations (21)-(23) τ is a complex function

$$\tau = R + i \, I$$

(24)

Finite element formulation of quasi-stationary heat conduction problems leads to a matrix equation of the form [8]

$$\underset{\sim}{K} \, \underset{\sim}{\phi} + \underset{\sim}{J} = 0$$

(25)

where

$$\underset{\sim}{K} = \begin{bmatrix} K_q & -\omega C_q \\ -\omega C_q & -K_q \end{bmatrix} ; \qquad \underset{\sim}{J} = \begin{bmatrix} FR, FI \end{bmatrix}^T ; \qquad \underset{\sim}{\phi} = \begin{bmatrix} R, I \end{bmatrix}^T$$

(26)

R_r, I_r (r=1,n) being the nodal parameters.
Typical matrix elements are

$$(K_{rs})_q = \Sigma \int_{\Omega_e} k_q \left(\frac{\partial N_r}{\partial x} \frac{\partial N_s}{\partial x} + \frac{\partial N_r}{\partial y} \frac{\partial N_s}{\partial y} \right) d\Omega + \Sigma \int_{\Gamma_e} \alpha_q \, N_r N_s \, d\Gamma ;$$

(27)

$$(C_{rs})_q = \Sigma \int_{\Omega_e} \varrho \, c_q \, N_r N_s \, d\Omega ;$$

(28)

$$FR_r = \Sigma \int_{\Gamma_e} \alpha_q \, R_a \, N_r \, d\Gamma ;$$

(29)

$$FI_r = -\Sigma \int_{\Gamma_e} \alpha_q \, I_a \, N_r \, d\Gamma$$

(30)

where (r,s=1,n).

Equation (25) represents again a symmetric-two degrees of freedom-system. Thus the same program utilized in the analysis of the heat and mass transfer problem can be used also for periodic heat conduction. Minor modifications include the output of data, where modulus $|\tau| = (R^2 + I^2)^{1/2}$ and phase $\varphi = \arctan(I/R)$ of temperature oscillations are given for each node.

ILLUSTRATIVE EXAMPLES

In the programs, two-dimensional isoparametric elements describe the various regions and these are also capable of incorporating curvilinear sides. The integrations in equations (16)-(19) and (27)-(30) are carried out numerically. Since the $\underset{\sim}{K}$ and $\underset{\sim}{C}$ matrices in equations (20) and (25) do not depend on the potentials, a Gaussian elimination technique is first used in the solution procedure to obtain a partial inverse. This inverse is then utilized to back substitute the variable vector occurring at each stage of the calculation. This way much computer time is saved when transient problems are dealt with.

The order of accuracy reached with the numerical calculations was evaluated by considering sections where one-dimensional distributions of potentials exist. In this case the analytical and the finite element solutions agree within 1% [3,7,8].

In the examples presented here, heat and mass transfer in the foundation basement of a cold store is considered first to demonstrate the applicability of the method in a most general case.

Then moisture movement in porous walls is investigated, neglecting the influence of temperature fields on mass transfer.

Finally, situations where diffusion of moisture does not influence the transfer of heat are considered. Reference is made to

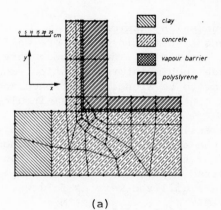

clay

concrete

vapour barrier

polystyrene

Fig. 1 - Heat and moisture transfer in the foundation basement of a cold store. (a) Mesh used. (b) Steady state potential distributions with a complete vapour barrier. (c) Steady state potential distributions with the vapour barrier removed from the floor.

(a)

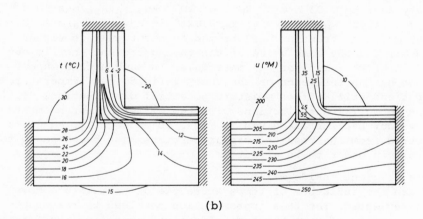

(b)

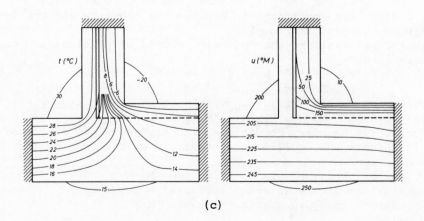

(c)

geometrically complex structures and solutions of the resulting
heat conduction problems are given for boundary conditions which
are periodic functions of time.

Heat and moisture transfer in a foundation basement

It is known that excessive mass transfer in cold store walls
can seriously damage their thermal insulation. Thus, vapour barriers are usually utilized to reduce such moisture migration. However, it is possible for the barrier to be wrongly positioned and
in such circumstances the moisture content in the thermal insulation can rise dangerously.

A steady state finite element analysis with the mesh shown
in Fig. 1a was used to demonstrate the applicability of the method
in such cases. Boundary conditions of the first kind and non conductive external surfaces are referred to, as indicated in the
figures. The physical property values utilized were as follows:
- concrete: k_q=0.32 W/m·K; k_m=1.4x10^{-8} kg$_{moisture}$/m·s·°M
- soil (clay): k_q=1.14 W/m·K; k_m=1.1x10^{-7} kg$_{moisture}$/m·s·°M
- polystyrene: k_q=0.03 W/m·K; k_m=1.05x10^{-9} kg$_{moisture}$/m·s·°M
- vapour barrier: k_q=1.0 W/m·K; k_m=2.8x10^{-12} kg$_{moisture}$/m·s·°M.

Any moisture transfer was assumed to occur only in vapour
form ($\varepsilon \cong 1$) and the same value of the thermogradient coefficient
(δ =0.5 °M/K) was used for all materials. In such a case vapour
diffusion has no influence on the thermal field, but temperature
distributions still affect moisture potential distributions [4,9].

The potential distributions with two different arrangements
of the vapour barrier are given in Figs. 1b and 1c [7]. As it can
be inferred from the results, vapour barriers must be extended to
the floors in order to be effective.

Moisture movement inside walls

Often the influence of temperature fields on moisture movement is neglected. Yet this approximation can lead to reasonably
accurate results if most moisture migration occurs in liquid form.
This is the case, for example, with drying of recently built
walls, a situation considered in Fig. 2. The geometry and the mesh
used are represented in Fig. 2a.

In the calculations the following values of physical properties were used:
- soil (clay): k_m=1.1x10^{-7} kg$_{moisture}$/m·s·°M;
  ρc_m=2.7 kg$_{moisture}$/m^3·°M
- concrete: k_m=8x10^{-8} kg$_{moisture}$/m·s·°M; ρc_m=0.26 kg$_{moisture}$/m^3·°M
- brick: k_m=6x10^{-8} kg$_{moisture}$/m·s·°M; ρc_m=2.16 kg$_{moisture}$/m^3·°M.

Convective boundary conditions with higher values of transfer coefficients and equilibrium potentials at the outside surfaces
α_m=5x10^{-6} kg$_{moisture}$/m^2·s·°M; u_a=100 °M
and lower values at the internal surfaces
α_m=3x10^{-6} kg$_{moisture}$/m^2·s·°M; u_a=50 °M
were considered.

A constant value of the mass transfer potential: u_w=200 °M

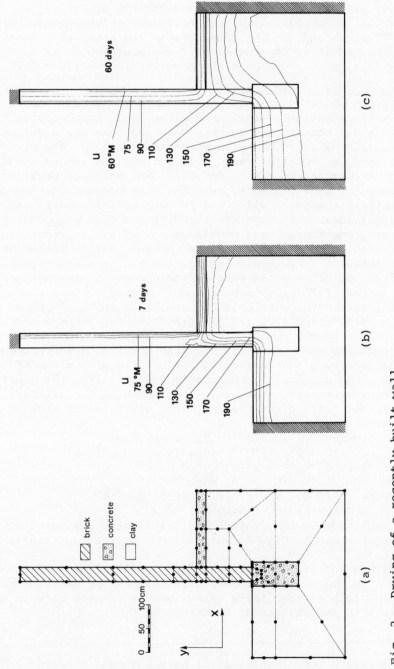

Fig. 2 – Drying of a recently built wall.
(a) Mesh used. (b) Potential distributions after seven days. (c) Potential distributions after two months.

was assumed at the bottom surface and constant stationary initial
conditions: u_i=200 °M were taken throughout the domain.

Potential distributions at two different values of time are
shown in Figs. 2b and 2c. The computations indicated that steady
state is reached after, approximately, two months even if most of
the variations in moisture content occur within the first week.

Thermal waves in a corner junction

Equipment selection for air conditioning systems is often
based on the assumption of maximum cooling loads occurring at the
same time in all conditioned spaces. This approach was developed
when all design calculations were performed manually. Now with
computers used for routine design calculations, it is practical
to make a more extensive design analysis. For example, neglecting
the influence of mass transfer, heat gains by conduction through
one-dimensional composite walls can be computed, for variable operating conditions, utilizing the transfer function method [11].

However, one-dimensional models cannot always represent adequately civil engineering structures. Prefabrication techniques
and the consequent large scale utilization of the same components
in different buildings often make a finite element analysis of
periodic thermal fields the most economic choice.

To illustrate the possibilities available in this type of a-
nalysis, amplitude and time lag distributions for temperature oscillations in a corner junction are investigated here.

The geometry considered and the mesh utilized are represented
in Fig. 3a. Far away boundaries are substituted with non conductive boundaries placed at a reasonable distance from the junction.

Convective heat transfer was assumed both at the external and
at the internal surface and the following estimates of physical
parameters were made:
- external surface: α_q=20 W/m$^2\cdot$K; $|\tau_a|$=20 K; φ_a=0;
  ω=7.27x10^{-5} rad/s (i.e. $2\pi/\omega\cong$24 h)
- internal surface: α_q=8 W/m$^2\cdot$K; $|\tau_a|$=0 K
- concrete: k_q=1.75 W/m$\cdot$K; ρc_q=1.85x10^6 J/m$^3\cdot$K
- polystyrene: k_q=0.039 W/m$\cdot$K; ρc_q=1.6x10^4 J/m$^3\cdot$K
- brick: k_q=0.7 W/m$\cdot$K; ρc_q=1.5x10^{-6} J/m$^3\cdot$K.

The amplitude of temperature oscillations at any point of
the domain is shown in Fig. 3b while time lags are represented in
Fig. 3c.

Amplitude and time lag of heat flux oscillations were also
calculated. In regions where one-dimensional fields develop and
a comparison is possible, there exists good agreement between numerical results and analytical solutions [8,11].

CONCLUSIONS

The approach to the solution of heat and mass transfer problems in buildings proposed in this paper has wide applications
and has been shown to give accurate results.

The versatility of the finite element method in dealing with

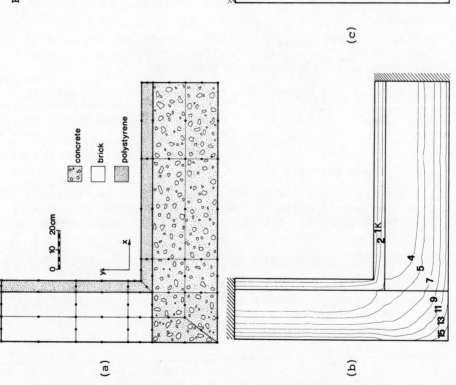

Fig. 3 – Thermal waves in a corner junction.
(a) Mesh used.
(b) Amplitude of temperature oscillations.
(c) Time lag of temperature oscillations.

concrete

brick

polystyrene

0 10 20cm

(a)

(b)

(c)

35

complicated geometries and physical property variations makes possible the solution of many practical problems in the field of
building thermophysics.

REFERENCES

1.   Luikov, A.V. 1935. On Thermal Diffusion of Moisture (in Rus-
      sian). Zh. Prikl. Khim. 8:1354-1358.

2.   Luikov, A.V. 1936. Moisture Gradients in the Drying Clay.
      Trans. Ceram. Soc. 35:123-129.

3.   Lewis, R.W., Comini, G., and Humpheson, C. 1975. Finite Ele-
      ment Application to Heat and Mass Transfer Problems in Porous
      Bodies (in Russian). Inzh. Fiz. Zh. 29:483-488.

4.   Luikov, A.V. 1966. Heat and Mass Transfer in Capillary-Porous
      Bodies, Pergamon, Oxford.

5.   Luikov, A.V., and Mikhailov, Y.A. 1965. Theory of Energy and
      Mass Transfer. Pergamon, Oxford.

6.   Bonacina, C., and Comini, G. 1971. Computer Calculation of
      Heat and Mass Transfer Phenomena. Proc. III Conf. on Drying,
      Budapest.

7.   Comini, G., and Lewis, R.W. 1976. A Numerical Solution of
      Two-Dimensional Problems Involving Heat and Mass Transfer.
      Int. J. Heat Mass Transfer 19:1387-1392.

8.   Mikhailov, M.D., Comini, G., Del Giudice, S., and Runchi, G.P.
      1977. Determination of Thermal Wave Distributions by the Fi-
      nite Element Method. Int. J. Heat Mass Transfer 20:195-200.

9.   Luikov, A.V. 1975. Systems of Differential Equations of Heat
      and Mass Transfer in Capillary-Porous Bodies (Review). Int.
      J. Heat Mass Transfer 18:1-14.

10.   Zienkiewicz, O.C., and Parekh, C.J. 1970. Transient Field
      Problems: Two-Dimensional and Three-Dimensional Analysis by
      Isoparametric Finite Elements. Int. J. num. Meth. Engng 2:
      61-71.

11.   ASHRAE 1972. Handbook of Fundamentals. ASHRAE, New York.

THERMAL RESISTANCE
OF STRUCTURAL MEMBERS
UNDER UNSTEADY
CONDITIONS

YU. N. KUZNETSOV

All Union Heat Engineering Institute
USSR

ABSTRACT

The method for constructing formulas for calculating thermal resistance of structural members under unsteady conditions has been proposed. For a number of practical cases some design formulas have been presented.

NOMENCLATURE

τ — time

$t, \bar{t}, \theta, \vartheta$ local, average, dimensionless, relative temperatures, respectively

q heat flux

R_t thermal resistance

$\bar{R}_t$ dimensionless thermal resistance

Fo Fourier parameter

Subscripts

s surface

f fluid

l limiting

The analysis of thermal characteristics of any structures is associated with the consideration of the interconnected unsteady processes of heat transfer in the structural members and the environments. Mathematically, it often requires the solution of the conjugated problems for the different types of differential equations.

A rather effective means to facilitate the solution of such difficult problem is the use of thermal resistances of the structures and heat transfer resistance which enables considerable simplification of the initial equations.

In this case the solution of the conjugated problems can be reduced to the solution of a number of non-conjugated problems.

It is reasonable to consider thermal resistance of the structural members under unsteady conditions provided the dynamic thermal characteristics of the medium are known. Let, in particular, the temperature field in the structure section of the area

ω with a boundary "s" under unsteady conditions be described by the following boundary problem for the heat conduction equation:

$$\rho c \frac{\partial t}{\partial \tau} = \frac{\partial}{\partial x}\left(\lambda \frac{\partial t}{\partial x}\right) + \frac{\partial}{\partial y}\left(\lambda \frac{\partial t}{\partial y}\right) \tag{1}$$

$$t/_{\tau=0} = t_0(x,y); \quad -\lambda \frac{\partial t}{\partial n}\bigg|_s = \alpha(\tau,x,y)\left(t/_s - t_f(\tau,x,y)\right) \tag{2}$$

In practice, the analysis of the dynamic thermal characteristics of the structural members requires some thermal characteristics of the object, such as the average temperature

$$\bar{t} = \frac{1}{\omega}\int_\omega t\, d\omega \tag{3}$$

the surface temperature, the surface temperature gradient and the like. This enables simplification of the problem involved.

Integration of equation (1) with respect to section ω with utilisation of Grin formula leads to an ordinary differential equation

$$\rho c \frac{dt}{d\tau} = \frac{1}{\omega}\int_s\left(-\lambda \frac{\partial t}{\partial n}\bigg|_s\right) ds \tag{4}$$

The boundary conditions and the theorem of the mean give:

$$\rho c \frac{d\bar{t}}{d\tau} = \frac{s}{\omega} K\left(\bar{t}_f - \bar{t}\right) \tag{5}$$

Here

$$K = \frac{-\lambda \frac{\partial t}{\partial n}\big|_s}{\bar{t}_f - \bar{t}} = \left(\frac{1}{\alpha} + R_t\right)^{-1} \tag{6}$$

and thermal resistance R_t of the structural member is

$$R_t = \left(\frac{-\lambda \frac{\partial t}{\partial n}\big|_s}{\bar{t}/_s - \bar{t}}\right)^{-1} \tag{7}$$

Engineering analysis of dynamic characteristics of structural members can be carried out rather simply by using the equations (5) – (7) if R_t is known, which characterises the structural member temperature field under unsteady conditions.

From equation (7) it follows that for the structural member it suffices to solve the boundary problem with known surface temperature gradient, provided thermal properties and initial temperature field are constant. More complicated cases can be derived from this approach rather simply.

Thus, the regularities of variation of R_t can be obtained by analysing the solution of the following boundary problem

$$\frac{\partial t}{\partial \tau} = a\mathcal{L}(t) \tag{8}$$

$$t/_{\tau=0} = t_0; \quad -\lambda \frac{\partial t}{\partial n}\bigg|_s = q^*(\tau) \tag{9}$$

or for a dimensionless form:

$$\frac{\partial \theta}{\partial Fo} = \mathcal{L}(\theta) \tag{10}$$

$$\theta/_{Fo=0} = 0; \quad \frac{\partial \theta}{\partial \mathcal{N}}\bigg|_s = q(Fo) \tag{11}$$

Here L = Laplas operator, $\theta = \dfrac{t-t_0}{t_0}$; $Fo = \dfrac{a\tau}{l_*^2}$; $q = \dfrac{q^* l_*}{\lambda t_0}$;

l_* = geometry scale; $\mathcal{N} = \dfrac{n}{l_*}$; $S = \dfrac{s}{l_*}$.

Respectively, for thermal resistance

$$\tilde{R}_t = \frac{R_t}{\lambda/2\ell_*} = 2\frac{\theta/s - \bar{\theta}}{q} \tag{12}$$

For the considered case, according to equation (4)

$$\frac{d\theta}{dFo} = \frac{S}{\Omega}q \; ; \quad \bar{\theta} = \frac{S}{\Omega}\int_0^{Fo} q\,dFo \tag{13}$$

Therefore, it is reasonable to introduce new variable ϑ

$$\vartheta = \frac{\theta - \bar{\theta}}{q} \tag{14}$$

From (10), (11) and with due account of (13) it follows that the field of the dimensionless relative temperature ϑ is described by the following system:

$$\frac{\partial\vartheta}{\partial Fo} + K_q(Fo)\vartheta = \mathcal{L}\vartheta - \frac{S}{\Omega} \tag{15}$$

$$\vartheta\big|_{Fo=0} = 0 \; ; \quad -\frac{\partial\vartheta}{\partial N}\Big|_s = 1 \tag{16}$$

In this case $$\tilde{R}_t = 2\vartheta/s \tag{17}$$

Equation (15) contains new parameter $K_q(Fo)$ which characterizes the effect of the variability in time of the boundary conditions, for the given case of heat flux, on the heat transfer process for the structural member.

By definition

$$K_q(F_q) = \frac{d\ln|q(Fo)|}{dFo} \tag{18}$$

The analogous parameter was used earlier for investigation of the unsteady processes of hydrodynamics and convection heat transfer [1] . There, detailed analysis of the effect of the parameters like those used in (18) was also carried out. Here, this parameter is used only for the analysis of some regularities of temperature field, and, consequently, of thermal resistance under unsteadly conditions.

From equation (15) it follows that the temperature field and thermal resistance at a given instant Fo are defined by the whole process hystory, i.e., by all values of $K_q(Fo)$ during (0,Fo). It means that the characteristics of   the unsteady thermal process depend on the law of variation of the boundary conditions in time. But with large Fo, the values of Kq wholly define the thermal conditions of the object irrespective of the law of variation of the boundary conditions. In this case, if there exists $\lim\limits_{Fo\to\infty}K_q(Fo) = \tilde{K}_q$, then the temperature field of body with increased Fo approaches limiting distribution ϑ_ℓ which is the solution of the problem

$$\mathcal{L}\vartheta_\ell - \tilde{K}_q \vartheta_\ell = \frac{S}{\Omega} \tag{19}$$

$$-\frac{\partial\vartheta_\ell}{\partial N}\Big|_s = 1 \tag{20}$$

It is evident that in this case $\tilde{R}_t \longrightarrow \tilde{R}_{t1} = 2\vartheta_\ell/s$.

According to asymptotic theory of functions [2] three different behaviours of $K_q(Fo)$ with large Fo are possible:
I) For the finite functions with the order lower unity, $\tilde{K}_q = 0$;
II) For the exponential functions, $\tilde{K}_q$ = const; III) For the infinite functions, $\tilde{K}_q = \infty$.

With the variation of $q(Fo)$, described by the function I which covers, for example, all polynominals, with the increased Fo, the temperature field ϑ approaches the distribution ϑ_ℓ which, according to equation (19), is described by Poisson equation.

In this case, temperature field θ_ℓ is of the following form:

$$\theta_\ell = \lim_{Fo \to \infty} \theta = \frac{S}{\Omega} \int_0^{Fo} q(Fo)dFo + \vartheta_\ell q(Fo) \tag{21}$$

For example, for the flat wall with the thickness of δ

$\ell_* = \delta$; $X = \frac{x}{\delta}$ and heated from one side only, we obtain:

$$\vartheta_\ell = \frac{x^2}{2} - \frac{1}{6}; \quad \widetilde{R}_{t\ell I} = \frac{R_{t\ell}}{\delta/2\lambda} = \frac{2}{3} \tag{22}$$

It should be noted that to calculate dynamic characteristics of structural members the steady-state values of R_t (for flat wall $R_t = \delta/2\lambda$) are usually used which may result in great errors even with large Fo (in the given example thermal resistances differ by 1.5 times).

Analogously, for the cylindrical wall with inner diameter r_o and outer diameter r_1 ($1* = ro$; $R = \frac{r}{ro}$; $R_1 = \frac{r_1}{ro}$) with thermal disturbance on the inner surface we obtain:

$$\vartheta_\ell = \frac{1}{R_1^2 - 1}\left[\frac{R^2}{2} - R_1^2\left(\ell n R - \frac{R_1^2}{R_1^2 - 1}\ell n R_1 + \frac{3}{4}\right) - \frac{1}{4}\right]$$

$$\widetilde{R}_{t\ell I} = \frac{R_{t\ell}}{\iota_0/2\lambda} = \frac{1}{R_1^2 - 1}\left[\frac{1}{4} + R_1^2\left(\frac{R_1^2}{R_1^2 - 1}\ell n R_1 - \frac{3}{4}\right)\right] \tag{23}$$

If we introduce $\widetilde{R}_{t1I} = \frac{R_{t\ell}}{(\iota_1 - \iota_0)/2\lambda} = \frac{\widetilde{R}_{t\ell I}}{R_1 - 1}$,

then at least with $R_1 \leqslant 1.5$ with an error of $< 2\%$, $\widehat{\widetilde{R}}_{t1I} = \frac{2}{3}$.

Accordingly, with the thermal disturbance on the outer surface

$$\vartheta_\ell = \frac{R_1}{R_1 - 1}\left[\frac{R^2}{2} - \ell n R + \frac{R_1^2}{R_1^2 - 1}\ell n R_1 - \frac{R_1^2}{8} - \frac{3}{8}\right]$$

$$\widetilde{R}_{t\ell I} = \frac{R_1}{R_1^2 - 1}\left[\frac{R_1^2}{8} + \frac{1}{2}\frac{1}{R_1^2 - 1}\ell n R_1 - \frac{3}{8}\right] \tag{24}$$

It should be noted that with $R_1 \leqslant 1.5$ the values of thermal resistances in both cases differ but insignificantly($< 1\%$).

With exponential variation of the boundary condition in time, the temperature field and R_t with large Fo depend on the limiting value of $\widetilde{K}_q$ only (for $q = \exp(\alpha$ Fo$)$ $K_q = \widetilde{K}_q = \alpha$)

In particular, for the above-considered case of the flat wall

$$\vartheta_\ell = \frac{1}{\sqrt{\widetilde{K}_q}}\frac{ch(\sqrt{\widetilde{K}_q}\,X)}{sh\sqrt{\widetilde{K}_q}} - \frac{1}{\widetilde{K}_q}$$

$$\widetilde{R}_{t\ell II} = 2\left(\frac{1}{\sqrt{\widetilde{K}_q}}cth\sqrt{\widetilde{K}_q} - \frac{1}{\widetilde{K}_q}\right) \quad ; \quad \widetilde{K}_q > 0 \tag{*}$$

$$\widetilde{R}_{t\ell II} = 2\left(\frac{1}{|\widetilde{K}_q|} - \frac{1}{\sqrt{|\widetilde{K}_q|}}ctg\sqrt{|\widetilde{K}_q|}\right); \quad \widetilde{K}_q < 0 \ .$$

Thus, the stabilized value of thermal resistance $\widehat{R}_{tlII}$ with the exponential variation of the heat flux differ from $\widehat{R}_{tlI}$. With the increased heat flux, $\widehat{R}_{tlII}$ is lower, while with the decreased value it is higher than $\widehat{R}_{tlI}$, (Fig.1). It should

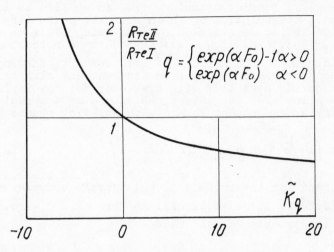

Fig.1 Stabilized values of thermal resistance.

be noted that if with $q = \exp(-\alpha \, Fo)$, α coincides with some eigenvalue of the corresponding boundary problem, then at this point there is discontinuity for $\widehat{R}_{tlII}(\widehat{K}_q)$.

Analogously, for cylindrical wall with the thermal disturbance on the inner surface:

$$\vartheta_\ell = \frac{U(\sqrt{\widetilde{K}_q}\,R)}{-U'(\sqrt{\widetilde{K}_+})} - \frac{2}{(R_1^2-1)\widetilde{K}_q} \; ; \; R_{t\ell_{ii}} = 2\,\vartheta_\ell\big/_{R=1} \; .$$

Here

$$U(\sqrt{\widetilde{K}_q}\,R) = I_0(\sqrt{\widetilde{K}_q}\,R)\cdot K_1(\sqrt{\widetilde{K}_q}\,R_1) + I_1(\sqrt{\widetilde{K}_q}\,R_1)\cdot K_0(\sqrt{\widetilde{K}_q}\cdot R).$$

Accordingly, with the thermal disturbance on the outer surface:

$$\vartheta_\ell = \frac{V(\sqrt{\widetilde{K}_q}\,R)}{V'(\sqrt{\widetilde{K}_q}\,R_1)} - \frac{2R_1}{(R_1^2-1)\widetilde{K}_q} \; ; \; R_{t\ell_{II}} = 2\,\vartheta_\ell\big/_{R=1} \; ;$$

$$V = I_0(\sqrt{\widetilde{K}_q}\,R)\,K_1(\sqrt{\widetilde{K}_q}) + I_1(\sqrt{\widetilde{K}_q})\,K_0(\sqrt{\widetilde{K}_q}\,R) \; .$$

The knowledge of the limiting values of the temperature field and thermal resistance enables those to be obtained at any moment irrespective of the thermal disturbance functions. It is evident that the regularities of the limiting heat transfer for the exponential functions should be used, since the above regularities differ from the steady-state ones.

In particular, if the heat flux obeys the following low $q(Fo)=\exp(i\beta \; Fo)$ then the complex function ϑ_ℓ , describing the limiting temperature field and being the solution of the steady-state problem (19)-(20) with $\widetilde{K}_q = i\beta$ is related to the influence function G(Fo, x, y), describing the temperature field with thermal pulse disturbance, can be written as follows:

$$\tilde{\mathcal{V}}_e(x,y) = \int_0^\infty \left[G(x,y,Fo) - \bar{G}(Fo) \right] \exp(-i\beta Fo) dFo$$

So, function $\tilde{\mathcal{V}}_e$ is Fourier transformation of the influence function for the difference between the local and average temperatures. Indeed, if Fourier transformation is applied to the problem (10)-(11), the solution of which with the pulse disturbance q(Fo) is influence function G, then using (14), we can arrive at system (19), (20) with $K_q = i\beta$ It should be noted that the module and argument of function $\tilde{\mathcal{V}}_e$ are equal to the amplitude and phase of the field temperature fluctuations with harmonic variation of the heat flux with frequency β . The influence function G allows for the determination of the temperature field of the body at any moment Fo and with any disturbance by using the following relationship:

$$\theta(Fo, x, y) = \int_0^{Fo} G(Fo - \tau, x, y) q(\tau) d\tau \tag{25}$$

For the considered case it is reasonably to intorduce the influence functions for the surface G_s and average temperatures $\bar{G}$ and their difference G_Δ respectively. In this case $G_\Delta = G_s - \bar{G}$.

From (13) it directly follows that

$$\bar{G} = \frac{S}{\Omega} \tag{26}$$

To arrive at G_Δ , it is sufficient to apply inverse Fourier transformation to function $\tilde{\mathcal{V}}_e / S$:

$$G_\Delta = \frac{1}{2\pi} \int_{-\infty}^\infty \tilde{\mathcal{V}}_e / S \exp(i\beta Fo) d\beta \tag{27}$$

Thus, the solution of sufficiently simple steady-state problem (19), (20) with subsequent utilization of (27) enables the determination of the influence function $G_\Delta(Fo)$ which provideds for the calculation of the thermal resistance of the structural member under unsteady conditions at any time and with any thermal distrubance q(Fo) from the formula that follows:

$$\tilde{R}_t(Fo) = \frac{1}{q(Fo)} \int_0^{Fo} G_\Delta(Fo - \tau) q(\tau) d\tau \tag{28}$$

The respective calculations carried out for the flat and cylindrical (R ≤ 1.5) walls enabled for the influence function G_Δ the following expression to be obtained:

$$G_\Delta(\tilde{Fo}) = \frac{\tilde{R}_{teI}}{2} \frac{a_2/Fo + a_3}{2 + \varphi(\tilde{Fo}) + S^{-1}(\tilde{Fo})} ,$$
$$\varphi(\tilde{Fo}) = a_1 \tilde{Fo}^{a_2} \exp(a_3 \tilde{Fo}) ; \quad a_1 = 0.0768; a_2 = -0.572; a_3 = -31.20 . \tag{29}$$

Here $\tilde{Fo} = \frac{a\tau}{\sigma^2}$ for the flat wall and $\tilde{Fo} = a\tau / (\imath_1 - \imath_0)^2$ for the cylindrical wall. $\tilde{R}_{t1I}$ can be obtained from (22)-(24), for the cylindrical wall with an error of <2% $\tilde{R}_{t1I} = 2/3(R_1 - 1)$. For illustration, Fig.2 presents the results of calculations of thermal resistances for the flat and cylindrical walls according to formulas (28), (29) for the cases of saltatory, linear and parabolic variation of $q(\widetilde{Fo})$.

If $q = A\,\widetilde{Fo}^m$, then $K_q(\widetilde{Fo}) = \dfrac{m}{\widetilde{Fo}}$; $\widetilde{K}_q = 0$.

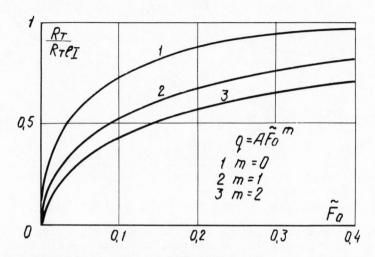

Fig.2 Unsteady values of thermal resistance.
The larger m, then with given $\widetilde{Fo}$ the greater $\widetilde{K}_q(\widetilde{Fo})$ differs from K_q and the greater $\widetilde{R}_t$ differs from the stady-state value of $\widetilde{R}_{t1I}$ which is identical for all m. With the increase of $\widetilde{Fo}$, $\widetilde{R}_t \rightarrow \widetilde{R}_{t1I}$ the faster, the lower m. If the stabilization time $\widetilde{Fo}_*$ of the thermal resistance is determined from $\widetilde{R}_t(\widetilde{Fo}_*) = 0,98\widetilde{R}_{t1}$, then with the jump of the heat flux $\widetilde{Fo} \cong 0.3$. In this case, for example, stabilization time $\mathcal{T}_* = (\widetilde{Fo}_* \, \delta^2/a)$ for a concrete wall 100 mm thick is $6.1 \cdot 10^3$ sec and increases proportionately to the square of the wall thickness. Analogously, for a steel wall this value is $0.4 . 10^3$ sec. The time period during which the unsteady value of the thermal resistance for the same conditions differs from the steady-state as high as more than 2 times $\mathcal{T}_{0,5}$ is $0.51 . 10^3$ and $0.02 . 10^3$ sec for the concrete and steel walls, respectively. With the increase in m, the stabilization time increases proportionately to m, for example, $\widetilde{Fo}_{0,5} = 0,25 + 0.06\,m$, which for a concrete wall with m = 1, and for steel wall with m = 2 gives $\mathcal{T}_{0,5} = 0.173 . 10^4$ and $\mathcal{T}_{0,5} = 0.8 . 10^3$.

A rather specific variation of the thermal resistance is observed for synusoidal fluctuations of the heat flux $q = A \sin$ (BFo) (Fig.3) $\widetilde{R}_t$ can assume negative values and periodically feature discontinuities. This is due to the phase shift of the fluctuations $q(Fo)$, δ (Fo) and Θ/S (Fo) . It should be noted that the parameter $K_q(Fo) = Bctg(BFo)$ and also varies in a likely manner.

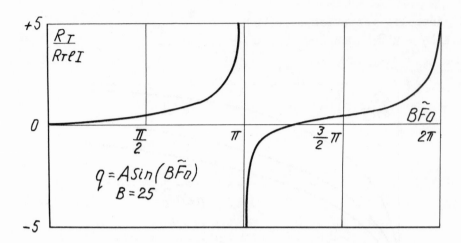

Fig.3   Thermal resistance with heat flux fluctuations .

Thus, the proposed method enables rather effective investigations of the regularities of thermal resistance of structural members under unsteady conditions during deep disturbances and design formulas to be obtained for calculating R_t under any

thermal disturbances which, in turn, allows for simple analysis of the dynamic characteristics of the structures involved.

References

1. Kuznetzov Ju.N. Preprints of papers presented at the IV
   Int.Heat Transfer Conf.v.II, Fc 4.3, Paris, 1970.
2. Evgrafov M.A. Asymptotic estimations and integral functions.
   Fizmatgiz, Moscow 1962.

SIMULTANEOUS HEAT AND MOISTURE TRANSFER IN POROUS WALL AND ANALYSIS OF INTERNAL CONDENSATION

MAMORU MATSUMOTO

Faculty of Engineering
Kobe University
Rokko Kobe, 657 Japan

ABSTRACT

The development of mathematic models of transient heat and mass transfer in a porous body of building are presented. System of this media is considered to be homogeneous multicomponent two phase heterogeneous mixture. Balance and constitutive equations are derived based on irreversible thermodynamics. Using these equations dynamics of internal condensation process are analized and compared with experimental results. The analytical results show good agreement with the experimental results.

NOMENCLATURE

A_i =a volume fraction of ith phase to bulk volume (m^3/m^3)
F_{ik} =external body force on kth component in ith phase (N/kg)
H_{ik} =euthalpy of kth component in ith phase (J/kg)
H_i =euthalpy of ith phase (J/kg)
$\quad = \Sigma \gamma_{ik} H_{ik}$
j_{ik} =mass flux of k in phase i relative to mass average velocity ($kg/m^2 s$)
J_{ik} =mass flux of kth component in ith phase relative to solid skeleton velocity ($kg/m^2 s$)
$J_{k(ij)}$ =mass flux of k at interface from phase i to j ($kg/m^3 s$)
$J_{(ij)} = \Sigma_k J_{k(ij)}$
k =permeability of liquid water (m^3/sN)
M_a =molecular weight of dry air (kg/mol)
M_w =molecular weight of water (kg/mol)
P =thermodynamic pressure (N/m^2)
P_v =partial vapour pressure (N/m^2)
P_a =partial pressure of dry air (N/m^2)
P_{vs} =saturated water vapour pressure (N/m^2)
Q_i =internal heat source in phase (w/m^3)
q_i =heat flux in phase i (w/m^2)
q =$\Sigma_k q_i$
q_{ij} =heat flux at interface from phase i to j (w/m^3)
R_w =universal gas constant of water (Nm/kgk)
R_a =universal gas constant of dry air (Nm/kgK)
S_i =eutropy of ith phase (J/kgk)
S_{ik} =eutropy of ith component in kth phase (J/kgk)
T =temperature (K)
t =time (s)
V_{ik} =velocity of k in ith phase (m/s)

V_i = mass average velocity of ith phase (m/s)
V^l = RwT/P
V^m = RaT/P
α_a = heat transfer coefficient (w/m^{2}s)
α'_μ = mass transfer coefficient related to chemical potential gradient (kg/m s J/kg)
α_T = moisture transfer coefficient related to temperature gradient (kg/m^{2}s K)
$\beta_{\shortmid\shortmid}$ = moisture transfer coefficient between phases (kg/m^{3}s J/kg)
β = $\beta_{\shortmid\shortmid}$/T
γ_{ik} = bulk density of kth component in ith phase (kg/m^3)
γ_i = bulk density of ith phase = $\Sigma\gamma_{ik}$ (kg/m^3)
θ = water content (m^3/m^3)
Φ = volume fraction of unit bulk volume occupied by humid air (m^3/m^3)
Φ_0 = true porosity (m^3/m^3)
ψ = volume fraction of unit pore volume occupied by liquid and/or adsorbed water (m^3/m^3)
λ = thermal conductivity (w/m K)
λ_μ = thermal conductivity related to chemical potential gradient (w/m J/kg)
λ' = total moisture conductivity = $\lambda'_{\mu g} + \lambda'_{\mu l}$ (kg/m s J/kg)
$\lambda'_{\mu g}$ = moisture conductivity of water vapour (kg/m s J/kg)
$\lambda'_{\mu l}$ = moisture conductivity of liquid water (kg/m s J/kg)
λ'_T = total moisture conductivity related to temperature gradient (kg/m s K) = $\lambda'_{Tg} + \lambda'_{Tl}$
λ'_{Tg} = water vapour conductivity related to temperature gradient (kg/m s K)
λ'_{Tl} = liquid water conductivity related to temperature gradient (kg/m s K)
ρ = true density (kg/m^3)
μ = chemical or generalized potential (J/kg)
μ_{a0} = chemical potential of pure dry air (J/kg)
μ_{v0} = chemical potential of pure water vapour (J/kg)
μ_a = $R_aT \ln p_a/P$ (J/kg)
μ_v = $R_wT \ln p_v/P$ (J/kg)
μ_w = $R_wT \ln p_v/P_{vs}$ (J/kg)
μ_{vg} = μ_v of water vapour (J/kg)
μ_{vl} = μ_v of liquid water (J/kg)
μ_{wg} = μ_w of water vapour (J/kg)
μ_{wl} = μ_w of liquid water (J/kg)
σ = entropy production (J/kg k m^3 s)
C_{ik} mass fraction of kth component in ith phase (kg/kg)

Subscript
i,j = ith and jth phase   k = kth component
a = dry air   v = water vapour
w = liquid and/or adsorbed water   l = liquid phase
g = gas phase   s = solid skeleton
l = gas phase   2 = liquid phase

I. INTRODUCTION

The recent progress of the digital computer allow us more detail prediction of the temperature and humidity fields in the building and its elements, especially of dynamics of the field. Attension is paid to thermal and hygric design of the elements with permitting the internal condensation without damage[1][2]. Those are over-simplified model. Numerical analysis by computer was reported[3] by

using deVries equation[4] under quisi-steady state of temperature
and under the isothermal condition[5]. Rate of temperature increase
and max. temperature in wall are essential for structure design and
design of antifirecovering of steel structure. Methods and
models accounting for effect of moisture flow and evaporation, and
effect of bulk flow of gas have been reported for the hygroscopic
moisture content[6][7]. Room air temperature and humidity(absolute)
have been analyzed[8][9] and calculated[10] using impulse responses
of room to heat and moisture supply which were calculated, based on
linear simultaneous heat and moisture transfer equation. These can
not be applied to cases of internal condensation.

For analysis of above mentioned problem, the common and sufficiently precise mathematic model of heat and mass transfer may
be required. Further, for the building engineering, there are
several peculiar feature such as transfer through multi-layers,
usually, (need of common potential), asymmetric boundary condition
and action of external pressure. Generally, the features are different
from drying engineering and soil science.

As well known there are many excellent investigation of combined
heat and mass transfer, such as Luikov [11], Krischer [12], deVries
and Philip [13]. Luikov have introduced irreversible thermodynamics
to the problem at the first time. In soil physics, Taylor and Cary
[14] presented phenomenological constitutive equations for heat and
mass flux based on irreversible thermodynamics for unsaturated soil.
The detail discussion have been reported [15] for saturated soil.

These works are based on local equilibrium of moisture between
phases. But, from the point of view above mentioned, they are not
sufficient. For the hydrodynamic analysis of the dispersed media
of poly phase, Nigmatulin [16] have derived the heat, mass momentum
and entropy balance equation based on continuum mechanics and
presented the constitutive equations. In this paper, along his
investigation, balance and constitutive equation in porous media are
derived with the assumption that porous media is heterogeneous mixture
of gas and liquid phases, each of which is homogeneous mixture [17],
without assumption of local equilibrium between phases. Liquid phase
is composed of liquid and/or adsorbed water, soluble salts and
dispersed solid skeleton. Gas phase is composed of humid air and
disperse of solid skeleton. It is assumed that dry air is innert
gas and salts are involatile. The concept of non-local equilibrium condition of moisture had shown but classical one [18][19]
[20][21]. The derived equations are expressed by form which contain
flux term [22] expressed by mass flux relative to solid skeleton,
for eliminating the effect of skeleton deformation. And then chemical
potentials   relative to pure water vapour or pure liquid water
are derived as mass transfer potential. At local equilibrium between
phases and uniform thermodynamic pressure, equations of mass and
energy balance derived here are equivalent to the equations by
Slegel et al.[23].

For internal condensation process, the over-simplified model of
moving boundary treatment have been presented and compared with
experiments [24]. The analysis of this process have been reported
by Kooi [25] using deVries-philip equation and measured diffusion
coefficient from his precise and reliable experiment. Comparison
with experimental result shows good agreement but there are no comparison with moisture distribution. Only moisture balance equation
were used and temperature fields did not analyzed. We are concerning
that modified values of moisture diffusion coefficient from measured
value show good agreement with experimental results. Here experi-

mental results of the internal condensation in gas concrete are compared with results from our combined heat and moisture transfer equation using chemical potential.

2. BALANCE AND CONSTITUTIVE EQUATION

We regard the building material as porous media. In the media there are solid skeleton of material, liquid including the solute and gas. The porous media composed of above components, is regarded as two phases multicomponent media. One phase is liquid in which liquid water, adsorbed water, solute of soluable material such as salt and solid particles of skeleton are dispersed. Solid or solid particles of skeleton are treated as solute. Another phase is a gas phase in which dry air, water vapour and solid particle of skeleton are dispersed. It is regarded that solid in gas phase is inert with water vapour. In each phase, it is regarded as homogeneous mixture and continuum system. Between phases there are interface. Interraction between phases take place at the interface. This means that porous material is heterogeneous mixture of homogeneous liquid and gas phase.

The macroscopic mass density γ_{ik} of kth component in ith phase is defined as the mass of κ per unit bulk volume, which are total volume of gas and liquid phases. The macroscopic mass density of phase i is γ_i.

$$\gamma_i = \Sigma \gamma_{ik} \tag{1}$$

Mass average velocity of phase i, V_i, is

$$V_i = \frac{1}{\gamma_i} \Sigma_\kappa \gamma_{ik} V_{ik} \tag{2}$$

where V_κ is the component velocity of kth component in ith phase. The conservation of mass for component k in ith phase with respect to the spacial coordicate system can be written

$$\frac{\partial \gamma_{ik}}{\partial t} + \nabla \gamma_{ik} V_i = -\nabla j_{ik} + J_{k(ji')} - J_{k(ij')} \tag{3}$$

where j_{ik} is the diffusion flux of kth component in ith phase and $j_{ik} = \gamma_{ik}(V_{ik} - V_i)$. $J_{k(ij)}$ and $J_{k(ji)}$ are mass transfer rates of k through interface, from phase i to j, and from j to i respectively, owing to phase transition. $J_{k(ij)} \geq 0$ and $J_{k(ji)} \geq 0$. The sum of equations (3) for all k components yields

$$\frac{\partial \gamma_i}{\partial t} + \nabla \gamma_i V_i = J_{(ji)} - J_{(ij)} \tag{4}$$

where $J_{(ji)} = \Sigma J_{k(ji)}$, $J_{(ij)} = \Sigma J_{k(ij)}$. Using the mass fraction of component κ in phase i, $C_{ik} = \gamma_{ik}/\gamma_i$, equations of mass conservation are

$$\gamma_i(\frac{\partial C_{ik}}{\partial t} + V_i \nabla C_{ik}) = - \nabla J_{ik} + J_{k(ji)} - J_{k(ij)} - C_{ik}(J_{(ji)} - J_{(ij)}) \tag{3'}$$

From the conservation of internal energy, energy balance equation [26] [27] is

$$\gamma_i(\frac{\partial H_i}{\partial t} + V_i \nabla H_i) = \gamma_i W_i - \nabla q_i - \nabla \Sigma j_{ik} H_{ik}$$
$$- \Sigma_\kappa (J_{k(ji)} X_{ik(ji)} - J_{k(ij)} X_{ik(ij)})$$
$$- q_{(ij)} + \gamma_i Q_i + \Sigma j_{ik} F_{ik} \tag{5}$$

where work done by internal force per unit bulk volume $\gamma_i W_i$ is

$$\gamma_i W_i = A_i(\frac{\partial P}{\partial t} + V_i \nabla P) + \nu_i f_{ij}(V_j - V_i) + \tau_i : e_i$$
$$+ J_{ji} \frac{(V_{ji} - V_i)^2}{2} - J_{ij} \frac{(V_{ij} - V_i)^2}{2} \tag{6}$$

$\chi_{ik(ji)}$ and $\chi_{ik(ij)}$ are the heat of fluxes from the ith phase to the material which undergoes transformation j→i and i→j, respectively, per unit mass of transition [28]. There are relations

$$\chi_{1k(21)} + \chi_{2k(21)} = H_1 - H_2 \quad \chi_{1k(12)} + \chi_{2k(12)} = H_2 - H_1$$

$q_{(ij)}$ is heat transfer through the interface from phase i to j per unit bulk volume. F_{ik} is the external force acting on component k in phase i per unit mass of component k. $f_{(ij)}$ is the rate at which phase i is gaining momentum of dissipation from phase j. ν_i is fraction of kinetic energy of mixture which can be dissipated as the result of interaction between phases and which transfer directly into the internal energy of the phase i($\Sigma \nu_i = 1$). τ_i is the viscous stress tensor in phase i and e_i is the deformation rate tensor in phase i.

From the entropy balance, entropy equation of the system composed of liquid and gas phases is

$$\Sigma \left(\frac{\partial \gamma_i S_i}{\partial t} + \nabla V_i \gamma_i S_i\right) = -\Sigma \nabla \left(\frac{J_{qi}}{T} + \Sigma S_{ik} J_{ik}\right) + \sigma \qquad (7)$$

where σ is entropy production rate and $J_{qi} = (q_i + J_{ik} H_{ik})$. By using Gibbs relation and eq(3') and (6), and neglecting the inertial term at interface, entropy production σ is

$$\sigma = \left[\frac{\gamma_1 Q_1}{T_1} + \frac{\gamma_2 Q_2}{T_2}\right] + \left[q_{12}\left(\frac{1}{T_1} - \frac{1}{T_2}\right) + \Sigma_k J_{k(12)}\left(\frac{\mu_{1k}}{T_1} - \frac{\mu_{2k}}{T_2}\right) + i_{k(12)}\left(\frac{1}{T_1} - \frac{1}{T_2}\right)\right.$$

$$+ \Sigma_k J_{k(21)}\left(\frac{\mu_{2k}}{T_2} - \frac{\mu_{1k}}{T_1}\right) + i_{k(21)}\left(\frac{1}{T_2} - \frac{1}{T_1}\right)\right] + \left[\frac{\tau_1 : e_1}{T_1} + \frac{\tau_2 : e_2}{T_2}\right]$$

$$- \left[f_{12}(V_2 - V_1)\left(\frac{\nu_1}{T_1} + \frac{\nu_2}{T_2}\right) + \left(q_1 \frac{1}{T_1} \nabla \ln T_1 + q_2 \frac{1}{T_2} \nabla \ln T_2\right)\right]$$

$$+ \frac{1}{T_1} \Sigma j_{1k}(\nabla_T \mu_{1k} - F_{1k}) + \frac{1}{T_2} \Sigma j_{2k}(\nabla_T \mu_{2k} - F_{2k})\right] \qquad (8)$$

where $i_{k(12)} = H_1 + \chi_{1k(12)} = H_2 - \chi_{2k(12)}$, $i_{k(21)} = H_1 - \chi_{1k(21)} = H_2 + \chi_{2k(21)}$. The first term of right hand side is a term of pure entropy production by external heat source. The other term are entropy production by flux of nonequilibrium process. The second term is of product of 0th order tensor and the third is of 2nd order tensor. The fourth term is of product of 1st order tensor.

Here We assume that (A) temperature and thermodynamic pressure are local equilibrium between phases, that is $T_1 = T_2$ and $P_1 = P_2$, (B) the deviation from the equilibrium state is not so much that there are linear relations between fluxes and driving forces[29]. By using Curie's principle constitutive equations are obtained. For vectors

$$q = q_1 + q_2 = -B_{00} \nabla \ln T - \sum_p^n B_{0p} \nabla_T \mu'_{1p} - \sum_p^m B_{0(n+p)} \nabla_T \mu'_{2p}$$

$$j_{1k} = -B_{k0} \nabla \ln T - \sum_p B_{kp} \nabla_T \mu'_{1p} - \sum_p^m B_{k(n+p)} \nabla_T \mu'_{2p} \qquad (9)$$

$$j_{2k} = -B_{(n+k)0} \nabla \ln T - \sum_p B_{(k+n)p} \nabla_T \mu'_{1p} - \sum_p B_{(k+n)(n+p)} \nabla_T \mu'_{2p}$$

$$f_{12} = \beta_{00} \frac{V_1 - V_2}{T}$$

For scalor fluxes $\qquad (10)$

$$J_{k(12)} = \sum_p \beta_{kp} \frac{\mu_{1p} - \mu_{2p}}{T}$$

For 2nd order tensor

$$\tau_i = \sum_p B_{ip} e_p \qquad (11)$$

There are relations that $\sum_k^n j_{1k} = 0$ and $\sum_k^m j_{2k} = 0$ and

$$\nabla_T \, \mu'_{ik} = \nabla_T \mu_{ik} - F_{ik} = \nabla_T \mu_{ik} - F_{ik} + S_{ik} \nabla T$$

Matrix $[B_{ij}]$, $[\beta_{ij}]$ and $[B_{ij}]$ must be positive definite and are expected to be symmetric matrix from Onsager reciprocal theory. In our problem of porous media, mass centre velocity V_i is very low. Then we can assume that entropy production by τ_i and f_{12} is negligible small.

The mass fluxe of the kth component in ith phase relative to the solid skeleton J_{ik} are

$$J_{ik} = \gamma_{ik} (V_{ik} - V_{is}) = j_{ik} - \frac{\gamma_{ik}}{\gamma_{is}} j_{is} \qquad (12)$$

where V_{is} is the velocity of the solid skeleton in ith phase and γ_{is} is the bulk density of solid skeleton in ith phase. In the porous material it can be treated $V_{1s} = V_{2s}$. Using Gibbs-Duhem theorem and neglecting the contribution of inertial term of each component we have

$$\sum_k \gamma_{ik} \nabla_T \bar{\mu}'_{ik} = \nabla P - \sum_k \gamma_{ik} F_{ik} + \sum_k \nabla_T \tau_{ik} + r_{ij} = 0 \qquad (13)$$

for ith phase. Here $\bar{\mu}_{ik}$ are generalized potential[30] and r are force per unit bulk volume acting to ith phase from jth phase. From the conservation of momentum, right hand side of eq(13) is equal to zero. By substituting eq(12) and (13) into eq(9) or eq(8) we obtain the alternative constitutive equation of the first order tensor

$$[J] = [C_{ij}] [F] \qquad (14)$$

where
$$[J] = [q, \, J_{11}, \, J_{12}, \cdots , J_{1(n-1)}, J_{21} \cdots , J_{2(m-1)}]^T$$
$$[F] = -[\nabla \ln T, \, \nabla_T \bar{\mu}'_{11}, \cdots , \nabla_T \bar{\mu}'_{1(n-1)}, \nabla_T \bar{\mu}'_{21}, \cdots , \nabla_T \bar{\mu}'_{2(m-1)}]^T$$
$[C_{ij}]$ is $(n+m-1) \times (n+m-1)$ matrix in which n is a number of components in gas phase and m is in liquid phase.

$[J]$ and $[F]$ are column vector of $[m+n-1]$ dimensions. Matrix $[C_{ij}]$ must be positive definite for positive entropy production and may be symmetric matrix. Eq(14) is the basic constitutive equation in this study.

3. DIFFUSION EQUATION

We can assume without generality that in the building problems, components of liquid phase are only two components, that is, water and solid dispersed media(skeleton) and of gas phase are three, that is, water, dry air and solid skeleton(if it may be imaginary). Since n=3 and m=2 in eq(14). Further, the generalized potentials including the internal stress term are approximately equal to the ordinary chemical potentials. In other words, states considered here are not so far from the mechanical equilibrium. Then the ordinary chemical potential is used in eq(14) in the following i.e. $\nabla_T \bar{\mu}'_{ik} = \nabla_T \mu'_{ik}$. In this case, linear constitutive equation(14) is reduced to following form

$$[J] = [q, \quad J_{1w}, \quad J_{1a}, \quad J_{2w}]^T \qquad (15)$$
$$[F] = -[\nabla \ln T, \quad \nabla_T \mu'_{1w}, \quad \nabla_T \mu'_{1a}, \quad \nabla_T \mu'_{2w}]^T$$

$[C_{ij}]$ in eq(14) is a 4x4 matrix $(i,j=0,1,2,3)$. Lower subscripts 1, 2, a and w mean gas phase, liquid phase, dry air and water, respectively. According to the rigidity of liquid water interfaces[31], cross effects between phases are considered to be weak, then coefficients C_{ij} related to those cross effects become to zero, i.e. $C_{13}= C_{31}= C_{23}= C_{32}= 0$.

For scalar flux

$$J_{w(\overline{12})} = \beta \, \frac{\mu_{1w} - \mu_{2w}}{T} = -J_{w(\overline{21})} \qquad (16)$$

In the following, $J_{w(\overline{12})}$ and $J_{w(\overline{21})}$ are abbreviated by J and $-J$, for the component occurring phase transformation is water only.

At the equilibrium between phases $\mu_{1w} = \mu_{2w}$

Of the porous media, We can assume that concentration of solid in gas phase do not affect the state of gas, then chemical potentials of gas are independent of solid concentration. The chemical potential of dry air μ_d and of water μ_{H_2O} are

$$\mu_d = \mu_{a0}(T,P) + \mu_a^{H_2O} \tag{17}$$

$$\mu_{H_2O} = \mu_{vo}(T,P) + \mu_v = \mu_{w0}(T,P) + \mu_w \tag{18}$$

where

$$\mu_a = R_a T \ln\frac{p_a}{P} \qquad \mu_v = R_w T \ln\frac{p_v}{P} \qquad \mu_w = R_w T \ln\frac{p_v}{p_{vs}}$$

μ_{a0}, μ_{v0} are the chemical potentials of pure dry air and pure water vapour, respectively. μ_{w0} is of pure liquid water. From Gibbs-Duhem equation

$$\nabla_T \mu_a' = \frac{M_w}{M_a} \frac{p_v}{P-p_v} \nabla_T \mu_v' \tag{19}$$

From definition of chemical potential

$$\nabla_T \mu_{H_2O}' = \nabla_T \mu_{vo} + \nabla_T \mu_v - F_w = \upsilon_v \nabla P + \nabla_T \mu_v - F_w \tag{20}$$

$$= \nabla_T \mu_{w0} + \nabla_T \mu_w - F_w = \upsilon_w \nabla P + \nabla_T \mu_w - F_w \tag{20'}$$

$$\nabla_T \mu_d' = \nabla_T \mu_{a0} + \nabla_T \mu_a - F_a \tag{21}$$

where $\upsilon_v = R_w T/P$ $\upsilon_a = R_a T/P$ $\upsilon_w = 1/\rho_w$. There are relations

$$\nabla_T \mu = \nabla\mu - \frac{\mu}{T}\nabla T \tag{22}$$

Substituting the relations (19),(20'),(21) and (22) into eq(14) and (15) we obtain

$$[J] = [D_{ij}][F] \tag{23}$$

$$[J] = [q, \quad J_{1w}, \quad J_{1a}, \quad J_{2w}] \tag{24}$$

$$[F] = -[\nabla T, \quad \nabla\mu_{wg}, \quad \nabla P, \quad \nabla\mu_{w\ell} - F_w] \tag{25}$$

μ_{wg} and $\mu_{w\ell}$ are μ_w of water vapour and liquid water. The external forces F_a and F_w in gas phase are omitted in the above equation because of weak effects on fluxes.

$$[D_{ij}] = \begin{bmatrix}
[\frac{C_{00}}{T} + (C_{01}+\xi C_{02})(W_s - \frac{\mu_{wg}}{T}) - C_{03}\frac{\mu_{w\ell}}{T})] & [C_{01}+\xi C_{02}] \cdot \\
[\frac{C_{10}}{T} + (C_{11}+\xi C_{12})(W_s - \frac{\mu_{wg}}{T})] & [C_{11}+\xi C_{12}] \cdot \\
[\frac{C_{20}}{T} + (C_{21}+\xi C_{22})(W_s - \frac{\mu_{wg}}{T})] & [C_{21}\pm\xi C_{22}] \\
[\frac{C_{30}}{T} - C_{33}\frac{\mu_{w\ell}}{T}] & 0
\end{bmatrix}$$

$$\begin{matrix}
[C_{01}\upsilon_v + C_{02}\upsilon_a + C_{03}\upsilon_w - (C_{01}+\xi C_{02})\upsilon_v] \cdot & C_{03} \\
[C_{11}\upsilon_v + C_{12}\upsilon_a - (C_{11}+\xi C_{12})\upsilon_v] \cdot & 0 \cdot \\
[C_{21}\upsilon_v + C_{22}\upsilon_a - (C_{21}+\xi C_{22})\upsilon_v] \cdot & 0 \cdot \\
C_{33}U_w & C_{33}
\end{matrix} \bigg] \tag{26}$$

where $W_s = \dfrac{R_w T}{p_{vs}}\dfrac{dp_{vs}}{dT}$ and $\xi = \dfrac{M_w}{M_a}\dfrac{P}{P-p_v}$.

Eq(23) is the diffusion equation expressed by the relative chemical potential μ_w to the pure liquid water. By same procedure as above, the diffusion equation by μ_v to the pure water vapour can

be derived.   As a matter of course, matrix $[D_{ij}]$ does not limited
to be positive definite and is not symmetric matrix.   If obtained
the all values of coefficients D_{ij}, we can resolve the values of the
original coefficients C_{ij} and the validity of the coefficients can
be inspected from the point of positive entropy production.

Eq(23) expressed by μ_w is quite equivalent to the same equation
expressed by potential μ_v relative to water vapour.   Theoretically,
it is a matter of choise which is employed.   On the other hand, of
coefficients D_{ij} and equivalent coefficients related to potential
μ_v, these values are depend extremely on the values of temperature,
relative chemical potential.   Since there are adequate or convenient
choise for use, according to the condition.   As well known, at higher
water content, eq(23) based on free water is adequate and at lower or
hygroscopic moisture content, equation based on pure vapour is pre-
fered.

The flux of the liquid water relative to the solid skeleton J_{2w}
in eq(23) is the extension of Darcy's law for unsaturated porous
media and can be applied to deformable solid body.   The flow equation by Klute[32] is equal to above in which temperature gradient
set zero.   Sum of the equation J_{1w} and J_{1a} is equivalent to and extension of Darcy's law in gas phase or of Poiseuille laminar flow.

4. HEAT AND MASS TRANSFER EQUATION

With the assumption of non-swelling body($V_s=0$), following equations of heat and mass conservation are yielded from eq(3), (4) and
(5).   Let Φ_0 be the true porocity of porous material and Φ be the
pore volume that is the fraction of a unit bulk volume of material
occupied by humid air. ψ is the volume fraction of true pore volume
occupied by liquid water.   They are related by

$$\Phi = \Phi_0 (1 - \psi) \qquad (27)$$

The water vapour, dry air, humid air(mixed gas of dry air and
water vapour) and liquid and/or adsorbed water conservation equations are

$$\frac{\partial \Phi \rho_v}{\partial t} + \nabla J_{1w} = \beta (\mu_{w\ell} - \mu_{wg}) \qquad (28)$$

$$\frac{\partial \Phi \rho_a}{\partial t} + \nabla J_{1a} = 0 \qquad (29)$$

$$\frac{\partial \Phi \rho_g}{\partial t} + \nabla (J_{1w} + J_{1a}) = \beta (\mu_{w\ell} - \mu_{wg}) \qquad (30)$$

$$\frac{\partial (\Phi_0 - \Phi) \rho_w}{\partial t} + \nabla J_{2w} = \beta (\mu_{wg} - \mu_{w\ell}) \qquad (31)$$

where ρ_v , ρ_a and ρ_g are the true density of water vapour, dry air
and humid air, respectively.

Mass centre velocity V is very low because of the reposed
solid skeleton which hold overcoming fraction of mass per bulk unit
volume.   Then we can omit the following term from eq(5), that is
work term by internal force as internal viscous dissipation. And
kinetic energy of phase chage and substantial derivative of pressure
are omitted.   Further, we neglect the contribution to heat flow,
of enthalpy fluxes by mass diffusion relative to mass centre in gas
phase. Then the heat balance equation(5) is redused to

$$C\gamma_s \frac{\partial T}{\partial t} + [C_g(J_{1w} + J_{1a}) + \Phi_0 \psi C_w \rho_w J_{2w}] \nabla T$$

$$= - \nabla q + J (H_1 - H_2) + Q \qquad (32)$$

$$J = \beta (\mu_{wg} - \mu_{w\ell})$$

where C is the specific heat of the total system

$$C = (\Phi \rho_g C_g + \Phi_0 \psi C_w \rho_w + (1-\Phi_0) \rho_s C_s)/ \gamma_s \qquad (33)$$

C_w and C_s are specific heat of liquid water and solid skeleton, respectively. C_g is isobaric specific heat of humid air. $(H_1 - H_2)$ is heat of vaporization of water or of adsorption of moisture(relative to the free water).

There are relations between moisture content and the relative chemical potential under the equilibrium state.

$$\psi \Phi_0 = \Phi_0 - \Phi = f(\mu_{w\ell}, T) \qquad (34)$$

This relation is valid at non-equilibrium state as far as mixture of liquid phase is homogeneous or, in other word, the states are not so far from the equilibrium state. In this paper, we omit the hysteresis effects, though those effects can be accout for the same ways but much complication in computation processes[33], as far as the relation is deterministic. Then relation(34) is unique.

The simultaneous equations (28), (29)or (30), (31) and (32) are basic equations for transient heat and mass transfer in the porous materials and applicable to various problems. In local equilibrium $\mu_{1w} = \mu_{2w}$. Sum of eq. (28) (31) yield the water balance equation. If $\nabla P=0$ i.e. there are no external pressure difference or no intense evaporation, eq. (29) and (30) are trivial.

The boundary conditions are

$$\alpha_\mu^{'} \Phi (\mu_w^S - \mu_{gw}) + \alpha_t^{'} \Phi (T^S - T) = J_{1w} - (J_{1w}+J_{1a})\frac{\rho_v}{\rho_g} \qquad (35)$$

$$\alpha_\mu^{'} (1- \Phi) (\mu_w^S - \mu_{we}) + \alpha_t^{'} (1- \Phi)(T^S - T) = J_{2w} \qquad (36)$$

$$\alpha (T^S - T) = q - (H_1 - H_2) T_{2w} \qquad (37)*$$

$$P = P^S \qquad (38)$$

where eq. (35) (36) and (37) are from water vapour, dry air, liquid water and heat conservation, respectively. Superscript s means the value of ambient air.

In local equilibrium between phases and uniform pressure, the heat and moisture balance equation reduce to**

$$(\rho_v - \rho_w) \frac{\partial \Phi}{\partial t} + \Phi \frac{\partial \rho_v}{\partial t} = \nabla (\lambda_\mu^{'} \nabla \mu_w + \lambda_T^{'} \nabla T) \qquad (39)$$

$$C\gamma_s \frac{\partial T}{\partial t} = \nabla \lambda \nabla T + (\lambda_{\mu\ell}^{'} \nabla \mu_w + \lambda_{T\ell}^{'} \nabla T) \nabla T + (H_1 - H_2)J$$

$$J = -\rho_v \frac{\partial \Phi}{\partial t} + \nabla(\lambda_{\mu g}^{'} \nabla_{\mu w} + \lambda_{Tg}^{'} \nabla T)$$

$$= -\rho_w \frac{\partial \Phi}{\partial t} - \nabla(\lambda_{\mu\ell}^{'} \nabla_{\mu w} + \lambda_{T\ell}^{'} \nabla T) \qquad (40)$$

where conductivities $\lambda, \lambda^{'}$ are difined as follows

$$q = \lambda \nabla T + \lambda_\mu \nabla \mu_w \fallingdotseq \lambda \nabla T$$

$$J_{1w} = \lambda_{\mu g}^{'} \nabla \mu_w + \lambda_{Tg}^{'} \nabla T$$

---

* heat transfer by mass diffusion is omitted.
** For simple expressions, external force term are omitted.

$$J_{2w} = \lambda'_{\mu\ell}\nabla\mu_w + \lambda'_{T\ell}\nabla T$$

$$J_{1w} + J_{2w} = \lambda'_\mu\nabla\mu_w + \lambda'_T\nabla T$$

and $\lambda'_\mu = \lambda'_{\mu\ell} + \lambda'_{\mu g}$ $\qquad\qquad \lambda'_T = \lambda'_{T\ell} + \lambda'_{Tg}$

In eq. (39) we assume ρ_w = constant. Coefficient of temperature gradient in heat balance equation, [$\lambda+\lambda'_{Tg}(H_1-H_2)$] is effective thermal conductivity as [34]. To solve eq. (39), following relations must be substituted in eq. (39).

$$\frac{\partial\Phi}{\partial t} = \frac{\partial\Phi}{\partial T}\frac{\partial T}{\partial t} + \frac{\partial\Phi}{\partial\mu_\ell}\frac{\partial\mu_\ell}{\partial t} \qquad\qquad \frac{\partial\rho_w}{\partial t} = \frac{\partial\rho_w}{\partial T}\frac{\partial T}{\partial t} + \frac{\partial\rho_w}{\partial\mu_\ell}\frac{\partial\mu_\ell}{\partial t}$$

Boundary conditions are eq. (37) and sum of eq. (35) and (36). Usually the second term of left had side of eq. (35) is almost equal to zero, then

$$\alpha'_\mu(\mu_w^S - \mu_w) + \alpha'_T(\mu_w^S - \mu_w) = J_{1w}$$

$$\alpha(T^S - T) = q - (H_1 - H_2)J_{2w} \qquad\qquad (41)$$

5. EXPERIMENT OF INTERNAL CONDENSATION

Internal moisture condensation in cellular concrete is investigated. Detail description of experimental arrangement are reported in ohter [35]. Speciments for experiment (10X10X2 cm) are put horizontally on a plate which is upper face of water bath regulated its temperature. This arrangement set in the climate room maintained at constant temperature and humidity. Front face of each specimen is exposed to climate room. At rear face there is no flow of moisture by moisture barrier but heat flows. Side surfaces of each speciemen are covered with alluminium foil bonded by epoxi resin as same as rear face, and specimens are buried in insulation board of foam-polystylene with same depth of specimen, for guarantee of one dimensional heat and moisture flows. Sensitivity of control are ± 0.05°C of water bath, ±0.1°C, ±1%RH in the climate room. Velocity of air main flow upon specimens is ca.1.5(m/s). Six specimens are put on water bath. One of these is used to weigh succeedingly total amount of water content. Other one is for measurement of temperature distribution through specimen along the heat flow by thermocouples of copper constantan, which insert in specimen at interval of 0.5cm. Remainders are for measurement of water content distribution. These specimens cut off to 4 piece of depth 0.5cm.

Procedure of experiment is as follows. Specimens are equilibriumed with climate room air of 20.3°C 75%RH, by exposing to the air for 30 days without vapour and heat barrier and 7 days on water bath with same arrangement of the experiment in which temperature of water bath is maintained constant 20.3°C. After the equilibrium, temperature of water bath are decrease so as to reach 12.2°C of rear surface temperatures of specimens and after attaining this value those surface are maintained at 12.2°C. Fig. 1 shows values of climate room air temperature and himidity, and of rear surface temperature during the experiment. Deviations of the condition from the set point at 12 and 41 days are due to trouble of humidity control of the climate room.

Experimental results of increase of water content, distribution of water contents and temperature distribution during experiment are shown in Fig. 2, 3 and 4 respectively.

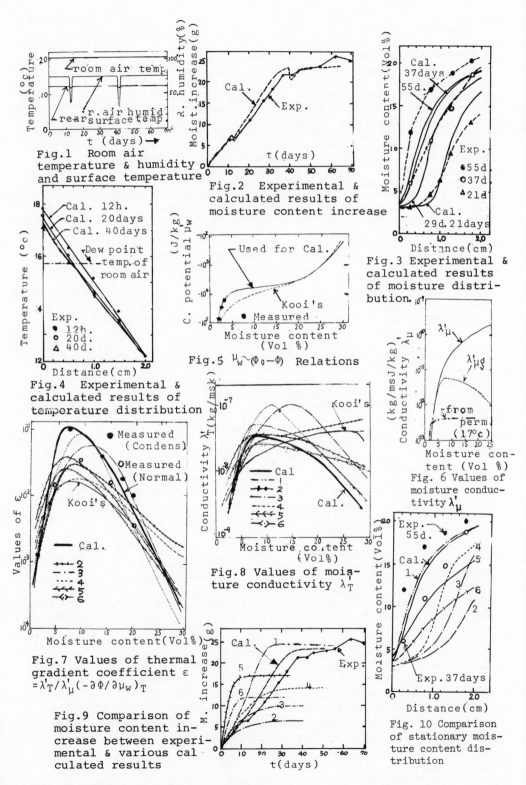

Fig.1 Room air temperature & humidity and surface temperature

Fig.2 Experimental & calculated results of moisture content increase

Fig.3 Experimental & calculated results of moisture distribution.

Fig.4 Experimental & calculated results of temperature distribution

Fig.5 $\mu_w \sim (\Phi_0 - \Phi)$ Relations

Fig.6 Values of moisture conductivity λ'_μ

Fig.7 Values of thermal gradient coefficient $\varepsilon = \lambda'_T / \lambda'_\mu (-\partial \Phi / \partial \mu_w)_T$

Fig.8 Values of moisture conductivity λ_T

Fig.9 Comparison of moisture content increase between experimental & various calculated results

Fig.10 Comparison of stationary moisture content distribution

55

6. COMPARISON WITH ANALYTICAL RESULTS

Dynamic process of moisture content and temperature in specimens are computed by using the eq. (39) with eq. (41) at front
surface, prescrived temperature and no flow of moisture (J_{lw}=0) at
rear surface, and by using value of experimental condition shown in
Fig. 1. The equations are solved numerically by finite difference
method. The scheme of finite differentiation is of "Crank-Nicolson"
type. Pure nonlinear term such as heat flow by mass flux are
linearized by application of Crank-Nicolson scheme only to main
variable as temperature and explicit method to another variable.
Increment of coordinate variable Δx is one tenth of specimens depth
(Δx=0.2cm) in the computation. Time increment Δt is used of 1/12(h).
The equation with these increment was stable and solutions of this
show good agreement with the solution from time increment Δt=1/120(h)
during initial several days. We did not discuss stability by eigenvalue of each inverse matrix.

Values of coefficients used in this computation are as follows.
Measured values are as ρ_s=486(kg/m^3) Φ_0=0.7 vapour permeability
$\lambda''_\mu = \lambda'_v/(\partial P_v/\partial \mu_v)=\lambda'_v/(\partial P_v/\partial T)$ at 75%RH=3.51×10^{-11}[kgm/Ns] and
$\varepsilon = \lambda'_v/\lambda''_\mu (-\partial \Phi/\partial \mu)_T$ shown in Fig. 7. α=15.2 (W/m^{2}K) from flow velocity
measurement, mass transfer coefficient $=\alpha'/(\partial P_\mu/\partial \mu)_T=\alpha'/(\partial P_v/\partial T)_\mu$
=8.16 [kg/Ns] from Lewis relation and α, estimated values are as
C_s=1326 [J/kgK], H_1-H_2=2.465×10^6 [J/kg], λ=1.1624(0.1+10.2(Φ_0-Φ)×10/
3.0×ρ_s) [W/mk]. Relation of equilibrium moisture conents Φ_0-Φ=f(μ,T)
are shown in Fig. 5 in which value in hygroscopic range are from
the measured value and in high water content from the report [36].
Moisture conductivity λ''_μ are shown in Fig. 6 which are estimated
values (one third of Kooi's value)[37] and discuss later. In Fig. 6
broken line is the calculated value from vapour permeability and
dotted line is moisture conductivity of gas phase $\lambda''_\mug$ which are
calculated with assumption that moisture flow by temperature
gradient is only by gas phase, and with λ''_v value [36]. From λ''_μ
and ε we calculate λ''_T which are shown in Fig. 8. In Fig. 7 dot
Mark ● are calculated from steady state of this experiment and
circle mark ○ are from ordinary our measurement of ε.

In Fig. 2, 3 and 4 calculated results are shown with experimental results. There are good agreement between them. Differences
between them in Fig. 3 may be due to measurement error of moisture
contents.

7. SENSITIVITY ANALYSIS

Estimated values of moisture conductivity λ''_μ are used for the above
calculation. Validity of these value are discussed as follows. In
following calculation, variation of climate room condition are
ignored and we use constant value of room air temperature and humidity.
Computations are performed by using values of λ''_μ being 1.5times of
value used in preceding calculation and fixed value of ε. Calculated
result of this case are shown in Fig. 9 and 10 by curve 1. *)
By curve 2 calculated result are shown by using λ''_μ (three times of
the values shown in Fig. 7) from Kooi's measurement of stationary
experiment (increasing moisture content) [37] and ε from our ordinary
experiment almost same as kooi's value. There are very large
deviation from experiment.

*) In these figures, solid line marked by "Calculated" is the
calculated results of the preceding section (ignoring the variation
of the room air humidity).

Using various values of ε shown in Fig. 8 by curve no. 1-6 we perform the same calculation with the fixed values of the moisture conductivity λ_μ' which are the measured values by Kooi(three times of the values shown in Fig. 6). The calculated results are shown in Fig.9 and 10. They show very large deviations from the experimental results and the "calculated" results from the preceding section. Those indicate the validity of the values of the parameter λ_μ' used in the preceding section. Further investigation must be made for identifying the values of λ_μ' and for the uniqueness of ε. For the purpose of this analysis, dynamic method may be effective.

8.CONCLUSIONS

Combined transient heat and mass transfer equations are derived for thermal and hygric dynamic analysis of the building and its elements, based on irreversible thermodynamics applying to the heterogeneous two phase flow. The equations can be applied to the case of non-local equilibrium between phases and to case of the existence of thermodynamic pressure gradients, as well as pure diffusion. The relative chemical potentials μ_w or μ_g are presented based on thermodynamics, though this concept had well but not strictly known such as capillaric potential. Use of this potential allow unified treatment of multilayer problems without any additional relation. From the view of the numerical analysis it is hoped that equations are close to linear system. Chemical potentials are not usually best from this point and other variables may be better according to condition analyzed. But introduction of other variable is to be regarded as purely mathematical transformations.

Experiment of internal condensation in cellular concrete is compared with numerical analysis based on the equations derived here with local equilibrium condition between phases. They show good agreement of total moisture content, moisture distribution and temperature distribution. This indicates the validity of this equations of this case. We think that the validity of the used values of parameter λ_μ', especially at high water content, must be discussed by other experiment, since the values are different from the values measured by stationary or /and isothermal conditions. And also the uniqueness of ε must be discussed. As a matter of course, values of moisture conductivity at outer drier parts play important role to internal condensation rates. Finally, parameter estimation from dynamic process may be effective for the identification of porous systems.

REFERENCES
[1] Glaser,H.   Kälte Technik 10: 86, 1958
[2] Vos, B.H.   Building Sci. 3: 191, 1969
[3] Sandberg,P.I. Inst. für Byggnads Tech. Lund Report 43   1973
[4] deVries, D.A. and Philip, J.R.   Trans. Amer. Geophys. Un.
      38:   222   1957
[5] Gertis, K. and Kiessl,K.   Deutsch. Ausschuss. für Stahlbeton
      258;   87   1975
[6] Harmathy, T.Z.   Industrial and Eng. Chem. Fundamental
      8;   92   1969
[7] Matsumoto, M. Prep. of Annual Meeting of Architectural Inst. of
      Japan   p177   1971   (in Japanese)
[8] Maeda, T.   Method of Analysis of Transient Room Air Humidity
      and Temperature in Building Composed of   Multi   Rooms, Consider-
      ing the Moisture Uptake of Surrrounding Wall, Report of the
      Design Committee, Thermal Eng. Arch. Inst. of Japan   1962
      (in Japanese)

[9]   Maeda, T.   Tras. of Arch. Inst. of Japan   89: 325   1963
[10] Maeda, T. and Matsumoto, M.   Prep. of the Annual Meeting
      of Arch. Inst. of Japan   P131   Aug. 1969 (in Japanese)
[11] Luikov, A.V.   Heat and Mass Transfer in Capillary porous Body
      Pergamon Pr.   1966
      Luikov, A.v.   Int. J.   Heat and Mass Trans.   18; 1,   1975
[12] Krischer, O.   Die Wissenschaftlichen Grundlagen der Trocknungs-
      technik 2Auf.   Springer V.1.   1963
[14] Cary, J.W.   Int. J of Heat and Mass Trans.   7:   531, 1964
      Taylor, S.A. and Cary,T.W.   Soil Sci. Soc. Amer. Proc. 28;
      167,   1964
[15] Groenvolt,P.H. and Bolt,G.H.   J. of Hydrology   7; 358, 1969
[16] Dorokhov,I.N., Kafarov, V.V. and Nigmatulin, R.I.
      Appl. math. and Mech.   39;   461,   1975
[17] Matsumoto, M.   The 6th Symposium of thermal problem, Environ.
      Eng.   Arch. Inst. of Japan   6:   15   1976 (in Japanese)
[18] Nordon, p. and David, D.   Int. J. Heat and Mass Trans.
      10:    853   1967
[19] Maeda, T. and Matsumoto, M.   Soc. of Heating,   Ventilating and
      Sanitary Eng.   Kinki Branch of Environ. Eng.   Report no.11:
      1, 1967   (in Japanese)
[20] Matsumoto,M. and Fujihana, R.   CIB/RILEM   The 2nd Int. Symp.
      on Moisture Problem in Building   2.1.2   1974
[21] Novak,L.T. and Coulman, G.A.   Canadian J. of Chem. Eng.
      53:    60,   1975
[22] Raats,P.A. and Klute, A.   Soil Sci. Amer. Proc.   32;   452,
      1968
[23] Slegel, D., Davis,L.and Boersma,L.   Heat and Mass Transfer
      in the Biosphere part 1 Ed. by deVries,D. and Afgan, N.
      Scripta Pr.   p87   1975
[24]   ibid.[2]
[25] Van der Kooi, J. Moisture Transport in Cellular concrete,
      Ph.D Thesis Eindhoven Univ.of Technology, Waltman, Delft, 1971
[26] Nigmatulin, R.I.   J. Appl. Math. and Mech.   34: 1033, 1970
[27] Bird,R.I.,Stewart, W.E. and Lightfoot,E.N.   Transport Phenomena
      Wiley Pr.   1960
[28]   ibid. [26]
[29] Luikov, A.V.   Int. J. Heat and Mass Trans.   9: 139, 1966
[30] Fitts,D.D.   Nonequilibrium thermodynamics   McGraw H. 1962
[31] Philip, J.R.   Trans. Amer. Geophys.Un.   38: 782,   1957
[32]   ibid. [22]
[33] Matsumoto, M. and Matsushita, T.   Proc. Kinki Branch Arch.
      Inst. of Japan   p69   May 1977   (in Japanese)
[34] deVries, D.A.   Heat and Mass Transfer in the Biosphere
      Ed. by deVries,D.A. and Afgan, N.   Scripta Pr.   p5   1975
[35] Matsumoto, M. and Akayama, A.   Proc. of Kinki Branch Arch.
      Inst. of Japan   p85   (NO. 423)   1977 (in Japanese)
[36]   ibid. [25]
[37]   ibid. [25]

FUNDAMENTALS OF MOISTURE
AND ENERGY FLOW
IN CAPILLARY-POROUS
BUILDING MATERIALS

JOHAN CLAESSON

Department of Mathematical Physics
University of Lund — Box 725
Lund, Sweden 22007

ABSTRACT

Basic physical features of combined energy and moisture flow in porous building materials for dynamic and stationary processes are discussed. The mathematical and physical structure of these dynamic processes is discussed in terms of local thermodynamical equilibrium, flows induced by gradients in intensive state variables, and conservation of energy, moisture and other components.

The conditions for thermodynamical equilibrium in a capillary-porous material containing moisture in liquid and gas phases, air, and possibly salts in the pores are presented as a prerequisite for the understanding of the dynamic flow processes. The peculiarities of pore-water tension are discussed. Moisture may flow in gas phases and in liquid phases while interacting in a complicated way. The specific problems arising from the complex character of the flow are discussed. The relation between the moisture flow and gradients in intensive state variables such as pore-water pressure or tension, vapor pressure, gas pressure and temperature is discussed.

Some problems concerning the energy flow are discussed. The energy flow is complicated by the mass flow in the pores and especially by continuous condensation and evaporation of water in the pores. Divisions of the energy flow into heat flow and energy transfer due to mass flow and the effects of condensation and evaporation are discussed.

The equilibrium relations and the transfer coefficients for porous materials often exhibit rather pronounced hysteresis. The causes of hysteresis and the complications due to hysteresis in the description, are dealt with.

INTRODUCTION

A building material like concrete or brick is penetrated by a complex pore system. The pore space is filled by air, water and possibly salts. Suitable parts of the pore volume are occupied by capillary condensed liquid water. The free pore walls are covered by layers of adsorbed water molecules. The remaining parts of the pore space contain a mixture of air and water-vapour. The phases of liquid water may contain salts.

Consider a small part or region of the porous material. The region is small compared with the overall dimensions of the considered piece of building material, but it is large compared with the minute single pores. The region contains a reasonably representative sample of the pore structure. The state of such a region is characterized by a number of variables like the moisture content w (kg H_2O/m^3), the temperature T, the water-vapour pressure p, and so on. We will consider such a region as a point.

The state will usually vary from point to point in the material. The distribution of states in the material will change with the time t during a dynamical process when moisture and energy flow through the porous material. The aim of the study of these processes is to be able to calculate how this distribution of states through the building material evolves with time. The states at an initial time and the conditions at the boundaries during the dynamical process have to be known.

MATHEMATICAL STRUCTURE

The mathematical description is based on the conservation equations for moisture and energy. We will, for the sake of notational simplicity, consider the one-dimensional case with flows and changes only in the direction of x. Let F_w and F_e denote the flows of moisture and energy respectively. The moisture flow $F_w(x,t)$ in a point x at a time t refers to the total mass of the waters molecules that pass the point x per unit time and unit area perpendicular to the flow direction x. The moisture flow F_w may have contributions from liquid flow, convective and diffusional gas flow and possibly migrations along the pore walls. The energy flow F_e refers to the total energy flow. The heat flow F_q is one contribution to F_e.

The moisture content, i.e. the total mass of water molecules in the myriads of liquid, gas and surface phases per unit volume of the porous material, is w. Let e denote the total energy content per unit volume. (It is preferable to use moisture and energy contents per unit mass of dry solid in situations where volume changes of the solid are of importance.)

The introduced quantities F_w, F_e, w and e are functions of x and t. The equations for conservation of moisture and energy are:

$$\frac{\partial w}{\partial t} = -\frac{\partial F_w}{\partial x} \qquad\qquad \frac{\partial e}{\partial t} = -\frac{\partial F_e}{\partial x} \qquad\qquad (1)$$

The left-hand members give the increase in moisture and energy content in a point (per unit time and unit volume). The right-hand members give the net influx to a point (per unit time and unit volume) of moisture and energy respectively. The net influx depends on the rate of change of the flow intensity along the x-axis.

We must now have suitable expressions for the flows F_w and F_e in order to be able to perform the calculations. These expressions serve to relate the flows to the present distribution of states through the material. The flows present the paramount problem. The flows in a point x at a time t depend on the present state at the point and on the slightly differing states in the vicinity of the point. Differently phrased, the flows depend on present gradients of state variables at the point and on the state at the point.

We must be able to specify the flows that arise for every distribution of moisture and energy that occurs during the dynamical process. The conservation equations (1) express, once one has got used to them, a rather trivial bookkeeping for moisture and energy. The difficulties and the main physical content lie in the actual expressions for the flows, in which these flows are related to the spacial rates of change and to the levels of various state variables.

It is often a bit difficult to discern the structure behind the great variety of alternative, more or less clearly formulated, conservation equations that are used. However, the mathematical structure contained in the conservation equations (1) is indeed simple. Suppose that the moisture and energy distributions $w(x,t)$ and $e(x,t)$ are known at a time t. Then the state is known at each point. A certain information concerning the preceding moisture history for each point may also be necessary due to hysteresis effects. The complete knowledge of the present state in a point also requires a number of formulas that relate pertinent state variables like the temperature T, vapour pressure p, and so on to the conservation variables w and e. The expressions for the flows then give $F_w(x,t)$ and $F_e(x,t)$ for each point x at the considered time t. There must be additional expressions that render it possible to specify the flows through the boundaries during the dynamical process. From formulas (1) we obtain the rate of change of w and e for each point. From this we get w(x,t+Δt) and e(x,t+Δt) for each point x at a slightly later time t+Δt. The procedure is repeated for the new time t+Δt, and so on. This mathematical structure of the calculations is illustrated in figure 1.

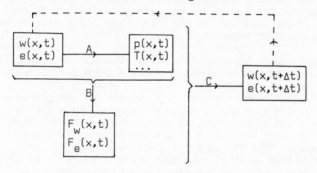

A: Equilibrium relations   B: Expressions for the flows
C: Conservation equations

Figure 1.   Mathematical structure of a dynamical process with moisture and energy flow in a porous building material.

A numerical simulation of a dynamical process may closely follow the procedure of figure 1. The continuous moisture and energy distributions are approximated by the values in a number of discrete points or cells. A set of equilibrium relations allows the calculation of other pertinent state variables for each cell. The flows of moisture and energy through the boundary between two consecutive cells are calculated from the differences in the state variables between the two cells and from the prevailing levels of the state variables with the aid of the expressions for the flows. The boundaries require additional formulas that give the flows through these. Direct moisture and energy balances then give new moisture and energy contents for each cell after a time increment Δt. This procedure is repeated for time-step after time-step.

Differences in concentrations of other substances and changes due to flows of these substances may be of importance in the dynamical process. There is an additional conservation equation for each new substance. Consider as an example a case when the air and a salt in the pores are of importance. Let F_a(kg air/m² s) denote the flow of air and a (kg air/m³) the air content. Let F_b and b be the corresponding quantities for the salt. Then we get two additional conservation equations for the air (a) and the salt (b):

$$\frac{\partial a}{\partial t} = -\frac{\partial F_a}{\partial x} \qquad\qquad \frac{\partial b}{\partial t} = -\frac{\partial F_b}{\partial x} \qquad\qquad (2)$$

The difficult problem is again the expressions that make it possible to calculate the four flows, when the state variables are known through the porous material.

THERMODYNAMICAL EQUILIBRIUM

It is of fundamental importance to base the studies of these dynamical processes on a thorough thermodynamical analysis of the behaviour of moisture and other substances in the pore system of a building material. The thermodynamics of the pore system with its different constituents provides the basis for all deeper physical insight in these processes. This thermodynamical system is in a way very complicated due to the complex pore structure in which liquid and gas phases are intertwined in a way that may seem completely intractable. However, the analysis of the conditions for thermodynamical equilibrium produces a amazingly large amount of valuable and useful information. The thermodynamical investigation also provides valuable insight into the behaviour in non-equilibrium, since we obtain the direction of the local flows. The intensities of the overall flows are given from measured or inferred transport and transfer coefficients.

Let us now consider a piece of a porous material in thermodynamical equilibrium. The pore space is occupied by liquid water phases and gas phases. We will always have a connected region in mind when the word phase is used.

A gas phase contains a mixture of air and water-vapour. All gas phases will in equilibrium be in exactly the same state. A gas phase deep inside the material will be in the same state as the surrounding humid air, even if these two phases are seperated from each other by a long sequence of other liquid and gas phases. Thus all gas phases have the same gas pressure p_g and the same water-vapour pressure p. Consider as an illustration a non-equilibrium situation, where a single gas bubble deep inside the material has a higher pressure p_g than that of the outside surrounding air. Air from the bubble will dissolve in neighbouring liquid phases and diffuse through the liquid out to the surrounding air. The gas bubble will eventually adjust its pressure to the equilibrium value or, if this is not feasible, vanish completely.

Let us first consider the case when the liquid phases do not contain any salts. Then all liquid water phases must in equilibrium be in exactly the same state. They have the same pressure p_ℓ. The pore-water pressure p_ℓ is in general different from the pressure p_g in the gas phases in the pores. The liquid and gas phases exchange water molecules. The chemical potential for water in the gas phases and in the liquid phases must in equilibrium be the same. This condition gives the so called Kelvin equation [1]. A very accurate expression for the condition of equal chemical potential is [2]:

$$p_\ell - p_s(T) = \frac{RT}{v_\ell} \ln \left(\frac{p}{p_s(T)} \right) \qquad (3)$$

Here $p_s(T)$ is the saturation vapour pressure of (pure) water at the temperature T. The temperature T is in degrees Kelvin, and R is the general gas constant. The quantity v_ℓ is the volume of one mole of liquid water. The factor RT/v_ℓ is equal to $1340 \cdot 10^5$ Pa at room temperature. Table I shows corresponding values of relative humidity p/p_s and pore-water pressure p_ℓ for T=290K.

$\frac{p}{p_s}$	0.8	0.9	0.99	0.999	1	1.001
p_ℓ (bar)	-300	-140	-13	-1.3	+0.02	+1

Table 1.   Corresponding values of relative humidity p/p_s and pore-water pressure
p_ℓ at room temperature in a case without salts. (1 bar=10^5 Pa)

There are two conspicuous facts concerning the pore-water pressure p_ℓ.
The first is that p_ℓ usually turns out to be negative. This means that there
is a state of tension in the liquid phase. The second remarkable fact is the
large numerical values of p_ℓ. A change $\Delta(p/p_s)=0.01$ in relative humidity corresponds to a change $\Delta p_\ell \approx 13 \cdot 10^5$ Pa in pore-water pressure or tension.
Equation (3) gives the condition for diffusional equilibrium between a
liquid water phase and neighbouring water-vapour in a gas phase. Consider now
a non-equilibrium situation with fixed relative humidity p/p_s. Water will
evaporate from a liquid phase, if the pore-water pressure is higher than the
equilibrium value according to equation (3). Conversely, water-vapour will condense on the liquid surfaces, if the pore-water pressure lies below the equilibrium value. Suppose as an example that the relative humidity is lowered from
saturation $p/p_s=1$ to $p/p_s=0.90$. Then, according to table 1, the pore-water
pressure must fall to $p_\ell=-130\cdot 10^5$ Pa in order to prevent evaporation and maintain equilibrium. All liquid water will eventually evaporate if the liquid
phase fails to establish this high tension or negative pressure. We know that
there is liquid water in many porous materials in equilibrium below saturation.
This means that there must exist a mechanism which has the ability to create and
maintain these high tensions or negative pressures in the pore water.
These states of high tension are caused by the surface tension of suitably
curved water surfaces. Consider a point on a water surface in a pore. Let κ
denote the so-called mean curvature of the surface in the point. The inverse of
κ gives a mean radius of curvature. It can be shown with the aid of differential
geometry [2] that the local net effect of the surface tension γ of the curved
water surface is a force in the normal direction to the surface. The magnitude
of the net force per unit area of water surface is $2\kappa\gamma$. This is an exact expression irrespective of the shape of the water surface. There is a pressure
p_g in the gas phase outside the liquid and a pressure p_ℓ in a liquid phase.
The pressure difference p_g-p_ℓ is balanced by the force $2\kappa\gamma$. This equation for
force balance at a point on a curved water surface is due to Laplace:

$$p_g - p_\ell = 2\kappa\gamma \tag{4}$$

The force equilibrium is illustrated in figure 2. The mean curvature κ must in
equilibrium have the same value at all points of all water surfaces in the pores,
since p_g and p_ℓ have the same values throughout. This is a strong condition on
the shapes of the water surfaces.
Coercive forces between the water molecules keep the liquid phase together
and prevent rupture. Coercive forces between the pore wall and the water in
the vicinity of the walls keep the liquid attached to the pore walls in spite
of the tension in the liquid.

Figure 2.   Cross-section through a pore.
The pressure difference p_g-p_ℓ
over the water surface is balanced by the force $2\kappa\gamma$ from
the surface tension γ of the
curved water surface with mean
curvature κ.

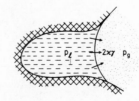

High pore-water tensions require heavily curved water surfaces. Take as an example a situation with $p_\ell = -130 \cdot 10^5$ Pa, which corresponds to $p/p_s = 0.90$. The so-called mean curvature κ in any point at a free surface of a liquid phase is given by (4). The corresponding mean radius of curvature $R = 1/\kappa$ becomes $R = 110$ Å $(1.1 \cdot 10^{-8}$ m).

The liquid phases in the pores of a building material usually contain salts. These influence the chemical equilibrium condition between liquid water and water-vapour considerably. There remains a lot of research to be done concerning the influence of salts on the dynamical flow processes. The condition for equilibrium is now instead of (3):

$$p_\ell - p_s(T) = \frac{RT}{v_\ell} \, \ell n \, (\frac{p}{p_s(T)}) + \frac{RT \, M_w}{v_\ell} \, \nu \phi m \qquad (5)$$

The difference from (3) is the last term, which is of osmotic character. Here M_w is the mole weight of water, ν the number of ions per dissolved salt molecule, and m the molality (i.e. the number of moles of salt per kilogram of water) of the salt in the considered liquid phase. The factor ϕ is the molal osmotic coefficient for the salt dissolved in water [3]. It is a function of the concentration m. A reasonable approximation is usually $\phi \approx 1$. Relation (5) is shown in figure 3 for sodium chloride. We note as a rule of thumb that a decrease by ten per cent in relative humidity $(\Delta(p/p_s) = -0.10)$ or in relative weight of salt in the salt solution $(\Delta m \approx -2)$ corresponds to a decrease of pore-water pressure p_ℓ of the order of $100 \cdot 10^5$ Pa.

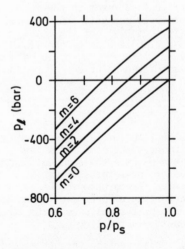

Figure 3. Relation between pore
water pressure p_ℓ and
relative humidity p/p_s
for different molali-
ties m of sodium chlo-
ride at room tempera-
ture. The concentration
m=2 means 12 weight
per cent salt in the
water.

The concentration m of the salt may be different in seperate water phases even in thermodynamical equilibrium. The pore-water pressure p_ℓ will then have different values in different water phases in the pores.

Equations (3), (4), and (5) lose their meaning in extremely small liquid phases. Consider water in a pore with linear dimensions of say 15 Å, which corresponds roughly to the thickness of five layers of water molecules. The influence of the solid walls will give a more complicated state of tension than that of a simple isotropic pressure or tension. The description requires the use of a full tension tensor. The equations for thermodynamical equilibrium are valid for a point in the liquid water as long as a simple isotropic state of tension, characterized by a hydraulic pressure or tension p_ℓ, prevails.

The equilibrium conditions are valid locally also during a dynamical flow process for which the states through the material change with time. A liquid phase and a neighbouring gas phase lie so close together that the local deviations from equilibrium ought to be exceedingly small, even during the most violent dynamical process.

MOISTURE FLOW

Water molecules may flow or diffuse in liquid phases, gas phases and along adsorbed water layers on the pore walls. We will in the following discussion disregard the possibility of migration of water in the adsorbed water layers on the pore walls.

The water of a liquid phase will move when there is a gradient in the local pore water pressure p_ℓ. A gradient in the local gas pressure p_g will cause a convective flow of the gas with its water-vapour, while a gradient in the water-vapour pressure p will cause a diffusion of vapour through the gas phase. Moisture flows are caused by gradients in p_ℓ, p_g and p. These are (excluding the possibility of surface migration) the three main direct physical causes for moisture movements in the porous material.

Convective gas flows require gradients in the gas pressure p_g. The main cause, besides externally applied pressure differences over the material, for changes in p_g is temperature changes. A temperature change in a gas phase of essentially constant volume entails a minor pressure change in accordance with the gas law. The pressure changes when the temperature is raised are mainly due to evaporation of water so that the vapour pressure increases considerably. The vapour pressure will roughly change as the saturation vapour pressure $p_s(T)$, as long as the material contains enough liquid water. A rapid solar heating of a humid wall to say 50°C may raise the vapour pressure to $p\approx0.25\cdot10^5$ Pa. This will give an increase of the gas pressure p_g with 25 per cent. An extreme heating of a wall of concrete during a fire to say 200°C may increase the gas pressure p_g in the pores from $1.0\cdot10^5$ Pa up to $16\cdot10^5$ Pa. The intensity of the convective gas flows through the material depends on the degree of free gas passages through the pore system. An increasing extension of the liquid phases will diminish the gas passages and lower the gas flow. In conclusion we have that a high moisture flow intensity in the form of convective gas flow requires considerable and rapid temperature changes (or externally applied pressures).

We will now limit the discussion to the case when gas pressure gradients and convective gas flows are negligible. There are water-vapour flows due to gradients in p and liquid flows due to gradients in p_ℓ. The flow pattern on the pore level may be extremely complicated. There is vapour diffusion in gas phases and convective flows through liquid phases. Water evaporates from and condensates on myriads of water surfaces in the pores. These flows in the phases and between phases affect each other. The heat of evaporation and condensation, the energy flows, and in particular the heat flows also affect the moisture flow pattern.

A particular water molecule may diffuse through a gas phase, condensate on a liquid surface, flow through the liquid phase, evaporate into a new gas phase, and so on. A separation of the moisture flow in a vapour and a liquid component is not a very precise concept. It is only the total moisture flow that has a precise measurable meaning.

The state in a point of the material is given by two independent variables, when we assume constant gas pressure p_g and neglect effects of salts. A natural choice, from a physical point of view, of independent variables is p and p_ℓ.

The macroscopic moisture flow is given by a phenomenological expression of the following type (in the one-dimensional case):

$$F_w = -D_1 \frac{\partial p}{\partial x} - D_2 \frac{\partial p_\ell}{\partial x} \qquad (6)$$

The two transport coefficients D_1 and D_2 are functions of the state variables. Complications due to hysteresis are discussed below. The coefficients D_1 and D_2 are necessarily positive.

Vapour diffusion will dominate at low moisture contents. Liquid flow will normally dominate at high moisture contents. The transport coefficients D_1 and D_2 will tend to zero for high respectively low values of w. The dynamical flow problem is much simpler in many respects in the limits of low and high moisture contents, when one type of moisture flow prevails. The really difficult situation is for intermediate moisture contents, when vapour diffusion and liquid flow interact and both play important roles.

I suspect that the coefficients D_1 and D_2 invariably will decrease respectively increase monotonously as functions of the moisture content w. Alternative moisture transport coefficients, which refer to other gradients, in particular to the gradient in w, tend to become rather irregular functions of w [4].

It is remarkable that there does not exist, as far as the author knows, any method to measure negative pore-water pressures (except through the measurement of relative humidities, which becomes problematic close to saturation.) The state of the moisture may be characterized by the water-vapour pressure p or the relative humidity p/p_s. The state of the moisture is alternatively specified by the pore-water pressure p_ℓ. From table 1 we have that the state of the moisture changes considerably over the region of high relative humidities. The pore-water pressure changes from $-13 \cdot 10^5$ Pa to $+1 \cdot 10^5$ Pa, when the relative humidity increases from $p/p_s = 0.99$ to saturation. Thus p_ℓ is the appropriate variable to characterize the state of the moisture at high relative humidities. The absence of methods to measure p_ℓ seems to have concealed the predominant importance of the pore-water pressure or tension at higher humidities.

We shall briefly indicate how negative pore-water pressures may in principle be measured with the use of osmotic effects. Let the moisture in the porous material be in contact through a semi-permeable membrane with a water solution that contains a solved substance A. The water may pass the membrane but not the substance A. The solved substance A will decrease the chemical potential of the water in the solution. The water solution can therefore at a positive pressure establish equilibrium with the moisture with its negative pore-water pressure. A knowledge of the concentration of A (and of the positive pressure) in the solution gives the pore-water pressure p_ℓ in the solution.

Other variables like p and T or p_ℓ and T may be used instead of p and p_ℓ as independent variables. This gives phenomenological expressions for F_w which are analogous to (6). Formula (3) and the rules for the differentiation of a composite function give the relations between the coefficients of the different descriptions. Another natural choice of independent variables is w and T. The special complications for this choice due to hysteresis are discussed below.

ENERGY AND HEAT FLOW

We need in a rigorous description of a dynamical moisture and heat flow process the total energy flow, not merely the heat flow. The heat flow coincides with the total energy flow in the case without mass flows. The difficulties arise when there are mass flows through the system. Let us limit the analysis to the situation when only water-vapour and liquid water flow through the pore system.

Let $F_{w\ell}$ and F_{wv} denote the liquid and vapour flows respectively. The total moisture flow F_w is equal to $F_{w\ell} + F_{wv}$ This splitting should be used with caution, since we cannot measure these two flows separately. The vapour and liquid carry their respective internal energies e_v' and e_ℓ' (J/kg). The considerable difference $e_v' - e_\ell'$ is equal to the latent heat of evaporation (at constant volume). This gives a convective contribution $e_v' F_{wv} + e_\ell' F_{w\ell}$ to the total energy flow F_e. There are also contributions from pressure work in the displacement of the masses. Finally there is a heat flow component F_q. Neglecting contributions from pressure work we have

$$F_e = F_q + e_v' F_{wv} + e_\ell' F_{w\ell} \tag{7}$$

This equation may be regarded as a definition of what we call the heat flow. The physical content of the formalism lies in the expression for F_q; for example when we postulate Fourier's law:

$$F_q = -\lambda \frac{\partial T}{\partial x} \tag{8}$$

Here λ is the heat conductivity. It will depend on the moisture content w.
The trouble with (7) is that we do not know the separate flows F_{wv} and $F_{w\ell}$. A more rigorous approach is to postulate directly for F_e a linear expression in gradients of state variables in the same way as we did for the moisture flow F_w. The energy transport coefficients in this linear expression should in principle be measured directly. This however poses great experimental difficulties.

Fortunately, the moisture flows in porous building materials are in most applications very small. The contribution to the energy flow from the moisture flow is then negligible compared to the ordinary heat flow. The energy flow F_e is with good accuracy equal to the heat flow F_q. The change in the energy content due to changes in the moisture content is then also negligible. The energy conservation equation becomes the usual heat conduction equation:

$$\frac{\partial e}{\partial t} \simeq C \frac{\partial T}{\partial t} \qquad F_e \simeq F_q = -\lambda \frac{\partial T}{\partial x} \qquad C \frac{\partial T}{\partial t} = \frac{\partial}{\partial x}\left(\lambda \frac{\partial T}{\partial x}\right) \tag{9}$$

Here C is the heat capacity (at constant moisture content) per unit volume of porous material (J/m^3 $^\circ$C). The coefficients C and λ will depend on the moisture content w.

Let us now briefly consider the general case when the influence of the moisture flow on the energy balance cannot be neglected. The difficulty stems from the fact that we do not know the separate vapour and liquid flows. We may alternatively say that the difficulty is due to the fact that we do not know the net rates of evaporation or condensation in each point of the material. The heat of evaporation affects the energy equation in an uncontrollable way.

It should be noted that is physically unsound and arbitrary to make direct assumptions about the distribution of heat generation and consumption due to condensation and evaporation within the material. The gradients in state variables, which arise during the dynamical process, induce certain mass flows in accordance with the transfer coefficients. The evaporation and condensation is then governed by the requirement that local equilibrium shall prevail. For example, an excess net inflow · of water vapour to a small region of the material will be accompanied by an appropriate rate of net condensation so that local thermodynamical equilibrium is maintained.

HYSTERESIS

The relation between moisture content w and the state of the moisture, given for example by p/p_s or p_ℓ, often exhibit pronounced hysteresis for building materials. The amount and distribution of the pore water in the pores depend not only on the actual state of the moisture but also on the previous moisture history. We get a new branch for the relation each time the direction of the change in state of the moisture is altered. The moisture content w depends on the state of the moisture and on all preceding turning points or states at which the direction of change was altered. Moisture transport coefficients may also exhibit this complicated type of dependence.

Special complications arise when the moisture flow is related to the gradient of the moisture content w. Let D_w denote the corresponding transport coefficient. The problems are due to the fact that moisture flows are caused by gradients of intensive state variables like p and p_ℓ, and not by gradients of w. The dependence of D_w on the previous moisture history will become very intricate because of the moisture hysteresis. Consider as an illustration an equilibrium situation where the moisture content w varies through the material due to hysteresis. There are not any flows, and the intensive state variables are constant through the material. The transport coefficient D_w must in this completely ordinary situation be zero, since the gradient of w is different from zero. The description with D_w as transport coefficient should only be used in monotonous moisture processes, where there is a single relation between the moisture content and the state of the moisture.

These hysteresis effects are caused by irreversible processes when the moisture content changes locally in a pore. Consider a situation when a part of a pore is filled by pore water in an irreversible process. This part of the pore cannot be emptied by retracing the process in the opposite direction, because of the irreversible character of the process. The pore has to be emptied in some other way during other conditions. The pore water can therefore be distributed in different ways in the pores depending on the previous moisture history.

There are many possible types of irreversible processes. A water surface in a pore may encounter an obstacle. An example is when two water surfaces meet and coalesce irreversibly to a single water surface in a new positions. Different types of thermodynamical instabilities also cause irreversibilites. The water surfaces in the pores may reach an unstable position. Then the surface moves swiftly to a new, stable position.

The various branches for the relation between the moisture content and the state of the moisture, as well as the corresponding different branches for moisture transport coefficients, have to be known when a dynamical process is numerically simulated. The turning points, where the direction of the change in the state of the moisture is altered, must be saved for each point in the material in order to be used in the subsequent calculations. Hysteresis effects cause considerable complications to dynamical moisture and energy flow processes.

Acknowledgement

This study has been sponsored by the Swedish Council for Building Research.

REFERENCES

1. Thomson, W. (Kelvin). On the equilibrium of vapour at a curved surface of liquid. Phil. Mag., p. 448, 1871.

2. Claesson, J.. Theory of microcapillarity. I Equilibrium and Stability. Thesis, Lund, 1977.

3. Robinson-Stokes. Electrolyte Solutions. Butterworth, London, 1955.

4. Krischer, O.. Die wissenschaftlichen Grundlagen der Trocknungstechnik. Springer-Verlag, Berlin, 1963.

MOISTURE TRANSFER
IN POROUS MEDIUM UNDER
A TEMPERATURE GRADIENT

J. P. GUPTA

Chemical Engineering Department
Indian Institute of Technology
Kanpur 208016, India

S. W. CHURCHILL

Chemical & Biochemical Engineering
University of Pennsylvania
Philadelphia, Pennsylvania, USA 19174

ABSTRACT

Most places on earth undergo diurnal as well as seasonal temperature changes. Besides, the internal heating and cooling of the buildings in cold and hot weather also results in temperature variation across the walls of the building. These temperature changes result in the movement of moisture in the soil and the building materials which can be significant when sub-freezing temperatures are encountered. Since this can affect the structural properties of the building material and its foundations, it is important to understand the mechanism of the moisture movement. A model, proposed earlier, based upon the movement due to vapour pressure gradient is briefly presented alongwith one based upon the surface tension gradient in the liquid. Both models, though widely apart in concept, give order of magnitude results. It is difficult to reject either model based upon the present results. More specific experiments need to be done to decide between the two models.

NOMENCLATURE

a   constant in Eq. (2)

b   constant in Eq. (2)

h   height of water film on sand particle

t   time

T   temperature

v_x velocity in the water film

x   direction coordinate for the heat and moisture transfer

71

Greek Symbols

α_1 thermal diffusivity of the frozen zone

μ viscosity of water at the freezing temperature

ρ density of water

σ surface tension of water

INTRODUCTION

Building materials and soils are porous and contain moisture
which migrates whenever a temperature gradient is imposed either
by the natural environmental conditions or by the heating or cooling system installed in the building to keep the inside temperature
at a comfort level while the outside temperature varies between the
extremes of cold and hot. The direction of moisture migration is
the same as that of heat, i.e. away from a heat source and towards
a heat sink. This moisture migration resulting in a redistribution
of moisture is likely to cause changes in the structural properties
of the building materials and soils. In some cases, even the foundations might be affected. Hence it is important to know about
this process of the moisture migration.

In case the temperature no where across the building material
falls below the freezing temperature of moisture, the migration of
moisture from the higher temperature side to the lower temperature
side results in a build-up of the moisture concentration at the
lower temperature end. This results in a migration of moisture in
the opposite direction. Hence, after sometime, an equilibrium
moisture distribution is reached where any further moisture migration in one direction due to temperature gradient is balanced by an
equal transfer in the opposite direction due to hydrostatic gradient. Unless the temperature gradient is very large, the net redistribution of moisture is not very significant and hence the resulting effects on the structural properties of the material are
likely to be less pronounced.

A situation of deeper concern arises when the temperature
profile of the medium does cross the freezing temperature, i.e.
when the temperature on one side of the building material or soil
is sub-freezing whereas that on the other side is above freezing.
The moisture migration is still from the higher temperature side
to the lower temperature side but the moisture reaching the location
of the freezing temperature isotherm immediately gets frozen and is
not available for migration in the reverse direction unlike the
case discussed in the preceding paragraph. Thus the lower temperature side continues to act as a sink both for heat and moisture.
This points to the fact that if such a sub-freezing temperature
gradient is maintained long enough, almost all the moisture from

the higher temperature side would have migrated towards the lower temperature side and be frozen there. Thus, one part of the material where the temperature is above freezing will be rather dry, whereas the other part of the material where the temperature is below freezing will have high moisture content in the form of ice. Furthermore, for this to happen, the overall temperature gradient does not need to be very high although a high value would result in a higher rate of moisture migration. It is evident that the structural property change in this case would be more pronounced than in the earlier case with above freezing temperatures. Since habitable areas where sub-freezing temperatures occur are vast and distributed all around the globe, the study of this phenomena is of great interest.

We have studied this phenomena of moisture migration under a subfreezing temperature gradient. The experimental part and a mathematical model have been described respectively in References (1) and (2). The relevant literature survey upto the recent past has been summarized in Reference (3). It does not however claim to be complete in all aspects. Before we go on any further, it would be proper to discuss the experimental set up briefly.

EXPERIMENTS

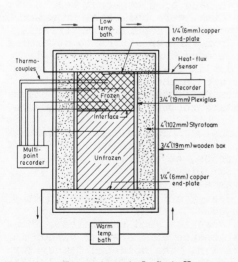

Fig 1:   Experimental Set-Up

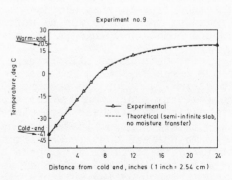

Fig 2:   Temperature Profile

A line sketch of the experimental set-up is shown in Figure 1. The cell was 24-inch (61-cm) long and one-foot square (30.5-cm square) in cross-section. The end plates were made of ¼-inch (6 mm) copper with a manifold of 25,¼-inch (6 mm) copper tubes

soldered on the outside for the circulation of liquid for controlling the temperature. The lateral sides and the bottom were made
of 3/4-inch (19-mm) Plexigalas sheets. A heat-flux sensor was
mounted on the inside surface of the cooled copper end-plate to
measure the amount of heat being withdrawn by the cold plate. The
whole cell was surrounded by a 4-inch (102 mm) thick Styrofoam
insulation and placed in a 3/4-inch (19 mm) thick wooden box. The
thickness of the insulation had been precalculated so as to have
negligible heat transfer from the sides. The cell was placed in a
horizontal position so that the sampling was easy and the gravity
effects, if any, were uniform.

The porous material used was Ottawa Binding Sand No. 80, a
naturally occurring sand from the Ottawa River in Illinois. The
particle size distribution used was above 270 mesh (U.S.). The
sand was wet uniformly and packed to a dry-density of 100 lb/cu.ft
(1.6 gm/cc.) to a depth of 3-inch (7.6 cm), the remaining cell
volume being filled with Styrofoam insulation to further cut down
on the heat gain from the surroundings. The copper end-plate with
the heat flux sensor was cooled to sub-freezing temperature by circulating a Freon TMC refrigerant through the manifold. The other
end-plate was kept at the initial starting temperature by using
another controlled temperature bath. Forty-eight copper-constantan
thermocouples were used to measure the cross-sectional and axial
temperature profiles. Preliminary experiments established that the
heat transfer was one-dimensional, thus proving the adequacy of the
insulation. At various times during the experiments, samples were
collected along the length of the cell both from the frozen and the
unfrozen zones to obtain the moisture concentration profiles by
gravimetric analysis and drying in an oven to a constant weight.
The next set of samples, at a later time, were collected from a
laterally different location. At the end of a particular experiment, extra samples were collected from various depths to determine
any possible gravitational effect but none was found. No ice-lens
formation was observed and a differential scanning calorimetric
experiment showed that all the moisture in the sand froze at 0°C.
This meant that there was no unfrozen moisture in the region below
0°C and hence no change in the moisture content of the frozen zone.

The details of all the experiments conducted with the values
of the parameters used and the results obtained are available
elsewhere (3). For the present purposes, we will refer only to
the relevant experiments which are Experiment Nos. 9, 10 and 11.
Information about their parameter values is in Table 1 below.
The moisture content in Experiment Nos. 9 and 11 was kept almost
the same whereas that in Experiment No. 10 was at a level which
saturated the pore or void space between the sand particles. This
was done to study the effect of the absence of the water vapour in
the void space on the moisture migration. The cold-end temperature
in Experiment Nos. 9 and 11 was taken to be different to study the
effect of temperature gradient on the moisture migration. Further,
Experiment No. 11 was carried out for a much longer period so as
to study the effects of a steady-state temperature gradient on the
moisture migration. For the same reasons, the samples were withdrawn at various times in all the experiments. These experiments
were done so as to be able not only to study the extent of the

TABLE 1:   EXPERIMENTAL PARAMETERS

Experiment Number	Initial Moisture, % weight	Cold-end Temp., °C	Warm-end and Initial Temp., °C	Sampled at, Hours	Duration of Experiment, Hours
9	9.9067	-41.0	20.5	1.5, 3, 7	7.0
10	16.3 (saturated)	-41.0	23.2	1.5, 4.3, 11.5	11.5
11	9.87	-17.9	32.5/ 26.0	2, 14, 22, 38, 62.5, 91.5, 116, 143, 208	208.0

moisture migration in each case but also to study the coupling, if any, between the heat and moisture transfer and to formulate a model to predict the results in other situations.

RESULTS

Temperature Profile

A typical temperature profile for Experiment No. 9 is shown in Figure 2. Also shown is a theoretical calculation using Neumann's analysis (4) for a semi-infinite slab without moisture transfer. The two profiles match very well. Hence the effects of moisture redistribution on the thermal properties seem to be negligible and one can solve for the temperature and the moisture profiles independently. The only interaction is the effect of the temperature gradient on the moisture transfer in the unfrozen zone.

Moisture Profiles

The composite moisture profiles for the three experiments are shown in Figures 3 to 6. Figure 3 shows that the interface values of the moisture content in the frozen and the unfrozen zones tend to reach asymptotic values. This is the case when the cold-end temperature has not affected the temperature profile near the warm-end meaning thereby that the experimental cell is behaving as a semi-infinite one.

Figure 4 shows that in Experiment No. 10 the moisture content did not change from the initial value before the experiment was started. There is no trend observable (unlike in Experiment 9) with regard to the moisture migration. This means that when there is no vapour space in the pores (being full with liquid water) there is no moisture migration.

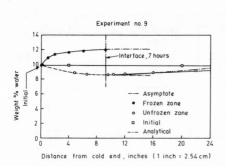

Fig 3:   Moisture Profile

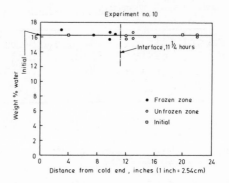

Fig 4: Moisture Profile

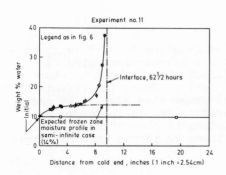

Fig 5:   Frozen Zone Moisture
         Profiles

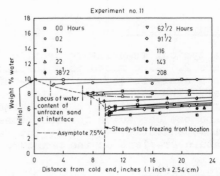

Fig 6:   Unfrozen Zone Moisture
         Profiles

Figures 5 and 6 show the moisture profiles in the frozen and the unfrozen zones for Experiment No. 11. Because of the long duration of the experiment and the steady state temperature profile reached after $62\frac{1}{2}$ hours, it was thought pertinent to show the profiles for all the sampling periods. Since there is no moisture migration within the frozen zone, the moisture profiles for all the periods overlap (Fig. 5). An asymptotic profile is shown for the

case when the cell behaves as a semi-infinite one and the region
near the warm-end is not affected. As time progresses the effect
is felt and the movement of the interface is slowed down which
ultimately achieves a steady location when the temperature gradient
becomes steady. However, the moisture keeps migrating towards the
interface which, after 208 hours, had a moisture content of about
37 percent, significantly more than what is required to fill the
pore space. The sand particles in this region had been pushed
apart a little due to the expansion of moisture on freezing to
accommodate the increasing moisture content. For the unfrozen
zone, the moisture profiles are shown in Figure 6. After the
steady temperature profile was reached, the moisture in the unfrozen zone kept on reducing. If significantly more time had been
allowed, the unfrozen zone would have probably become dry.

A mathematical
model for moisture
migration in the
vapour phase was
formulated (2,3). It
was based upon the
assumption that the
liquid and vapour in
a partially filled
pore (Fig. 7) are in
equilibrium at the
prevailing temper-
ature (5). The
vapour pressure was
assumed to vary
linearly with
temperature in
Reference (2) and
exponentially with
temperature (Antoin's
relation) in
Reference (3). The
derivation of the
equations and final
analytical results

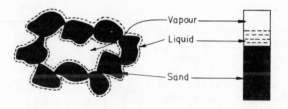

Fig 7: Sand Particles & Pore Space

for the moisture profile and the interface concentration are given
in the above references and are not repeated here in order to save
space. The analytical unfrozen zone moisture profile has been
plotted in Figure 3. It shows a close correspondence to the experimental results. It could also be fortuitous because the values
of vapour diffusivity obtained from these experiments were order
of magnitude higher than expected whereas the values of liquid
diffusivity thus obtained compared favourably with the ones obtained
by independent experiments (3). The results showed that the unfrozen zone profile **was** only a function of $x/2\sqrt{\alpha_1 t}$ and hence a
single plot gives the unfrozen zone moisture profile for all x and
t (3).

The model does predict a number of other observations as
listed below:

i) For the saturated sand, since there is no vapour space, the concentration of vapour is zero and hence no moisture migration which is confirmed in Figure 4

ii) If the initial temperature is the freezing temperature, there is no vapour pressure gradient and hence no moisture migration

iii) If the cold-end temperature is lowered further, resulting in a higher temperature gradient and a faster rate of freezing, the step rise from the unfrozen to the frozen zone concentration at the interface is less (Figs. 3 and 5).

After the above model was presented (2,3), it was suggested (6) that another possible reason of moisture movement could be due to the surface tension gradient in the liquid phase instead of the vapour pressure gradient proposed by us in our model. The movement due to the surface tension gradient caused by an applied temperature gradient is called the 'thermocapillary motion' by Levich (7). He has developed the theory fully for a steady state temperature gradient above the freezing temperature for a thin liquid film in a shallow pan. In such a case, the liquid movement due to the thermocapillary motion is balanced by a reverse flow in the opposite direction due to the hydrostatic pressure. In our case, however, due to the sub-freezing temperature at one end, the moisture freezes and there is no return velocity. The velocity profile in this case is linear.

We have adapted the formulation of Levich to our case. The film of water on the sand particles is a thin one and hence the convection currents would be negligible (Fig. 7).(See Levich (7) for further assumptions). The Navier-Stokes equation for velocity in the x-direction, neglecting higher order differentials, is

$$\frac{\partial^2 v_x}{\partial y^2} = 0 \qquad (1)$$

or

$$v_x = a + by \qquad (2)$$

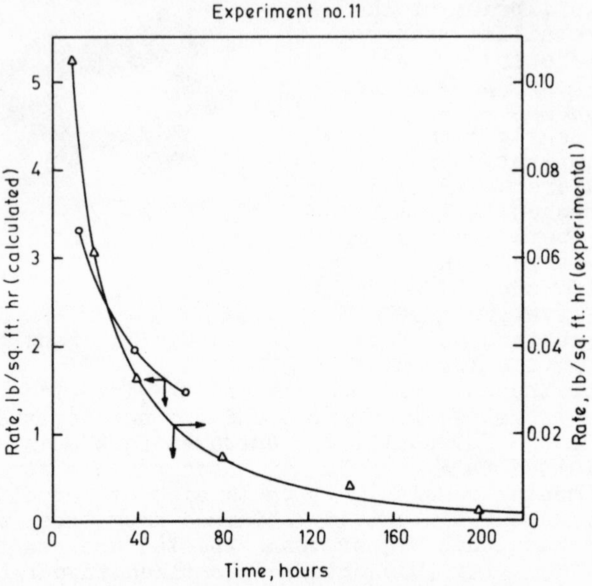

Fig 8:   Rate of Moisture Migration

Boundary conditions are:

(i) No slip at the sand-water interface

$$v_x \Big|_{y = h} = 0 \tag{3}$$

(ii) Continuity of shear stress at the air-water interface

$$\mu \frac{\partial v_x}{\partial y} \Big|_{y = 0} = \frac{\partial \sigma}{\partial T} \text{ grad } T \tag{4}$$

The substitution of Eqs. (3 and 4) into Eq. (2) gives

$$v_x = \frac{1}{\mu} \frac{\partial \sigma}{\partial T} (y - h) \text{ grad } T \tag{5}$$

The total flow rate per unit perpendicular length of the film is

$$\rho \int_0^h v_x \, dy$$

Substitution of Eq. (5) in it gives

$$= - \frac{\rho h^2}{2\mu} \frac{\partial \sigma}{\partial T} \text{ grad } T \tag{6}$$

(The flow rate is positive since $\frac{\partial \sigma}{\partial T}$ is negative).

Calculations of the instantaneous flow rate were carried out using the results of Experiment No. 11 both when steady state temperature profile had been reached (62½ hours) and earlier when it had not been reached (14 and 38½ hours) using the temperature gradient in the unfrozen region at the interface. The height of the liquid film was calculated by assuming the available water to be equally distributed in the unfrozen region (Fig. 6). Though there was a known gradation of the sand particles (3), 80 mesh size (0.177 mm) was found to be the average diameter for these calculations and the particles were assumed spherical. The void space, the voidage between the 'spherical' sand particles, was taken to be the usual 40 percent as in spherical packing. The dry density of sand was approximately 100 lbs/ft³ (1.6 gm/cc) whereas the particle density was 142 lbs/ft³ (2.29 gm/cc). The calculations for the moisture transfer rate gave values upto 70 times more than the experimentally measured values (Fig. 8). This discrepancy could probably be due in part to the assumptions regarding the sand particles being uniform spheres, voidage being 40% and the like. Also, probably some of the water film on the sand is 'bound water' thus reducing the actual height of the film the motion of which causes the migration of moisture towards the cold-end. The height of the film appears with an exponent of two (Eq. 6) and hence any change in its value will have large effect on moisture transfer. The difference between the calculated and the measured moisture migration rates is not so much as to cause an outright rejection of the 'thermocapillary motion' model. This model also explains two of the three situations explained by the earlier model:

i) When the pore space is full there is no portion of the film exposed to air and hence the surface tension concept does not come into the picture. One can say that $\partial\sigma/\partial T = 0$ in Eq. (6) giving the flow rate to be zero. There is however one draw back. There is surface tension even at the water-sand inter-face and hence, unless it is neutralized by the attractive forces of the sand for water molecules at the sand-water inter-face, the above conclusion is doubtful (8).

ii) When the initial temperature is same as the freezing temper-ature, grad T is zero in Eq. (6) and hence the flow rate is zero.

For the third situation of lower cold-end temperature, this model says that since grad T is higher, the rate of migration should be higher which was experimentally observed also.

CONCLUSIONS

       There are two possible mechanisms for the moisture migration in building materials and soils caused by an imposed temperature gradient. One is based upon the diffusion of water vapours through the void space due to a vapour pressure gradient whereas the other is based upon the thermocapillary motion of the liquid film due to a surface tension gradient. Although the latter gave results which were upto 70 times more than the observed values, yet it cannot be ruled out because the discrepancy could be due to the simplificat-ions made in the model or 'bound water' on the sand particles. It should also be pointed out that the vapour phase diffusivity value obtained in the earlier model was order of magnitude more than the expected value which would have underpredicted the result. Hence neither model is totally satisfactory yet, although both do explain qualitatively all the observed facts. It is likely that a combin-ation of both the models might balance out the over-prediction of one and the under-prediction of the other. Equally likely is the possibility of some other model involving convection also. More specific experiments need to be carried out to resolve this matter. It would then be possible to predict quantitatively the extent of the moisture migration in the building materials based upon the extreme conditions of temperature that such materials are likely to encounter in specific geographical locations, and then to calcu-late their structural properties to be used in the design of the building so that an optimum use of the building material can be made.

REFERENCES

1. Gupta, J.P., and Churchill, S.W., "Heat and moisture transfer in wet sand during freezing", presented at the Winter Annual Meeting ASME, Washington D.C., November 30. Published in Environmental and Geophysical Heat Transfer, HTD-4, p. 99, American Society of Mechanical Engineers, New York (1971).

2. Gupta, J.P., and Churchill, S.W., "A model for the migration of moisture during the freezing of wet sand", AIChE Preprint 13, presented at the 13th National Heat Transfer Conference,

Denver, August 6-9, 1972. Published in the <u>Chemical Enginee-</u>
<u>ring Progress Symposium Series No. 131</u>, 69 : 192 (1973). In
the latter, $\rho(T)$, ρ_0 and ρ_f in Eqs. (14 to 20) should actually
be $p(T)$, p_0 and p_f.

3. Gupta, J.P., <u>Moisture migration and heat transfer in wet sand</u>
    <u>during freezing</u>, Ph.D. Thesis, University of Pennsylvania,
    Philadelphia (1971).

4. Neumann, F., in Riemann-Weber: <u>Die partiellen Differential-</u>
    <u>Gleichungen der mathematischen Physik</u>, 5th ed., Vol. 2,
    F. Vieweg and Sohn, Braunschweig, 121 (1912).

5. Tung, L.N. and Drickmayer, H.G., "Diffusion through an inter-
    face - binary system", <u>J. Chem. Phy.</u>, 20 : 6 (1952).

6. Graves, D., University of Pennsylvania, Philadelphia, Personal
    Communication.

7. Levich, V.G., "Physicochemical hydrodynamics", 2nd ed.,
    Prentice Hall, Princeton (1962).

8. Nigam, P.C., I.I.T. Kanpur, Personal Communication.

AN ENGINEERING SYSTEM FOR DETERMINING THERMAL AND MOISTURE TRANSFER PROPERTIES OF STRUCTURAL MATERIALS

ALLAN SHAW

Department of Mechanical Engineering
The University of Adelaide
Adelaide, South Australia 5001

ABSTRACT

The materials used in building design are an important determining factor of the capital and energy costs of heating and cooling buildings. A laboratory chamber for the study of the thermal and moisture transfer properties of structural materials is proposed. The purpose of the design is to simulate both the "inside" and "outside" temperature and humidity on the two sides of a test panel of structural material to be analysed. Any desired inside pair and any desired outside pair of temperature-humidity conditions over a wide climatic range can be investigated. The system can accommodate most structural materials from a glass sheet to a concrete wall.

A unique low energy inexpensive engineering system is described which will obtain results to close tolerances.

NOMENCLATURE

dbt   dry bulb temperature

eh   specific enthalpy of air at entering wet bulb temperature condition to direct expansion coil (kJ/kg of dry air.)

h_r specific enthalpy of refrigerant

$\dot{m}_a$ mass rate of flow of dry air

$\dot{m}_r$ mass rate of flow of refrigerant

W   humidity ratio (kg/kg)

INTRODUCTION

Recent years have seen a growing demand by research scientists both in the mechanical and life sciences for two-variable, temperature-humidity controlled chambers which will operate stably over a wide range.

A search of the literature on the engineering aspects of two-variable control over a climatic range reveals discouraging recommendations. The general consensus is not to design temperature-humidity, wide climatic range systems. It has been said that due to the inter-dependence of the temperature and humidity properties, there is an inbuilt instability to this type of design. It is felt that since the time constant of the dehumidifying process is different from the humidifying process and different again from heating and cooling, design solutions become too complicated and expensive.

Another discouraging aspect is that, though there is a very large demand, the user's specifications differ markedly from each other depending on the area of interest and the degree of sophistication of the research program. Manufacturers not seeing a mass production approach have avoided entering this area.

A fundamental design for determining thermal and moisture transfer properties of structural materials is presented here. However this purpose is only a particular application. This design is applicable to the broad field of climate simulation.

THE AIMS

The specific aims of the design may be expressed as a set of specifications.

(1) System must be designed to maintain its temperature and humidity operating conditions to close tolerances and with low temperature gradients.

(2) System must have the capacity to offset all sensible and latent heat loads.

(3) System must have the capability to operate over a wide range of temperature and humidity conditions, above and below ambient conditions.

(4) System shall have the capacity to automatically changeover temperature and humidity operating settings as per some prescribed program.

(5) System shall have the capacity to function continuously over long periods of usage. Momentary interruptions such as are caused by defrost cycles are not admissible.

(6) System must be easily operable by non-engineers.

(7) System must be designed to fail safe.

(8) System to be inexpensive in capital costs.

(9) System to have low energy requirements.

THE HARDWARE ARRANGEMENT

The chamber [1] would consist of two compartments separated by a test frame designed so that a test panel of variable thickness may be inserted and withdrawn. (See Fig. 1.)

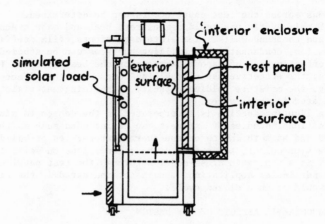

Fig. 1. Schematic
cross-section
through test
panel.

Thus a sheet of glass or a cavity brick wall may serve as a subject for
investigation. On one side of the test panel would be a light weight but well
insulated interior enclosure. The surface of the test panel facing this
enclosure would represent the still air interior surface of a building wall and
the volume of this enclosure would represent interior building conditions. It
is assumed that the laboratory which will house this facility would be air-
conditioned and set at the desired interior building condition of the particular
investigation. Thus with the enclosure in place there would be negligible heat
and moisture transfer between the "inside" building condition within the
enclosure and the ambient condition of the laboratory in which the enclosure
sits since they are both at the same conditions at the start of a test.

The design is concerned with simulating outside air conditions on the
"exterior" surface of the panel under investigation. This side would be
subjected to an air movement at a pre-established air velocity. For example
should the panel be tested for winter conditions it may be desired that the
air movement simulate a 24 to 32 kilometre per hour wind velocity. Depending
on the purpose of the user, this "outside air" chamber may be designed to
include simulation of a solar light and heat load facing the test panel. It
is the objective of this design to simulate precisely within this outside
chamber any desired weather condition over a climatic range. In addition
provision may be made to include automatic change-over from one set of
temperature and humidity conditions to any other combination of temperature
and humidity as per a stipulated program. Most commonly it is expected that
change-over between "day" and "night" conditions will be desired.

The "inside condition" section as well as the structural panel itself will
be instrumented to indicate the changes and rate of change of temperature
and humidity. By observing these changes, the point at which steady state

conditions across the test panel occurs can be determined.

With this system one could demonstrate heat and mass transfer through simple and compound walls. Temperature gradients, film coefficients, vapour transmission, condensation on building surfaces can be studied. Studies on the effectiveness of vapour barriers or the best location for vapour barriers, the effectiveness of caulking compounds, the effect of building porosity, the moisture holding capacity of building materials could be investigated in this system.

If a lighting section is incorporated in the design to simulate sunlight many additional studies of interest can be included such as the effect of altitude and azimuth angles on the thermal properties of building structures, the effect of colour, the effect of haze conditions on solar transmission. Figure 1 is a schematic cross-section through the test panel of the chamber. In this particular application, though not indicated in the figure, the air system would be in a closed cycle.

THE THERMODYNAMIC ASPECTS OF THE DESIGN

Consider a conventional vapour compression refrigeration cycle operating continuously, using a thermostatic expansion valve and having a fixed condensing temperature and a constant air flow rate. The cycle employs a single direct expansion coil with a selected surface whose performance can be determined. Assume some specified climatic range to be the shaded area indicated on a psychrometric chart. This range may cover a dry bulb temperature from about 4°C to 43°C and a humidity ratio of from 0.005 to 0.025. Assume this area is divided by five equally spaced constant enthalpy lines as indicated in figure 2 on a psychrometric chart.

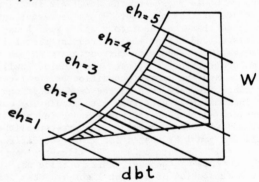

Fig. 2. Climatic Range

Now, in the context of the constraints set forth above let us assume that air enters the direct expansion coil at an enthalpy of "3". In figure 3 are indicated on a capacity versus suction temperature, pressure diagram the condensing unit and the direct expansion coil performance curves.

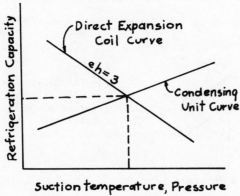

Fig. 3. Balance
point for two
steady flow
streams.

The intersection of the condensing unit curve with the curve at the direct
expansion coil would result in a simultaneous solution which would represent
the balance point for two steady flow streams wherein the heat and mass
transfer from the air stream at entering enthalpy "3" can be equated to the
heat transferred to the refrigerant.

Thus with reference to figures 4 and 5,

$$\dot{m}_a \Delta h = \dot{m}_r \Delta h_r$$

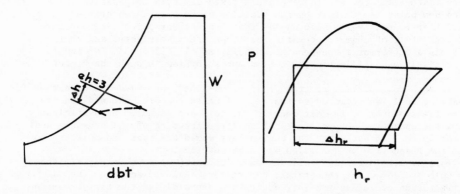

Fig. 4. Entering and leaving
conditions of air at direct
expansion coil.

Fig. 5. Refrigerant cycle with
Refrigerant effect = Δh_r

Now let us repeat this same process for all of the five entering enthalpy
conditions, still maintaining the same condensing temperature, the same air
flow rate and the same components in the refrigeration cycle.

The results on the refrigeration capacity vs suction temperature, pressure
diagram would indicate 5 fixed refrigeration capacities each related to 5
fixed suction temperatures and pressures and 5 fixed balance points.

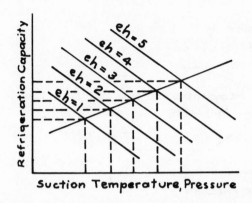

Fig. 6. Balance points
for five different
entering air enthalpy
conditions.

Up to this point only one component having a control action has been
indicated, the totally self-contained performance of a thermostatic expansion
value.

Now let us consider that the system is required to cover a range as
depicted on the psychrometric chart of figure 2 with a specific enthalpy
varying from h'1' to h'5'. To this system we add a heater and a humidifier
in the air stream downstream of the direct expansion coil.

Two proportional controllers with sensing elements within the chamber of
the climate simulator act to maintain a fixed dry bulb temperature and a
fixed dew point temperature by the addition of heat to a reheater and to a
humidifier respectively. Let us say that 5 different pairs, (temperature
and humidity), of chamber operating settings are investigated and that each
pair has a different corresponding enthalpy of '1', '2', '3', '4' and '5'
respectively when it leaves the test chamber to flow through the direct
expansion coil.

In each case precise control would be obtained, provided the direct
expansion coil was selected so that in all cases the dry bulb and the dew
point temperatures of the air leaving the coil are found to be sufficiently
below the operating set points to allow for offsetting of all sensible and
latent loads. In each case the controllers would add heat and moisture so
that the respective eh condition would be obtained.

This system is a self realizing one. On start up or change-over, the
dry bulb and dew point temperature controllers begin to act when the
refrigeration cycle has sensibly cooled and dehumidified the air stream to
a point below the respective operating settings. When this occurs the
combination of the sensible heat load (assuming there is one) and the sensible
heat added by the dry bulb temperature controller in the reheater plus the
latent heat load and the water vapour introduced into the air stream by the
dew point temperature controller together obtain the fixed entering enthalpy
corresponding to the selected pair of operating settings ('1' to '5').

Thus the refrigeration cycle is driven to a steady flow state having a
fixed suction temperature and a fixed capacity as the controllers add heat
and moisture to satisfy the desired operating settings.

Now let us assume a load change takes place within the controlled air
section so as to increase the sensible heat load. This occurrence would have
no effect on the steady flow nature of the refrigeration cycle since the dry
bulb temperature controller would act to reduce the sensible heat addition to

the reheater an equivalent amount to the increase in sensible heat load and both the refrigeration capacity and suction temperature would remain fixed.

Inter-action between the dry bulb temperature controller and the dew point temperature controller is non-existant in that though the temperature of the surface of the water in the humidifier (assuming a pan-humidifier was used) will add some sensible heat – this will not effect the steady flow nature of the refrigeration cycle since the surface temperature of the water, remains constant – and were it to rise, the reheater would be relieved of an equivalent addition of sensible heat.

Described above is a simple system of wide range temperature-humidity control that will maintain conditions to close tolerances, perform stably and handle both sensible and latent loads.

There remains the need to so select the refrigeration capacity as to avoid unnecessary sensible cooling and dehumidifying and a very inexpensive system of climate simulation exists with minimal components and minimal controls.

The system as described will have a degree of energy wasted at high entering enthalpy settings. This energy expenditure can be negligible for small refrigeration systems below 3 KW capacity. By modification the higher expenditure of energy for the higher entering enthalpy conditions can be reduced. One solution (unlikely to be employed commercially) is to use an open type compressor and to reduce the compressor speed.

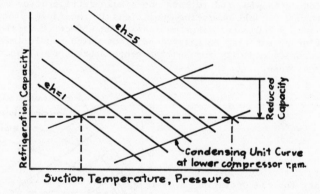

Fig. 7. Reduction of Refrigeration Capacity with reduced compressor speed.

Thus as indicated in figure 7, the refrigeration capacity for entering enthalpy 5 could be reduced to that of entering enthalpy 1.

Another more practical way of reducing the energy requirements for high entering enthalpy settings is to include a face and bypass damper in the system which can be made to act completely automatically to maintain a fixed refrigeration capacity. In this case an open compressor would not be required.

To explain let us examine what occurs when the air flow rate of figure 3 is increased and decreased. The resultant effect would be as indicated in figure 8.

Fig. 8. Effect of
Increase and Decrease
of Air Flow Rate.

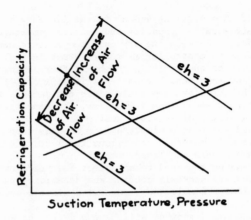

It is thus possible by decreasing the air flow rate or bypassing some
of the air around the direct expansion coil to decrease the refrigeration
capacity and suction temperature to the condition of eh = '1' for all
entering enthalpies and to have one single refrigeration capacity and suction
temperature for all operating conditions in the range.

Figure 9 indicates diagramatically the effect of varying degrees of bypass
on different entering enthalpy conditions which result in a single fixed
refrigeration capacity and suction temperature for all operating settings
within a range.

Fig. 9. Face and Bypass
Dampers Effecting Constant
Capacity for all Operating
Conditions.

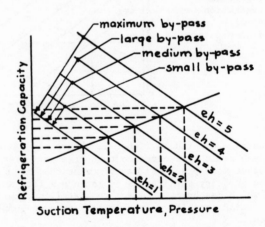

The steady flow state of the refrigeration system establishes a fixed
energy datum level at the direct expansion coil surfaces when the control
system approaches any desired pair of temperature-humidity settings in the
design range. The system is very amenable to adjustments and arrangements
which though maintaining the fixed energy datum will permit this datum to

change. This was illustrated by describing the effect of introducing a face and bypass damper.

This is basic to the design approach taken here. Apart from the simplicity of the system - the stability and close tolerance performance is a function of the fixed energy datum. This datum would be common to numerous applications of this design and permits a mass production approach in manufacturing.

This range system using a single direct expansion coil for offsetting both the sensible and latent heat loads will of necessity experience some energy penalties. There will be occasion when the sensible cooling will have to be more than is actually required because of the need to offset the latent heat load with sufficient dehumidification. This will occur during operating settings which simulate high dry bulb temperatures combined with arid conditions. At these times the direct expansion coil will be approaching dry performance. (See curve 1 in figure 10). On the other hand there may be occasion when dehumidification will be more than is actually required because of the need to offset the sensible heat load. This will occur during operating settings having high relative humidities. (See curve 2 in figure 10). However, if the direct expansion coil is carefully selected [2] the energy penalty can be minimized. The alternatives - the use of absorbents or the use of several air paths with one coil performing as a sensible cooler and another coil as a dehumidifier pose problems which increase cost of manufacture, require more controls, increase the size of the system, the expense of maintenance, and makes the system more difficult to operate.

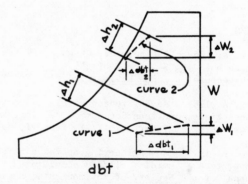

Fig. 10. Energy expense where both the cooling and dehumidifying function is carried out by a single direct expansion coil.

It should be noted that the specifications in this case, as well as for most applications of climate simulation do not involve rapid load changes. Though the design presented here can cope with rapid load changes it takes full advantage of the case where it does not occur. Once the operating set points are reached the control action becomes almost negligible. This points to one of the major differences with air conditioning design problems where load changes do occur but where the requirement of range is almost entirely absent. In this design once operating settings are reached if the system is switched from automatic to manual control, the desired conditions would hold for hours with only minor drift associated with small load changes.

The expected performance of this system would be similar to that of the phytotron unit at the Waite Agricultural Research Institute of the University of Adelaide [3].

ACKNOWLEDGEMENTS

Carrier Air Conditioning Pty. Ltd. Australia contributed the direct expansion coils for research in this area of design. Their help is gratefully acknowledged.

CONCLUSIONS

An engineering system for determining thermal and moisture transfer properties of structural materials has been described. This system has the capacity to satisfy the requirements of scientists seeking economically viable climate simulators of high quality performance. The system overcomes the instabilities arising from simultaneous control of temperature and humidity and the problems associated with reproducing climatic conditions over wide ranges.

The performance of the system depends on establishing a steady flow refrigeration cycle having constant direct expansion coil temperatures which provides a fixed energy datum. Apart from the self-contained action of a conventional thermostatic expansion valve there are no controllers necessary for the cooling and dehumidifying processes. Controllers add heat and moisture to the fixed energy datum to obtain stable close tolerance performance.

REFERENCES

1.   Shaw, A. 1967 "A Laboratory Chamber for Study of Heat and Mass Transfer in the Building Science Course"
      Architectural Science Review. Vol. 10, No. 3, pp. 90-92.

2.   Shaw, A. 1965 "An Environmental Plant Growth Chamber"
      Transactions, American Society of Heating, Refrigerating and Air Conditioning Engineers, Vol. 71, Part II, pp. 102-122.

3.   Shaw, A. 1975 "An engineering approach to the manufacture of temperature - humidity climate simulators".
      Journal of the Society of Environmental Engineers
      Vol. 14-1 (Issue 64) March 1975, pp. 31-35.

EFFECTS OF INFILTRATION
ON HEAT TRANSFER
THROUGH VERTICAL SLOT
POROUS INSULATION

P. J. BURNS AND C. L. TIEN

University of California
Berkeley, California, USA

ABSTRACT

The present paper reports the results of an analytical investigation of convective heat transfer through vertical slot porous insulation with infiltration. The objective is to understand the effect of infiltration upon the heat transfer characteristics of building-wall insulation. An equivalent porous covering material is used to simulate porous openings, holes or cracks. The influence of the modified Rayleigh number, the size and location of the opening, the infiltration due to wind pressures, and the pertinent parameters characterizing the covering material is analysed. It is demonstrated that infiltration may significantly increase the heat transfer through vertical slot porous insulation.

NOMENCLATURE

A   aspect ratio, H/d

c_p specific heat (KJ/kg-°K)

d   horizontal distance between the hot and cold walls (m)

Da   Darcy number, K/d^2

g   acceleration of gravity (m/sec^2)

H   vertical distance between the horizontal walls (m)

K   permeability (m^2)

k_o thermal conductivity of the stagnant medium (W/m-°K)

M   number of grid spaces in the horizontal direction

N   number of grid spaces in the vertical direction

Nu   Nusselt number, $\frac{1}{A} \int_0^A (U_w \theta - \frac{\partial \theta}{\partial x}) \, dy$

$\bar{P}$ pressure (N/m^2)

P   non-dimensional perturbation pressure, $\dfrac{K_2}{\mu\alpha}\,[\overline{P} - P_{\infty_1} + \rho_1 gy]$

P_{∞_1} pressure in the left external environment at y = 0

p^* absolute pressure at the location (x = 0, y = 0)

Ra^* modified Rayleigh number defined in equation (8)

Res   criterion for numerical convergence in equation (16)

T   temperature (°K)

ΔT   temperature difference between hot and cold walls (°K)

t   thickness of a covering material

$\overline{u}$ horizontal velocity (m/sec)

u   non-dimensional horizontal velocity, $\overline{u}d/\alpha$

U_w non-dimensional horizontal blowing or suction velocity,
           u(x = 0, 1;y)

$\overline{v}$ vertical velocity (m/sec)

v   non-dimensional vertical velocity, $\overline{v}d/\alpha$

$\overline{V_w}$ wind velocity (m/sec)

$\overline{x}$ horizontal length coordinate (m)

x   non-dimensional horizontal length coordinate, $\overline{x}/d$

$\overline{y}$ vertical length coordinate (m)

y   non-dimensional vertical length coordinate, $\overline{y}/d$

Greek Symbols

α   thermal diffusivity, $k_0/\rho_1 c_p (m^2/sec)$

β   coefficient of cubical expansion (1/°K)

θ   non-dimensional temperature difference, $(T - T_3)/(T_1 - T_3)$

ν   kinematic viscosity (m^2/sec)

ρ   density of the fluid (kg/m^3)

τ   any non-dimensional numerical variable, either φ or Ψ, associated
           with equation (16)

φ   non-dimensional perturbation temperature, θ - 1 + x

Ψ non-dimensional stream function, $u = \dfrac{\partial \Psi}{\partial y}$ and $v = -\dfrac{\partial \Psi}{\partial x}$

Subscripts

1   hot boundary covering material or hot wall

2   interior porous medium

3   cold boundary covering material or cold wall

INTRODUCTION

A porous material is introduced into an enclosure to decrease the convective and radiative transfer of heat. Although radiation may contribute significantly to the overall energy transfer, at room temperature the effect is small [1-3], and is lumped into the experimentally determined thermal conductivity. The differential energy balance must then be formulated in terms of the stagnant thermal conductivity which is dependent upon the local geometry in a very complex fashion [4]. The correspondingly complicated flow system is described in a macroscopic sense by Darcy's law [5].

Many published works have dealt with free convection in porous media heated from below. An excellent current review is presented by Bankvall [6]. Yet the conjugate problem of heating from the side has received little attention. Experimental [6-11] and analytical [6,7,9,10,12,13] results are available for a variety of configurations and insulation materials.

Existing analytical studies are concerned primarily with very idealized physical systems, although some recent authors [14] have published some results for more realistic flow situations. The anisotropic permeability was shown to exert little influence upon the heat transferred, except where it is of crucial importance in the definition of the modified Rayleigh number. Temperature gradients on the vertical boundaries may affect the heat transfer significantly by eliminating the usually small heat fluxes at the top of the hot wall and the bottom of the cold wall. Small leakage velocities tend to increase the overall energy needed to maintain the enclosure at steady state.

Leakage velocities, in practice, are the result of the imbalance of external pressure forces acting on a wall as well as the action of the internal flow and density field. Holes or cracks are prevalent in walls due to present construction practices. Holes exist at switches, electrical outlets, baseboards, ceilings, etc. Most such openings offer a very restricted flow path. Therefore, it is reasonable to model these holes analytically by assuming an effective area of flow. Even such materials as sheet rock, plaster, brick and paint are indeed permeable. With deterioration or faulty installation procedures, something akin to a porous covering material will exist.

The physical model to be considered is shown in Fig. 1. A rectangular, two-dimensional enclosure is envisioned. Although many cracks elicit three-dimensional flow patterns, some cracks, such as along baseboard and ceiling moldings can certainly be modeled as two-dimensional. The steady-state case is considered. The enclosure is of width, d, and height, H. The hot wall is on the left, and the cold wall is on the right side of the enclosure while the horizontal boundaries are assumed to be insulating. As the non-dimensional height, H/d, is increased, the horizontal boundary conditions will gradually become less significant. The left and right covering materials, subscripted as 1 and 3, respectively, are of heights h_1 and h_3 and thicknesses t_1 and t_3. Their locations are given by the non-dimensional variables y_1 and y_3. The interior fibrous medium is denoted by the subscript, 2.

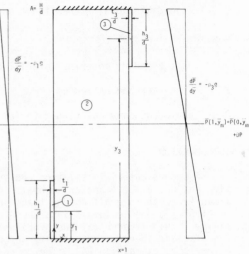

The external pressures are taken as hydrostatic at the temperature of the respective environment. The different pressures existing at differing heights across the enclosure will act to force flow through the medium. Fig. 2 shows the flow situation existing when the walls are perforated uniformly, and the flow is due to the forced convection alone [15]. The stack action only is depicted in Fig. 2(a) with the neutral zone at the mid-height. The inside is assumed hotter than the outside. Flow is "out" over the top half of the enclo- sure and "in" over the bottom half of the wall. Exhaust fans or internal chimneys would decrease the absolute magnitude of the internal pressure, thus raising or lowering the neutral zone by shifting one of the curves horizontally. The pressures due to wind forces are

Fig. 1   The physical system.

sketched in Fig. 2(b), with pressures of equal magnitude acting on the windward and leeward sides. Flow is "in" on the windward side and "out" on the leeward side. In actuality, both situations act together, and there results a compli- cated balance such as is suggested in Fig. 2(c).

The above situation is an oversimplification of an extremely complex phys- ical phenomenon. The pattern of pressure fields existing on both sides of a wall depends on all acting pressure forces and the size and distribution of all openings, both external and internal. The pressure field in a single room is coupled with the pressure distribution throughout the entire building. The pressure forces due to the wind are continually changing, and dynamic balances including acceleration are required. Each building presents a unique problem, and rational approximations must be introduced to render the problem tractable. In this light, the main focus of this paper will be upon determining qualitative trends in the hope that design procedures may be improved.

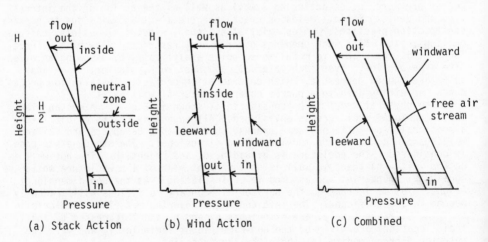

Fig. 2   Forced convection effects due to uniform perforations.

ANALYTICAL FORMULATION

The porous medium is assumed to be an effective continuum. This is generally valid for systems where the Darcy number (K/d^2) is much less than one.
Darcy's law then adequately describes the transfer of momentum provided the
Reynolds number based upon pore diameter is less than one [16]. Both of these
criteria are satisfied for almost all porous insulation assemblages wherein the
motion is due mainly to a gradient in the body force resulting from the transfer of heat. The Boussinesq approximation of constant properties except in the
body force term is made, limiting the overall temperature difference. For such
a system as has been described, the continuity equation and the equations governing the transfer of momentum and energy are, in dimensional form, as follows
[5]:

$$\frac{\partial \bar{u}}{\partial \bar{x}} + \frac{\partial \bar{v}}{\partial \bar{y}} = 0 \tag{1}$$

$$\bar{u} = -\frac{K}{\mu}\frac{\partial \bar{P}}{\partial \bar{x}} \tag{2}$$

$$\bar{v} = -\frac{K}{\mu}\left[\frac{\partial \bar{P}}{\partial \bar{y}} + \rho g\right] \tag{3}$$

$$\bar{u}\frac{\partial T}{\partial \bar{x}} + \bar{v}\frac{\partial T}{\partial \bar{v}} = \alpha\left[\frac{\partial^2 T}{\partial \bar{x}^2} + \frac{\partial^2 T}{\partial \bar{y}^2}\right]. \tag{4}$$

The penalty for the apparent good fortune of the linearity and the lower
order of the equations is not directly evident upon inspection, but lies in
the ambiguity of the permeability and the stagnant thermal conductivity, which
both depend on the geometrical arrangement of the fibers. Further experimental
and theoretical investigation of these parameters is needed to apply the results to existing physical systems, although much fine work has been accomplished to date [5-8].

Although the transfer of water vapor by diffusion and convection is significant regarding material deterioration, it is forthwith neglected as the
effect is of second order in determining the overall heat transfer. Also, the
no-slip boundary condition may be dropped when the Darcy number is very small
because the influence of the internal flow resistance is then greater than the
influence of the diffusion of vorticity from a boundary.

The variables, $\bar{x}$, $\bar{y}$, $\bar{u}$, $\bar{v}$, $\bar{P}$, and T are non-dimensionalized as follows:

$$\frac{\bar{x}}{d}, \frac{\bar{y}}{d}, \frac{\bar{u}d}{\alpha}, \frac{\bar{v}d}{\alpha}, \frac{(\bar{P}-P_\infty+\rho_1 g\bar{y})K_2}{\mu\alpha}, \frac{T-T_3}{T_1-T_3} \tag{5}$$

to yield the un-barred quantities, x, y, u, v, P and θ, respectively. The
pressure has been perturbed from the hydrostatic pressure existing in the hot
infinite medium. A non-dimensional stream function is defined, and equations
(2 and 3) are cross-differentiated to eliminate the pressure. The resulting
system of equations is as follows:

$$\frac{\partial^2 \psi}{\partial x^2} + \frac{\partial^2 \psi}{\partial y^2} = -Ra^*\frac{\partial \theta}{\partial x} \tag{6}$$

$$\frac{\partial \Psi}{\partial y} \frac{\partial \theta}{\partial x} - \frac{\partial \Psi}{\partial x} \frac{\partial \theta}{\partial y} = \frac{\partial^2 \theta}{\partial x^2} + \frac{\partial^2 \theta}{\partial y^2} \ . \tag{7}$$

The non-dimensional parameter appearing in the coupling source term of equation (6) is the modified Rayleigh number, explicitly:

$$Ra^* = \frac{g\beta\Delta TKd}{\nu\alpha} \tag{8}$$

For ordinary fibrous insulation systems, the modified Rayleigh number is limited to less than about 50, typically being less than 10. Qualitatively, the modified Rayleigh number is the ratio of a buoyant force to a drag force.

The pressures may be determined to within a reference pressure by the direct integration of equations (2 and 3), which, in dimensionless form, are:

$$u = -\frac{\partial P}{\partial x} \tag{9}$$

$$v = -\frac{\partial P}{\partial y} - Ra^* (1 - \theta) \tag{10}$$

Once the velocities and temperatures are known, the pressures may be calculated by integrating equation (9) or (10) along lines of constant value of either y or x.

The reference pressure may be determined by applying an overall continuity balance, as follows:

$$\int u_1 dy = \int u_3 dy \tag{11}$$

Equation (11) can be formulated in terms of the pressure by applying a momentum balance to the covering materials. The flow is assumed to be one-dimensional through the covering materials. The velocity of flow through the covering materials may then be described by:

$$u_1 = -\frac{K_1}{K_2} \frac{d}{t_1} \Delta P_1 \tag{12}$$

$$u_3 = -\frac{K_3}{K_2} \frac{d}{t_3} \Delta P_3 \tag{13}$$

where ΔP is the difference in the non-dimensional external pressure and the internal pressure as calculated from equations (9) and (10). This difference is positive in the positive sense of the horizontal coordinate axis.

The resulting flow system is one that balances the flow through the covering materials and the internal flow, in accordance with the imposed pressure and temperature boundary conditions. Energy must be supplied to heat the fluid as well as to accelerate it and push it through the fibrous medium and the covering materials. Heat is also lost by conduction through the boundaries. The energy supplied to maintain the steady state is characterized by the Nusselt number, explicitly:

$$Nu = \frac{1}{A} \int_0^A (U_w \theta - \frac{\partial \theta}{\partial x}) \ dy \tag{14}$$

The problem has been reduced from three regions to one region by the introduction of the covering materials, which couple the boundary conditions to the external pressures, assumed to vary hydrostatically conforming to the density at the external temperature. The problem may be parametrically stated as follows:

$$Nu = Nu(Ra^*, A, \frac{K_1 d}{K_2 t_1}, \frac{K_3 d}{K_2 t_3}, y_1, h_1, y_3, h_3) \qquad (15)$$

The last four parameters are indicative of the size and location of the covering materials, while the third and fourth parameters describe their construction and the first two are apposite to the fibrous medium and the transferred fluid.

METHOD OF SOLUTION

Equations (6 and 7) are coupled by virtue of the source term in the equation of motion and the convection component of the energy equation. In addition, the boundary conditions on the stream function are coupled to the interior flow field. The extremely complex nature of the system renders analytical solution difficult, especially when the boundary conditions are not uniform. In view of these observations, the problem was solved numerically on a CDC 7600 computer employing non-conservative central difference representations [17] of the governing equations.

The calculation scheme proceeded as follows: (1) the stream function and the perturbation temperature were initialized to their conduction values, i.e. the values obtained under the asymptotic limit of the aspect ratio becoming infinitely large; (2) new stream function and perturbation temperatures were calculated from equation (6) and a modified version of equation (7); (3) relative pressures were calculated to within a reference pressure, p^*, by integrating equations (9 and 10) numerically; (4) the reference pressure, p^*, was calculated using equation (11) and injection velocities and new boundary values for the stream function were determined; and (5) steps (2) through (4) were repeated until convergence was achieved.

The solutions were assumed to be converged when the following criterion was satisfied:

$$\left| \frac{\tau_{new} - \tau_{old}}{\tau_{new}} \right|_{max} < Res \qquad (16)$$

where the subscript "max" denotes the maximum value over all the grid points and the symbol τ denotes any dependent variable, either the stream function or the perturbation temperature. The value of Res was tested, and a value of 5×10^{-5} was found to yield sufficient convergence as there was less than a one percent change in the Nusselt number when Res was decreased from 5×10^{-4} to 5×10^{-5}.

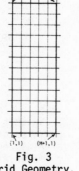

Fig. 3
Grid Geometry.

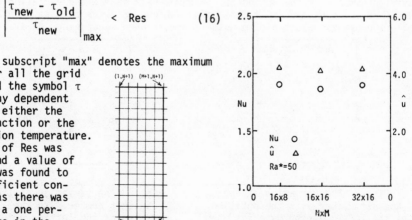

Fig. 4 The variation of the Nusselt number with the grid size.

The value of 5×10^{-5} was therefore used for all runs reported hereinafter.

The grid system employed in solving this problem is shown in Fig. 3. The accuracy of the scheme was determined by decreasing the grid size and observing the change in the global quantities of the Nusselt number and the average wall infiltration velocity. The variation of these quantities with the number of grid points N x M is shown in Fig. 4. Based upon these calculations, a grid size (N x M) of 16 x 8 was used for all examples set forth succeedingly. During these runs, the field values were observed to shift as much as ten percent at a particular point, so the profiles lack the high degree of accuracy attributed to the global quantities.

RESULTS AND DISCUSSION

All calculations were performed for an aspect ratio of ten, corresponding approximately to that existing in ordinary stud walls. In all cases presented graphically, the covering materials encompassed one-fourth of the area of the walls. Such a large area was considered so that a sufficient number of grid points existed in the region of infiltration. The large number of cases calculated prohibited a fine mesh. In fact, the average infiltration velocity is not expected to vary strongly with the size of the covering material, and this has been substantiated numerically by halving the areas of infiltration and observing little change in the average leakage velocity. Modified Rayleigh numbers of 10, 30 and 50, only, were considered. The vertical location of the covering materials is denoted by y_1/A and y_3/A for the location on the left and right walls, respectively, about which the covering material is centered. The pressures are equated at the mid-height of the wall except where otherwise noted. The quantities $K_1 d/K_2 t_1$ and $K_3 d/K_2 t_3$, symbolized by $K_i d/K_2 t_i$, are both held constant at a value of 1×10^{-2}, except in Fig. 18, where they are allowed to vary. The value 1×10^{-2} corresponds approximately to that of a dense piece of cardboard about one millimeter in thickness. These values are inversely related to the resistance to flow through the covering materials.

The variation of the Nusselt number with the position of the covering materials is shown in Figs. 5-8. The right covering material is centered about the seven-eighth, five-eighth, three-eighth, and one-eighth position in Figs. 5, 6, 7 and 8, respectively, while the position of the left covering material is varying as the abscissa in the Figures. Results are proffered for modified Rayleigh numbers of 10, 30 and 50. Figures 9-12 depict the variation of the mean infiltration velocity in the same fashion. For the purpose of comparison, Table 1 provides values of the Nusselt number when no openings exist.

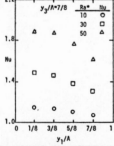

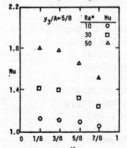

Fig. 5 The variation of the Nusselt number with the position of the left covering material when the right covering material is fixed at the 7/8 position.

Fig. 6 The variation of the Nusselt number with the position of the left covering material when the right covering material is fixed at the 5/8 position.

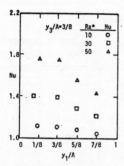

Fig. 7 The variation of the Nusselt number with the position of the left covering material when the right covering material is fixed at the 3/8 position.

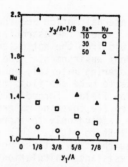

Fig. 8 The variation of the Nusselt number with the position of the left covering material when the right covering material is fixed at the 1/8 position.

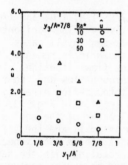

Fig. 9 The variation of the average infiltration velocity with the position of the left covering material when the right covering material is fixed at the 7/8 position.

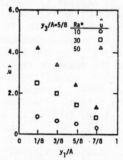

Fig. 10 The variation of the average infiltration velocity with the position of the left covering material when the right covering material is fixed at the 5/8 position.

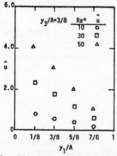

Fig. 11 The variation of the average infiltration velocity with the position of the left covering material when the right covering material is fixed at the 3/8 position.

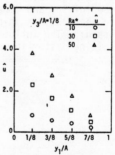

Fig. 12 The variation of the average infiltration velocity with the position of the left covering material when the right covering material is fixed at the 1/8 position.

The internal motion and the resulting internal pressure field compels the assemblage to act as a pump, displacing fluid from the hot external environment to the cold external environment. The hot fluid enters through the left covering material, and the strong buoyancy force tends to make the fluid rise, thus enhancing the primary flow field. Forced flow, due to the different pressures on the covering materials, is of secondary importance, acting only to increase or decrease the bulk flow from left to right. This may be demonstrated by calculating the flow rate using an approximate one-dimensional resistance calculation, and assuming the flow is due to the external pressure differences alone. There is never a flow reversal as would be expected from forced flow considerations when $y_1/A = 7/8$ and $y_3/A = 1/8$, although the flow rate is very small. With an increase in the aspect ratio, flow reversal would eventually occur as the pressure differences due to height would then become greater, and would no longer be negated by the internal motion.

Table 1.
Nusselt numbers
for no leakage

Ra^*	Nu
10	1.01
30	1.11
50	1.27

Streamlines of the cases for y_1/A, y_3/A of 1/8, 7/8; 1/8, 1/8, and 7/8, 1/8 are shown sketched in Figs. 13, 14 and 15, respectively. The horizontal and vertical scales are unequal, as the aspect ratio is ten in all cases. The modified Rayleigh number is 50. The streamlines are shown in increments of 2, except for the bounding streamline which adjusts to the flow situation. The zero streamline identifies the boundary between the through-flow and the recirculating flow. As observed in Fig. 14, where the through-flow directly opposes the natural flow, the recirculating region is small. In Fig. 13, the inflow is large, tending to push the fluid toward the center of the enclosure before it rises. In Fig. 15, the through-flow region is small and the infiltration does not dominate the flow field. In most cases, there is a tendency towards a relatively stagnant core, of slow velocity, rotating with the natural flow. This is similar to the conditions prevailing when no porous medium is present to inhibit the flow [18-20]. The strong buoyancy effects are readily apparent from the flow patterns. In addition, the preponderance of the flow occurs close to the walls. Some of the finer details sketched are not justified due to the coarse grid employed during calculations.

The influence of the flow upon the temperature distribution may be determined by viewing Fig. 16, which is a sketch of the isotherms for a modified Rayleigh number of 50, and impermeable boundaries. In the core region, the isotherms are vertical, as in pure conduction, or as occurs when the flow is "fully developed." The influence of the horizontal walls is to turn the fluid; thus the resulting distortion of the isotherms due to the normal flow. This

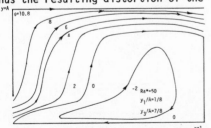

Fig. 13 Non-dimensional streamlines when the left and right covering materials are fixed at the 1/8 and 7/8 positions, respectively.

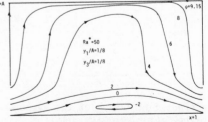

Fig. 14 Non-dimensional streamlines when the locations of the left and right covering materials are fixed at the 1/8 position.

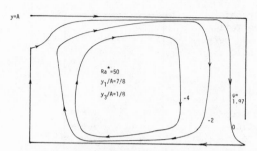

Fig. 15 Non-dimensional streamlines when the locations of the left and right covering materials are fixed at the 7/8 and 1/8 positions, respectively.

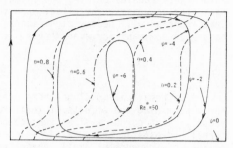

Fig. 16 Non-dimensional stream function and isotherms when no infiltration is allowed.

is very similar to the "conduction regime" observed by Eckert and Carlson [20] at lower temperature differences. The flow is likely to remain in this regime for nearly the entire range of practical applications due to the large drag force inherent in porous media.

To determine the effect of pressure differentials due to wind, the pressure difference across the mid-height of the enclosure was changed from zero to that corresponding to a wind of about 15 mph. As a result of many measurements, it was determined [15] that the pressure difference, $\Delta \overline{P}$, resulting from the action of the wind could be approximated by:

$$\Delta \overline{P} \;=\; 0.64 \; (\tfrac{1}{2} \, \rho \, \overline{V_w}^2) \qquad\qquad (17)$$

In non-dimensional form, the resulting pressure difference is about 100. Both positive and negative pressure differences were considered, modeling the effects on the windward and leeward sides of buildings. The representative variable is identified as DP, the pressure of the right external environment relative to the left external environment at the mid-height. Alternatively, this could be viewed as the displacement of the neutral zone due to excess supply air, internal ventilation, etc. The vertical variation is yet hydrostatic. The manner with which the Nusselt number and the mean infiltration velocity vary with the non-dimensional pressure difference, DP, is shown in Fig. 17. The effect is small; the trends are as expected from the enhancement or the abatement due to the forced convection.

The variation of the Nusselt number and the mean infiltration velocity with the non-dimensional permeability-thickness parameter of the covering materials is illustrated in Fig. 18. As the abscissa varies in the positive sense, the resistance to flow shifts from being dominated by the covering materials to being about the same order of magnitude as the interior porous medium. The ascendency of this parameter is apparent from the results. Albeit predictable, this is most unfortunate due to the lack of catholic characterization of this parameter for physical systems. Some experimental work is necessary to assess values of this parameter in a general sense. Numerical stability became exceedingly difficult to achieve at the high values of this parameter, with the result that not many

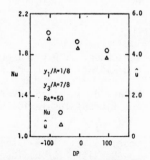

Fig. 17 The variation of the Nusselt number and the mean infiltration velocity with the non-dimensional pressure difference.

values were considered. As the value of this parameter increases, the effect upon the total heat transfer will become significant, and perhaps even dominant for even very small holes.

SUMMARY

A mathematical model has been proposed for determining the effects of infiltration velocities on porous systems subjected to external hydrostatic pressure variations. The salient consequence of small leakage velocities has been demonstrated. The domineering effect of the permeability-thickness parameter of the covering materials has been illustrated. The need for further work in the area of characterizing this parameter is manifest. For the particular system studied, any combination of locations has been shown to affect the heat transfer adversely. The importance of the free convection upon the infiltration has been indicated. Wind effects and specific

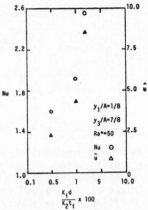

Fig. 18 The variation of Nusselt number and the mean infiltration velocity with the parameter characterizing the covering materials.

locations have been seen to be of a secondary nature in determining the overall energy transfer. The size of the inlet or exhaust opening has been postulated to exert a tertiary influence upon the mean infiltration velocity. Buoyancy forces due to height differences were shown to induce more flow than those due to the temperature difference alone.

The following algebraic expression correlates the data fairly well:

$$Nu = 1 + Ra^*[0.015 + 4.5 \times 10^{-4} \frac{y_1}{A} \left(1 - 21(\frac{y_1}{A})^2\right) - 0.075(\frac{y_3}{A} - \frac{1}{2}) - 1.8 \times 10^{-5}DP$$
$$+ 1.3 (\frac{K_i d}{K_2 t_i} - 10^{-2}) + 20 (\frac{K_i d}{K_2 t_i} - 10^{-2})^2]$$

ACKNOWLEDGEMENT

This research was supported through a Grant on "Research in Building Insulation" by the Center for Building Technology, National Bureau of Standards (NBS). Many discussions with Dr. T. Kusuda have been most helpful to the present work.

REFERENCES

1. Verschoor, J. D., and Greebler, P. 1952. Heat Transfer by Gas Conduction and Radiation in Fibrous Insulation. Trans. ASME. 74: 962.

2. Mumaw, J. R. 1968. Variations of the Thermal Conductivity Coefficient for Fibrous Insulation Materials. M.S. Thesis. The Ohio State University.

3. Lopez, E. L. 1969. Techniques for Improving the Thermal Performance of Low-Density Fibrous Insulation. Prog. in Aero. and Astro. 23: 153.

4. Kaganer, F. M. G. 1969. Thermal Insulation in Cryogenic Engineering. Translated from Russian by A. Moscona. Israel Program for Scientific Translations, Jerusalem.

5. Scheidegger, A. E. 1974. The Physics of Flow through Porous Media. University of Toronto Press.

6. Bankvall, C. G. 1972. Natural Convective Heat Transfer in Insulated Structures. Lund Inst. of Tech. Report 38.

7. Bankvall, C. G. 1973. Heat Transfer in Fibrous Materials. J. of Testing and Evaluation 3: 235.

8. Fournier, D. and Klarsfeld, S. 1974. Some Recent Experimental Data on Glass Fibre Insulating Materials and Their Use for a Reliable Design of Insulations at Low Temperatures. ASTM STP 544: 223.

9. Holst, P. H. and Aziz, K. 1972. A Theoretical and Experimental Study of Natural Convection in a Confined Porous Medium. Can. J. Chem. Engr. 50: 232.

10. Bories, S. A. and Combarnous, M. A. 1973. Natural Convection in a Sloping Porous Layer. J. Fluid Mech. 57: 63.

11. Kaneko, T., Mohtadi, M. F., and Aziz, K. 1974. An Experimental Study of Natural Convection in Inclined Porous Media. Int. J. Heat Mass Transfer 17: 485.

12. Chan, B. K. C., Ivey, C. M., and Barry, J. M. 1970. Natural Convection in Enclosed Porous Media with Rectangular Boundaries. J. Heat Transfer 92: 21.

13. Weber, J. E. 1974. Convection in a Porous Medium with Horizontal and Vertical Temperature Gradients. Int. J. Heat Mass Transfer 17: 241.

14. Burns, P. J., Chow, L. C., and Tien, C. L. (in press) Convection in a Vertical Slot Filled with Porous Insulation. Int. J. Heat Mass Transfer.

15. ASHRAE Handbook of Fundamentals. 1967. 405.

16. Muskat, M. 1937. The Flow of Homogeneous Fluids through Porous Media. J. W. Edwards.

17. Torrance, K. E. 1968. Comparison of Finite-Difference Computations of Natural Convection. J. Research of the National Bureau of Standards. B. Mathematical Sciences 72B: 281.

18. Batchelor, G. K. 1954. Heat Transfer by Free Convection Across a Closed Cavity Between Vertical Boundaries at Different Temperatures. Quart. J. Appl. Math. 12: 209.

19. Ostrach, S. 1972. Natural Convection in Enclosures. Adv. Heat Transfer 8: 161.

20. Eckert, E. R. G. and Carlson, W. O. 1961. Natural Convection in an Air Layer Enclosed Between Two Vertical Plates with Different Temperatures. Int. J. Heat Mass Transfer 2: 106.

THE EFFECT OF CONVECTIVE HEAT EXCHANGE ON THERMAL-INSULATING PROPERTIES OF PERMEABLE POROUS INTERLAYERS

V. A. BRAILOVSKAYA, G. B. PETRAZHITSKY, AND V. I. POLEZHAEV

Moscow High Technical School
Moscow, USSR

INTRODUCTION

An increase in efficiency of modern engineering systems and designs depends considerably on our knowledge of numerical regularities of the convective heat and mass exchange processes. Investigation of convective heat transfer in porous interlayers is particularly important in reliably determining characteristics and increasing the qualities of thermal insulation particularly in civil engineering, where the convective component may significantly deteriorate thermal-insulating properties of insulation elements.

In this respect it is of interest to obtain a complex of local and integral characteristics of heat exchange in circular layers of isotropic porous medium, the liquid being filtered by lifting power under non-isothermal conditions.

FORMULATION OF PROBLEM.

A limiting model of stationary two-dimensional thermal convection in a finely divided medium with fine pores, based on the Darcy linear law, is used:

$$\vec{V} = -\frac{K}{\mu}\, grad\, p \quad , \qquad (I)$$

where P - is the pressure, μ - is the coefficient of dynamic viscosity of the liquid filling up pores, K - is the permeability coefficient of medium, $\vec{V}$ - is the rate of filtration.

The cylindrical surfaces bounding the porous material are impermeable and kept at constant temperatures T_1 and T_2.

The system of equations for steady-state natural convection in a porous medium with taking account of (1) and of approximation of lifting forces in the Boussinesc approximation has the form:

$$\frac{\mu u}{K} = -\frac{\partial p}{\partial x}$$

$$-\frac{\mu \upsilon}{K} = -\frac{\partial p}{\partial y} + \rho g \beta (T - T_0)$$

$$\frac{\partial u}{\partial x} + \frac{\partial \upsilon}{\partial y} = 0 \tag{2}$$

$$\rho C_p \left(u \frac{\partial T}{\partial x} + \upsilon \frac{\partial T}{\partial y} \right) = \lambda^* \left(\frac{\partial^2 T}{\partial x^2} + \frac{\partial^2 T}{\partial y^2} \right) \quad ,$$

where ρ - is the density, β - is the volume expansion coefficient , C_p — is the specific heat of the liquid filling up the pores, u and υ — are the transmissibility rate projections onto the axes x and y , λ^* — is the heat conductivity of the porous medium without taking account of liquid motion, and $\Delta T = T - T_0$ — is the difference between local and certain characteristic temperatures.

After inserting the function of current ψ , determined from conditions, $u = \partial \psi / \partial y$, $v = -\partial \psi / \partial x$, of reducing system (2) to the dimensionless form and turning to the polar coordinate system, we get:

$$\Delta \psi = -Ra^* \left(\frac{\partial \theta}{\partial z} \cos \varphi - \frac{1}{z} \frac{\partial \theta}{\partial \varphi} \sin \varphi \right) \tag{3}$$

$$\Delta \theta = \frac{1}{z} \left(\frac{\partial \psi}{\partial \varphi} \frac{\partial \theta}{\partial z} - \frac{\partial \psi}{\partial z} \frac{\partial \theta}{\partial \varphi} \right) \quad ,$$

where ψ and θ — are the dimensionless functions of current and temperature, respectively, and $Ra^* = g \beta \delta K \rho^2 C_p \Delta T / \mu \lambda^*$ is the Rayleigh filtration number, $\delta = R_2 - R_1$ - is the width of gap between cylindrical surfaces of radii R_1 and R_2.

Steady-state solutions are found by iterating with respect to a ceratin parameter τ analogous to time. At the initial moment of time filtration ($\psi = 0$) is supposedly and there is a regime of heat conductivity, according to which temperature in the circular channel is distributed by the law $\theta = \ln \frac{z}{z_1} / \ln \frac{z_2}{z_1}$, provided $\theta = 0$ at $z = z_1$ and $\theta = 1$ at $z = z_2$.

Assuming the pattern of flow to be symmetrical relative to the vertical axis of symmetry, the solution of system (3) is obtained for one half of the circular region $(-\pi/2 \leqslant \varphi \leqslant \pi/2)$ and at the boundary $\varphi = \pm \pi/2$ we assume $\partial \theta / \partial \varphi = 0$ at $z_1 \leqslant z \leqslant z_2$.

In practice, of particular interest is the value of the mean heat transfer through the layer and its dependence on the Rayleigh criterion and on the geometry of cavity. As in the work on convection of homogeneous liquid interlayers [1], the average heat transfer is characterized by the coefficient of convection

$$\mathcal{E}_k = \overline{Nu}\ z_i\ \ell n\ z_2/z_1\ ,$$

where

$$\overline{Nu} = \frac{1}{\pi}\int_{-\pi/2}^{\pi/2}(\partial\theta/\partial z_i)\,d\varphi\ ,$$

which shows an excess of the mean heat flow in convection conditions over the flow in the regime of pure heat conductivity.

Heat exchange in an annular porous interlayer was investigated on the basis of numerical solution of system (3) by the scheme developed earlier for solving the problems of natural convection of a homogeneous liquid [1] (explicit scheme obtained by the method balance). The control of accuracy of calculations was performed by correlating the solutions on various lattices and by verifiying discrepancies of integral heat balances.

RESULTS OF SOLUTION

Calculations were carried out for various convection conditions at $10 < Ra^* < 1000$ and at the ratio of radii of external and internal cylinders $1.01 \leq z_2/z_1 \leq 8$.

With an increase in the Rayleigh filtration number at constant geometry of the annulus the coefficient of convection, i.e. the heat transfer through the layer increases, ehich impairs the thermal-insulating properties of the porous interlayer. Dependence of $\mathcal{E}_k$ on the Rayleigh filtration number at various z_2/z_1 is shown in Fig.1. It may be easily seen that these

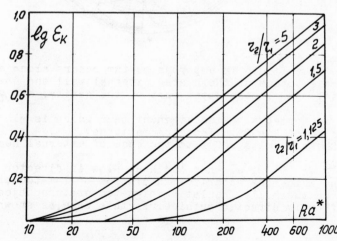

Fig.1.

dependence asymptotically tend to $\mathcal{E}_K = 1$ at some "critical" Rayleigh number which is different for each relative width of the gap. The loss z_2/z_1 , the larger "critical" Ra^* ,i.e.

the larger the values of Ra^* , the more considerable of the convective component contribution. From these data, assuming particular numerical parameters, we can determine the mean heat flow through a shell, and also, for each porous interlayer characterized by the z_2/z_1 ratio to obtained the values of

Ra^* , starting with which the role of convection in heat exchange cannot be neglected.

A typical picture of developed natural convection in the region under consideration is shown in Fig.2. The natural filt-

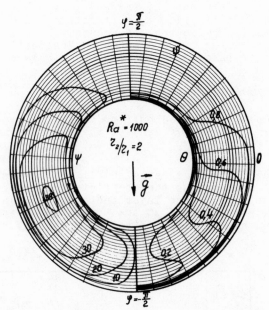

Fig.2.

ration of a liquid in a porous medium occurs along crescent-shaped trajectories up the heated external wall and down the cold internal one. In the upper part of the cavity, at $Ra^* = 1000$

there is formed a heated stagnant band which is also seen on the left side of Fig.2 in isotherm patterns.

These results refer to the case of external heating

$(T_1 < T_2)$ i.e. when the heat flow is directed inside the region, which characterizes insulation of low-temperature systems. On thermally insulating high-temperature objects, the heat flow is directed outside. In this case,as shown in Fig.3

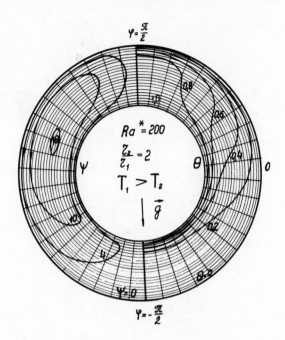

Fig.3.

the centre of crescent trajectories of flow is located above the
horizontal axis of symmetry, whereas the cold stagnant band is
in the lower part of cavity. At constant properties of medium the
pattern of distribution of ψ and θ at $T_1 > T_2$

is symmetrical relative to the horizontal axis of symmetry of
the similar pattern for the case of $T_1 < T_2$ at equal

other parameters. Convection in annulus is characterized by the
fact that the motion of liquid results in temperature stratification: in the upper part of the layer on the central line

$(z = z_1 + 1/2 ; -\pi/2 \leqslant \varphi \leqslant \pi/2)$ the temperature is higher, and
in the lower part it is lower than the average temperature of
$\theta_m = 1/2$. The regime of the developed convection

is characterized by the appearance of a region with the inverse
temperature gradient (Fig.2), as in the case of the homogeneous
liquid convection. [1] .

Temperature stratification affects considerably the development of boundary layers and, consequently, the distribution
of heat flows over cold and heated surfaces. Distribution of
Nusselt local numbers along the internal and external boundaries
of the region at different values of the Rayleigh filtration number is presented in Fig.4. It is seen that the maximum local heat
flows are realized in the initial region of the liquid flows
washing the cold and hot walls. Then, as the boundary layers
grow, the Nusselt local numbers decrease, reaching, at $\varphi = \pi/2$

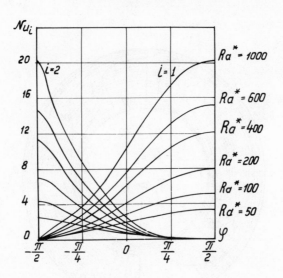

Fig.4.

on the hot and at $\varphi = -\pi/2$ on the cold surfaces, the values less than those for the corresponding flows in conditions of pure heat conductivity. As follows from graphs in Fig.4, this non-uniformity of heat flows at the boundaries increases with

Ra^* . The same tendency takes place with increasing

z_2/z_1 ; therefore, the data only on the average values of convective heat transfer ($\overline{Nu}$ or $\mathcal{E}_K$) may be insuffi-

cient, for instance, to calculate the heating of structures protected with porous insulation.

Significant non-uniformity of local heat flows on the walls of interlayers can be explained by the specific properties of a porous medium described in this case by the Darcy law, the medium exhibiting resistance to mixing and strengthening considerably the effect of temperature stratification as compared with convection in a homogeneous liquid.

Similar properties of convection in a porous medium for flat interlayers were described in [2] . It should be noted that the data on the effect of convection upon the thermal--insulating properties of porous interlayers are practically not available in reference literature, stadards, and specifications.

The data obtained on the dependence of the average heat flow through the porous interlayers on the Rayleigh filtration number at different dimensions of a cavity may be used to determine the possibility of arising convection, its quantitative effect on the thermal characteristics of porous materials and the choice of ways of convection suppression by varying the parameters inserted into dimensionless criteria.

REFERENCES

1. Petrazhitsky G.B., Bekneva E.V., Brailovskaya V.A., and
   Stankevich N.M. Calculation of flow and heat exchange in free
   liquid motion in a horizontal annulus. VestnikL L'vovskogo
   politekhnicheskogo instituta, No.46 "Voprosy elektro- i tep-
   loenergetiki", 1970.
2. Vlasyuk M.P., and Polezhaev V.I. Natural convection and heat
   transfer in permeable porous materials. Preprint IPM AN SSSR,
   No77,1975.

RADIATIVE ENERGY TRANSFER EFFECTS ON FIBROUS INSULATING MATERIALS

E. ÖZIL

Middle East Technical University
Ankara, Turkey

R. BIRKEBAK

College of Engineering
University of Kentucky
Lexington, Kentucky, USA 40506

ABSTRACT

The results of an experimental study are presented and correlated with a theoretical model for the combined modes of heat transfer through a fibrous insulating material.

The study involves the effects of environmental thermal radiation exchange on the heat transfer in a fibrous insulating material which has one free and one bounded surface. Such conditions are found for insulation used in attics of houses.

Experimental results are presented in terms of temperature profiles and net radiative heat flux in the fibrous material as a function of environmental radiation temperature, air temperature, base or substrate temperature and base heating rate.

NOMENCLATURE

A, B constants in equations (6) and (7)

d fiber diameter (micrometers)

F photon mean free path function in equation (3)

f^1 radiation function, equation (5)

k thermal conductivity (w/m-°K)

L bed depth (mm)

N_f fiber bed parameter, equation (4)

n_f number of fibers per unit area $(1/mm^2)$

q heat flux (w/m^2)

T absolute temperature (°K)

T^1 temperature difference ratio, equation (1)

y position coordinate normal to base (mm)

Subscripts

a   air

e   environment

eff   effective

f   fiber

o   base or skin

r   radiative

s   surface

Greek Alphabet

ε emittance

η dimensionless coordinate (y/L)

ϕ azimuthal angle

θ polar angle

θ_x dimensionless temperature (T_x/T_o)

ρ density (Kg/m^3)

σ Stefan-Boltzmann constant $(5.6697 \times 10^{-8} w/m^2 \ °K^4)$

INTRODUCTION

Solid fibers and a gas, usually air, make up the bed of a fibrous insulating material. This bed is contained either between two bounding surfaces or one free and one bounding surface. The striking difference between the two types is that the latter is exposed to ambient conditions and the former is not.

Usually, the insulative properties and the energy transfer characteristics of fibrous insulating materials are studied using a guarded hot-plate method where the material is physically bounded on both sides. However, this method fails to predict the energy transfer effects for insulating materials with a free surface simply because the boundary conditions do not match. And most of the applications of the fibrous insulating materials (common attic insulations, etc.) and all the biological insulating materials (furs, feathered integuments, etc.) are of this type.

As a result of the addition of the heat exchange with the environment to internal heat exchange, the energy transfer in fibrous insulating materials with one free surface becomes a more general and a much more complex problem. If we examine this general case we will observe that the energy transfer is normal to the plane(s) of the fibrous bed surfaces and includes transfer of heat by conduction, radiation, and sometimes by convection.

The relative importance of all three forms of heat transfer depends on several factors like geometrical arrangement of the fibers in the bed, dimensions of the fibers, the bed depth, and ambient conditions. Usually it is accepted that convection within the bed is negligible. The effects of conduction through the solid material is usually small [1-7].

Several analytical methods have been tried in order to solve the problem of combined heat transfer in fibrous materials. The presence of both the external and the internal radiation makes the energy equation a non-linear and an integrodifferential one for which there is no closed form of solution available. An extensive review of the works on this problem can be found in reference [8].

DESCRIPTION OF THE TEST MATERIAL AND THE TEST EQUIPMENT

A. The Test Material

Artificial fur samples supplied by Norwood Mills, Inc., Janesville, Wisconsin, were used for the tests. The reason for this choice is that the artificial fur resembles closely some types of real furs in appearance, effective densities, and insulative properties. It also is a well organized fibrous material which we can precisely describe and represents a common insulating material although it may be less dense than the average material.

Black artificial fur, type 755, was used for the experiments and some of its characteristics were supplied by the manufacturer as

(1) The material is composed of 85% Verel (Modacrylic) and 15% Acrylic,
(2) the finished fiber length is 38 mm, and
(3) the weight per lineal yard (1 yard long, 59 in. across) is 44 ounces.

Other physical properties were measured by Özil [8] and Wilson [9]. Artificial fur is woven with a thread mesh that serves as skin. The fibers are collectively placed in stitches and sewn throughout the mesh. The number of the fibers per unit area, n_f, was about 36 ± 1.8 fibers per mm^2. The square root of the average of the squares of the fiber diameters is 46.3 μm. The effective bed density to that of a fiber, ρ_{eff}/ρ_f is 0.082.

The thermal properties of the fiber were measured by Özil [8] and Birkebak [10]. The thermal conductivity, k, was found to be about 0.2 w/m°K. The emissivity, ε, of the artificial fur is approximately 0.9. We must point out here that the apparent emissivity of the fiber layer changes with depth. The value 0.9 represents an overall apparent emissivity. A discussion for the apparent emissivities is presented in [8].

B. Test Equipment

A schematic diagram of the test equipment is given in Fig. (1). The test samples were placed into an environmental chamber whose wall temperature was controlled by a constant temperature bath. The range of the bath temperature was from 250 - 350°K. The stainless steel cylindrical chamber was sprayed black inside and was 260 mm long and 185 mm in diameter. It had a hemispherical cap and several ports of different sizes. These ports were used for electrical and thermocouple feed-throughs, for the test plate support system, and one port contained a thermopile detector which was used to measure the radiative fluxes. The whole chamber was insulated by a 50 mm thick fiberglas insulation layer. For vacuum measurements the environmental chamber was connected to a turbomolecular pump.

The test plate basically was a 12.5 mm thick plexiglas piece with a length and width of 200 and 100 mm, respectively. Sandwiched between two layers of a double-back tape, were six individually controlled heaters on the test plate. The samples were laid on the uppermost layer of the tape and eighteen copper-constantan, Awg. 30, thermocouples and two heat flux sensors were placed underneath the skin of the samples. Fig. (2) shows the test plate and the positions of the heaters, thermocouples, and the two heat flux sensors.

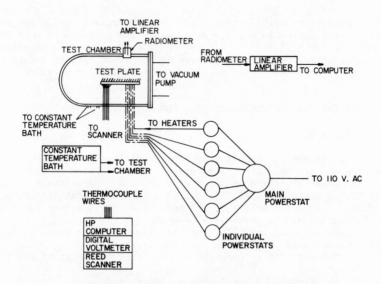

Fig. (1)   General Schematic Diagram of the Experimental Set Up

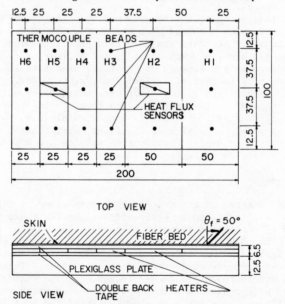

Fig. (2)   The Test Plate and the Positions of the Thermocouples

A thermocouple gradiometer, Fig. (3), was placed into the samples in order to record the temperature distribution within the fiber bed and above the sample.   The gradiometer was aranged in a ladder-like fashion.

All the thermocouples, heat flux sensors, and the detector were connected to 2114B Hewlett-Packard mini-computer through a 2402A HP Integrating Digital Voltmeter and 2912-40 channel Reed Scanner for the data acquisition.

A more detailed description of the test equipment can be found in ref. [8].

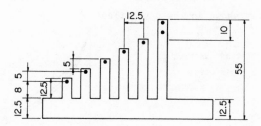

Fig. (3) The Gradiometer

Test Procedure

All of the measurements were made at steady state conditions. The computer was programmed to check whether or not the steady state conditions prevailed.

Two sample thicknesses were employed in the experiments; 20 and 25 mm. Before a set of runs, the sample was groomed so that the desired thickness was obtained. The gradiometer then was placed into the fiber bed making sure that the gradiometer did not change or disturb the bed's overall structure.

The test plate was then put into the environmental chamber and the appropriate connections to the heaters and the computer were made. The test plate and the chamber walls were heated and kept at the desired temperatures with the aid of the heaters and the constant temperature bath.

When steady state conditions were reached all the necessary measurements were made and the calculations were done by the computer. The temperature within the fiber bed, the wall, air and detector temperatures, and the heat fluxes were printed as output.

Again ref. [8] gives more details about the test procedure and the calculations.

Test Results and Discussion

The following non-dimensionalizations are utilized in this paper:

$$\eta = \frac{y}{L} , \quad T' = \frac{T - T_e}{T_o - T_e} \tag{1}$$

and

$$\theta = \frac{T}{T_o} , \quad \theta_e = \frac{T_e}{T_o} , \quad \theta_a = \frac{T_a}{T_o} \tag{2}$$

where y is the position coordinate in the vertical direction, L is the fiber bed depth, and T is the temperature (°K). Subscript e is for the environment, o is for the backskin, and a is for air.

Measured temperature profiles are presented in Figures (4-6). Figures (4) and (5) present data for two different skin temperatures for varying θ_e and θ_a, while the results presented in Figure (6) are for different base heating rates and values of θ_e, θ_a and T_o.

The T' distributions are used to show more clearly the effects of heat conduction in the fiber bed. We see from Figures (4-6) that the profiles are curve-linear which is caused by the interaction of the radiation exchange within and at the free surface of the bed. An increase in the air temperature T_a decreases the convective loss from the surface and the temperatures within the fiber bed must increase in order to dissipate the heat by thermal radiation.

This effect is clearly shown in the θ profiles.

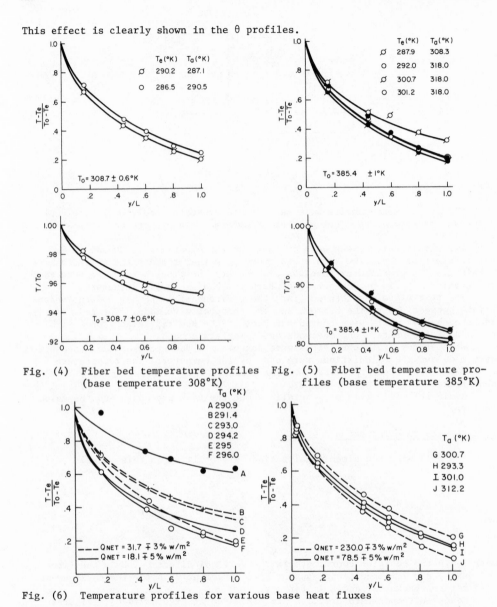

Fig. (4) Fiber bed temperature profiles Fig. (5) Fiber bed temperature pro-
          (base temperature 308°K)   files (base temperature 385°K)

Fig. (6) Temperature profiles for various base heat fluxes

The temperature distributions, T's, as a function of T_e and T_a are presented in Figure (6) for various base heating rates. It is seen that increasing the air temperature when $T_a < T_o$, causes a decrease in T' at $\eta = 1.0$, which
means a decrease in the convective heat loss from the fiber bed. Concomitant
with an increasing T_a is an increase also in T_o, the base or skin temperature.
The effect is more pronounced for the lower heat fluxes. As a matter of fact,
for higher heat fluxes, a fourfold increase in the heat input shows only
slight changes in the temperature profiles with the air temperature. Low heat
fluxes represent lower radiative losses and consequently convection and radiation are of equal importance in the energy transfer process. With increase in
heat input the radiative heat losses become more significant and the effect of

convection or air temperatures on the temperature distribution becomes smaller. The T' distribution is then almost strictly a function of T_e.

The variation of the net radiative heat flux from the fiber bed as a function of the environmental radiation temperature is shown in Figure (7). Shown in the figure are three curves for three different θ_a ratios. Again, as expected, the radiant heat losses increase with decreasing θ_e's. Additionally decreasing the air temperature results in a decrease in the radiant heat loss portion, a sign of the presence of convection.

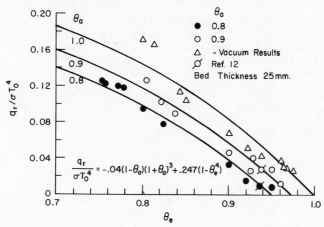

Fig. (7) Net Radiative Heat Flux

If we can assume for the moment that conduction is the dominant energy transfer process in the fiber bed, the temperature terms in the radiation energy equations can be simplified to give the following for the transfer of diffuse radiation in fiber beds according to David and Birkebak [7].

$$q_{r-y} = (2\sigma/\pi)\,(T_o^4 - T^4)\int_0^\pi \int_0^{\pi/2} e^{-\eta N_f F/\cos\theta}\,\cos\theta\,\sin\theta d\theta d\phi$$

$$+ (2\sigma/\pi)(T^4 - T_e^4)\int_0^\pi \int_0^{\pi/2} e^{-(1-\eta)N_f F/\cos\theta}\,\cos\theta\,\sin\theta d\theta d\phi$$

$$- (8\sigma T^3/N_f \pi)L(\partial T/\partial y)\int_0^\pi \int_0^{\pi/2} f'\,\cos^2\theta\sin\theta\phi\phi \qquad (3)$$

where $N_f = {}^{n_f d_f \varepsilon L/}\cos\theta_f = 4(\rho_{eff}/\rho_f)\varepsilon L/\pi d_f$ $\qquad (4)$

and $\quad f' = \left[2 - \exp(-\eta N_f F/\cos\theta) - \exp[-(1-\eta)N_f F/\cos\theta]\right]/F$

$$-N_f\left[\frac{(1-\eta)}{\cos\theta}\exp(-[1-\eta]N_f F/\cos\theta) + \frac{\eta}{\cos\theta}\exp(-N_f \eta F/\cos\theta)\right] \qquad (5)$$

In these equations, σ is Stefan-Boltzmann constant, F is a photon mean free path function and is a function of the geometry of the fiber in the bed [5-7, 11], N_f is a measure of the optical thickness of the bed, and q_{r-y} is the net flux of diffuse radiant energy in the y-direction.

In Eq. (3), the first term is the net radiation from the base or skin through a plane located at η. The second term is similar to the first and gives the net radiant transfer passing through the plane at η toward the environmental radiation sink at T_e. The third term, that involving the temperature gradient $\partial T/\partial y$, is the amount of radiation passing through the η plane because of emission and scattering by the fibers.

For our particular needs at present, eq. (3) can be simplified to the following form

$$q_r \cong A\sigma\bar{T}^3\Delta T_r + B\sigma(T_0^4 - T_e^4) \tag{6}$$

We must now interpret $\bar{T}^3$ and ΔT_r. Equation (6) does not involve the interaction of conduction and radiation in the fiber bed. We believe, that for this simplified first order model or approach that the interaction or coupling influence can be partially achieved by introducing in some way the air temperature T_a into eq. (6). There are several suitable choices for expression $\bar{T}^3$ and ΔT_r in terms of T_a, T_e, T_s and T_0 and we can put forth arguments justifying their validity. Our choice in this paper is based on what temperature would be most readily available. We have assumed the mean temperature $\bar{T}$ to be approximated by $T_0(1+\theta_a)/2$ and $\Delta T_r \cong T_0 - T_a = (1-\theta_a)T_0$. With these assumptions eq. (6) becomes

$$q_r/\sigma T_0^4 = \frac{A}{8}(1-\theta_a)(1+\theta_a)^3 + B(1-\theta_e^4) \tag{7}$$

Our results were least square fitted to eq. (7) and the constants $A/8$ and B are -0.04 and 0.247, respectively. The solid lines on Figure (7) were calculated using eq. (7) and we see that the first order model approach does quite well in correlating the data.

Also presented in Fig. (7) are data from Birkebak et al. [12] and measurements obtained for our fiber bed under vacuum conditions. The vacuum results are presented here only for general interest. These results are somewhat comparable to the case of $\theta_a \cong 1$. However, the absence of air alters the temperature profile greatly and therefore the base and radiative heat transfer processes.

In a similar manner the base heat flux was correlated. Our base heat transfer results are shown in Figure (8) and were correlated by assuming that q_0 was equal to an overall conductance times the difference between (T_0-T_a) plus the net radiative flux. The resulting equation after least square fitting of the data is

$$\frac{q_0}{\sigma T_0^4} = \frac{1.66}{\sigma T_0^3}(1-\theta_a) - 0.04(1-\theta_a)(1+\theta_a)^3 + 0.247(1-\theta_e^4) \tag{8}$$

The agreement of this simple model and the data is quite acceptable.

CONCLUSIONS

Changes in the dimensionless environmental radiation temperature θ_e are shown to greatly effect the overall net radiative heat transfer for the type of fiber bed studied.

We see from Fig. (7) that a net radiative heat flux into the fiber bed can occur when $\theta_a < \theta_e$ for $\theta_e < 1.0$.

In general, for fiber beds of the type studied with large N_f (in our case $\cong 57$) and $\rho_{eff}/\rho_f \cong 0.05$ to .1 (in our case 0.082) that gas conduction in the bed is the most important mechanism for heat transfer [1-9]. However, the presence of the fibers alters both the radiative and conductive heat transfers in the bed and these mechanisms of heat transfer are greatly effective by changes in θ_e, θ_a and T_0.

The equations developed from Davis and Birkebak's [7] analysis correlate the data reasonably well. If as according to Davis [6] such beds as the one studied are conduction dominated, then eq. (7) should work; and, indeed, it does correlate our results.

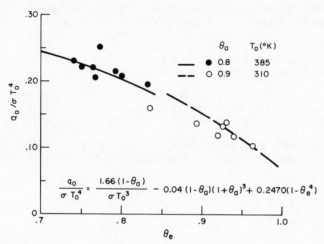

Fig. (8) Base Heat Flux

ACKNOWLEDGEMENTS

The financial support from the National Science Foundation on grants GB-15579 and ENG 75-21938 is gratefully acknowledged. Dr. Özil would also like to acknowledge the support received from the Department of Mechanical Engineering at the University of Kentucky.

REFERENCES

1. Finck, J.L., "Mechanism of Heat Flow in Fibrous Materials," Bureau of Standards Journal of Research, Vol. 5, 1930, pp. 973-984.

2. Allcut, E.A., "An Analysis of Heat Transfer Through Thermal Insulating Materials," Proc. General Discussion on Heat Transfer, IME, ASME, Mst. of Mech. Engrg., London, 1951, pp. 232-235.

3. Hager, N.E. and Steere, R.C., "Radiant Heat Transfer in Fibrous Thermal Insulation," J. Applied Physics, Vol. 31, 1960, pp. 39-50.

4. Verschoor, J.D. and Graebler, P., "Heat Transfer by Gas Conduction and Radiation in Fibrous Insulations," Trans. ASME, Vol. 74, 1952, pp. 961-968.

5. Davis, L.B. and Birkebak, R.C., "An Analysis of Energy Transfer Through Fur, Feathers, and Other Fibrous Materials," HTL-TR-5, University of Kentucky, Mechanical Engineering, 1971.

6. Davis, L.B., "Energy Transfer in Fur," Ph.D. dissertation, University of Kentucky, Dept. of Mechanical Engineering, 1972.

7. Davis, L.B. and Birkebak, R.C., "On the Transfer of Energy in Layers of Fur," Biophysical Journal, Vol. 14, 1974, pp. 249-268.

8. Özil, E., "Experimental and Theoretical Study of Combined Heat Transfer in Biological Insulating Materials," Ph.D. dissertation, University of Kentucky, Dept. of Mechanical Engineering, 1976.

9. Wilson, D.L., "Experimental Study of Heat Transfer in Artificial Fur," Masters Thesis, University of Kentucky, Dept. of Mechanical Engineering, 1976.

10. Birkebak, R.C., unpublished thermal conductivity data, High Temperature and Thermal Radiation Laboratory, University of Kentucky.

11. Birkebak, R.C. and Davis, L.B., "Radiative Thermal Conductivities of Materials with Various Fiber Orientations," XIII International Conference on Thermal Conductivity, University of Missouri-Rolla, 1974, pp. 243-251.

12. Birkebak, Richard, Birkebak, Roland, and Warner, D.W., "Total Emittance of Animal Integuments," J. Heat Transfer, Vol. 86, 1964, pp. 287-288.

ANALYSIS OF HEAT TRANSFER THROUGH WALLS WITH VERTICAL AIR LAYERS AND WALL LEAKAGE

V. BYSTROV AND V. GALAKTIONOV

Moscow Power Engineering Institute
Moscow, USSR

ABSTRACT

In this report heat transfer from the warm air to the cold one across the vertical air layer is considered under mixed natural-forced convection. Correlations between heat transfer coefficients and constructive characteristics, conditions of heat transfer from the side of the warm and cold air, a value of the leakaging air are obtained by means of a numerical solution of the system of equations.

Nomenclature

C_p	specific heat
G	a value of the leakaging air
h	a height of the layer
K	heat transfer coefficient
Ra	Rayleigh number
Re	Reynolds number
t	temperature
u	air velocity in the layer
α	heat transfer coefficient
δ_ℓ	a thickness of the air layer
ν	kinematic viscosity
ϱ	density

Subscripts

δ	basic value
T	warm
X	cold
cn	mean value

125

Heat transfer process across constructions with air layers
is a complex of interacting phenomena and its analysis has a
certain interest. So these sort of constructions one can attri-
bute various wall constructions with hollow inclusions and so on.
Heat transfer phenomena interacting between each other determine the heat transfer coefficient across the considered
construction. Finally, the coefficient is dependent on constructive and thermophysical parameters, namely: a thickness, thermal
conductivity and emissivity of surfaces of considered constructive materials, a thickness of the air layer and outside conditions.

Figure 1 is a sketch of the heat transfer from the warm air
to the cold one.

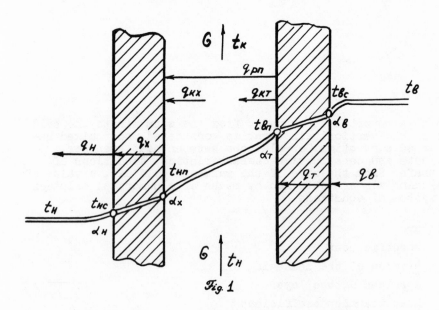

$\mathcal{F}ig.$ 1

The cold air leakaging to the layer becomes warm in it because of a contact   with a warm surface and leaves the layer
having a temperature t_k .

Under passing of the leakaging air, intensity of heat transfer in consequence of natural-forced convection is dependent on
heat transfer coefficients near the warm and cold walls.

Heat transfer on outside surfaces is characterized by effective heat transfer coefficients taking into account both convection and radiation.

The system of equations describing the heat transfer process
consists of nine equations. In order to close the system it needs
to have equations for a calculation of heat transfer coefficients
in layers under mixed convection.

With a purpose of obtaining of these equations experimental

study of heat transfer under mixed convection was carried out at the model of a flat vertical channel with different temperatures of opposite sides. The study of heat transfer and temperature distributions was carried out with a help of an interferometer and thermocouples placed into the channel walls.

As a result interpolative equations were obtained allowing to make a calculation both local and mean heat transfer coefficients at regions:

$$c \leq Re \leq 6 \cdot 10^5 \; ; \quad 1.2 \cdot 10^5 \leq Ra \leq 7.05 \cdot 10^5$$
$$c \leq x/\delta \leq 10$$
$$\alpha_T , \alpha_x = f(\lambda, \delta_\varepsilon, Re, Ra) \tag{1}$$

The abovementioned system of equations is solved by a method of a threestaged iterative procedure and correlations are obtained between the heat transfer coefficient K and constructive characteristics, conditions of heat transfer of the warm and cold air, a value of the leakaging air.

Because it is more suitable to present obtained results through dimensionless parameters, firstly the basic version of heat transfer is calculated at the following conditions:

$$\alpha_\varepsilon = 8.7 \; W/M^2 K \; ; \; \alpha_H = 23.3 \; W/M^2 K \; ; \; t_\varepsilon = 20°C \; , \; t_H = -30°C \; ; \; \lambda_T = \lambda_x = 0.81 \; W/MK$$
$$\delta_T = \delta_x = 25 \cdot 10^{-4} \; M \; ; \; \delta_\varepsilon = 0.15 \; M \; ; \; (h/\delta_\varepsilon)_\varepsilon = 10 \; ; \; G_\varepsilon = 0.011 \; Kg/M^2 s$$

At these conditions the basic heat transfer coefficient K_δ is equal to 1,49 $W/M^2 K$.

In order to find out the heat transfer coefficient dependence on intensity of heat transfer of the cold air for the every regime of a leakage characterized by a relative value of the air $\bar{G} = G/G_\delta$, which changed from 0,25 up to 4, the values of $\bar{K} = K/K_\delta$ are calculated at the variation of a relative heat transfer coefficient $\bar{\alpha}_H = \alpha_H /(\alpha_H)_\delta$ from 2,5 to 1. Because at $\bar{\alpha}_H$ = const $\bar{K}$ is not practically dependent on $\bar{G}$ so finally it makes a sense to present $\bar{K}$ as a function only of an argument $\bar{\alpha}_H$ (Fig.2).

Analytic approximation of a calculate equation is
$$y = A \cdot x / (B + x) \tag{2}$$
where A and B are constant.

The values A and B are equal, correspondingly, to 1,17 and 0,17. Finally, the calculate equation is of the form

$$\bar{K} = 1.17 \cdot \bar{\alpha}_H / (0.17 + \bar{\alpha}_H) \tag{3}$$

This equation describes the data with the error no more than 1%.

Analysis of the appearing values shows that the changing of a heat transfer regime at an increase of a heat transfer coefficient near the outside cold wall leads to a reduction of temperatures t_{HC} , t_{HII} , $t_{\varepsilon II}$, $t_{\varepsilon c}$, small reduction of temperatures t_K and t_{cII} and an increase of α_x and α_T .

In order to solve a problem about the influence of a heat transfer coefficient of the warm air on $\bar{K}$, the value $\bar{\alpha}_\varepsilon = \alpha_\varepsilon /(\alpha_\varepsilon)_\delta$

is varied from 0,33 to 2,7 at a changing of G from 0,25 up to 4.
As a result of a computation, the value of $\bar{K}$ can be given as

$$\bar{K} = 2,18 \cdot \bar{\alpha}_\varepsilon / (1,18 + \bar{\alpha}_\varepsilon) \qquad (4)$$

Analysis of obtained values shows that an increase $\bar{\alpha}_\varepsilon$ leads
to an increase of temperatures $t_{нс}$, $t_{нп}$, $t_{сп}$ and a considerable increase of temperatures $t_{сп}$, $t_{сс}$ and $t_к$.

At the same time there is a small increase of the heat transfer coefficient from the leakaging air to the cold wall and more
considerable increase of the coefficient from the warm wall to
the air.

In this report the most frequently occuring case is discussed
when the values $\hbar = \lambda_{1}\delta$ for the warm and cold walls are
equal to each other.

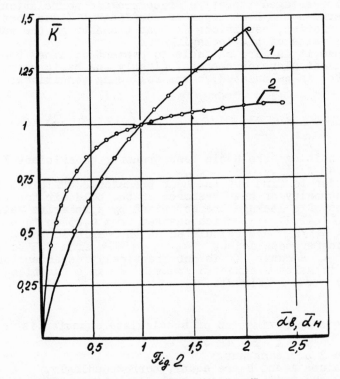

Fig.2

The value of $\bar{K}$ as a function only of $\hbar$ is shown in Fig.3.

Similar to (3) and (4) the obtained correlation can be approximated by the following equation

$$\bar{K} = 1,03 \bar{\hbar} / (0,03 + \bar{\hbar}) \qquad (5)$$

A value of a relative temperature drop is introduced for convenience of an analysis at the study of the intensity of heat
transfer across the considered construction depending on the temperature of the outside and inside air.

$$\bar{\Delta t}_н = \frac{(t_\varepsilon)_\delta - t_н}{(t_\varepsilon)_\delta - (t_н)_\delta} \qquad (6)$$

$$\overline{\Delta t_{\mathscr{B}}} = \frac{t_{\mathscr{B}} - (t_H)_\delta}{(t_{\mathscr{B}})_\delta - (t_H)_\delta} \tag{7}$$

The values of $\overline{K}$ as functions of some parameters are shown in Fig. 4 and 5. Apparently, $\overline{K}$ is dependent not only on values of $\overline{\Delta t_H}$ and $\overline{\Delta t_{\mathscr{B}}}$, but $\overline{G}$ also.

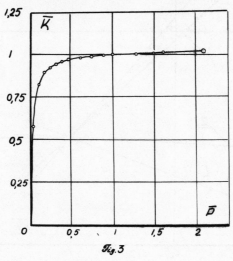

Fig. 3

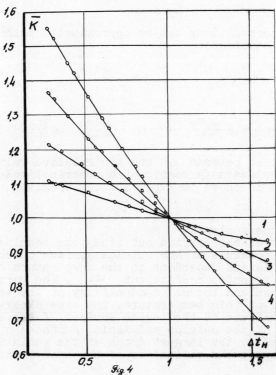

Fig. 4

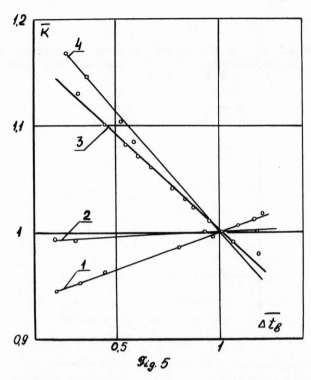

Fig. 5

Mentioned correlations can be approximated mathematically by the following equations:

$$\bar{K} = \left(1 - \frac{0,17\bar{G}}{3,53 + \bar{G}} \right) \; \bar{\Delta t}_{H}^{\;0,04 - \frac{0,81\bar{G}}{1,05 + \bar{G}}} \tag{8}$$

$$\bar{K} = 0,89 + \frac{0,5\bar{G}}{2,53 + \bar{G}} + \left(0,11 - \frac{0,5\bar{G}}{2,42 + \bar{G}} \right) \bar{\Delta t}_{\beta} \tag{9}$$

The correlation between $\bar{K}$ and the relative value of the leakaging air obtained as a result of a numerical solution of the basic version is shown in Fig.6 and can be given as

$$\bar{K} = 1,92 - 1,28\bar{G} / (0,39 + \bar{G}) \tag{10}$$

Results of the analyses.

As a result of the carried out study the conclusion can be done that the temperature of the outside surface of the cold wall is more sufficient dependent on the temperature of the outer air and the heat transfer coefficient $\mathcal{A}_{H}$. The variations of the layer thickness and thermal conductivity of the walls influence least of all on this temperature. The same observations are true of the temperature of the inside surface of the cold wall. The temperatures of the outside and inside surfaces of the warm wall are subjected to the largest drops at the variations of the cold and warm air and the variations of the heat transfer coefficient α_{ξ} .

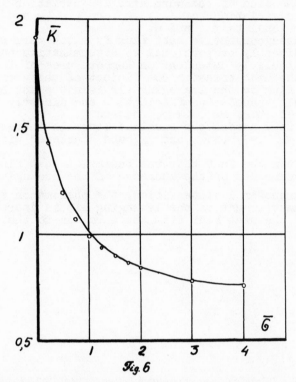

Fig. 6

Qualitative variations of the examined temperatures are analogous according to the most parameters except for thermal conductivity.

The increase of thermal conductivity of walls leads to the increase of temperatures t_{HC} and $t_{\xi\pi}$ at the simultaneous reduction of $t_{\pi\pi}$ and $t_{\xi c}$. The variation of these temperatures is directly proportional to the variations of α_{ξ}, Δt_{ξ} and inverse proportional to α_H , Δt_H , δ_{ξ} and G.

The temperature of the air leaving the layer t_K and the mean temperature of the air in the layer $t_{c\pi}$ are sufficiently dependent on variations of parameters α_{ξ} , Δt_H and Δt_{ξ} , besides of it $t_{c\pi}$ is a function of G. Qualitative variations of the both temperatures are identical and they are increased with the increase of α_{ξ} , μ , Δt_{ξ} , δ_{ξ} or with the reduction of α_H , Δt_H and G.

As a rule, the heat transfer coefficient near the cold wall α_K is a stable value and slightly varies with the variation α_H , α_{ξ} , μ , Δt_H , Δt_{ξ} and G. The parameter δ_{ξ} is an exception because the layer thickness and the ratio h/δ_{ξ} are determinating in respect of a flow regime near the cold wall. The value of α_T is more sensitive to variations of the most parameters and almost constant only in according to G. In all cases α_K and α_T are directly proportional to variations of the examined parameters and only an increase of δ_{ξ} leads to a reduction of α_K .

Radiation heat flux $q_{\rho\pi}$ is directly proportional to the temperature drop between inside surfaces of warm and cold walls and

therefore it is more sufficiently dependent on heat transfer conditions from the side of the warm air. Its variation is directly
proportional to variations of α_H , α_δ , μ , $\bar{\Delta t}_H$ and $\bar{\Delta t}_\delta$ and
inverse proportional to δ_ξ and G.

In all cases convective heat flux from the warm wall to the
layer air is directly proportional to determinative parameters
and it is particularly dependent on temperatures of the warm and
cold air and the heat transfer coefficient of the warm air.

The heat flux to the leakaging air is not practically dependent on the heat transfer coefficient α_H and directly proportional
to variations of α_δ , μ , $\bar{\Delta t}_H$, $\bar{\Delta t}_\delta$ and G.

The heat flux across the examined layer is increased with a
rise of α_H , α_δ , μ , $\bar{\Delta t}_H$ and $\bar{\Delta t}_\delta$ and decreased with a rise of
δ_ξ and G.

Proceed from the fact that the radiation heat flux between
the walls is decreased at the increase of the leakaging air va-

lue and heat transfer is intensified, the phenomenon takes place
at which the heat amount to the leakaging air is increased and
simultaneously the heat flux from the warm air to the cold one
is decreased.

References

1. Разумов Н.Н. "Графо-аналитический метод исследования и
   расчета воздухообмена в зданиях любой объемно-пространствен-
   ной композиции".
   Докторская диссертация, М.,1969.

2. Шкловер А.М. "Основы строительной теплотехники".
   Госстройиздат, М., 1956.

VENTILATION AND AIR MOVEMENT INSIDE BUILDINGS

VENTILATION OF BUILDINGS: A REVIEW WITH EMPHASIS ON THE EFFECTS OF WIND

A. F. E. WISE

Department of the Environment
London SW1 P 3EB, England

ABSTRACT

In this review of ventilation, particular attention is given to the natural ventilation of housing in which wind is the dominant force. Meteorological data in a form of presentation used in the UK are outlined. Air movement around buildings is described as a basis for information on wind pressure distributions. A simple theoretical treatment indicates when buoyancy may be more important than wind in determining ventilation rates; and ventilation in practice is discussed with reference to physical principles and human factors. Research needs are emphasised in a final discussion.

NOMENCLATURE

A Area of ventilation opening m^2

Ar Archimedes number (defined in text)

C_p Pressure coefficient (defined in text)

g Acceleration due to gravity m/s^2

h Vertical distance between openings m

p Static pressure at a surface N/m^2

p_a Static pressure in the free wind N/m^2

Q_B Volume flow rate of fresh air due to buoyancy m^3/s

Q_T Total volume flow rate of fresh air m^3/s

U Reference wind speed (measure at 10 m in open country) m/s

Δ Difference between two values of the same variable

ρ Density of air kg/m^3

θ Temperature of air $^\circ C$

See also Fig. 5 and Table 1

INTRODUCTION

The brochure for the Seminar announced a lecture on air movement around buildings. This subject in itself, although interesting as background to the main theme of the Seminar, plays only an indirect role in relation to the principal heat transfer processes and thus in the environmental design of buildings, ventilation and energy conservation. It therefore seemed appropriate to widen the scope of the paper to include not only an outline of air movement aspects but also to touch on other more direct ways in which the wind influences environmental design and particularly the ventilation of buildings. Ventilation is, of course, one of the key issues in energy conservation. Whereas in the past it was necessary to ensure that not less than some minimum level of ventilation was provided to meet certain basic requirements, now it is important to ensure that no more than the minimum is supplied with a view to minimising energy consumption. In the UK some 20–40% of the heat input to the average house may be required to account for the ventilation loss.

Such considerations suggest that a fresh look be taken both at the natural and mechanical ventilation of buildings. The present paper is offered as a contribution to such a review, emphasising particularly the role that the wind plays in the natural ventilation of dwellings. It does not provide new solutions, rather it draws attention again to some perhaps well-known aspects of the subject and urges that the gaps in knowledge that emerge be tackled through research. Comprehensive coverage was not feasible in the space available – for example, the basic subject of ventilation requirements is not considered – and the review is limited to the following. Meteorological data are considered and a form of presentation that has been found useful for environmental design in the UK is outlined. Buildings in the wind are discussed, first to provide a background on air movement and then to summarise information on wind pressure distributions, an important factor in ventilation. The buoyancy of the air also plays its part in ventilation and the paper considers when buoyancy effects are likely to be more important than the effects of wind. The paper continues with an account of some results of natural ventilation measurements in occupied buildings and points to the practical problems of interpreting data obtained under field conditions. The need for further research is emphasised in a final discussion

Before embarking on the general account a reminder of the principal factors that influence natural ventilation may be in order. In special circumstances turbulent fluctuations of wind velocity and presure may be the determining factors in natural ventilation – as, for example, with a room ventilated by one open window – but normally the mean pressures created by wind and temperature difference (buoyancy) are the main forces at work. The airflow into and through a building depends on the overall pressure distribution and on the effective areas along the air flow paths. The natural ventilation rate even for complex buildings may be calculated from computer programmes incorporating the many non-linear equations involved. For present purposes, where the principles involved are of more concern, it will be sufficient to refer to a simple idealised model of a building (Fig. 1) and to recall

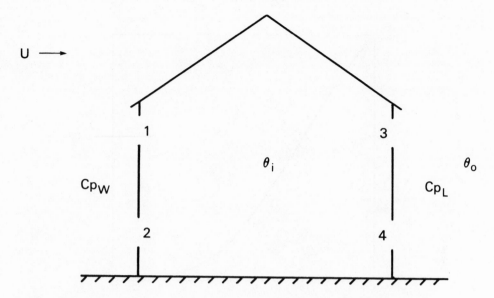

Fig. 1 Simple model of a building for ventilation calculations.

that the ventilation rates due to wind and temperature difference acting separately are given from equations of the form:

Wind: $\qquad Q_w = C_d \; AU \; (2 \, \Delta c_p)^{\frac{1}{2}}$ 1

Buoyancy: $\qquad Q_B = C_d \; A \; \left(\dfrac{4 \Delta \theta \, gh}{\theta}\right)^{\frac{1}{2}}$ 2

FREQUENCY DISTRIBUTION OF WIND SPEED

Fundamental to an understanding of natural ventilation is some knowledge of the frequency distribution of wind speeds for the area concerned. Such information is obtained from meteorological records. In the United Kingdom, for example, wind speed and direction are continuously measured and recorded by anemographs at more than 100 meteorological stations. From the detailed traces certain information is extracted and published monthly viz. the mean wind speed and direction and the maximum gust speed occurring during each hour. Frequency tables of hourly mean wind speed and direction based upon such data have been published by the Meteorological Office for 35 sites in the UK and this information is basic to ventilation design.

For a selected site the frequency with which any given wind speed is exceeded can be obtained from cumulative frequency plots of the kind shown in Fig. 2, an example for Kew, London. It has been shown, moreover, that the cumulative frequency distribution of wind speed for all 35 sites (1) can be expressed as a single curve (Fig. 3) based on the long term mean speed U_{50} ie the speed exceeded for 50% of the time at any given site. These recent studies have, moreover, led to the development of the mean wind speed map shown in Fig. 4. This information provides a basis for assessing the distribution of ventilation rates in a building or determining a reference wind speed for

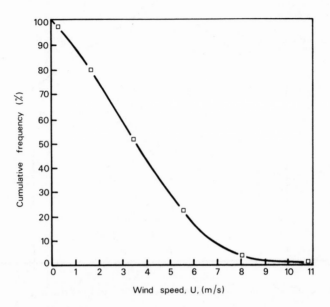

Fig. 2 Cumulative frequency of speeds greater than U for all wind directions
at Kew, 1950-1959

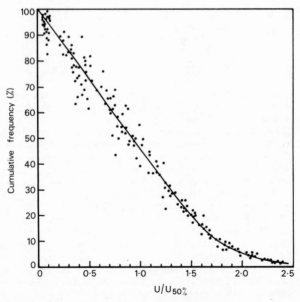

Fig. 3 Cumulative frequency of U/U_{50} for 35 sites in the UK.

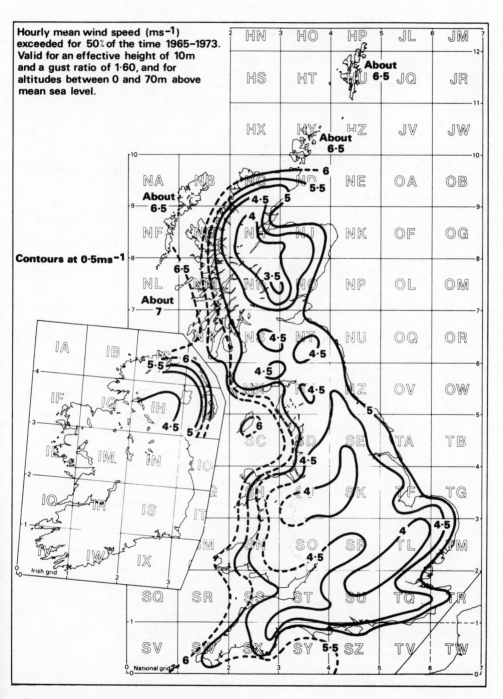

Hourly mean wind speed (ms^{-1})
exceeded for 50% of the time 1965–1973.
Valid for an effective height of 10m
and a gust ratio of 1·60, and for
altitudes between 0 and 70m above
mean sea level.

Contours at 0·5ms^{-1}

Fig. 4. Contours of U_{50} for the UK

139

computing the ventilation rate to be set against some chosen standard.
Consider, for example, a site near London for which the mean wind speed (Fig 4)
is 4.5 m/s. Fig. 3 provides a means of determining the distribution of speeds
around this mean and hence of assessing ventilation. Fig. 3 shows that for a
cumulative frequency of 70%, for example, U/U_{50} is 0.6 and hence U is obtained
as 2.7 m/s. Similarly for other frequencies a speed may be calculated, giving
a frequency distribution as a basis for assessing ventilation rates.

The estimation of wind speed may be refined in several ways. The values
determined as above relate to a standard height of 10m in open country where
most meteorological stations are located. Thus a correction for height above
ground may be appropriate in some circumstances, and also a correction from
rural to urban conditions if the building in question is located in an area
of extensive development. For such purposes it is necessary to adjust mean
speeds in accordance with information on velocity profiles in the boundary
layer over different types of terrain. The power law variation due to
Davenport (2) is simple and convenient and its use in making such corrections
has been discussed (3). A more detailed study might also require an
examination of the influence of wind direction and of seasonal effects. It
has been found, nevertheless, that the basic information given in Fig. 3 and
4 is adequate for the broad assessment of ventilation for a wide range of
practical conditions.

AIR MOVEMENT AROUND BUILDINGS

Impetus to the study of air flow around buildings has been given in
recent years by environmental designers who sought reliable information on how
to achieve comfortable conditions for pedestrians in open air situations.
Experience has shown that unpleasantly windy conditions often occur around
buildings much taller than their neightbours, for example, isolated tall
buildings or groups of buildings surrounded by streets of 2 and 3 storey
housing, or sometimes buildings of medium height – perhaps 20 or 30m. high –
in open country. Various data are required for the solution of such problems
including meteorological information (the data already set out have proved
useful), aerodynamic results and human requirements (3) but for present
purposes an outline of the aerodynamic findings will be sufficient. Several
investigators have studied the flow patterns around single and groups of
model buildings in wind tunnels and have reported on the following general
lines as to the main mechanisms involved.

A flow approaching a building standing well above its surroundings – a
two-storey house well exposed, for example, or a tower block dominating an
area – is diverted both over the top and around the sides. The flow separates
at sharp front edges and a wake of separated flow is formed at the rear.
Boundary layer conditions, with a vertical gradient in the local dynamic
pressure, pertain in the approaching flow and this is of significance in
determining the details of the flow pattern. T ically a stagnation point on
the windward face of the building occurs at some 70 to 80% of the height above
the ground. A flow of air down the front face below the stagnation point is
induced and this is of fundamental importance for environmental design.
Essentially it means that faster moving air from higher levels is drawn
downwards to ground level. This has the effect both of increasing the speeds
of the accelerated flow round the sides of the building near the base, and of
feeding a reverse flow in front of the building. Excepting for narrow
buildings there is a tendency to form an organised vortex, and the presence of
a lower building upstream, and the height of the latter, also the angle of the
wind to the face, are all important. The greatest speeds occur at the sides
of and beneath a tall building and some generalisation is possible (3). The
following results relate to the symbols in Fig. 5. For the conditions specified

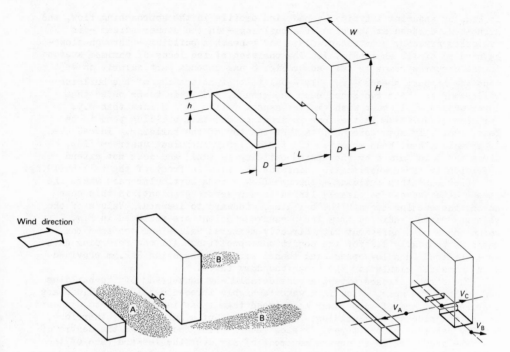

Fig. 5 Definition sketch for air movement round buildings.

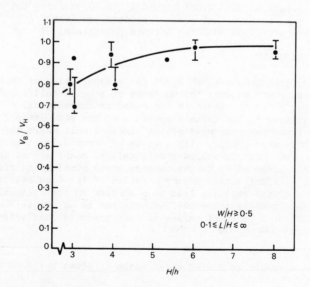

Fig. 6 Speed ratios in corner streams.

in Fig. 6, assuming a typical urban wind profile in the approaching flow, the
highest wind speed at the side of a building — in the corner stream — is
typically given by a ratio* of 0.9; and beneath a building — through-flow —
by a ratio of 1.2 as in Fig. 7. The position of the locus of maximum speeds
in the corner streams is defined by Fig. 8 and depends particularly on $W^{\frac{1}{2}}$.
Thus the highest speeds — roughly equal to that at the top of the building —
will occur in these two zones near the ground, and it is to be noted that
lower speeds will occur with smaller aspect ratios ie W/H less than 0.5.
For comparison speeds in the zone in front of the tall building tend to be
less, typically about one half that at the top of the building. Indeed even
lower values have been reported (2) for a highly turbulent upstream flow,
where the wake formed by the upwind building is small and does not extend
sufficiently far downstream to reinforce the flow in front of the tall building.

Flow beneath a building — through-flow — is a particular case where the
speeds attained can be clearly linked to a pressure gradient, in this case
across the opening beneath the building, windward to leeward. Values of the
velocity ratio predicted from the pressure gradient are included in Fig. 7
which shows good agreement with directly measured values and confirms the
ratio as typically 1.2 for the conditions specified. All the foregoing data
were obtained in a low speed wind tunnel with cross section 2x2.5m provided
with means for simulating the urban boundary layer.

The corner streams extend a considerable way downstream of the building
and enclose the wake. Flow in a wake is highly three-dimensional. Winds are
turbulent but are of much lower mean speed than those in the zones to windward
and at the side of the building. The wake area of a building is, therefore,
usually regarded as sheltered. A slow reverse flow occurs near the centre of
the wake and there is an upward movement of air over the leeward face of the
building due to vertical gradients of negative pressure. According to a recent
review (4) the turbulence effects do not extend beyond 15 to 20 building
heights downstream and can be much less for a tall, narrow building. The
effects on mean velocity do not extend beyond 15 to 20 building heights except
when the angle of flow is such that corner vortices are formed over the roof.
The effects may then extend 80 building heights downstream.

WIND PRESSURE DISTRIBUTION

The foregoing paragraphs set the scene for a description of the typical
distribution of mean pressure over the surfaces of a model building in a
boundary layer flow. With the pressure expressed as conventional
coefficients by dividing by the dynamic pressure of the wind at roof level,
Fig. 9 shows that pressures over most of the windward wall where the wind is
slowed down are above atmospheric, with a value of about 0.75 at the
stagnation point. The contours follow a well defined shape around this centre,
and we may note small areas close to the corners where accelerating flow
produces suction. The deflection and acceleration of the flow over the roof
and round the sides of the building lead to a suction on the wide walls, roof
and leeward wall, and mean pressure coefficients may be as low as − 2.0.
Table 1 brings together a range of values of mean pressure coefficient for a
variety of conditions (following the text).

---

* The results are expressed as a wind speed ratio R_H given by:

$$R_H = \frac{\text{wind speed at pedestrian height}}{\text{free wind speed at height H, } V_H}$$

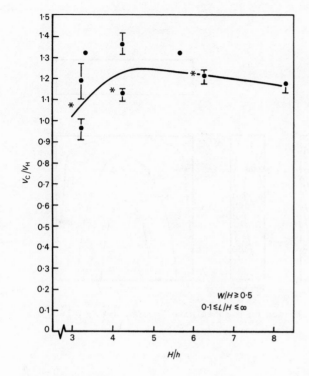

Fig. 7 Speed ratios in through flow
* predicted.
● measured.

Fig. 8 Corner streams (a) typical speeds for a model with H=0.4m, W=0.4m, L=0.3m (b) dependence of Y on W and H.

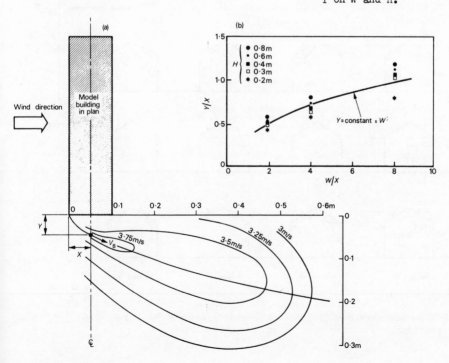

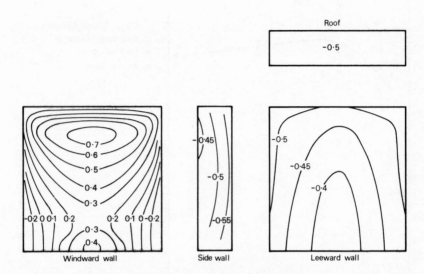

Fig. 9 Typical pressure distribution on a model with H=0.4m, W=0.4m, L=0.3m

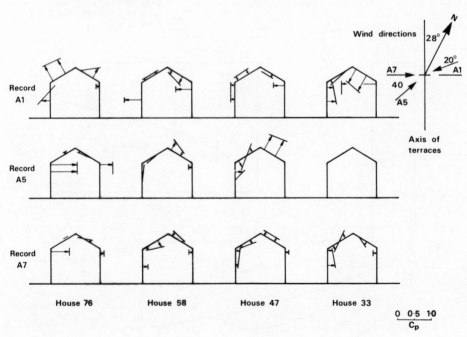

Fig. 10 Mean pressure coefficients on 2-storey houses for three wind directions.

The results in Table 1* represent a maximum condition where the coefficient of pressure difference across the building with any wind direction amounts to about 1.0 in the direction of the wind. As might be expected, buildings sheltered by others from the full force of the wind exhibit different characteristics. From extensive full-scale measurements of wind pressures on 2-storey houses on an estate of 93 houses near London (5), useful information on the effects of shelter was obtained and is summarised in Fig. 10. The results shown were obtained in houses roughly at the centre of long terraces; the axes of the terraces are at right angles to the page. To the left of the buildings was a stretch of open country, to the right of them the urban terrain. In the lower example, the wind was blowing square on to the terraces, in the middle example it was blowing at 40° to the normal and in the upper it was at 20° to the normal in the opposite direction. In these examples, there is generally one zone of substantial pressure ie. on the windward wall of the windward house with a coefficient of 0.6-0.8. On the other houses there are small pressures only (although there is an unexplained anomoly in the upper example), indicating the generally substantial degree of shelter afforded by the windward row of houses. It will be seen that with houses sheltered from the full force of the wind, the effective pressure difference may be as low as 0.1, as compared to 1.0 for an exposed building.

There seems to be a lack of information about the conditions under which the effective pressure difference lies between these two extremes. Lacking also is reliable information on what proportion of buildings may be regarded as exposed, what proportion sheltered, and what proportion can be regarded as in some intermediate category as regards wind pressure. This seems to be an area for further wind tunnel study linked to carefully planned field surveys.

WIND VERSUS BUOYANCY

Although the present review is concerned primarily with wind and its effect on ventilation, an interpretation of ventilation in practice requires a means of assessing when wind is a dominant factor and when it plays only a secondary role in ventilation. The following paragraphs outline the general principles underlying the interaction of mean wind pressure and temperature difference, identified at the outset as the primary forces governing natural ventilation, by reference to a simple, idealised situation. The general form of the equations relating ventilation rate to the factors that determine these natural forces have been set out at (1) and (2) with reference to Fig. 1, a schematic representation of a simple building.

To develop the argument it is useful to introduce a dimensionless term, the Archimedes number, which represents the ratio between buoyancy and inertia forces in the air:-

$$Ar = \frac{(\Delta \theta gh)^{\frac{1}{2}}}{(\bar{\theta} U^2)} \qquad \dots\dots\dots\dots\dots 3$$

Equation 2 giving the ventilation rate due to temperature difference may then be written in demensionless terms as:-

$$\frac{Q_B}{C_d AU\ Ar} = 2 \qquad \dots\dots\dots\dots\dots 4$$

*(See page 153 for Table 1).

Equation 1 may also be written in dimensionless terms as:-

$$\frac{Q_W}{C_d \, AU} = (2 \, \Delta c_p)^{\frac{1}{2}} \qquad \qquad \cdots\cdots\cdots\cdots\cdots \; 5$$

In practice both wind and buoyancy are likely to act simultaneously, especially in the heating season, and in a recent paper (6) a colleague at the Building Research Establishment has calculated the flow rate Q_T that would be expected to occur under these circumstances. The results are plotted in dimensionless form in Fig. 11 which also shows lines representing wind and buoyancy acting alone.

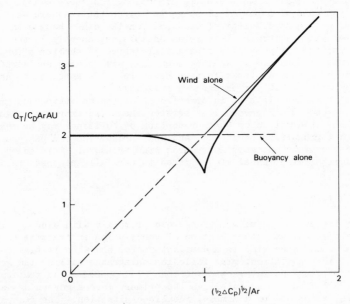

Fig. 11 A comparison of total ventilation rate with the rates due to buoyancy and wind alone.

This particular example relates to the schematic representation in Fig. 1 but similar general forms of curve are obtained when the distribution and size of opening areas are varied.

For immediate purposes it will suffice to point out, with reference to Fig. 1, that buoyancy and not wind dominates when:-

$$\frac{(\frac{1}{2} \, \Delta c_p)^{\frac{1}{2}}}{Ar} < 1 \qquad \qquad \cdots\cdots\cdots\cdots\cdots \; 6$$

This may also be seen from an inspection of equations 4 and 5 for when Q_W is less than Q_B. As a simple numerical example consider the two extremes already identified of an exposed dwelling of 2 storeys when Δc_p may be taken as 1.0, and a sheltered dwelling when Δc_p may be as low as 0.1. Suppose that the height between the centre of the openings in Fig. 1 is 3m and that the mean temperature difference, indoors to outdoors, is $10^{\circ}C$. Considering a range of wind speeds from 1.5 to 7.5 m/s and with $\overline{\theta}$ = 300°K the results in Table 2 were obtained.

Table 2. Wind versus buoyancy for a simple example

	Δc_p	$(\frac{1}{2}\Delta c_p)^{\frac{1}{2}}$	Archimedes Number, Ar at different wind speeds			
			U = 1.5	U = 2.5	U = 5.0	U = 7.5
Exposed dwelling	1.0	0.7	0.7	0.4	0.2	0.1
Sheltered dwelling	0.1	0.2				

Thus with an exposed dwelling wind dominates under these conditions over most of the wind speed range whereas with a sheltered dwelling buoyancy is seen to dominate, with $(\frac{1}{2}\Delta c_p)^{\frac{1}{2}}$ less than Ar over much of the range. In this simple case with buoyancy dominant, air will flow in to the building at the lower level and out at the upper level. With wind dominant there will be a flow across the building from windward to leeward. In an intermediate situation the flow pattern will derive from a combination of these character-istics. This example, is of course, merely illustrative but the principles behind this simple theory are fundamental to an interpretation and under-standing of ventilation in practice.

NATURAL VENTILATION IN PRACTICE

Natural ventilation in practice is complicated by the effects of various factors beyond those already outlined and these may be regarded as including at least three types:-

1. Local meteorological conditions varying due to location, topography, degree of development and so on.

2. Building design and detail much more complex than in the simple scheme considered.

3. Human factors assuming importance, with occupants free to control their living conditions by the controls available to them – windows, doors, heating, thermostats and so on.

Perhaps the most extensive field investigations involving such factors were carried out some 25 years ago in the UK (7, 8) and in this work it was found possible to interpret many of the results in terms of the simple physical principles outlined in the previous section. Figs. 12 and 13 reproduced from this early work illustrate, for example, a series of measure-ments in a two-storey house under exposed, winter conditions in which the wind was fairly constant in direction. Using the form of theory outlined here and relating it as closely as possible to his experimental situation, the author calculated the wind speed below which buoyancy dominated as 3.5 miles per hour ie 1.5 m/s. Fig. 12 shows the relation between ventilation rate and temperature difference. Fig. 13 shows the relationship between rate and wind speed, with the two closely proportional above 1.5 m/s. Below 1.5 m/s wind

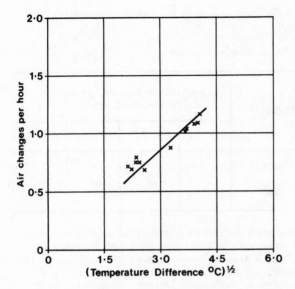

Fig. 12 The effect of
temperature difference on
ventilation rate at low
wind speeds.

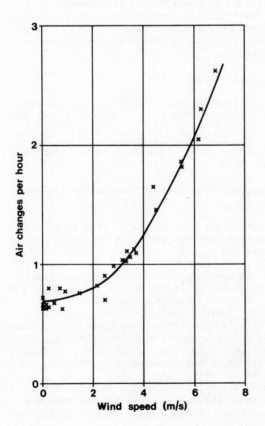

Fig. 13 The effect of
wind speed on ventilation
rate.

has little effect on ventilation rate and the results are seen to be consistent with the principles outlined. The research covered similar investigations under a variety of conditions and took the subject a stage further by examining in some detail the air leakage paths available in dwellings and the ways in which occupants controlled ventilation by the adjustment of doors, windows and so on. These are, of course, key questions and the research showed the ways in which the use of windows was related to wind speed and external air temperatures. The control possible by means of windows was found to be very coarse and air leakage was found often to be excessive. Improvements in design together with the provision of ventilators separate from the windows were recommended.

These and other studies in the literature provide a good foundation of knowledge of ventilation, both theoretical and experimental, but more needs to be done to extend the range of data over the full range of practical conditions as a basis for environmental design and energy conservation. In the UK after a period of years in which little was done ventilation research has once again assumed importance. Work has begun again at the Building Research Establishment (eg. 6) and substantial studies are being carried out by the various nationalised fuel industries (eg 9), also by universities. The work is being tackled both by laboratory research and by studies in occupied buildings since the key to an advance in this field lies in a co-ordinated programme of field and laboratory work. Already it is being found that dwellings in the UK are much tighter than 20 years ago, largely as a result of the weather stripping of many windows – see, for example, Table 3 which illustrates the effect of this treatment – and improvements in their quality and the absence of flues. Thus in a range of houses it was recently found (6) that ventilation rates at mean wind speed varied from

Table 3.   Equivalent orfice areas for some closed wood and metal windows from air leakage tests.

Side hung casements	
No weatherstripping	170 to 635
Weatherstripped	40 to 190
Top hung vent lights	
No weatherstripping	230 and 800 (2 examples)
Weatherstripped	85 (1 example)
Horizontally pivoted	
No weatherstripping	510 to 1080
Weatherstripped	4 to 300
Note: Areas in mm^2 at 75 N/M^2 pressure difference	

0.3 to 1.3 air changes per hour whereas 25 years ago a figure of 2 or more was likely. Reduced ventilation has, however, contributed to an increase in condensation in some circumstances and considerable maintenance expenditure has been necessary as a result. Part of the problem is that the control achievable by windows remains much too coarse and further improvement in this respect seem needed. From recent tests on 3 houses, for example, it was reported that opening two windows on opposing walls to only their first fixable position resulted in an increase in room ventilation rate of at least two-fold and, on average, four-fold (6). Further investigations seem to be required of ventilators separate from windows and also of the possibilities for mechanical ventilation linked to well-designed window/ventilator arrangements, in which the human factors are properly explored. Research is also in progress in other countries and it is to be hoped that the Seminar will stimulate further initiatives. Much remains to be done, especially by field work, if the assessment of ventilation in practice is to be put on a thoroughly sound basis, paving the way for practical design procedures and the proper integration of ventilation with building planning, design and heating.

WIND AND MECHANICAL VENTILATION

In many buildings there can be an interaction between natural and mechanical ventilation, the magnitude of which depends primarily on the acting pressures. Whilst the effect is often negligible it can be significant where low-powered mechanical ventilation is in use, for example in domestic housing projects. It is possible to assess the likely effect of wind in relation to the flow rate expected from a fan by considering the fan characteristic and circuit resistance in relation to the expected wind pressure differential. In many situations wind pressure will be trivial compared to the fan pressure but with the simple mechanical ventilation systems used in many multi-storey housing projects, wind pressure may sometimes be significant compared to a fan pressure of say 10 or 20mm water gauge.

A study of the effects of wind was made some years ago in the UK as part of a study of extract ventilation systems for internal bathrooms in multi-storey flats (10). Such systems, once constructed, require balancing ie adjustment of grilles to achieve equal flows into all extract points, and as an example an assessment was done of the effects of pressures due to wind and buoyancy on this process in an 11-storey block of flats. The system is shown in Fig. 14 (a), whilst Fig. 14, (b) and (c), shows how variations in natural forces could influence the distribution of areas obtained with two examples of grilles in practical use, one of 6500 mm^2 and the other of 1950 mm^2 when fully open. The wind suction was assumed in calculations to be one half of the dynamic pressure of the wind, with a speed of 9 m/s at 10m above the ground and a variation of speed with height. These results and related findings in occupied buildings led to design recommendations (10) to make such systems reasonably independent of natural forces by advocating certain minimum pressures related to grilles and fans for various practical situations. A more recent paper (6) has considered the contribution of mechanical ventilation as an aid in energy conservation.

IN CONCLUSION

Wind is only one of a number of factors that influence the ventilation of buildings, nevertheless it is important as the dominant force in many situations. Its contribution to ventilation has, therefore, been studied, both theoretically and experimentally, in several investigations reported in the literature. Further studies would, however, be useful to extend the range and detail of information on meteorological factors relevant to ventilation; to extend the data on pressure distributions around buildings by wind tunnel study linked to carefully chosen field experiments; and to explore further the interactions of wind and buoyancy in determining natural ventilation.

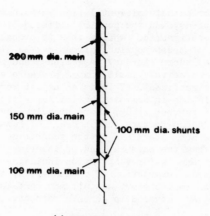

(a) Diagram of system

200 mm dia. main

150 mm dia. main

100 mm dia. shunts

100 mm dia. main

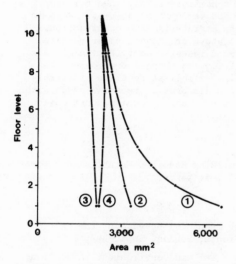

(b) Area of 6,500 mm² when fully open

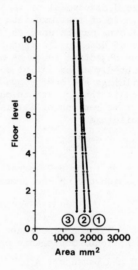

(c) Area of 1950 mm² when fully open

Conditions

①.No external effects

②.Chimney effect due to 10°C temperature difference

③.Wind suction at fan outlet acting with ② above

④.② and ③ above with suction in the internal rooms due to wind

Fig. 14 The effect of external factors on ventilation balancing.

151

In most of the work reported natural ventilation rates have been explained in terms of average wind speeds, and the possible effects of rapid changes in the wind environment have been ignored. Whilst this is probably adequate for most situations, the effects of pressure fluctuations may be significant in certain situations; for example, where buoyancy is dominant and pressure difference across a building is comparatively small, rapid pressure variations associated with turbulence may be significant. This is an aspect worthy of study. The use of mechanical ventilation in domestic buildings, allied to natural ventilation, also merits further study. More broadly, the investigation of ventilation in practice needs to take full account of human factors and the ways in which conditions are determined by the general design and use of buildings, as distinct from the physical and engineering factors emphasised here. The general requirements for ventilation including such aspects as odour and pollution control, prevention of condensation, provision of air for combustion appliances, thermal comfort and air movement are outside the scope of the present paper but are covered in certain respects in other papers to be given at the Seminar.

Natural ventilation by its very nature is less a matter for precise design than for a broad assessment of the range of conditions likely in a building under the climate of its immediate locality and taking account of the ways in which the building may be used as well as its engineering and construction. A wider range of data on the factors set out above are nevertheless required if this latter objective is to be met. A broad research contribution of this kind is undoubtedly an essential requirement for the overall environmental design of buildings and energy conservation. The fullest benefits are likely to be realised if the scientists can develop a dialogue on research needs with those whose prime responsibility is towards the planning, design, construction and use of buildings.

ACKNOWLEDGEMENT

I am indebted to Dr P R Warren of the Building Research Establishment for helpful discussion during the preparation of this paper.

REFERENCES

1. Caton P G F "Standardised maps of hourly mean wind speed over the UK" 4th International Conference on Wind Effects, London, September, 1975.

2. Isyumov N and Davenport A G "The ground level wind environment in built-up areas" 4th International Conference on Wind Effects, London, September 1975.

3. Penwarden A D and Wise A F E "Wind Environment Around Buildings", London HMSO, 1975.

4. Peterka J A and Cermak J E "Turbulence in building wakes" 4th International Conference on Wind Effects, London, September 1975.

5. Eaton, K J and Mayne, J R "The measurement of wind pressures on two-storey houses at Aylesbury". Building Research Establishment Current Paper 70/74, Watford, England.

6. Warren, P R. "Natural infiltration rates and their magnitude in houses" Building Research Establishment PD 28/76, Watford, England.

7. Dick, J B "The fundamentals of natural ventilation of houses". JIHVE, London, June, 1950, 123-134

8. Dick, J B, "Ventilation research in occupied houses". JIHVE, London, October, 1951, 306-326.

9. Brundrett, G W "Ventilation: a behavioural approach". Electricity Council Research Centre, Chester, England, February 1976.

10. Wise A F E and Curtis M "Ventilation of internal bathrooms in multi-storey flats". Building Research Establishment Current Paper E9, 1965, Watford, England.

Building height ratio	Building plan ratio	Side elevation	Plan	Wind angle α	C_{pe} for surface A	B	C	D	Local C_{pe}
$\frac{h}{w} < \frac{1}{2}$	$1 < \frac{\ell}{w} < \frac{3}{2}$			$0°$	+0·7	-0·2	-0·5	-0·5	-0·8
				$90°$	-0·5	-0·5	+0·7	-0·2	
	$\frac{3}{2} < \frac{\ell}{w} < 4$			$0°$	+0·7	-0·25	-0·6	-0·6	-1·0
				$90°$	-0·5	-0·5	+0·7	-0·1	
$\frac{1}{2} < \frac{h}{w} < \frac{3}{2}$	$1 < \frac{\ell}{w} < \frac{3}{2}$			$0°$	+0·7	-0·25	-0·6	-0·6	-1·1
				$90°$	-0·6	-0·6	+0·7	-0·25	
	$\frac{3}{2} < \frac{\ell}{w} < 4$			$0°$	+0·7	-0·3	-0·7	-0·7	-1·1
				$90°$	-0·5	-0·5	+0·7	-0·1	
$\frac{3}{2} < \frac{h}{w} < 6$	$1 < \frac{\ell}{w} < \frac{3}{2}$			$0°$	+0·8	-0·25	-0·8	-0·8	-1·2
				$90°$	-0·8	-0·8	+0·8	-0·25	
	$\frac{3}{2} < \frac{\ell}{w} < 4$			$0°$	+0·7	-0·4	-0·7	-0·7	-1·2
				$90°$	-0·5	-0·5	+0·8	-0·1	

ℓ- length of major face of building. w- width of building (length of minor face)

Table 1. Pressure coefficients C_p for vertical walls of rectangular clad buildings.

HUMAN DISCOMFORT IN INDOOR ENVIRONMENTS DUE TO LOCAL HEATING OR COOLING OF THE BODY

P. O. FANGER

Technical University of Denmark
DK 2800 Lyngby, Denmark

ABSTRACT

Due to comprehensive research over the last ten years it is possible today to predict those combinations of the environmental parameters (air temperature, mean radiant temperature, air velocity and humidity) which will provide thermal neutrality for man at a given activity and clothing.

However, thermal neutrality for the body in general is not always sufficient to provide thermal comfort for man. It is a further requirement that no discomfort is created due to local heating or cooling of the body. This may be caused by an asymmetric radiant field, a local convective cooling of the body (draught), by contact with a warm or cool floor or by a vertical air temperature gradient.

In the paper a review is given of recent research on discomfort due to local heating or cooling of the body. Limits for avoiding local discomfort are established and discussed.

INTRODUCTION

In a modern industrial society man spends the greater part of his life indoors. A large proportion of the population spends 23 out of 24 hours in an artificial climate - at home, at the workplace, or during transportation.

During recent decades this has resulted in growing understanding of and interest in studying the influence of indoor climate on man, thus enabling suitable requirements to be established which should be aimed at in practice.

At the same time an increasing number of complaints about unsatisfactory indoor climate suggest that man has become more critical regarding the environment to which he is subjected. It seems that he is most inclined to complain about the indoor climate of his workplace (offices, industrial premises, shops, schools, etc.) where he is compelled to spend his time in environments which he himself can control only to a very limited degree. Field studies indicate that in practice many of these complaints can be traced to an unsatisfactory thermal environment.

About one third of the world's energy consumption is used to provide thermal comfort for man. It is no wonder, therefore, that efforts towards energy conservation in recent years have led to an increased interest in man's comfort conditions in order to

155

assess the human response to different conservation strategies
|1,2,3,4|.
        In practice, local heating or cooling of the body is a fre-
quent reason for discomfort. In the present paper a review is given
of recent research on this subject.

DEFINITION OF COMFORT

        In agreement with ASHRAE's Standard 55-74 |5|, thermal com-
fort for a person is here defined as "that condition of mind which
expresses satisfaction with the thermal environment". A first requirement for comfort is that a person feels thermally neutral for
the body as a whole, i.e., that he does not know whether he would
prefer a higher or lower ambient temperature level. Man's thermal
neutrality depends on his clothing and activity and furthermore on
the following environmental parameters:

> Mean air temperature around the human body
> Mean radiant temperature*in relation to the body
> Mean air velocity around the body
> Water vapor pressure in the ambient air.

        The thermal effect of clothing, activity and environmental
parameters on man has been studied quite intensively during the
last decade |6,7,8,9,10| and all combinations of the parameters
which will provide thermal neutrality can be predicted from the
comfort equation and the corresponding comfort diagrams |6|.
        But thermal neutrality as predicted by the comfort equation
is not the only condition for thermal comfort. A person may feel
thermally neutral for the body as a whole, but he might not be
comfortable if one part of the body is warm and another cold. It
is therefore a further requirement for thermal comfort that no
local warm or cold discomfort exists at any part of the human body.
Such local discomfort may be caused by an asymmetric radiant field,
by a local convective cooling (draught), by contact with a warm or
cool floor, by a vertical air temperature gradient, or by a non-
uniformity of the clothing. Each of these cases will be dealt with
in the following and limits for avoiding local discomfort will be
discussed.

ASYMMETRIC RADIATION

        Asymmetric radiant fields in spaces may be caused by cold
windows, noninsulated walls, or by warm or cool panels in the wall
or in the ceiling.
        Limits for asymmetric radiation from vertical surfaces were
studied by Olesen et al. |11|. Accepting 5% of sedentary, thermally
neutral subjects feeling uncomfortable due to asymmetry, he recommended the following formula for estimating the limits of acceptable temperature differences between a vertical surface and the
mean radiant temperature:

---

*Mean radiant temperature is that uniform temperature of the am-
bient surfaces which would provide the same radiant heat loss from
the human body as in the real environment.

$$-2.4 - 1.8I_{cl} < \Delta t_w F_{p-w} < 3.9 + 1.8I_{cl}$$

where I_{cl} = clo-value of clothing

F_{p-w} = angle factor between sedentary person and the vertical radiant area. Can be found from Fanger's angle factor diagrams for sedentary persons |6|.

Δt_w = temperature difference (K) between radiant source and mean radiant temperature in relation to the person.

This formula which applies both for surfaces at the side, in front, and behind the subject is in reasonable agreement with the results obtained in similar experiments by McNall and Biddison |12| and by McIntyre and Griffiths |13|.

The formula indicates that a higher degree of asymmetry is permissible when wearing heavy clothing than when light clothing is worn. The formula allows larger temperature deviations of vertical surfaces than those which occur normally in residential or commercial buildings. But it should be noted that in Olesen's, as well as in the experiments by McNall and Biddison and by McIntyre and Griffiths, the increased (decreased) temperature of the heated (cooled) vertical surface was balanced by a decrement (increment) of the other surfaces to maintain a constant mean radiant temperature. This is somewhat unrealistic, as in practice the air temperature is usually changed to balance a low or high temperature of a surface. Further studies with normally clothed sedentary subjects are recommended to establish limits for acceptable asymmetric radiation, when the air temperature is modified to balance the effect of warm or cool vertical surfaces.

Recent (as yet unpublished) studies at the Technical University of Denmark indicate that man is much more sensitive to thermal asymmetry when exposed to overhead warm radiation (e.g. from heated ceilings). Accepting 5% feeling uncomfortable (due to warm head or cool feet), the following formula was found for sedentary, normally clothed subjects:

$$\Delta t_w \cdot F_{p-w} < 2$$

This result agrees quite well with the limit recommended by Chrenko |14|, although he predicts a higher percentage of people feeling uncomfortable.

DRAUGHT

Draught is defined as an unwanted local convective cooling of the body. It is perhaps the most common reason for complaints in ventilated spaces. As mentioned earlier, the mean air velocity around the body influences the ambient temperature necessary for thermal neutrality for the body as a whole |15,16,17,18|. However, in spite of thermal neutrality, local velocities can provide an unwanted cooling (= draught) of some parts of the body; the neck and the ankles seem to be the most sensitive parts of normally clothed persons. Unfortunately very few experimental results on this subject have been published.

However, extensive studies (as yet unpublished) on this problem have recently been performed at the Technical University of Denmark. More than one hundred college-age students have been involved in experiments where subjects were exposed at the neck

and the ankles to fluctuating and uniform air flows with different mean velocities, with different amplitudes and frequencies of
the fluctuating velocity, and with different air temperatures.
Based on the subjective reactions of the subjects, it has been
possible to establish a mathematical model which predicts the percentage of uncomfortable persons (due to draught) as a function
of the above-mentioned factors.

It was found that frequencies around 0.3-0.5 Hz were most
uncomfortable but that the most typical frequencies in practice
seem to be lower than 0.1 Hz. The maximum velocity is also important, people accepting a lower mean velocity the higher the max.
velocity. As an example of the results the diagram in Fig. 1

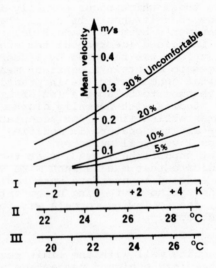

Fig. 1 The percentage of people predicted to feel
       uncomfortable due to draught as a function
       of the local mean velocity and air temperature.
       (Frequency of fluctuating velocity = 0.1 Hz.,
       max velocity = twice the mean velocity).

       Abscissa I   : Air temperature minus neutral
                      air temperature for the body

       Abscissa II : Air temperature during typical
                      summer conditions (clothing:
                      0.6 clo, activity: 1-1.2 met)

       Abscissa III : Air temperature during typical
                      winter conditions (clothing:
                      1 clo, activity: 1-1.2 met).

is shown. The diagram applies for a frequency equal to 0.1 Hz of the fluctuating velocity and a max velocity twice as high as the mean velocity. These values are believed to be quite typical for many cases in practice, but further field studies are recommended to study the characteristics of fluctuating velocities in different types of spaces in practice. It is also recommended to perform further laboratory studies where subjects are exposed to the more random velocity fluctuations which exist in spaces in practice.

Fig. 1 shows the percentage of people feeling uncomfortable as a function of the mean velocity and the difference between the local air temperature and the air temperature which is felt to be neutral for the body as a whole. This neutral temperature depends on the clothing and two scales of the abscissa have been calculated for typical clothing during summer and winter. If the temperature during winter is kept lower than neutral to save energy it is especially important to keep the mean velocity at a low level by proper design of the air distribution in the space. If low velocities are not maintained in the occupied zone it is likely that a higher ambient temperature will be required (= higher energy consumption) to decrease complaints of draught.

Fig. 1 is based on experiments where the subjects were exposed to an air flow from a horizontal duct situated very close to the back of the neck. In practice the natural convection current rising along the human body will interfere with the air flow in the space. The part of the curves below 0.1 m/s are therefore probably somewhat unrealistic and should be used with caution.

The subjects were found to be significantly less draught-sensitive on their (unclothed) ankles than on the neck. No difference was found between the draught sensitivity of men and women.

FLOOR TEMPERATURE

Due to the direct contact between the feet and the floor, local discomfort of the feet can often be caused by a too high or too low floor temperature. Studies on comfort limits for floor temperatures have recently been performed at the Technical University of Denmark by Olesen |19|, who found the following main results.

For floors occupied by <u>people with bare feet</u> (in swimming halls, gymnasiums, dressing rooms, bathrooms, bedrooms, etc.) the flooring material is important. Based on the results of experiments comprising 16 subjects and based on heat transfer theory, Olesen found the optimal temperatures and recommended the temperature intervals given in Table 1 for a number of typical flooring materials. For 10 minutes' occupancy about 10% of persons can be expected to experience discomfort at the optimal floor temperature while less than 15% can be expected to be uncomfortable within the recommended temperature interval. To save energy, flooring materials with a low contact coefficient (cork, wood, carpets) or heated floors should be chosen, to eliminate a desire for higher ambient temperatures caused by cold feet |20|.

For floors occupied by <u>people with footwear</u> (normal indoor footwear) the flooring material is without significance. Olesen found, based on his own experiments and a re-analysis of the results of Nevins |21,22,23|, an optimal temperature of 25 deg C for sedentary and 23 deg C for standing or walking persons. At the optimal temperature 6% of the occupants felt warm or cold discomfort at the feet. If one accepts up to 8% uncomfortable, the floor temperature

Table 1. Comfortable Temperatures of Floors Occupied by
          People with Bare Feet.

Flooring Material	Optimal Floor Temperature for		Recommended Floor Temp. Interval
	1 min Occupancy ^{0}C	10 min Occupancy ^{0}C	^{0}C
Pinewood Floor	25	25	22.5 - 28
Oakwood Floor	26	26	24.5 - 28
PVC-Sheet with Felt Underlay on Concrete	28	27	25.5 - 28
Hard Linoleum on Wood	28	26	24 - 28
5 mm Tesselated Floor on Gas Concrete	29	27	26 - 28.5
Concrete Floor	28.5	27	26 - 28.5
Marble	30	29	28 - 29.5

should be within the interval 22-30 deg C for sedentary and 20-28
deg C for standing or walking persons. At floor temperatures below
20-22 deg C the percentage of people experiencing cold feet in-
creases rapidly |19|. Although heavier clothing can provide thermal
neutrality at a lower ambient temperature, there is a risk of cold
discomfort at the feet, if they are not protected correspondingly
by well-insulated footwear. If normal light footwear is worn,
higher ambient temperatures (= higher energy consumption) will be
required to counteract the coldness at the feet.

VERTICAL AIR TEMPERATURE GRADIENTS

In most spaces in buildings the air temperature is not constant
from the floor to the ceiling; it normally increases with the height
above the floor. If this gradient is sufficiently large, local warm
discomfort can occur at the head, and/or cold discomfort can occur
at the feet, although the body as a whole is thermally neutral.
Little information on this subject has been published but preliminary results from studies at the Technical University of Denmark
by Schøler |24|, and results by McNair |25| and Eriksson |26| indi-
cate that the risk of local discomfort is negligible provided that
the air temperature difference between head and feet level is less
than 2-3K.

NON-UNIFORMITY OF THE CLOTHING

The clo-value is an expression for the mean thermal resistance of a clothing ensemble over the entire body. But if the clothing is very non-uniformly distributed over the body, it is likely that local warm and cold discomfort can occur at different parts of the skin although the body as a whole is thermally neutral. No systematic studies of this phenomenon have been performed, but McIntyre and Griffiths |27| found that although the general thermal sensation of sedentary subjects at 15 and 19 deg C was altered when they put on an extra sweater, this did not decrease the local cold discomfort on the hands and feet.

REFERENCES

1.   Fanger, P. O. 1975. Comfort Criteria and Energy Consumption. ASHRAE Special Bulletin, International Day, Atlantic City, pp 78-86.

2.   Gagge, A. P. and Nevins, R. G. 1976. Effect of Energy Conservation Guidelines on Comfort, Acceptability and Health. Report to FEA, Pierce Foundation Laboratory.

3.   Nevins, R. G. 1975. Energy Conservation Strategies and Human Comfort. Proceedings of the 6th International Congress of Climatistics "CLIMA 2000".

4.   Woods, J. E. 1975. Climatological Effects on Thermal Comfort and Energy Utilization in Residences and Offices. Proceedings of the Seventh International Biometeorological Congress, Maryland.

5.   American Soceity of Heating, Refrigerating and Air-Conditioning Engineers 1974. Thermal Comfort Conditions. ASHRAE Standard 55-74. New York.

6.   Fanger, P. O. 1970. Thermal Comfort. Danish Technical Press, Copenhagen. Republished by McGraw-Hill Book Company, New York, 1973, 244 pp.

7.   Nevins, R. G., Rohles, F. H., Springer, W., and Feyerherm, A. M. 1966. Temperature-Humidity Chart for Thermal Comfort of Seated Persons. ASHRAE Trans. 72, 1.

8.   McNall, P. E., Jr., Jaax, J., Rohles, F. H., Nevins, R. G., and Springer, W. 1967. Thermal Comfort (Thermally Neutral) Conditions for Three Levels of Activity. ASHRAE Trans. 73, 1.

9.   Rohles, R. H., Jr., Woods, J. E., and Nevins, R. G. 1973. The Influence of Clothing and Temperature on Sedentary Comfort. ASHRAE Trans.79, 2:71-80.

10.   Gagge, A. P., Stolwijk, J. A. J., and Nishi, Y. 1971. An Effective Temperature Scale Based on a Simple Model of Human Physiological Regulatory Response. ASHRAE Trans. 77, 1.

11. Olesen, S., Fanger, P. O., Jensen, P. B., and Nielsen, O. J. 1972. Comfort Limits for Man Exposed to Asymmetric Thermal Radiation. Proceedings of CIB Symposium on Thermal Comfort, Building Research Station, London.

12. McNall, P. E., Jr., and Biddison, R. E. 1970. Thermal and Comfort Sensations of Sedentary Persons Exposed to Asymmetric Radiant Fields. ASHRAE Trans. 76, 1.

13. McIntyre, D. A., and Griffiths, I. D. 1975. The Effect of Uniform and Asymmetric Thermal Radiation on Comfort. Proceedings of the 6th International Congress of Climatistics "CLIMA 2000".

14. Chrenko, F. A. 1953. Heated Ceilings and Comfort. Journ of the Inst. of Heating and Ventilating Engineers, 20:375-396, and 21:145-154.

15. Burton, D. R., Robeson, K. A., and Nevins, R. G. 1975. The Effect of Temperature on Preferred Air Velocity for Sedentary Subjects Dressed in Shorts. ASHRAE Trans. 81, 2:157-168.

16. Olesen, S., Bassing, J. J., and Fanger, P. O. 1972. Physiological Comfort Conditions at Sixteen Combinations of Activity, Clothing, Air Velocity and Ambient Temperature. ASHRAE Trans. 78, 2:199-206.

17. Ostergaard, J., Fanger, P. O., Olesen, S., and Madsen, Th. Lund. 1974. The Effect of Man's Comfort of a Uniform Air Flow from Different Directions. ASHRAE Trans. 80, 2:142-157.

18. Rohles, F. H., Jr., Woods, J. E., and Nevins, R. G. 1974. The Effects of Air Movement and Temperature on the Thermal Sensations of Sedentary Man. ASHRAE Trans. 80, 1.

19. Olesen, B. W. 1975. Termiske Komfortkrav til Gulve, (Thermal Comfort Requirements for Floors). Ph.D.-Thesis, Laboratory of Heating & Air-Conditioning, Technical University of Denmark.

20. Olesen, B. W. 1977. Thermal Comfort Requirements for Floors Occupied by People with Bare Feet. ASHRAE Trans. 83, 2 (in press).

21. Nevins, R. G., Michaels, K. B., and Feyerherm, A. M. 1964. The Effect of Floor Surface Temperature on Comfort. Part II: College Age Females. ASHRAE Trans. 70.

22. Nevins, R. G., Michaels, K. B., and Feyerherm, A. M. 1964. The Effect of Floor Surface Temperature on Comfort. Part I: College Age Males. ASHRAE Trans. 70.

23. Nevins, R. G., and Feyerherm, A. M. 1967. Effect of Floor Surface Temperature on Comfort. Part IV: Cold Floors. ASHRAE Trans. 73, 2.

24. Schøler, M. 1976. Vertikale temperaturgradienters indflydelse på menneskets termiske komfort (The Influence of Vertical Temperature Gradients on Human Comfort). M.S.-Thesis, Laboratory of Heating & Air-Conditioning, Technical University of Denmark.

25. McNair, H. P. 1973-74. A Preliminary/Further Study of the Subjective Effects of Vertical Air Temperature Gradients. British Gas Corporation, Project 552, London.

26. Eriksson, H.-A. 1975. Värme och Ventilation i Traktorhytter. Jorsbruks-tekniska Institut, S-25 Ultuna, Sweden.

27. McIntyre, D. A., and Griffiths, I. D. 1975. The Effects of Added Clothing on Warmth and Comfort. Ergonomics, 18, 2:205-211.

NATURAL VENTILATION
IN WELL-INSULATED HOUSES

D. J. NEVRALA AND D. W. ETHERIDGE

British Gas Corporation — Heating Division
London SW6 3HN, England

INTRODUCTION

Natural ventilation provides fresh air for a multitude of purposes, but it is also a source of heat loss from a building. In the past ventilation by natural means, either by infiltration through cracks or through purpose made openings, had been taken for granted. The only complaint at times had been of draughts; the energy loss caused by excessive ventilation rates had not attracted attention because the ventilation heat loss had been only up to 20% of the total heat loss from a typical house in the United Kingdom.

The energy crisis and the ensuing rise in fuel prices resulted in a range of energy saving measures, the most prominent being the adoption of higher insulation levels for new housing and it is probable that this trend will continue in the future. The application of high insulation levels disturbs the energy balance in a number of ways, one of the most important being the ratio of design fabric and ventilation heat loss, in a well insulated house. The ventilation heat loss can account for up to one half of the total loss under these circumstances. It has been realised that the law of diminishing returns applies to any further increase in insulation levels and therefore attention now has to be focussed on means of reducing the other component of energy loss - ventilation.

Measures have been introduced in prototype well-insulated dwellings to reduce the ventilation rate by specifying "tight" window designs or double glazing, and conventional ventilation heat loss calculations predict a substantially reduced design heat loss. However, experience has shown that where heating systems have not been fortuitously oversized, complaints of discomfort due to low temperature and draughts have arisen, which in turn suggests increased ventilation.

This paper presents an analysis of the mechanics of natural ventilation and describes a computer-based method developed by British Gas Corporation at Watson House for predicting ventilation patterns in houses. Calculations using the method are used to illustrate the basic reasons why natural ventilation is likely to cause problems in heating well-insulated dwellings. A detailed discussion of these problems is then given, particular attention being paid to the way in which ventilation could influence the sizing of appliances and the indoor thermal environment. Results of computer-based simulations of the thermal behaviour of a well-insulated dwelling are presented.

AIR INFILTRATION IN HOUSES

The natural ventilation of a house arises as a result of pressure differences generated across open areas by the action of the external wind (wind effect) and by bouyancy (stack effect). The open areas can take a wide variety of forms, and it is convenient to divide them into three groups. In the first group we have purpose-provided openings such as air vents, flues, chimneys and open windows. The second group consists of the cracks in and around room components (i.e. doors and windows). These cracks are such that they are readily identifiable and their dimensions can in general be measured. They have the important property that the spaces between which they communicate are also identifiable. In the third group, we have the open areas which remain when the two other types of crack have been sealed. This group is referred to here as "background leakage areas". Background leakage areas are often not visible as discrete cracks and the spaces between which they communicate are often not identifiable. Within this group will be cracks around electrical fittings and around the joints between ceilings and walls and also the porosity of room surfaces, which for certain types of floor can be very large.

Work carried out by the Government's Building Research Establishment[1] has shown that the infiltration of air into a house through background leakage areas can be very large. That is, when a house is pressurised by a large fan, and the cracks around doors and windows are sealed, a large flow out of the house remains and this is not due solely to the presence of purpose-provided openings. We have carried out tests to gain an idea of the magnitude of the background leakage area of individual rooms compared with the open areas of the components in the room. Both types of open area have been measured with the pressure/flow technique[2]. Basically, a known pressure is applied across the area and the resulting flow rate is measured. By making use of the crack flow equations[2], the open areas can be obtained from the pressure/flow measurements. The values obtained in this way are accurate for doors and windows, but are less accurate for the background leakage areas which are considered as the areas of hypothetical equivalent cracks. Nevertheless, it is reasonable to consider the areas of these hypothetical cracks as an approximate measure of the background leakage areas, since they have been found to give a fairly reasonable description of the flows associated with them[2].
Table 1 shows the results of tests carried out in a modern detached four-bedroom test house. The background leakage areas of the hall and the landing were not measured. For the tests, the lower suspended floor of the house was sealed with polythene sheet to simulate a concrete floor.

Table 1 - Comparison between background leakage areas of rooms and the sum of the open areas of the doors and windows in the rooms

ROOM	TOTAL OPEN AREA OF ROOM COMPONENTS m^2	BACKGROUND LEAKAGE AREA m^2
Lounge/Dining	0.073	0.031
Bedroom 1	0.025	0.037
Bedroom 2	0.020	0.031
Bedroom 3	0.026	0.036
Bedroom 4	0.022	0.032
Bathroom	0.022	0.025
W.C.	0.018	0.019
Kitchen	0.041	0.027

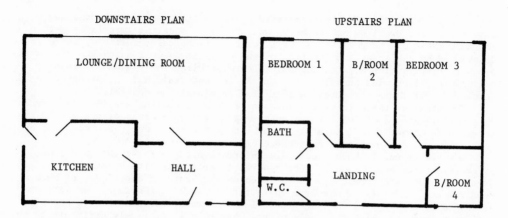

The main point to note from the results is that the background leakage areas of the rooms are of similar magnitude to the areas of the cracks around doors and windows in the rooms. Although the smaller background leakage areas occur for the smaller rooms (bathroom and W.C.), the largest background area does not occur for the largest rooms (lounge/dining room). Another point to note is that neither the room component areas nor the background leakage areas can be summed to give a total for the house. This is because some doors and some room surfaces are common to two rooms.

Account should be taken of background leakage areas when considering the ventilation of dwellings. This is likely to be more important for low-energy houses than for normal houses, since "weather-stripping" (e.g. reducing the open areas of cracks by application of foam strip) of the windows and doors will tend to increase the proportion of the total ventilation associated with the background leakage areas. By the same token, the effect on the air change rate of a house due to weather-stripping is effectively reduced by the presence of the background leakage areas. The problem of background leakage areas becomes very apparent when one is concerned with the prediction of ventilation of multi-room dwellings with a mathematical model. This is because one would ideally like to know not only the magnitude of the background leakage areas, but also the manner in which they communicate, both between rooms and between the exterior and the interior. Such information could be extremely difficult to obtain in full and for most purposes approximations will probably have to be employed. In the next Section a mathematical model which is being developed at Watson House is briefly described and the treatment of background leakage areas discussed.

PREDICTION OF VENTILATION

In essence the mathematical model$^{(3)}$ consists of the continuity equation coupled with the crack flow equations$^{(2)}$. The latter equations are semi-empirical in nature, having been derived from laboratory tests but with a good theoretical backing. Measurements in a test house have shown that the equations describe accurately the flow through the cracks of real doors and windows. When a dwelling is being studied, the individual open areas are each allotted a crack flow equation and this system of equations is solved with the continuity equation, once the required meteorological conditions and dwelling characteristics are specified.

The required meteorological conditions are wind speed, wind direction and external temperature. In addition the internal temperature must be known so as to account for stack effect.

The pressure distribution generated by the action of the wind over the exterior of the dwelling has to be known. Strictly speaking this is as much a characteristic of the dwelling as it is of meteorological conditions, since it is determined by the shape and location of the dwelling as well as by the wind. The pressure distribution and its dependence on wind direction can be obtained conveniently from wind tunnel tests, but it is essential that the atmospheric boundary layer characteristics be simulated as closely as possible.

The dwelling characteristics which have to be specified are the geometry and physical dimensions of the various open areas. For open areas of the second group referred to above, this information can be obtained by direct measurement or by a combination of mensuration and pressure/extract measurements. (2)

For background leakage areas this information can only be obtained from pressure/extract measurements. For general calculations with hypothetical dwellings, estimates would have to be made on the basis of experience of such measurements, and this is one area where further work is required. As well as estimates of the size of the background leakage areas, one also has to specify the spaces between which they communicate, but this is extremely difficult to do. However, the problem can be considerably simplified if one restricts the predictions to total air change rates of dwellings for the case with all internal doors open. For this case the internal open areas can be neglected and one is only concerned with the proportion of the background leakage areas which communicate with the exterior. By carrying out a fairly simple test on the house an estimate of this proportion can be made. This has been done for the four-bedroom test house referred to above.

Briefly, a large fan was used to supply air to the house at a known rate and the pressure inside the house relative to the outside was measured. Calculations were than carried out with the model using different distributions of background leakage area, and the distribution giving the best agreement with the measured pressure difference was found. For the house in question, this distribution was such that 50% of the background leakage areas of the downstairs rooms communicated through the external walls. In the upstairs rooms, 40% communicated through the external walls and 25% through the ceiling into the loft space.

ILLUSTRATIVE PREDICTIONS OF GENERAL TRENDS

The prediction method described above can be used in its present stage of development for illustrating some of the problems that could arise with the ventilation of low-energy dwellings. Calculations for this purpose are presented below. The calculations relate to a hypothetical detached house with four bedrooms, although the open areas of the various room components and the pressure distributions are values that have been measured for other houses. The size and distribution of the background leakage areas correspond approximately to those determined for the test house referred to above.

Fig.1 shows the predicted variation of air change rate with wind speed for a westerly wind and for a range of values of the internal/external temperature difference, ΔT. The air change rate is defined as the volume flow rate of fresh air entering the house divided by the volume of the house. It can be seen that above a certain wind speed the predicted air change rate is independent of ΔT. This is to be expected, because the pressures generated by the wind increase as the square of the wind speed, so that as the wind speed increases, the pressures generated by the stack effect eventually become negligible. Below the critical wind speed, the predicted air change rate is virtually independent of wind speed and depends only on ΔT. In the prediction method the pressures generated by the stack effect are simply added to the pressures generated by the wind. As a result of this, one would expect the predicted behaviour, because on the windward face the positive pressure due to the wind will tend to increase the inflow to the house on the lower floor, but it will tend to oppose the outflow on the upper

floor. The net effect will therefore tend to be small. The same argument
applies to the leeward face. Hence the nett result of the wind effect and stack
effect will be a tendency for the total air change rate to remain constant, until
the pressures generated by the wind become sufficiently large. The predicted
behaviour in Fig.1 can be observed in some recent ventilation measurements carried
out in a terraced house of new construction (see Fig.2). The wind direction is
not the same for each measurement and the ▲T values range from 14°C to 20°C,
nevertheless the predicted behaviour is still evident.

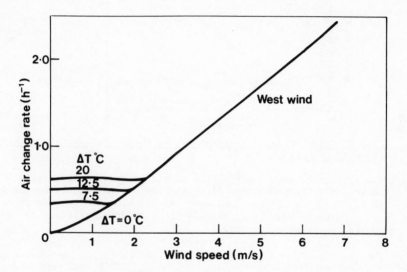

Figure 1 - Predicted variation of whole-house air change rate with wind speed and
           temperature difference

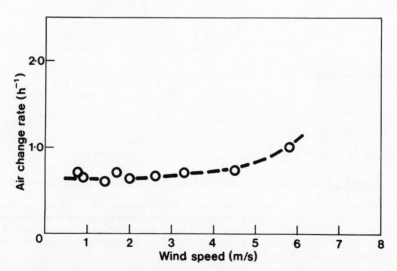

Figure 2 - Whole-house air change rates measured in a terraced house, illus-
           trating the importance of stack effect

Fig.3 illustrates the large changes in predicted ventilation which can occur when the wind direction changes and when the wind speed is fairly high. The two wind directions shown represent the extreme cases and fluctuations in the direction of the real wind would tend to reduce their difference in practice. However, it still seems probable that certain changes of wind direction could lead to a doubling of the air change rate.

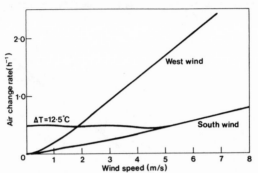

Figure 3 – Predicted variation of whole-house air change rate with wind speed for
         two wind directions

Fig.4 shows the frequency distribution of wind direction for a site in London[4] and it can be seen that the wind lies for a significant time in all the possible directions. Moreover the frequency distribution of the wind speed, also given in Fig.4, indicates that large changes in ventilation rate due to changes in wind speed are likely to occur often.

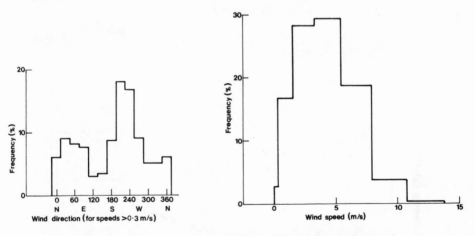

Figure 4 – Frequency distributions of wind direction and wind speed at Kew, London

Fig.5 shows the effect of neglecting the background leakage areas on the predicted air change rates. It is clear that the background leakage areas should be taken account of, since for this particular case they account for about 50% of the total air change rate. This is similar to the measurements for the semi-detached house given by Skinner[1].

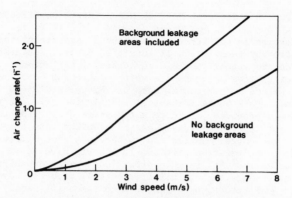

Figure 5 - Predicted variation of whole-house air change rate with wind speed,
illustrating the importance of background leakage areas

Fig.6 shows the flowpaths taken by the ventilation air for one particular case (internal doors closed) where the wind effect is dominant over the stack effect. It can be seen that there is a strong tendency for the air to enter through one side of the house and exit through the others. The figure also shows the distribution of the background areas and their contribution to the ventilation rate.

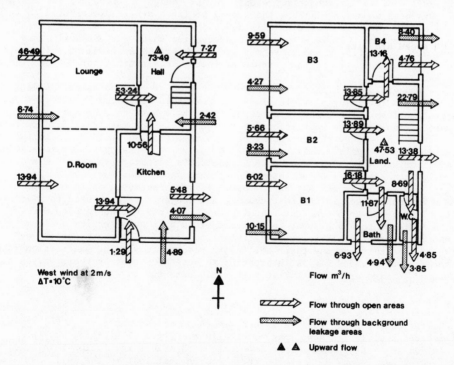

Figure 6 - Predicted distribution of ventilation flows within a house (internal
doors closed), illustrating passage of air from one side of the
house to the other

IMPACT OF VENTILATION ON DESIGN AND PERFORMANCE OF SPACE CONDITIONING SYSTEMS

Natural ventilation or infiltration may have a significant influence on the performance of heating systems in a well insulated house. The discussion in this paper is focussed on the situation where mechanical ventilation is not provided. However, the following brief comments are worth mentioning here. Mechanical ventilation is one method of overcoming some of the uncertainties inherent in natural ventilation. Whole house mechanical ventilation, i.e. where both a supply and extract are provided, could permit heat recovery from the extracted air, but the extra capital cost involved has to be taken into account. In the United Kingdom, for mechanical ventilation to become a cost-effective energy conservation measure, the relative price of fuel would have to rise substantially[5]. Our studies indicate that in situations where mechanical systems are used, unless precautions are taken, infiltration will probably have a significant role, imposing a large fluctuating component onto the basic ventilation rate.

It is convenient to consider the impact of natural ventilation under two headings, i.e. in terms of total whole-house air change rates and in terms of the air supply to individual rooms (see following). Under the first heading the magnitude of the ventilation rate and heat losses are considered, and in particular the effects of variation of ventilation rates on the design of plant for the whole house. Under the second heading the way in which the ventilation paths in the house affect the sizing of emitters is discussed. Before proceeding to this however, some other points are worthy of mention.

Fresh air requirements[6] of a typical house having a floor area of $90m^2$ are in the region of 45 1/s, which represents an air change of circa 0.8 volumes per hour. Of equal importance is how the supply of fresh air is distributed throughout the house. The greatest need will be in the living room, especially if smokers are present. It is a reasonable assumption that occupants will take action, e.g. by opening windows, to ensure an acceptable environment. It would therefore be unrealistic in the pursuit of energy conservation not to take these needs into account.

The manner in which air enters a room can also be of importance. A well insulated room retaining a single glazed window could be susceptible to low level draughts problems. The relative price of fuel would have to rise substantially for double glazing to be cost-effective in the British climate[5]. Cold currents generated by the window surface combined with infiltration could dominate the room air movement pattern in a well insulated enclosure, especially if for reasons of economy, radiators were positioned on inside walls. Subjective tests in a controlled temperature room at Watson House have shown that low level cold draught can result in overall discomfort. For the tests, an optimum temperature of 23^oC was maintained in the test chamber and subjects were exposed to simulated low level draughts at a constant velocity of 0.2m/s and varied temperatures. Results indicate that a draught temperature of 19^oC in an otherwise optimum thermal environment would cause dissatisfaction of 30% of the subjects. The siting of heat emitters and their ability to respond to varying ventilation rates is therefore important. It is interesting to note here that by their very nature, background leakage areas are unlikely to cause detectable draughts.

Whole-House Air Supply Rate

Energy conservation considerations are the most obvious motives for trying to minimise the ventilation heat loss of dwellings. Fig.7 illustrates how the ventilation heat loss becomes more important with higher insulation levels, constituting nearly one half of the total design heat loss in the well insulated house. The ventilation heat loss was calculated on the basis of empirical values in the 1970 IHVE Guide[7] Table A4.8 and is therefore identical for all three cases. A semi-detached house has been chosen as an example because it is the most numerous type of dwelling in the United Kingdom. Houses - detached, semi-detached and terraced - account for 78% of all existing dwellings and 70%[8] of new dwellings and this trend is likely to continue.

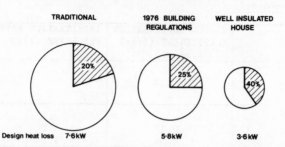

Figure 7 - <u>Ventilation heat loss as percentage of total design heat loss of a typical semi-detached house of 90m² floor area for three insulation levels.</u> (The specific fabric heat losses are: Traditional 1.4W/m³ °C; 1976 Building Regulations 1.0W/m³ °C; Well-insulated 0.5W/m³ °C.)

The design heat loss is based on a mean internal temperature of 19.3°C, external temperature -1°C and a mean ventilation rate of 1.02 air changes per hour.

Higher levels of insulation have resulted in smaller heat emitters or other sources of heat being required, resulting in increased sensitivity to any imposed load such as ventilation. Two factors combine to reduce the heat emitter size even further. The first is an economic pressure to save capital cost and as a result, emitters are not being oversized as had been the practice. The second stems from an attempt to tackle the largest single remaining component of the overall heat loss - the ventilation heat loss. By specifying "tight" or "weatherstripped" windows without fully taking into account the fresh air requirements of the occupants significant reductions in the size of the installed plant can apparently be made. Hence, the variation in air change rate which can arise as a result of variations in wind speed and direction (Fig.3) could lead to large changes in the heat demand of the house.

In practice, a building services engineer in the United Kingdom will probably use the IHVE Guide Section A4 to calculate the infiltration heat loss of weatherstripped windows. The method takes into account the actual crack length and is based on recommended window infiltration coefficients (1/ms at 1N/m²). Results of calculations of air change rates for a typical living room, as well as the usual recommended design air change rate* (Table A4.8), are summarised in Table 2. The recommended design value is identical for both the weatherstripped and the non-weatherstripped situation and is of the same order as the non-weatherstripped crack length value. The air change rate for the weatherstripped case shows a reduction by a factor of five. These values are applicable to single rooms and therefore to sizing of heat emitters, and the Guide suggests that approximately only one half of the sum of the room values should be taken into account when sizing the central plant (outside air enters only the windward rooms).

Although the crack length method may be satisfactory for office buildings, factories or even flats, or where mechanical ventilation is employed, it is obvious that serious difficulties can arise if used for houses. The reason for the divergence has been discussed above, i.e. the significant influence of background leakage open areas and of the needs of occupants.

To examine the sensitivity of the thermal response of houses (and individual rooms) to ventilation rates different from those used in the heat loss calculations, a series of computer simulations have been carried out. A well insulated semi-detached house of traditional construction described in Table 3 was used for the analysis. To simulate a real life situation as near as possible clear and cloudy

*To be used when the window characteristics and building plan are not known or when calculation is impracticable or inappropriate.

Table 2 - Air change rates of a typical living room - empirical and crack length
          method. Volume = 5.2x3.5x2.3 = 42m³. Crack length 10m; Surburban site
          (normal exposure)

	WINDOW INFILTRATION COEFFICIENT (1/ms) AT $1N/m^2$	INFILTRATION RATE PER METRE (1/s)	AIR CHANGE RATE (h^{-1})
Empirical IHVE Guide Value			
Non-weatherstripped	-	-	1
Weatherstripped	-	-	1
Crack Length Method			
Non-weatherstripped	0.25	0.2	0.86
Weatherstripped	0.05	1.0	0.17

design days, having a mean temperature of $-1^{\circ}C$, based on an analysis of 20 year
weather data were used. Incidental gains and system operation are tabulated in
Table 4. The computer program used for the analysis has been developed on the
basis of Rouvel's work[9].

Table 3 - Construction and heat loss of well insulated semi-detached house used in
          computer simulation studies

Construction

Internal Wall: Ground Floor: Internal Floors/Ceiling:

Plaster - 16mm   Carpet - 10mm   Plasterboard - 10mm
Lightweight block - 110mm   Timber - 20mm   Air Gap - 50mm
Plaster - 16mm   Glass Fibre - 25mm   Timber - 20mm
                              Concrete - 1000mm
External Wall:
                              Roof:
Plasterboard - 16mm
Lightweight Block - 110mm   Plasterboard - 10mm
Polystyrene - 50mm   Glass Fibre - 100mm
Brick - 115mm   Tiles - 10mm

Heat Loss

	AREA m^2	'U' $W/m^2 \, ^{\circ}C$	'UA' $W/ \, ^{\circ}C$
Roof	46.6	0.30	14.0
Wall	83.8	0.40	33.5
Floor	46.6	0.45	21.0
Window	14.1	2.50	35.3
Specific Fabric Heat Loss			103.8

        As an example the results of two computer simulations are shown in Fig.8,
where the energy requirement and internal and external temperatures of a well
insulated house (Table 3) on a cloudy design day are given. In both simulations
the heating system was operated intermittently and a conventional plant size ratio
of p=1.2 (i.e. 20% was added to the design heat loss) was used. The design heat
loss was based on a calculated air change rate of 0.2 volumes per hour; the Guide
recommendation to halve the sum of room values having been disregarded, as it
would have led to an extremely low value. In the first simulation the air change
rate was kept to the design value of 0.2 volumes per hour and the temperature
chart shows that the house required six hours to reach the design temperature of
$20^{\circ}C$ (test house results confirm this behaviour for similar plant size ratios[10]).
In the second simulation the air change rate has been changed to one volume per
hour (while all the other parameters remain the same) with the result that the

internal temperature never reaches the design value although the plant operated at full capacity. The maximum temperature at the end of the heating period reaches only 18°C; the mean over the heating period being slightly below 16°C. Such internal temperatures would inevitably lead to severe complaints.

Table 4 - System operation used in computer simulation studies

Occupancy (i.e. time period when internal design
 temperatures should be maintained)   8.00 - 23.00 hrs
Internal Design Temperature   20°C
Lights   300W
     a.m. - on : 8.00 hrs   p.m. - on : 1 hr before sunset
     a.m. - off : 1 hr after sunrise   p.m. - off : 23.00 hrs
Other Heat Gains
     A weighted average of gains from people, domestic appliances, DHW, etc.
     - day : 600W   - night : (all occupants presumed : 667 W
                       in house (2 adults, 2 children))
Intermittent Operation
     Heating system
     - on : 6.00 hrs - off : 23.00 hrs.

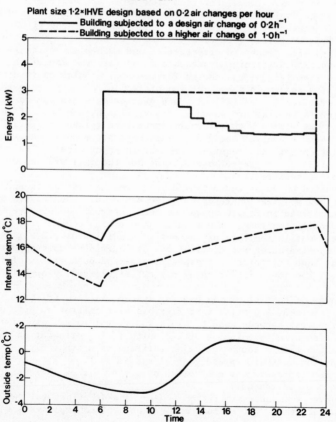

CLOUDY DAY

Figure 8 - Energy consumptions and internal temperatures of a well-insulated house on a cloudy design day (T_{MEAN} = -1°C).

Preliminary studies using the Watson House ventilation computer program indicate that the mean air change ratio of a "tight" or "weatherstripped" house in design conditions is higher than conventional calculations would suggest (0.1-0.2 volumes per hour) and would be in the region of 0.5 plus, depending on the design and general workmanship of the house. The above example (Fig.8) confirms that air infiltration through background areas (dependent on workmanship) has to be taken into account if complaints of discomfort are to be avoided.

Other simulations have shown that systems designed to operate continuously (plant size ratio=1) would also fail to reach internal design temperatures. A system designed for an air change of 0.2 volumes per hour when subjected to an air change of one volume per hour would achieve only an internal temperature of 17-18°C.

The results of the study indicate that solar and miscellaneous internal heat gains cannot be relied upon to compensate for higher than design ventilation rates

Distribution of Air to Individual Rooms

The flow rates and flow paths in Fig.6 indicate that under certain conditions the bulk of the fresh air may be entering a house through one particular side. If this is so, then the overall air change rate for the house may be irrelevant to the conditions in individual rooms, although retaining its importance in determining the boiler size. Individual room flow rates are all the more important because with higher values the probability of dissatisfaction caused by draughts is greater. In this context it should be remembered that the situation in a home is different from an office situation. Whereas in an office situation a percentage of occupants will always be dissatisfied and such a situation is accepted in the home the space conditioning system has to satisfy the individual occupant.

A typical room (Fig.9) was chosen for computer simulation studies of the consequences of a change in the direction of flow of air. As an example the internal temperatures for an inflowing and outflowing situation on a cloudy design day and continuous heating are shown in Fig.9. The heat output of the emitter in this situation was varied with external temperature (graph in Fig.9) as an external compensator control would do. The temperature chart shows that for the design ventilation rate the required temperature is attained. The outflowing situation results in a degree of overheating but the inflowing condition would create thermal dissatisfaction.

The situation on sunny and cloudy days as well as for north and south facing rooms for various ventilation rates was also explored. The results of the simulations are summarised in Fig.10 where the mean internal temperature is plotted against the ventilation rate. The results show a high sensitivity of internal temperature to ventilation rate at the design condition of 0.2 air changes per hour. The graph indicates the desirability of individual room temperature control and sufficient heat emitter output especially as the peak values of internal temperature for the south facing sunny day are significantly higher than the mean values.

Other modes of control of the heating system, except individual room temperature control combined with sufficient flexible heat emitter output, would result in a worse disparity between conditions in individual rooms, especially if the inflowing condition were to coincide with a north facing orientation and the outflowing condition with a south facing orientation on a sunny day. This is evidently true for the most widespread control system in the U.K. where the operation of the whole heating system is controlled by a single room thermostat either in the living room or in the hall.

The above computer simulation studies show that natural ventilation is one of the factors that will influence the choice of control systems for well insulated houses.

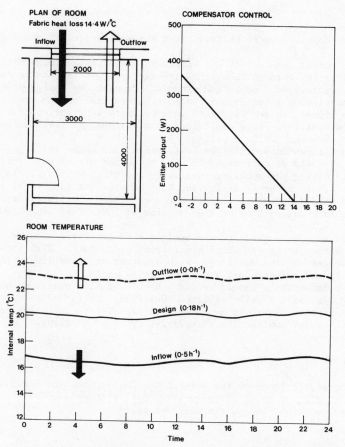

Figure 9 – Internal room temperatures for an inflow and outflow situation on a cloudy design day (T_{MEAN} = -1°C). Heat output varied with external temperature to simulate an external compensator control.

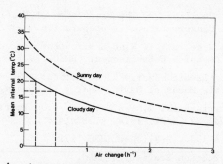

Figure 10 – Effect of air change rate on the mean internal temperature for sunny and cloudy design days (T_{MEAN} = -1°C), continuous heating, simulated external compensator control.   Sunny day = Sunny south facing room. Cloudy day = Cloudy south and north facing room and sunny north facing room.

CONCLUSIONS

The major conclusions to be drawn from the work described above are as follows.

When estimating the overall air supply to dwellings, allowance should be made for air infiltration through background leakage areas, which can be as large as that through cracks around doors and windows. Present day design procedures for estimating ventilation rates do not consider background leakage areas and consequently these procedures are likely to underestimate ventilation heat losses. For well-insulated dwellings, this could lead to a serious underestimation of the total heat requirement of the dwelling.

Large variations in ventilation heat losses, and hence total heat requirement can occur as a result of variations in wind speed and direction. This could lead to problems in sizing the appliance for well-insulated dwellings. In addition, a problem is likely to occur in sizing the heat emitters of individual rooms, as a result of the wide range of infiltration routes of the ventilation air. One solution to this problem might be the adoption of individual room temperature control.

Attention should be given to the avoidance of low-level draughts because it has been found that these can cause significant discomfort. With the small emitters of well-insulated dwellings, such draughts are more likely to occur, particularly if the emitters are placed on inside walls.

Careful consideration should be given to ventilation phenomena when designing space heating systems for well-insulated dwellings. It is believed that prediction methods, such as that described here, will be useful for predicting general trends and also possibly for application to specific types of dwelling.

ACKNOWLEDGEMENTS

The authors wish to thank the British Gas Corporation for permission to publish this paper, and their colleagues Mr. S.L. Pimbert and Mr. P. Phillips for their valuable contributions to the work described.

REFERENCES

1.   Skinner, N. Natural Infiltration Routes and their Magnitudes in Houses. Pt.2. Conference on Controlled Ventilation, Aston University, England, 24 Sept. 1975.

2.   Etheridge, D.W. Crack Flow Equations and Scale Effect. To be published in Building and Environment.

3.   Etheridge, D.W. & Phillips, P. The Prediction of Ventilation Rates and Implications to Energy Conservation. In preparation.

4.   Penwarden, A.D. & Wise, A.F.E. Wind Environment around Buildings. Dept. of Environment, Building Research Establishment, HMSO, 1975.

5.   A BRE Working Party Report. Energy Conservation: A Study of Energy Consumption in Buildings and Possible Means of Saving Energy in Housing. CP 56/75 June 1975.

6.   Tipping, J.C., Harris-Bass, J.N. & Nevrala, D.J. Ventilation and Design Considerations. BSE, Vol.42, Sept. 1974, p.132-141.

7.   IHVE Guide, Book A 1970. Published IHVE 1971.

8.   Housing and Construction Statistics HMSO.

9.   Rouvel, L. Berechnung des warmetechnischen Verhaltens von Räumen bei dynamishen Wärmelasten, Special supplement "FFE Berichte No.2" to Brennstoff-Warme Kraft 24 No.6 (1972).

10.   Nevrala, D.J. Heat Services for Future Housing Pt.1 The Insulated House Design Requirements. To be published in BSE.

VENTILATION OF FLATS

H. TRÜMPER AND H. BLEY

Lehrbereich Technischer Ausbau an der Architekturfakultät der Universität Karlsruhe
75 Karlsruhe, Englerstrasse 7, Federal Republic of Germany

Ventilations of Flats

A sufficient and undisturbed ventilation offers better living conditions in flats.
So far flat ventilation has been considered necessary for oxygen supply, however, it is even
more important for removal of smell and heat. Tenants usully consider opening of windows as
sufficient for removal of smell, but continued opening is regarded as molesting because of
draught. Additionally opening of windows results in high energy losses.

Figure 1 represents the plan of a three-bedroom-apartment, as usually built in
multiple family-dwelling. For rationalization in the years between 1950/60 flats became smaller
at the frontside and more extended with the neighbouring flats. As a consequence it was no
longer possible to provide the internal space with natural lighting.

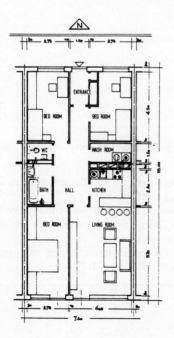

Figure 1

Plan of a typical 88 m^2 floor area
in a dwelling house ;
bath, WC and wash room without
outside windows.

Therefore these areas have been used as bath, WC, and utilityrooms, frequented only for a short time. In figure 1 even the kitchen is included in this interior area but enlighted by the livingroom. According to the building regulations of the GFR, kitchens are not permitted without windows. The rather inexpensive construction of these buildings however results problems concerning flat ventilation.

Contrary to figure 1 the cross-section of a flat shown in figure 2 has a kitchen and a bath on the outside. Such flats of an older type, are built before 1950 and don't offer good living quality because of the insufficient ventilation. Depending on winddirections and pressure the air is just transferred back and forth in different directions and not removed from the rooms.

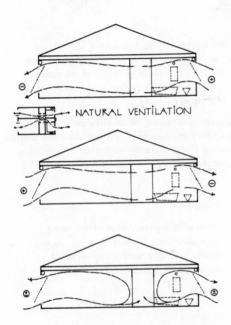

NATURAL VENTILATION

Figure 2

Systematic description of natural ventilation in the flat of figure 1, however bath, and WC are situated at the external wall with outside windows.

Bathrooms in the interior area without windows, already provided as early as 1930, were ventilated by ducts, as to be seen in figure 3.

In the GFR these baths and WC's have to be built with a ventilation by ducts accordir DIN 18017, which has been revised several times since 1950.
For buildings with more than four stories a main duct-ventilation, as developed in the Netherlands, is used often. Main duct ventilation, however, does not operate properly in the top story of a flat roofed houses because of insufficient shaft height.
The next step in the development of main duct ventilation provided statisfying conditions for the tenants by additional use of fans, according to DIN 18017 part 3.

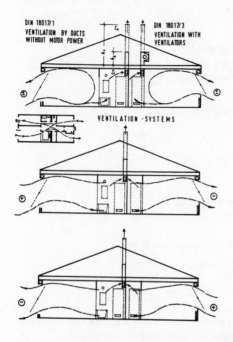

Figure 3

Systematic description of ventilation in the flat of figure 1 by ducts without motor power according to DIN 18017 part 2 and in amplification with ventilators according to DIN 18017 part 3.

Controlled ventilation has been developed in the Skandinavian countries, especially in Sweden. Nowadys in these countries it belongs to standard living comfort. Following these standards in the Federal Republic of Germany the "VDI-direction 2088" for flat ventilation has been developed, after numerous buildings have been provided with this kind of ventilation.

Controlled ventilation of an appartment, according to figure 1 is shown on figure 4.

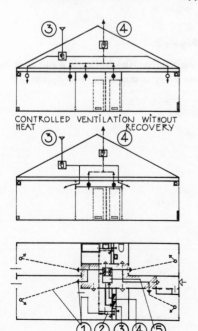

Figure 4

Ventilation in the flat of picture 1 according to VDI direction 2088.

The numbers have the following meaning :

1) Air inlet by tubes in the concrete ceiling and air
   intake at the window or through suspended ceiling
   in the central hall and slots on top of the bedroom
   and livigroom doors.

2) Exhaust valves in kitchen, bath, WC, and utility room.

3) Ambient air intake

4) Exhaust exit

5) Central unit with fans

The required air volume for ventilation by VDI-direction 2088 :

for kitchen	80 - 120 m^3/h
for bath	40 m^3/h (minimum air exchange four times)
WC	20 m^3/h (" " " " ")
	140 - 180 m^3/h

Heat requirements can be reduced by the amount of air losses because continous and uncontrolled ventilation through window gaps is replaced by controlled ventilation. During nighttime the recommended flow rates can be reduced as well as the flow rates for kitchen during minimum load hours. In order to avoid draught, it is necessary to provide an air heater. From an energy point of view this system is not much favorable compared with a conventional space heating system with radiators and main duct, figure 3. Further improvement of the system according to figure 4 was made in Sweden and France combining heat recovery with controlled ventilation, figure 5.

Figure 5

Ventilation like in figure 4 but with heat recovery and change-over system.

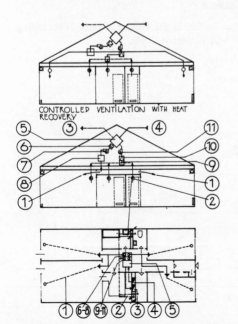

CONTROLLED VENTILATION WITH HEAT RECOVERY

The numbers in figure 5 have the following meaning :

    5) recuperator

    6) inlet air filter

    7) inlet air fan

    8) inlet air flap valve

    9) exhaust air flap valve

    10) exhaust air filter

    11) exhaust air fan

    12) air heater

    13) and 14) motor flap valves.

In this system 50 to 70 % of the energy in the exhaust air can be transfered to the ingoing air resulting in a total annual energy saving of about 10 - 20 %.

Incorporating a secondary air heater, the ventilation system can also provide the space heating. Preferably the inlet air is heated up to 15° or 16° C and additional electric heating is controlled by room thermostats, following quickly any changes of internal heat gains or solar radiation.

A rather important advantage of this system is the simplified settlement of heating accounts. A flat rate is charged for the basic central heating, whereas the additional electric heating account can be settled individually, encouraging the tenants to save energy.

The central airheating system provides an average temperature of $15 - 16^\circ$ C for all apartments, therefore no tenant can make profitable abuse of his neighbours' well heated walls and ceilings, whilst he saves his expenses.

CHANGE-OVER-SYSTEM

The change over flap valves (8) and (9), as shown in figure 5 are operating as follows :
Depending on daytime and heatload in the rooms these flap valves can reduce the exhaust air in bath and WC and raise in the kitchen area or regulate the air between living- and bedrooms. During nighttime the total reduced air flow can be directed to the bedrooms only.

The so called "Aldes"-system, especially used in France, heats the inlet flow only by exchanging heat with the exhaust air ; fresh air supply is only heated by human beings and electrical equipment, so that the heat economy in these houses is covered predominantly by internal heating.

Apartment energy economy :

Floor area	about	88 m^2
volume	about	220 m^2
Q_{Tr} (t_A = - 15°C)		3,300 W
Q_L (165 m^3/h)		2,100 W
Q_{tot}		5,400 W

$$Q_{sp} = \frac{5,400}{88} = \frac{60 \text{ W}}{\text{m}^2}$$

The inlet flow rate, related to rooms with a direct fresh air supply, results in an air exchange of 1.0/h, however related to the total volume of the apartment the air exchange is only 0.75/h.

Heat balance for room-heating and ventilation
with different versions.

CASE 1 - Standard version (central room heating and ventilation by ducts)

$$Q_{Tr} = 3.300 \text{ W} = 61 \%$$
$$Q_{Vent} = 2.100 \text{ W} = 39 \% \quad 160 \text{ m}^3/\text{h}$$
$$= 5.400 \text{ W} = 100 \%$$

With a central room heating and a ventilation with outside windows or by ducts heat recovery out of the exhaust air volume is quite impossible and an use of the internal heat gains is nearly excluded.

CASE 2 - Ventilation with heat recovery and central room heating

$$Q_{Tr} = 3.300 \text{ W} = \quad \text{(heating unit)}$$
$$Q_{Vent} = 2.100 \text{ W} = \quad (165 \text{ m}^3/\text{h})$$

For a system like in figure 4 and 5 leak air rate must be calculated although when the windows and doors are quite tight.

exhaust air volume = 160 m^3/h
leak air volume   = 50 m^3/h
ingoing our volume = 210 m^3/h

Q_{Vent} = 2.645 W (t$_{out}$ = -15°C, 210 m^3/h)

Q_{leak} =   630 W

$Q_{exhaust\ air}$ = 2.015 W

Recovery (recuperative/figure 5)

Q_{Re} = 60 % of $Q_{exhaust\ air}$ = 1.210 W

Loss

$Q_{outgoing\ air}$ =   805 W

Q_{leak} =   630 W

$Q_{loss\ total}$ = 1.435 W

Heat requirement for room heating and ventilation

Q_{Tr} = 3.360 W

Q_{Vent} = 1.435 W

Requirement   = 4.795 W

In comparison with case 1 a reduction of about 12 %.

CASE 3 - System like case 2 by consideration of internal heat gains

The internal heat gains depend on the installation of equipment and can be very great. Out of the exhaust air volume is partly a heat recovery possible. By raising the air temperature of the out-going air in the kitchen up to about 30°C this temperature would be about 10 ^{o}C higher than the room temperature. So a recuperative heat recovery with a part of about 65 % = 390 W is possible.

Summary for case 3

Q_{II} = 4.795 W

$Q_{Rec.\ II}$ = - 390 W
rest.   = 4.405 W

In relation to case 1 there is a whole heat requirement of only 82%. This is not a data for the whole time, because this described over-temperature is only during the main load time.

CASE 4 - Use of the solar radiation

It is a known fact, that a south orientated window is a propitious and free solar collector during heating season. But the basis for this is, the ingetting solar radiation energy can be used. The experience shows, that in rooms with south orientated windows a use of the whole ingetting solar energy is not possible on sunny days in winter-time and especially during spring and autumn. By ventilation with windows a part of this energy must be dissipated. Mostly the room heating system is too inactive, so that a use of the solar energy is not possible, and mostly the solar radiation is kept outside of the building by venetion blind or solar shading. The solar data is known and in the following is described how with a right disposition in the plan of figure 1 an optimal use is possible.

The curves in the upper part of figure 6 inform about the data of the middle total solar radiation on h horizontal and v south orientated faces. The flat in figure 1 has a brutto window area of 12.3 m^2 and by deducting the reducing parts there remains a net area of 10.0 m^2. By the hypothesis of reducing irradiation on 0.7 there resist an area 7.0 m^2 which can be used as a radiation area of 100%.

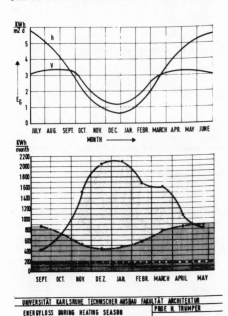

Figure 6

Ventilation like in fugure 5 but with auxiliary installation for extra air change between the south and north located rooms and for the transport of the solar heat gain.

On the lower part of figure 6 are the following curves represented :
Upper curve : The monthly heat requirement for the flat in figure 1
Middle curve : Irridated solar energy

A planimetry of the month october until april shows the following relations expressed as percentage

heat release = 100 % (without night draw down)
solar energy = 28 %
internal heat = 12 %
balance = 60 % of the whole heat requirement
of case 2

The dimension of the usable irridiated solar energy shows, that the inactive common heating systems as well as the dimension of the described ventilation systems with heat recovery are in able to use the irridiated solar energy for all rooms of a dwelling.

The system in figure 7 is a completion of the system in figure 5. By the installation of two change-over valves is an additional airway created. These airways called U_1 and U_2 between south orientated and north orientated rooms in a dwelling allow a transport of warmed up air from the southern rooms to the nothern parts. With quick adjustable fresh air heating systems the secondary air heater can be taken out of service and the assailed radiation heat on the southern part can be used because at the same time the heating equipment of quick adjustable electric heaters can be switched off.
So the irridiated solar energy can be extensivly used for room heating contrary to case 1 - 3.

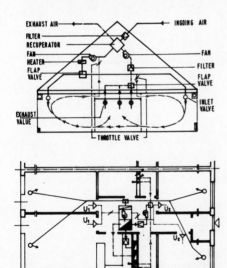

Figure 7

Upper part :
Presentation of total solar radiation an south orientated vertical and horizontal faces.
(BBC 1975 haft 8/9)

Conclusions

This essay should describe that until now common inactive heating systems like warm water heating without special room temperature control systems are not suitable for using directly irridiated solar energy in the rooms for example. Further it is not possible heat energy to transport into other rooms of a dwelling from the southern parts, which have the highest irridiation during heating season. Only with a ventilation system like the described fresh air heating system it is possible to realise a good room heating with a fair deduction system as well as saving energy supply system and the use of the irridiated solar energy.

VENTILATION THROUGH OPENINGS ON ONE WALL ONLY

P. R. WARREN

Building Research Establishment
Garston, Hertfordshire, United Kingdom

ABSTRACT

The main mechanisms giving rise to the natural ventilation of spaces with openings to outside air in one wall only are reviewed. Expressions which enable the magnitude of the ventilation rates due to each of these mechanisms to be calculated are derived from theoretical and wind tunnel studies. The results of measurements on a full-scale building are reported and compared with predictions.

NOMENCLATURE

A	area of opening (subscripts 1,2 see below) (m^2)
b	breadth of rectangular opening (m)
c	concentration (subscripts i,e)
f_1, f_2	functions of η
G	mass flow rate (kg/s)
g_1, g_2	functions of η_r
H	vertical distance between centres of two openings (m)
h	height of rectangular opening (m)
I	integral, defined in the text
k, k_p	wave number (= n/U_R ; = n_p/U_R) (m^{-1})
n, n_p	frequency, frequency at peak value of pressure spectrum (Hz)
p	pressure (subscripts 1,2,R) (N/m^2)
$p', \hat{p}$	fluctuating component of pressure; rms value of p' (N/m^2)
U	mean velocity (subscripts L,R,T see below) (m/s)
u	velocity in x-direction within mixing layer (m/s)
V	volume (m^3)
v	velocity in y-direction within mixing layer (m/s)
Ar	Archimedes number ($= \Delta\theta gh/\theta U^2$)
Cd	discharge coefficient
Cp	pressure coefficient (= $(p-p_o)/\frac{1}{2}\rho_o U_T^2$)
F	flow number (=Q/AU) (subscripts L,R,T indicate U_L etc)
Gr	Grashof number (= $\Delta\theta gh^3/\theta\nu^2$)
M	aspect ratio of rectangular opening (= h/b)
Re	Reynolds number (= UR/ν)
α	arbitrary constant relating to width of mixing layer

189

β   angle of flow direction in plane parallel to wall
γ   angle of wind direction (=0, perpendicular to wall)
Δ   difference between two values of same variable
ε   ratio of opening area (= A_1/A_2)
η, η_r similarity variable; value of η in plane of opening
θ   absolute temperature (deg K)
ν   kinematic viscosity (m^2/s)
ρ   density (kg/m^3)
φ   angle of opening of window 'vane'

SUBSCRIPTS

1,2   identification number of opening

i,e   value p and v at inner and outer edges of mixing layer

L,R,T   value of U, and dimensionless quantities Ar, F which depend upon U,
            local to the opening (L), at a chosen reference point remote from
            the building (R), and at a reference point in the free wind at a
            height equal to that of the building (T).

o   value of p and ρ taken for reference in defining Cp

INTRODUCTION

     Under the action of the steady pressures generated by wind and temperature
differences, air flows through openings, both purpose-made and adventitious, in
the outer skin and between spaces within a building. The general principles
underlying this form of ventilation were stated many years ago (1,2) and have
subsequently been developed and incorporated in computer-based methods for the
prediction of flow through buildings consisting of complex arrangements of
spaces (3,4). However there are situations when 'through' ventilation of this
type cannot occur, in particular when a space within a building is well-sealed
with respect to the rest of the building and has openings in one external wall
only. Typical examples are cellular offices or classrooms in which internal
doors are kept closed for reasons of privacy or noise, and for which the only
large openings available for ventilation in summer are in the one external
wall. Langdon and Loudon (5) have shown that such rooms tend to have higher
summer temperatures and give rise to greater user dissatisfaction than through-
ventilated rooms exposed to the same conditions of solar gain etc. The purpose
of the work described in part in this paper is to provide a basis for the
choice and sizing of openings to give the design summertime ventilation rate
in rooms of this type.
     The available agencies which can give rise to 'single-sided' ventilation
are:
  (i) Buoyancy, generally as a result of temperature differences
 (ii) Wind.
     Molecular diffusion could be included for the sake of completeness, but
a simple analysis indicates that its contribution to any exchange of air between
inside and outside of a space will be so small in comparison with other mechanisms that it may be neglected. Aspects of buoyancy and wind only will therefore
be considered.

BUOYANCY

Single Opening
     Consider an enclosed space, containing fluid at a density Δρ above the
fluid outside the space density ρ. The space is connected to outside by a
single rectangular opening of height h, and breadth b.   The difference in

weight will cause a flow into the enclosed space at the lower part of the opening and out at the upper part. The mass flow rate, G, is given by the following general expression,

$$G = \text{function } (b,h,\Delta\rho,\rho,\nu,g) \tag{1}$$

In continuing discussion the fluid will be taken to be air. Making the assumption that air is a perfect gas, then,

$$\frac{\Delta\rho}{\rho} = \frac{\Delta\theta}{\theta}$$

By dimensional analysis equation (1) may be written as,

$$\frac{G}{\rho bh} \left[\frac{\theta}{gh\Delta\theta}\right]^{\frac{1}{2}} = \text{function } \left[\left(\Delta\theta gh^3/_{\theta\nu}2\right), \left(\frac{h}{b}\right)\right] \tag{2}$$

The first independent dimensionless group is the Grashof Number, Gr, and the second is the aspect ratio, M, of the opening. If the temperature differences are small in comparison with the absolute temperatures, equation (2) may be written in terms of volume flow rate, and θ may be regarded as the mean between the internal and external absolute temperatures. The volume flow rate, Q, is therefore given by,

$$Q = A.\text{function}(Gr,M) . \left(\frac{g\Delta\theta h}{\theta}\right)^{\frac{1}{2}} \tag{3}$$

If the following assumptions are made,
 (i) viscous forces are small in comparison with those due to buoyancy, ie Gr is large, and
(ii) the vertical component of velocity at the opening is small in comparison with horizontal components ie the flow consists of two-dimensional horizontal layers,
then, as Shaw (6) and Brown and Solvason (7) have shown,

$$\text{function}(Gr,M) = \frac{1}{3}Cd . \tag{4}$$

Cd is the discharge coefficient for the opening, which if the latter is sharp-edged will approach the theoretical value of 0.61 at high Grashof numbers. Experimental results by Shaw confirm that this value is appropriate for openings typical of the size of open windows, for a range of $\Delta\theta$ less than 20 deg C.

Single Opening with a Vane
    The previous section applies to sliding windows. Many windows, however incorporate some form of opening component which opens at an angle to the plane of the wall, for instance, centre-pivoted and sidemounted casement windows. For convenience this component will be termed a 'vane'. It is not readily possible to deal with the flow that results from the presence of the vane, but its effect may be taken into account by including an additional factor in equation (3) as follows,

$$Q = \frac{1}{3}ACdJ(\phi) \left(\frac{g\Delta\theta h}{\theta}\right)^{\frac{1}{2}} \tag{5}$$

$J(\phi)$ must be determined experimentally.
    In order to obtain an indication of the form of $J(\phi)$, models constructed from plywood of horizontally centre-pivoted and side-mounted casement windows, of three aspect ratios, 0.86, 1.72 and 2.28, and width 0.5 m were set in the side of an otherwise totally sealed chamber of volume 18.5 m³. The temperature

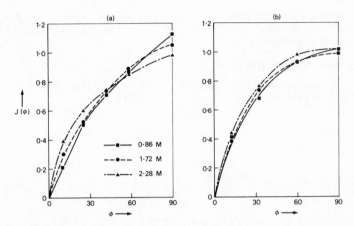

Figure 1 Variation of dimensionless flow rate, J(φ), with angle of opening, φ, for (a) side-mounted casement, and (b) centre-pivoted windows

was raised above that of the surrounding air by two oil-filled electrical heaters. Suitably shielded thermocouples were used to record temperatures within, and outside, the chamber. The ventilation rate was measured using a standard tracer gas decay method (8,9), using nitrous oxide, and an infra-red gas analyser to measure the gas concentration. The variation of $J(\phi)$ with ϕ is shown by Figure 1. The variation of $J(\phi)$ with aspect ratio, M, is small and largely within the expected experimental error. It is of interest to note that for both types of window that $J(\phi)$, and hence ventilation rate, is increased little by opening the window beyond approximately 60°.

Multiple Openings

Brown and Solvason (7) have noted that the analysis for a single opening can be extended to deal with openings of various dimensions separated by different vertical distances. The simplest arrangement consists of two openings of areas, A_1 and A_2, with their centres at a height H apart. The flow rate is then given by,

$$Q = \sqrt{2}A Cd \left[\frac{\varepsilon}{(1 + \varepsilon)(1 + \varepsilon^2)^{\frac{1}{2}}} \right] \left[\frac{g\Delta\theta H}{\theta} \right]^{\frac{1}{2}} \qquad (6)$$

If the areas are equal, ε equals 1 and equation (6) reduces to,

$$Q = \tfrac{1}{2}A Cd \left[\frac{g\Delta\theta H}{\theta} \right]^{\frac{1}{2}} \qquad (7)$$

WIND

The structure of the wind is complex, depending inter alia on the stability of the atmosphere and the roughness of the terrain. In general in the lower atmosphere the wind is turbulent and the mean value of its speed, its turbulence characteristics, and to a certain extent its direction vary with height. The resulting flow when the wind interacts with a bluff body, such as a building, is inevitably highly complex and can only be determined by wind tunnel or full-scale studies. Certain broad characteristics may, however be noted. Air deflected across the windward faces separates at sharp corners creating regions of reversed flow. Downstream a wake is formed. The surface of the building is subject to fluctuating pressures related not only to the turbulence in the wind upstream of the building but to the turbulence created by the building itself. Close to the surface the flow will be parallel to the surface, except at any stagnation points, although of course its speed and direction will fluctuate, particularly in regions of recirculation and on the leeward faces which will be affected by large scale eddies shed in the wake. The surface pressures may be charaterised by three quantities; a mean value, and the variance and spectrum of the fluctuating component. Under given conditions the mean pressure, the pressure fluctuations and the surface flow can give rise to an exchange of external and internal air in a space with openings on one wall.

Mean Pressure Distribution

This is the simplest situation to consider and applies when more than one opening is available. A difference in mean pressure at the openings will create a flow through the space. Again consider the simple situation of two openings, of areas A_1 and A_2 then the flow rate may readily be shown to be,

$$Q = ACdU_T(Cp_1 - Cp_2)^{\frac{1}{2}} \left[\frac{\varepsilon}{(1+\varepsilon)(1+\varepsilon^2)^{\frac{1}{2}}} \right] \qquad (8)$$

where $A = A_1 + A_2$. On rearrangement,

$$F_T = Cd\Delta Cp^{\frac{1}{2}} \left[\frac{\varepsilon}{(1+\varepsilon)(1+\varepsilon^2)^{\frac{1}{2}}} \right] \qquad (9)$$

If the areas are equal then $\varepsilon = 1$, and equation (9) reduces to,

$$F_T = 0.35Cd\Delta Cp^{\frac{1}{2}} \qquad (10)$$

Pressure Fluctuations

Given an opening of area A, variation in the external pressure at the opening will cause a change in internal pressure within the space. Since the space is sealed the density of air within the space will also vary about a mean value, ρ_o. If the process is considered to be adiabatic the rate of change of mass of the air within the space and in consequence the rate at which air is transferred through the opening is given by,

$$G = \frac{V\rho_o}{p_i\gamma} \cdot \frac{dp_i}{dt} \qquad (11)$$

where p_i is the pressure within the space. In order to make a simple estimate of the magnitude of the flow rate the following assumptions are made.

i) All of the pressure variation occurs at a single frequency, n. The amplitude of the fluctuating component of the external pressure at the opening, p_e', is therefore given by,

$$p_e' = \sqrt{2} \cdot \tfrac{1}{2}\rho_o U_T^2 \cdot \widehat{Cp}_e \qquad (12)$$

ii) The mass of external air which enters during each cycle mixes perfectly, and hence the outgoing air on the other half of the cycle contains 'contaminated' air from within the space.

iii) That the internal pressure follows the external pressure without any
lag or attenuation, ie the effects of inertia are ignored.

The use of these assumptions, together with equation (11) leads to the following
expression for an equivalent volume flow rate,

$$Q = \sqrt{2}\left[\frac{Cp\rho_o}{\gamma p_o}\right] \cdot nU_T^2 V \tag{13}$$

If n is taken to be equal to the peak frequency from the pressure spectrum,
n_p, use may be made of the relation,

$$n_p/U_T = k_p$$

where the wave number, k_p, will be independent of U_T, and, hence,

$$Q = \frac{1.41 V Cp\rho_o}{\gamma p_o} \quad k_p U_T^3 \tag{14}$$

It will be noted that the flow rate is independent of area and apparently
increases with velocity to the third power, both of which are contrary to
expectation and general observation. Cockroft and Robinson (10) report a
similar theoretical relationship using a more complex analysis of real
turbulence. They measured the turbulence characteristics in the flow upstream
of their test enclosure, which consisted of a sealed box, of dimensions
1.2 x 1.2 x 2.4 m, with a single square opening set up at approximately the
stagnation point, and related these by simple analysis to the fluctuating
pressure at the opening. Their measurements however indicated a linear
relationship between the ventilation flow rate and both air speed and opening
area.

Turbulent Diffusion
        Although the presence of turbulence at an opening is likely to provide a
mechanism, through diffusion, for the exchange of air within the space with
outside air, the complexity of the flows around buildings makes theoretical
analysis difficult. However the typical window dimension is generally very
much smaller than the typical dimension of a building. In consequence the flow
in the vicinity of an opening will be parallel to the plane of the wall.
Neglecting, initially, the fact that it is turbulent and considering the mean
flow at the opening itself, the flow resembles the 'mixing' layer formed when
a uniform stream is exposed to a region of zero velocity at one edge. Although
it is only a two-dimensional approximation it is suggested that this flow,
which is comparatively well understood be used as a first step in the determination of the turbulent exchange of fluid across the opening.
        Figure 2 shows the assumed pattern of flow. Outside of the mixing layer
the free stream has a uniform velocity equal to the mean velocity at the
opening, U_L. At the inner edge of the layer the longitudinal velocity, u,
is zero. At the magnitude of Reynolds number based on the opening dimensions
the flow will be turbulent within the layer. A number of theoretical solutions
based upon the classical phenomenonological theories of turbulence have been
derived for the distribution of mean velocity, u, and concentration of any
species, c, within the layer, notably by Tollmien (11) and Görtler (12). In
general good agreement with experimental results can be obtained with any of

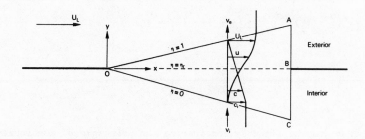

Figure 2 Schematic layout and nomenclature of two-dimensional mixing layer

these solutions since they each contain at least one arbitrary constant. All
are based upon the assumption of similar profiles for velocity and concentration.
Thus u and c may be expressed as functions of suitably chosen similarity
variable, η, where

$$\frac{u}{U_L} = f_1(\eta) \quad ; \quad \frac{c}{c_i} = f_2(\eta) \quad \text{and} \quad \eta = \frac{(y - y_i)}{\alpha x}$$

y_i is the value of y at the inner edge of the layer and is a constant which is
a measure of the rate at which the mixing layer spreads, and is defined so
that $\eta = 1$, when $y = y_e$ at the outer edge of the layer. c_i is the concentration
of some notional contaminant uniformly mixed with the air in the space. Referring
to Figure 2 the line OB is associated with a particular value of η, η_r. The
rate at which contaminant is transported across OB is equal to the rate at which
it is transported across, AB, which is given by,

$$\int_0^y uc\,dy = \alpha b \int_r^1 uc\,d\eta = \alpha b U_L c_i g_1(\eta_r),$$

where,

$$g_1(\eta) = \int_r^1 f_1(\eta) f_2(\eta)\,d\eta$$

Thus the enclosed space has a ventilation flow rate given by,

$$Q = \alpha A U_L g_1(\eta_r) \tag{15}$$

or, in dimensionless terms,

$$F_L = \alpha g_1(\eta_r) \tag{16}$$

Applying the principles of continuity of mass and momentum to the fluid
within the control volume OABCO leads to the following equation for the lateral
velocities, v_i and v_e, at the edges of the layer;

$$\frac{v_e}{\alpha U_L} = (1 - I_2) - \eta_r \quad ; \quad \frac{v_i}{\alpha U_L} = (I_1 - I_2) \qquad (17)$$

where

$$I_1 = \int_0^1 f_1(\eta) \, d\eta \quad ; \quad I_2 = \int_0^1 f_1(\eta)^2 \, d\eta$$

The value of η_r may be fixed by considering the principle of continuity applied to the enclosed space. If it is completely sealed OB is a mean streamline and there can be no net transfer of fluid across it, and $\eta_r = \eta_{ro}$, say. Thus the flow returning into the space across BC must be equal to the entrained flow along the edge OC, thus,

$$\int_0^b v_i \, dx = \alpha b U_L \int_0^{\eta_{ro}} f_1(\eta) \, d\eta = \alpha b U_L g_2(\eta_{ro}) \qquad (18)$$

where,

$$g_2(\eta_r) \text{ is defined as } \int_0^{\eta_r} f_1(\eta) d\eta$$

Thus,

$$g_2(\eta_{ro}) = \frac{v_i}{\alpha U_L} = (I_1 - I_2)$$

and hence η_{ro} may be determined and the position of the layer fixed.

In order to proceed further it is necessary to propose suitable expressions for the functions $f_1(\eta)$ and $f_2(\eta)$. There is little to choose for this present purpose between the various theoretical solutions and for ease of manipulation the functions proposed by Abramovitch (13) will be used:

$$f_1(\eta) = \left[1 - (1-\eta)^{3/2} \right]^2 \quad ; \quad f_2(\eta) = (1-\eta)$$

The functions $g_1(\eta_r)$ and $g_2(\eta_r)$ and the integrals I_1 and I_2 may now be determined, leading to,

$$\eta_{ro} = 0.62 \quad ; \quad v_i = 0.134 \, \alpha U_L \quad ; \quad g_1(\eta_{ro}) = 0.056.$$

It remains to assign a suitable value for the constant α. Liepmann and Laufer (14) determined experimentally a value of 0.23 for a uniform steady free stream, almost free of turbulence. Abramovitch (13) quotes values in the range 0.2 to 0.3, and also notes that α is increased by the presence of turbulence in the free stream. Bearman (15) examined the base pressure behind flat plates normal to a uniform stream with varying levels of turbulence. He suggested that the reduction in base pressure with increasing level of the turbulence, was due to thickening, and consequent enhanced entrainment into the mixing layers springing from the edges of the plate. In the context of buildings the free stream will be highly turbulent, possessing a wide range of eddy sizes, and it is reasonable to expect that the layer would be considerably thickened, with higher ventilation rates as a consequence. However it is also probable that the profiles of concentration and velocity will be distorted making the functions chosen for $f_1(\eta)$ and $f_2(\eta)$ less valid. Further, the above analysis was based on the assumptions of two-dimensional flow whereas there will almost certainly be very substantial three-dimensional effects in practice due to the low aspect ratio of most windows and to the fact that the flow may well be at an angle to the sides of the opening rather than perpendicular as assumed in this analysis. However despite these limitations, in the absence of further knowledge, use of the two-dimensional model of the flow will be continued and these various effects will be assumed to be included within the variation of the parameter α.

Some information on the effect of high levels of free stream turbulence may be derived from the results of Brown and Solvason (7) who included within their studies of heat transfer through rectangular openings in vertical, insulated partitions, additional, but limited, experiments to determine the effect of a flow parallel to the partition. The flow was created by conventional air-conditioning fans and would be expected to be highly turbulent. For high velocities the rate of heat transfer was independent of Grashof Number indicating that it was dependent on the flow only. These results have been employed to deduce values of F_L for the three sizes of openings used. Using the previously determined value for η_{ro} of 0.62, and $g_1(\eta_{ro})$ of 0.056, in conjunction with equation (16) values of α have been determined and are listed with their corresponding values of F_L in Table 1.

TABLE 1   Values of F_L and determined from Brown & Solvason (7)

Opening Dimensions (m)	F_L	α
0.15 x 0.15	0.130	1.84
0.23 x 0.23	0.096	1.71
0.30 x 0.30	0.090	1.60

Single Opening with a Vane

The mechanisms due to wind discussed so far have been applicable to plane openings. A vane will interact with the local airflow creating an exchange of air across the opening. Again the complexi nature of the flow precludes theoretical analysis. Using dimensional analysis the main independent variables may be identified.

$$F_L = \text{function } (\phi, \beta, M, Re_L).$$

The sense of ϕ and β is indicated in Figure 3.

Comparison of Mechanisms

Substitution of typical values into the expressions derived in previous sections (except for 'vane' effect which is discussed later) of typical values of the main variables indicates that they are all approximately of the same magnitude, except for the effect of pressure fluctuation. It is useful therefore to consider the simultaneous action of the main mechanisms. In this paper, however only one will be considered.

Turbulent Diffusion and Temperature Difference

If the additional variable, U_R, a reference wind speed is included with the other variables in equation (1) a similar application of dimensional

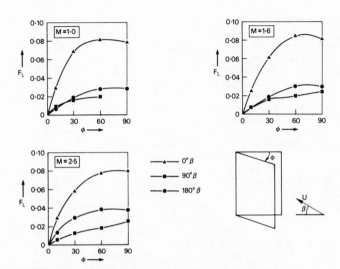

Figure 3 The variation of dimensionless flow rate, F_L, with angle of opening, ϕ, and flow direction, β, for side-mounted casement windows of aspect ratios, M 1.0, 1.6 and 2.5

TABLE 2   A comparison of the orders of magnitude of mechanisms for ventilation of spaces with openings in one wall only

Mechanism	Q(l/s)	Assumptions made in calculating, Q
TEMP. DIFFERENCE		
(1) single opening	40–120	$A=1.0$ m^2;$1<\Delta\theta<10^{\circ}$C;h=1.0 m;$\theta=290^{\circ}$K
(2) two openings	70–20	$A=1.0$ m^2;$1<\Delta\theta<10^{\circ}$C;H=1.5 m;$0=290^{\circ}$K
Pressure fluctuation	0.1–2.0	$A=1.0$ m^2;$2<U<10$ m/s;Cp=0.2;V=40 m^3
Mean pressure diff.	140–680	$A=1.0$ m^2;$2<U<10$ m/s;Cp=0.1;
Turbulent diffusion	200–1000	$A=1.0$ m^2;$2<U<10$ m/s;$F_L=1.0$;
'Vane' effect	200–1000	$A=1.0$ m^2;$2<U<10$ m/s;$F_L=1.0$;$\phi=60^{\circ}$

analysis yields the following,

$$F_R = \text{function } (M, Re_R, Ar_R) \tag{19}$$

If we assume that the aspect ratio, M, and the Reynolds number, Re_R will not be important. In the case of the Reynolds number this is effectively reiterating assumptions that have already been made when discussing the separate action of these two mechanisms. Thus,

$$F_R = \text{function } (Ar_R) \tag{20}$$

When temperature difference is the predominant effect, ie when Ar_R is large the ventilation flow rate will tend to that predicted by equation (5), which on division by U_R gives,

$$F_R = \frac{1}{3}J(\phi)Cd.(Ar_R)^{\frac{1}{2}} \tag{21}$$

When wind dominates, $Ar_R \to 0$, and F_R becomes independent of Ar_R. This suggests a useful basis of judgement in the analysis of measurements made in full size buildings, where control over the main variables, wind speed, direction and temperature difference is not possible. To isolate those results mainly influenced by wind the results may be plotted in the form F_R against $Ar_R^{\frac{1}{2}}$. Measurements dominated by temperature will tend to lie along the straight line defined by equation (21), whereas those largely due to wind will lie between the F_R axis and the line, equation (21).

WIND TUNNEL STUDIES

Experimental Arrangement

The working section of a small open jet wind-tunnel, described by Sexton (16), was altered so that it was totally enclosed with height 0.8 m and width 1.1 m. A small test chamber was constructed to have one wall in common with the side wall of the wind-tunnel. Its dimensions were 0.5 x 0.5 m in section. The depth away from the tunnel wall could be varied between 0.5 and 1.0 m, but for the experiments described here it was kept at 0.5 m, giving a cubic chamber. The wall common to both the tunnel and the chamber was so constructed that panels containing openings of various types could be inserted flush with the tunnel wall and sealed to prevent any leakage. These panels were square, and so enabled in the case of model casement windows, three settings of the angle β, 0.90 and 180°. Ventilation rates in the chamber were measured using conventional tracer gas techniques, using infra-red gas analyser to monitor gas concentration.

Plane Openings

The first column of Table 3 lists the sizes of plane openings which were tested. In the case of the slot openings the longer edge was arranged perpendicular to the flow. Measurements of ventilation rate were made for 5 or 6 values of tunnel speed, in the range 1.5 to 5 m/s. The results for each opening tested were invariably linear indicating the absence of scale effect and the relative unimportance of Reynolds number. The results are listed in Table 3. It is apparent that the square openings have consistently higher values of F_L than those for the slots, which lie much closer to the theoretical value for a non-turbulent free stream of 0.013. One likely explanation is that the slot is much closer to the two-dimensional assumption made in deriving the theoretical values, whereas the square would be expected to generate a fairly three-dimensional flow, giving rise to enhanced exchange of air between the chamber and the tunnel. The results for the slots show a peak at width of 54 mm followed by a steady reduction thereafter as the width increases. The peak may be due to upstream boundary layer on the tunnel wall or possibly some form of resonance. The steady reduction can in part be explained by the induced rotational flow set up in the chamber as the slot width increases. The longitudinal velocity created at the inner edge of the mixing layer can be shown theoretically to reduce entrainment, and as a consequence the exchange of air.

It was not possible to model the turbulence likely to occur in full scale situations, but the effect of increasing the turbulence in the wind tunnel stream was investigated. A biplane grid consisting of 46 x 7 mm slats (set with their larger side perpendicular to the flow), in a square pattern at 148 mm centres was inserted in the working section 1.5 m upstream of the centre of the openings in the chamber wall. The grid was sized to give a lateral scale of turbulence $L_y(u)$ at the test chamber opening of 25 mm, using the results of Baines and Peterson (17). The intensity of the longitudinal component of turbulence was measured as 9.0%. The results for the measurements are given in column 3 of Table 3 and may be compared with those without enhanced turbulence. The overall effect is to increase the value of F_L by some 60% for the slots and approximately 40% for the square openings, reflecting the higher contribution of three-dimensional effects to the exchange rate in the case of the latter.

TABLE 3   The effect of a turbulence-producing
grid on a flow number F_L

Opening type and size	F_L	
	No grid	Grid
SQUARE		
0.127 x 0.127 m	0.025	0.035
0.178 x 0.178 m	0.026	0.035
0.254 x 0.254 m	0.023	0.034
SLOT		
0.025 x 0.508 m	0.014	0.023
0.051 x 0.508 m	0.019	0.029
0.102 x 0.508 m	0.017	0.025
0.178 x 0.058 m	0.015	0.022
0.254 x 0.508 m	0.014	0.022
0.330 x 0.508 m	0.015	0.022
0.406 x 0.508 m	0.013	0.023
Free stream turbulence – u/U_L	0.8% (meas)	9.0% (meas)
$L_y(u)$	–	24 mm (calc)

Single Opening with a Vane
      The effect of an opening vane was studied by inserting small models of
openable casement windows in the test chamber wall. Three model windows were
used, each of area 0.026 m^2, but with aspect ratios of 1.0, 1.6 and 2.5.
Measurements of ventilation rate were made for a range of wind tunnel speeds
from 1.0 to 5.0 m/s, for the three values of β, 0, 90 and 180°, and, at each
of these, for values of window opening, φ, of 10, 30, 60 and 90°. Again, as for
the plain opening, the measured ventilation rates varied linearly with flow
speed indicating the absence of Reynolds Number effects and giving confidence
inthe application of the results to the full scale. Figure 3 summarises the
results. Substantially higher values of flow number occur for β = 0° than for
β = 90° or 180°. A possible explanation is that at β = 0° the exchange of air
is entirely due to the three-dimensional flow created by the presence of the
vane, whereas for the other angles the exchange depends to a certain extent
upon turbulent mixing processes akin to those previously discussed for the plane
openings. This would certainly be expected to be the case for β = 90°, φ = 90°,
when the vane should have little effect, and indeed, the values of F_L are close
to those found for the square plane openings, of approximately 0.025. It is
possible therefore that in a flow with higher levels of turbulence values of F_L
for β = 90 and 180° would be considerably increased. It is of interest to note
that F_L is close to its maximum value for φ ⪌60°, and that F_L and, in consequence
ventilation rates are most sensitive to the angle of window opening at low values
of this parameter.

Experimental Arrangement
      Field measurements of ventilation rate were made in two test rooms in the
single storey building shown in plan in Figure 4a. Both rooms were sealed.
Room A was fitted with a pair of vertically sliding windows, and Room B with

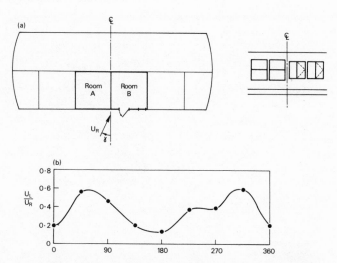

Figure 4(a) Plan view and window arrangement of the experimental building
(b) Variation of local flow velocity at the windows of room A with wind direction, γ

two side-mounted casement windows, shown diagramatically in Figure 4a. Ventilation rates were measured using standard tracer gas decay techniques, with
nitrous oxide as the tracer and an infra-red gas analyser to monitor gas
concentration. Simultaneously with each measurement of ventilation rate the
wind speed, U_R, at a height of 10 m, wind direction, γ, and temperatures inside
and outside of the rooms were recorded. The expected accuracies of measurement
were $\pm 10\%$ for ventilation rate, and ± 0.5 m/s for wind speed and $\pm 0.2^\circ$ C for the
temperatures.

Vertically sliding window - Room A
    With a window opening of 0.15 m, giving a single rectangular area of 0.15 m^2,
measurements were made for a wide range of U_R, γ, and $\Delta\theta$. In order to compare
the full scale results with the values predicted for particular mechanisms it is
necessary, where possible, to separate the results dominated by particular
mechanisms from each other. In order to separate wind and temperature the
procedure suggested in a previous section has been adopted and the results have
been plotted in the form F_R v $Ar_R^{\frac{1}{2}}$ in Figure 5a. By inspection those results
predominantly affected by wind were isolated, and plotted in Figure 5a, against
wind direction, γ.
    Measurements were also made with the same open area, but arranged in the
form of top and bottom openings of equal area, and separated by a vertical
distance of 1.16 m. These results were also analysed by plotting them in the
form F_R v $Ar_R^{\frac{1}{2}}$, in Figure 5b. As might be expected there are rather more results
in this instance clearly due mainly to temperature difference. The results
predominantly affected by wind are plotted against wind direction in Figure 5b.

Side-mounted casement window - Room B
    Measurements were made with a single casement window, of aspect ratio 1.9,
area when fully open of 0.60 m^2, and height 1.06 m, at the following values of
ϕ; 4, 12, 20, 27, 34 and 67°. Again the results due mainly to wind have been
separated from the remainder and plotted against wind direction. Figure 6 shows
the results.

Discussion of Results
    In order to relate the full scale results to the theoretical and model

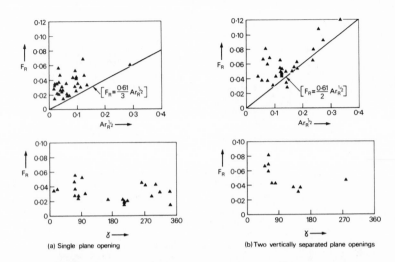

Figure 5 Variation of dimensionless flow rate, F_R, with Archimedes Number, $Ar_R^{1/2}$, and wind direction, γ, for (a) a single plane opening, and (b) two vertically separated plane openings

studies it is necessary to relate the reference wind speed U_R, and wind direction γ, to the local flow speed U_L and direction, β, in the region of the window. In order to do this a 1:25 scale model of the building was constructed and set up on the wind-tunnel turntable. A graded rod array was used to simulate the wind speed gradient, giving a 1/7 power law profile for mean velocity, but no attempt was made to simulate correctly atmospheric turbulence. It was not anticipated, however that the mean values of U_L and β would be affected by this omission. The local airspeed at the position of the sliding windows was measured using a DISA 55A01 hot wire anemometer, carefully aligned to the flow direction β, determined beforehand with a wool tuft. The wind tunnel equivalent to U_R was measured at the appropriate scaled height, upstream of the model. The ratio U_L/U_R so obtained is shown in Figure 4b as a function of wind direction γ. Using this, values of F_L have been determined for the results dominated by wind. The mean for the single opening is 0.105 (standard deviation; 0.030, sample size; 29) and for the double opening is 0.115 (standard deviation; 0.025, sample size; 13). The difference between these is not statistically significant. In view of the errors both in the full scale measurements and in the wind-tunnel determination the large spread in the calculated value of F_L is not unexpected. However the mean values are much higher than the theoretical values derived from the simple mixing layer model and close to those derived from Brown and Solvason (7). On the evidence of the wind-tunnel tests on the model test chamber the major reasons for this difference are the three-dimensional nature of real flows and the presence of high levels of turbulence. Figure 7 shows F_L against $Ar_L^{1/2}$ for <u>all</u> of the results for the double opening. Also included is the straight line representing ventilation due to temperature difference alone. At high values of $Ar_L^{1/2}$ the results group around this line; at low values the results deviate and tend to become independent of the Archimedes number, just as suggested earlier when the simultaneous action of wind and temperature difference was discussed. The result also implies that given the appropriate data, or design conditions, whichever of the two effects is calculated to give the largest ventilation rate may be assumed to be acting alone.

The results for the single side-mounted casement window have been similarly expressed in terms of the local windspeed U_L. However the wind-tunnel studies β must be taken into account. Direct comparison is restricted by the experimental arrangement to the three values of β; 0, 90 and 180°. From the 1:25 scale model these were found to occur for the following ranges of γ:

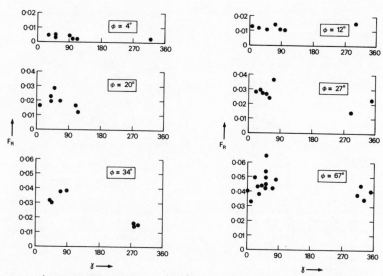

Figure 6 Variation of dimensionless flow rate, F_R, with wind direction, for a single side-mounted casement window at angles of opening, 0, of 4, 12, 20, 27, 34 and 67°

(i) $\beta = 0°$; $20 < \gamma < 70$
(ii) $\beta = 90°$; $150 < \gamma < 210$
(iii) $\beta = 180°$; $290 < \gamma < 340$

Unfortunately the results are not evenly distributed with respect to γ and the majority lie within range (i) corresponding to $\beta = 0°$. However the results within this range have been averaged for each of the six values of ϕ and plotted for comparison, in Figure 8, against the wind-tunnel result for a side-mounted casement with an aspect ratio of 1.6, the nearest to that of the full size window. The agreement is remarkably good and lends confidence to the use of the wind tunnel technique for further investigations, provided that the ventilation is occurring due to the presence of the vane alone and inde-pendent of turbulence levels in the free stream. This point is underlined by consideration of the limited number of results within range (iii). At $\phi = 67°$ the average of these results gives a value for F_L of 0.07, greatly in excess of the value from the wind-tunnel studies of 0.03. This is in accord with the previous suggestion that free-stream turbulence could make a contribution for angles of β other than $0°$, in respect of this type of window. In fact the value of F_L is very close to the value for $\beta = 0°$ and taking this, together with the full scale results for the plane openings, indicates a value of F_L of approximately 0.1 which applies both to sliding windows and side-mounted casements.

To make use of this result for design purposes also requires the means of relating the surface air speed U_L to an appropriate reference wind speed, such as that at 10 m available from Meteorological Office measurements. There is a dearth of information on U_L and in order to provide some guidance, limited tests were made using a simple cube model in the wind tunnel. Air speeds at two points, one third and two thirds of the height of the cube on the centre-line of one of the faces were measured and related to a reference air speed measured upstream at a height equal to that of the model. The ratio of U_L to U_T varied from 0.2 to 0.8 as the orientation relative to the flow was altered. For design purposes it would seem wise to use the lowest value.

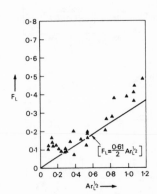

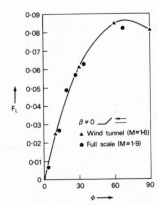

Figure 7 Variation of dimensionless flow rate, F_L, with Archimedes Number, $Ar_L^{1/2}$, for two vertically separated plane openings

Figure 8 Comparison of full-scale and wind tunnel model results for the variation of dimensionless flow rate, F_L, with angle of window opening, ϕ, for a side-mounted casement window

This then leads to the following simple formula for the ventilation flow rate for a single opening when the main mechanism is wind;

$$Q = 0.02 \, A \, U_T \tag{22}$$

The action of mean pressure difference if more than one opening is available will be to increase or give a higher flow rate than that given by equation (22). This equation will there tend to give the minimum likely rate and therefore the most suitable for design purposes since this fixes a maximum required area which can always be reduced.

CONCLUDING REMARKS

1   An analysis of the mechanisms that can give rise to the natural venti-
    lation of rooms with openings on one side only indicates that the following
    are the most important:
    a) Difference between internal and external air temperature
    b) Turbulent diffusion
    c) The interaction of an opening vane with the local air flow
    d) Mean pressure difference acting across openings, if more than one
        is present.

2   Measurements on a full scale building have lead to a simple formula
    which gives the minimum ventilation rate due to (b)and (c) above.

3   There is evidence from the full scale measurements that when temperature
    and wind are acting simultaneously the ventilation rate will be equal to
    the larger of the two individual calculated.

4   A simple model technique for investigating the performance of windows
    using a wind-tunnel has been developed and gives excellent results for
    exchange processes dominated by opening vane. The technique needs to
    be improved to deal with situations in which turbulence also plays a
    part.

5   A simple theoretical model has been developed to predict exchange rates
    due to turbulence. Although some aspects are oversimplified it is
    intended to use this model to assist in the analysis of the complex
    interactions between the mechanisms listed above.

6   The work described in this paper represents part of a larger study which
has included measurements on two further full scale buildings, studies
on combinations of windows and extension of the analysis to cover combinations of the main mechanisms, and which will be published later.

ACKNOWLEDGEMENTS

     The work forms part of the research programme of the Building Research
Establishment of the Department of the Environment and this paper is published
by permission of the Director. The author wishes to acknowledge the assistance
of his colleagues who made contributions to the work described in this paper,
in particular, Mr M Jenkins who was responsible for the full scale measurements,
Mr B C Webb who carried out the laboratory studies on buoyancy, and Mrs L Parkins
who carried out the wind tunnel measurements and assisted in many other ways
including the analysis of the experimental data.

REFERENCES

1   Shaw, W.N.   1907. Air Currents and the Laws of Ventilation. C.U.P.

2   Dick, J.B.   1950. The Fundamentals of Natural Ventilation of Houses.
     J.Inst.Heat.Vent. Engrs. $\underline{18}$:123-134.

3   Jackman, P.J.   1969. A Study of the Natural Ventilation of Tall Office
     Buildings. H.V.R.A. Laboratory Report No. 53.

4   Honma, H. 1975. Ventilation of Dwellings and its Disturbance. FAIBO
     Grafiska. Stockholm.

5   Langdon, F.J. and Loudon, A.G. 1970. Discomfort in Schools from Over-
     heatin in Summer. Building Research Station Current Paper, CP19/70.

6   Shaw, B.H. 1972. Heat and Mass Transfer by Natural Convection and
     combined Natural Convection and Forced Air Flow through large Rectangular
     Openings in a Vertical Partition. Inst.Mech.Engrs. Conference on
     Combined Forced and Natural Convection. Manchester, September 1971,
     Conference Volume C.819.

7   Brown, W.G. and Solvason, K.R. 1962. Natural Convection through
     Rectangular Openings in Partitions - Vertical Partitions. Int. J. Heat
     and Mass Transfer. $\underline{5}$:859-868.

8   Dick, J.B. 1950. Measurement of Ventilation Using Tracer Gas Technique.
     Heating, Piping & Air Conditioning. May:131-137.

9   Hitchin, E.R. and Wilson, C.B. 1967. A Review of Experimental Techniques
     for the Investigation of Natural Ventilation in Buildings. Building
     Science. $\underline{2}$:59-82.

10   Cockroft, J.P. and Robinson, P. 1976. Ventilation of an Enclosure
     through a Single Opening. Building and Environment. $\underline{11}$:29-35.

11   Tollmien, W. 1945. Calculation of Turbulent Expansion Processes. NACA
     TM No.1085.

12   Görtler, H. 1942 Berechnung von aufgaben der Frein Turbulenz auf eines
     Neuen Nahrengsansatzes. ZAAM. $\underline{22}$:244-254.

13   Abramovitch, G.N. 1963. The Theory of Turbulent Jets. M.I.T. Press,
     Cambridge, Massachusetts.

14   Liepmann, H.W. and Laufer, J. 1947. Investigation of Free Turbulent
     Mixing. NACA TN 1257.

15   Bearman, P.W.   1971.   An Investigation of the Forces on Flat Plates
      Normal to a Turbulent Flow.   J. Fluid. Mech.   46:177-198.

16   Sexton, D.E.   1968.   A Simple Wind Tunnel for Studying Air-Flow round
      Buildings.   Building Research Station Current Paper CP69/68.

17   Baines, W.D. and Peterson, E.G.   1949.   An Investigation of Flow through
      Screens.   Iowa Institute of Hydraulics, July 1949.

PREVENTION OF UNWANTED AIRFLOW BETWEEN AREAS HAVING DIFFERENT HYGIENIC STANDARDS IN HOSPITALS

H. ESDORN, K. GIESE, AND M. SCHMIDT

Hermann-Rietschel-Institut für Heizungs- und Klimatechnik
Technische Universität Berlin
Berlin, Germany

1. Introduction

Today many production and working procedures require environmental conditions which are not given without special measures being taken. Apart from particular requirements with regard to air temperature and humidity, the need for air to be free from particles, germs, gases or vapors is of increasing importance. To meet demands for special air standards, air conditioning systems are necessary for the areas concerned. One problem are the connections with surrounding areas, because of the necessity to prevent the transmission of air between the areas when doors are opened. This problem is of especial importance in hospitals because hygienic standards must be observed in order to reduce the transmission of contaminants and thereby to lower the danger of infection.

As part of a research programme by the department of environmental engineering at the Technical University Berlin on hospital design (financed by the Deutsche Forschungsgemeinschaft), an experimental and analytical investigation of ventilation systems for the reduction of transmission of airborne contaminants between areas of different hygienic standard was carried out at the Hermann Rietschel Institut for heating and air-conditioning engineering.

2. Definitions

An air lock is a facility to connect two areas of different air standards, which reduces the transmission of air between the areas. An air lock consists of at least one chamber with two doors which cannot be open simultaneously. According to the degree to which they are equipped with ventilation systems, one has to distinguish between passive and active air locks [1], [2]. In passive air lock systems the air lock chamber itself has no ventilation system, while at least one of the two areas has a ventilation system which keeps a certain pressure in the area

207

and thereby fixes the direction of flow through the air lock
(see Fig.1); In active air lock systems the chambers are equipped
with supply - and/or exhaust air systems. The two areas need not
be kept under fixed pressures (see Fig.2). For active air locks
one has to distinguish between pressurized air locks (volume
rate of supply air produced by the ventilation system $\dot{V}_S$ bigger
than the volume rate of exhaust air produced by the ventilation
system $\dot{V}_E$) and depressurized air locks ($V_E > V_S$). Air locks with
balanced volume rates for supply and exhaust air ($\dot{V}_E = \dot{V}_S$) are
called pressure-balanced air locks. The air-flow through the
door joints produced by the ventilation system is called air
barrier flow. Air locks can also be fitted with systems for air-
scavenging (in addition to the scavenging caused by the air
barrier flow). The pressure distribution of an air lock and the
two areas determines not only the value but also the direction
of the reduction of air transmission. If the separation of the two
areas is equal in both directions, the system should be called
non-directed, otherwise directed.

3. Description of Investigations

The transmission of air between two areas through an air lock
depends on a large number of parameters with mutual interference.
The parameters can be divided into those related to the design,
which are

> Volume of the air lock chamber,
> Design of the doors,
> Number of chambers,
> Pressure distribution,
> Scavenging air distribution,

and those related to the operation of an air lock

> door opening time,
> duration of dwell,
> passing frequency,
> passing direction.

The influence of the parameters on the air transmission was
investigated partly by experiment and partly by computer
calculation. The experiments were limited to those aspects which
can not be calculated sufficiently. Mainly these were all factors
which influence the transmission of air through a single door,
like the design of the door, the passage of persons through it
and the volume rate of flow through the open door resulting
from the difference in pressure on the closed door. The
investigation of the resulting effect of an air lock based on
the transmitted volume through a single door was done
numerically.

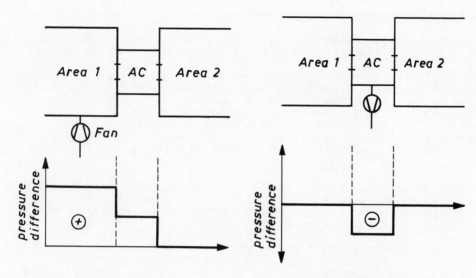

Fig.1   Passive Single-
          Chamber Air Lock
          (AC-Air lock chamber)

Fig.2   Active Depressurized
          Single-Chamber
          Air Lock

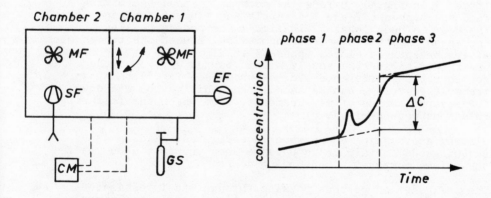

Fig.3   Plan View of Testing
          Arrangement
          MF - Mixing fan
          SF - Supply fan
          EF - Exhaust fan
          GS - Tracer Gas Supply
          CM - Tracer Gas
                Concentration Meter

Fig.4   Progress of Tracer
          Gas Concentration in
          Chamber 2

3.1. Experimental Investigations

The transmission of air through a door with persons passing through was investigated on a full scale model of an air lock.

The transmitted volume was measured by use of a tracer gas method. Nitrous oxide (N_2O) was taken as tracer gas because it has a good absorption in the infrared range and can therefore be measured even in low concentrations by the use of an infrared analizer. In the atmospheric air, there is no nitrous oxide so that one need not make reference measurements in the supply air. The method of testing will be explained with the aid of figs.3 and 4:

While the door is closed a constant volume rate of flow $\dot{V}$ is produced by the supply fan (SF) and the exhaust fan (EF) with the direction from chamber 2 towards chamber 1. Chamber 1 is supplied at the same time with a constant rate of tracer gas. A mixing fan (MF) in chamber 1 mixes the tracer gas with the air. When a steady state is reached there is a constant tracer gas concentration in Chamber 1. A flow of tracer gas through the door joint into chamber 2 is prevented by the counterflowing air barrier. If the final gas concentration is reached in chamber 1 the counterflowing air $\dot{V}$ and the tracer gas are turned off and these inlets are closed. The two mixing fans homogenize the concentration distribution, but because of the room air movement there is a small air flow through the door joint and thereby a slow concentration increase in chamber 2 (phase 1 in fig.4). Before the door is opened, the mixing fans are turned off and the supply and exhaust fans on. While the door is open (phase 2 in fig.4) the person passing through causes an intensified air transmission between the two chambers. After the door is closed (phase 3 in fig.4) the mixing fans are turned on and the supply air is turned off. From the concentration difference $\triangle c$ measured (see fig.4) the transmitted volume can be calculated. This method was used for several series of measurements, from which the following relations were derived.

Sliding and turning doors were investigated. Sliding doors are advantageous for air locks because they transmit about half the volume compared with turning doors.

The influence of persons passing through was investigated in several tests with varying directions and in additional tests without persons passing through. A clear relationship between the transmitted volume and the relative direction between the passing and the flow of the air barrier could not be derived from the data measured. For the sliding door the persons passing through produced a higher transmitted volume compared with the tests without persons. This could not be measured for turning doors, which is due to the dominating influence of the opening and closing process of the turning door on the room air movement.

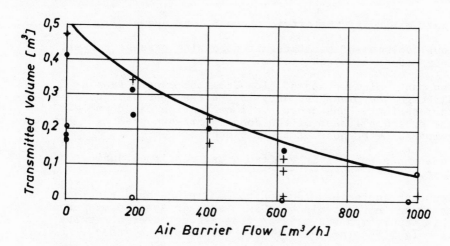

Fig.5   Air Volume Transmitted through Sliding Door per Passage
         (Door-opening time 8 sec.)
         • person passing from unclean to clean area
         + person passing from clean to unclean area
         o without person passing

The door opening times were varried between 8 and 30 seconds.
With longer opening times one got higher but more scattered
volume rates.

The volume rate of flow of the air barrier through the open door
was varied between 0 and 1000m³/h. Fig.5 shows a relationship
between the volume transmitted through a sliding door and
the volume rate of flow of the air barrier. With a sufficient
volume rate of flow a considerable reduction of the transmission
of air is possible against the direction of the air barrier flow
while persons are passing through a door.

For equal pressures and temperatures on both sides of a sliding
door the transmitted air volume produced by a single passage at
normal walking speed is about 0,5m³. Pressure differences varying
with the height due to different temperatures on the two sides
of the door produce a higher transmission rate if there is no
air barrier.

3.2. Numerical Investigations

A numerical model was developed to simulate the operation of air
locks. The results of the tests of which an account has already
been given were used to describe the air transmission at the
doors.

While developing the algorithm the following simplifying
assumptions were made:

a) Ideal mixing of the air in the air lock chambers.

b) Equal volumes of air transmitted during passage for all doors of an air lock system.

c) Constant volume rates of flow from the supply fans although the pressures vary in the chambers because of the opening of the doors. This assumption can be made if the pressure losses of the door joints are small compared with those of the air duct system.

d) Constant pressures in the two areas mentioned during operation of the air lock.

e) Constant contaminant concentration in one area (unclean area) and no contamination in the other area (clean area). This assumption was made to keep the calculated results free from the influence of the volume of the areas.

As the criterion for the quality of an air lock the volume rate of air flow from the unclean area to the clean area was calculated (average contaminant flow $\dot{X}$).

Although the effect of air lock systems depends on a large number of parameters the investigations were limited to the most important parameters in order to make the effect of the specific influences clearer. Except where otherwise stated, the calculations were made for the following conditions

a) The volume rate of flow through the door joint at a closed door, based on the tests of sliding doors, was calculated as [1]

$$\dot{V} = 0.0084 \cdot \Delta p^{0.61}$$

with

$\dot{V}$ - Volume rate of flow $\lfloor m^3/s \rfloor$.

Δp - Pressure difference $\lfloor Pa \rfloor$

b) The door opening time was 8 sec., the duration of dwell, i.e. the time a person is in a chamber while both doors are closed, was 2 sec. These times enable a person to pass through the air lock at slow walking speed.

---

1) Depends largely on the design of the door.

c) The Volume of the chambers was set $8.75m^3$ in accordance with the full scale model used for the tests.

d) The transmitted air volume for all doors was set to $0.5m^3$ per passage. This is a maximum value for sliding doors (see fig.5).

e) The calculations were made for a passage frequency of 20 per hour.

3.2.1. Calculated Results for Single Chamber Air Locks

3.2.1.1. Influence of the Door Design
======================================

As the test results had shown, the door design and the volume rate of air barrier flow have a large influence on the transmitted volume at a door. This influence on the resulting effect of the air lock systems is shown for two examples in fig.6. The average contaminant flow $\dot{X}$ is plotted against the transmitted volume of a single door. The worse separation effect of either pressurized or depressurized active air locks without additional air scavenging compared with the passive air lock is caused by the fact that in pressurized or depressurized air locks either contaminated air from the air lock is blown into the clean area or additional contaminated air is sucked into the air lock from the unclean area. In passive systems the pressure differences always cause a flow in the right direction. The magnitude of the possible concentration difference between two areas can be shown by an example:

An operating theatre with a volume of $100m^3$ is fitted with a ventilation system which produces 30 air changes per hour. The operating theatre can be entered through a passive single chamber air lock (see Fig.1), which is passed through 20 times per hour. If the air lock is fitted with sliding doors the contaminant concentration in the operating theatre is about 0.03% of the concentration in the adjacent area. If the air lock is fitted with turning doors (about double the transmitted volume) the final concentration in the operating theatre is about 0.1%, more than three times the value mentioned before.

3.2.1.2. Influence of the Frequency of Passage
==

The frequency of passage has an influence on the effect of air locks for two reasons. Firstly, there is only a transmission between the two areas because of the opening of the doors; secondly, the time for scavenging of the chambers is determined

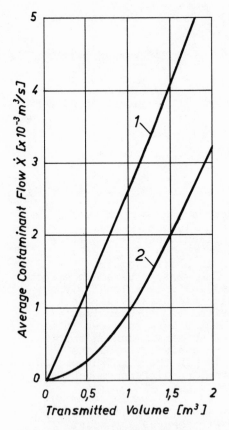

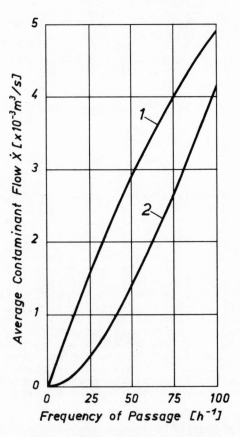

Fig.6 Average Contaminant
     Flow Related to the
     Volume Transmitted at
     a Single Door

Fig.7 Average Contaminant
     Flow Related to the
     Frequency of Passage

1 - Active single-chamber air lock without additional air
    scavenging
2 - Passive single-chamber air lock

by the frequency. The contaminant level in the chambers at the
beginning of a passage, which has an influence on the average
contaminant flow, depends on the time between two passages in
which the chambers will be scavenged by the air barrier flow.
In fig.7 the average contaminant flow Ẋ is plotted against the
frequency of passage for the same two air lock systems mentioned
in fig.6. The difference between active and passive air locks
is caused by the facts mentioned under 3.2.1.2.

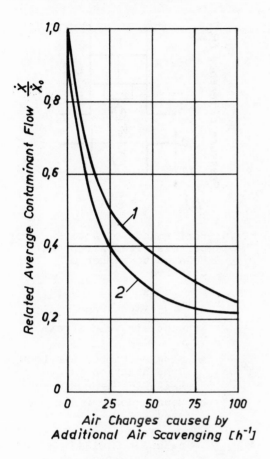

Fig.8   Contaminant Flow Influenced
         by Additional Air Scavenging

        $\dot{X}$ -   Average contaminant
               flow with additional
               air scavenging

        $\dot{X}_o$-   Average contaminant
               flow without additional
               air scavenging

        1 -   Active single-chamber
               air lock (pressurized,
               depressurized) without
               exterior pressure
               difference

        2 -   Active pressure-balanced
               single-chamber air-lock
               with exterior pressure
               difference

3.2.1.3. Influence of Air
         Scavenging
==========================

The effect of air lock
systems can be improved
by additional air-
scavenging in the chambers,
which leads to a continuous
dilution of the contamination. The determinative
factors are the number of
air changes by the
scavenging air in the
chambers and the position
of the scavenged chamber in
multi chamber air locks if
not all the chambers are
scavenged. In fig.8 the
influence of air scavenging
is shown for the above-
mentioned air lock systems.

3.2.2. Calculated Results
       for Multi-Chamber
       Air Locks

The possibilities of
improving single-chamber
air locks have been shown
above. If all these steps
are not sufficient to
produce a desired separa-
tion, multi-chamber air
locks must be used. These
systems allow a further
reduction of the contaminant
flow by increasing the number of chambers and by
producing suitable pressure
distributions. The influence
of the number of chambers
is shown in fig.9 with the
above mentioned example
of an operating theatre.
For single-, double- and
triple-chamber air locks
pressure difference diagrams
are given as well as the
final concentration in the
operating theatre

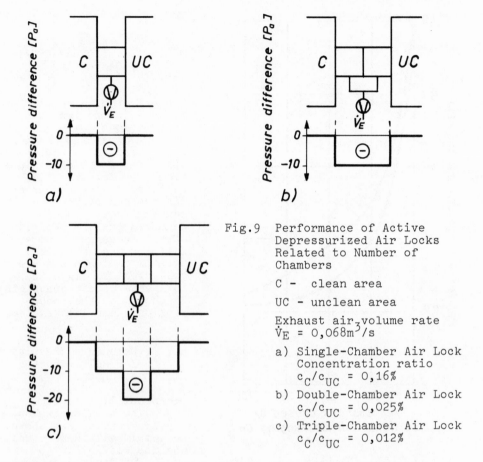

Fig.9 Performance of Active
       Depressurized Air Locks
       Related to Number of
       Chambers

C -   clean area

UC - unclean area

Exhaust air volume rate
$\dot{V}_E$ = 0,068m^3/s

a) Single-Chamber Air Lock
   Concentration ratio
   c_C/c_{UC} = 0,16%

b) Double-Chamber Air Lock
   c_C/c_{UC} = 0,025%

c) Triple-Chamber Air Lock
   c_C/c_{UC} = 0,012%

("clean area") related to that in the adjacent area ("unclean
area"). The influence of different pressure distributions is
shown in fig.10 for two examples of three-chamber air locks.
The given concentration values are for the above-mentioned example.

4. Summary

The tests on the transmission of air through doors have shown
that per passage through a sliding door of about 2m^2 with volume
rates of flow for the air barrier between 0 and 1000m^3/h a volume
of air of 0.5 - 0.1m^3 is transmitted from the unclean to the clean
area. Turning doors produce about double the transmitted volume.

The numerical investigations of various arrangements of air lock
systems on the basis of the maximum transmitted volumes at the
doors yielded the following results:

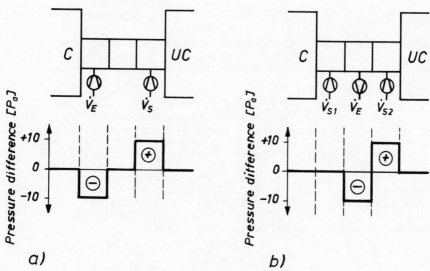

Fig.10 Performance of Multi-Chamber Air Locks Related to
        Pressure Distribution

  a) Supply air volume rate $\dot{V}_S$ = 0,068m^3/s
   Exhaust air volume rate $\dot{V}_E$ = 0,068m^3/s
   Concentration ratio c_C/c_{UC}= 0,006%

  b) Supply air volume rate $\dot{V}_{S1}$= 0,034m^3/s
            $\dot{V}_{S2}$= 0,086m^3/s
   Exhaust air volume rate $\dot{V}_E$ = 0,086m^3/s
   Concentration ratio c_C/c_{UC}= 0,0006%

1. Single-Chamber Air Locks

Active pressure-balanced air locks with a difference in
pressure from the clean to the unclean area produce the best
separation effect. In air locks without additional air-
scavenging the separation effect of passive systems with
exterior pressure distribution is better than that of either
active pressurized or depressurized air locks. Active single
chamber air locks with additional air scanvenging can produce
equal or better separations. The better separation effect of
passive and active pressure-balanced air locks compared with
either pressurized or depressurized air locks is caused by
the air flow through both doors from the clean to the unclean
area where as in active pressurized and depressurized air locks
at one of the two doors there is an air flow from the unclean
to the clean area.

2. Multi-Chamber Air Locks

Multi chamber air locks can produce, by a suitable pressure
distribution (directed air locks) and by additional air
scavenging, any desired separation effect.

3. Clean areas which are fitted with ventilation systems for
instance to provide desired environmental conditions, and
in which it is possible to keep a safe pressure difference
in relation to the unclean area under all circumstances of
operation and weather should be equipped with passive single-
chamber air locks. If it is impossible to maintain the
necessary positive pressure in the clean area, for instance
because of improper sealing, active air locks must be used.
For high requirements of separation active multi-chamber air
locks, if necessary with additional air-scavenging in the
chambers, must be employed.

5. References

[1] Esdorn, H.: Klimatisierung von Krankenhäusern,
     Deutsche Bauzeitung (1973), H.2, S.154

[2] Esdorn, H.: Luftströmung und Druckhaltung in
     Krankenhäusern
     Gesundheits-Ingenieur (1973), H.10, S.289

HEAT AND MASS TRANSFER
OF INTERNAL SURFACES
IN GUARDING CONSTRUCTIONS

P. M. BRDLICK

The Moscow Technical Institute for Forestry
Moscow, USSR

The calculation reliability of guarding construction thermal resistance is determined by the accuracy of the heat transfer coefficient calculation on it's internal surfaces. The internal aerodynamic regime of buildings depends on it's functional purposes. In the living accomodation the natural convection regime is relized, but in the most of production premises and housing for live-stock the mixed convection appears.

I. THE NATURAL CONVECTION REGIME

Under natural convection conditions the heat transfer depends on the flow regime and on the surface orientation.

1. Impenetrable Vertical Surface

The laminar regime of natural convection on the vertical surface may exist up to $Gr_{cr} \leq 10^9$. The heat transfer calculation may carry out using the Squire equation. This formula agrees with experimental data and numerical calculation satisfactory (Fig.1).

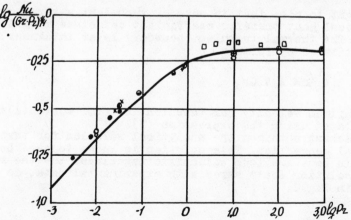

Fig.1 Comparison of Squire's (1) solution with numerical

solutions and experiments;
numerical solution: O- (2); ● -(3); ◑- (4); ◕- (5); ◉- (6);
experiment: ▽ - (7); +- (8); △ - (9); □ - (10); ×,■ - (4).

$$Nu_x = 0,508 \ Pr^{1/2} \ (0,952 + Pr)^{-1/4} Gr_x^{1/4} . \tag{1}$$

For air (Pr= 0.72) the expression(1) reduces to

$$Nu_x = 0.38 Gr_x^{1/4} - \text{local value}, \tag{2}$$

$$Nu_\ell = 0.507 Gr_\ell^{1/4} - \text{average value}. \tag{3}$$

The thermal boundary layer thickness for air is estimated by formula:

$$\delta_T / x = 5,25 Gr_x^{-1/4} \tag{4}$$

and the maximum value of velocity

$$U max = 0,592 \ ^{\nu}/x.' \ Gr_x^{1/2} \tag{5}$$

is reached at the distance from the wall $Y \ max = \delta_T /3$. (6)

   The turbulent regime at the vertical surface is realized when $Gr > 10^8$. Then Nusselt number is calculated as

$$Nu_x = 0,2(Gr_x \cdot Pr)^{1/3} (\frac{Pr^{2/3}}{2,14 + Pr^{2/3}})^{1/3} . \tag{6}$$

   The formula (6) generalizes experimental data in the range of Prandtl number Pr =0.005 ÷ 150. For air

$$Nu_x = 0,116 \ Gr_x^{1/3} . \tag{7}$$

It ought to note that in case of turbulent natural convection the local heat transfer coefficient coincides with its average value. The thermal turbulent boundary layer thickness for air is estimated by formula

$$\delta_T / x = 4,1 \ Gr_x^{-1/6} . \tag{8}$$

The maximum velocity for turbulent natural convection may be calculated using the expression (5).
   Eckert obtained the analytical solution for turbulent natural convection. This solution is known to have been published in some American scientific magazines. But the results of this solution don't agree with experimental data, as it may be seen in Fig.2.

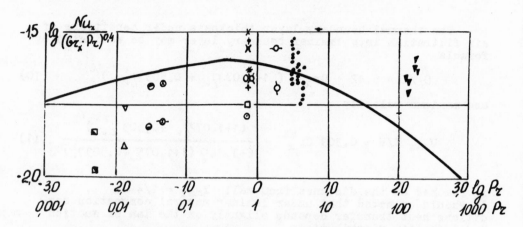

Fig.2. Comparison Eckert solution with the experiment.

2. Penetrable Vertical Surface

The approximate analytical solution for uniformly suction and injection is obtained by the method of relative conformity. In the conditions of laminar boundary layer this solution leads to expression

$$Nu_x / Nu_{x_o} = 1 - 0,645 \, \rho_w / \rho_\infty \cdot \eta_v , \qquad (9)$$

where Nu_{x_o} - local Nusselt number at impenetrable surface,

$\eta_v = Re_w \, (\, Gr_x \, /4)^{-1/4}$, $Re_w = Vw \cdot X/\nu$. Positive filtration velocity is injection velocity and negative-suction velocity. The solution (9) is compared with experimental data (fig.3) for $\eta_v = 1.8 - 1.8$. The experimental heat transfer coefficients are determined from temperature fields. This temperature fields were obtained using interferometer.

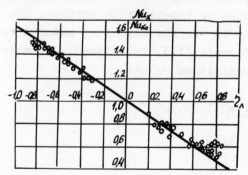

Fig.3.

The thermal boundary layer thickness under conditions of air filtration into laminar boundary layer may be estimated by formula

$$\delta_T/X = 5,42 \cdot Gr_x^{-1/4} \ (\ 1+1,07\,\eta_v + 0,503\,\eta_v^2\ )^{1/4} \qquad (10)$$

and maximum velocity

$$U_{max}\ X/\nu = 0,309\ Gr_x^{1/2} \cdot \frac{(1+1,07\,\eta_v + 0,503\,\eta_v^2\ )^{1/2}}{[6+3,84\,\eta_v(1+1,07\,\eta_v+0,503\,\eta_v^2)^{1/4}]} (11)$$

is reached at the distance from wall $Y_{max} = 1/4\ \delta_T$.
It should be noted that under laminar natural convection conditions heat transfer depends slightly on the law of suction and injection distributions.
    At great injection velocity ($\eta_v > 1$) the deformation of temperature profile is observed. Such temperature distribution has S-formed character. At $\eta_{cr} \approx 1,5$ heat flux at the wall reduces to zero.
The experiments show that at great suction velocity ($\eta_v < -1,5$) the heat transfer coefficient approaches to double heat transfer value for impenetrable surface. The influence of air injection and suction on the length of transition zone (from laminar to turbulent regime) was established experimentally.
The beginning of transition region is determined as

$$(\ Gr_x\ )^{trans.} = (Gr_x)_0^{trans}[1-0,3\,\theta_w\,x\ /\nu\cdot(Gr_x^{trans.}/4)^{-1/4}].\ (12)$$

The beginning of developed turbulent stream is estimated as

$$(Gr_x^{turb.} = (Gr_x)_0^{turb.}[1- 0,2\cdot\theta_w\,x/\nu\ (Gr_x^{turb.}/4)^{-1/4}].\quad (13)$$

In formulas (12),(13(,$(Gr_x^{trans})_0 = 10^7$ and $(Gr_x)_0^{turb} = 10^8$.
Index "O" attributes to impenetrable surface. The influence of wall penetrability on the air turbulent natural convection may be estimated by formula (14). This formula is obtained from analytically approximation solution

$$Nu_x/Nu_{x_o} = (1 - 0,705\ \eta_t^{1/3}\ )^2, \qquad (14)$$

where $\eta_t = 13.55\ Re_w^2/\ Gr_x$.

3. Water Vapor Evaporation and Codensation from the Vapor-Air Mixture

The influence of condensation and evaporation on heat transfer coefficient must be taken into account in calculations of guarding construction thermal resistance. The water vapor condensation and evaporation from vapor-air mixture are the heat and mass transfer problems for binary boundary layer.

In binary boundary layer the lifting forces appear to be a result of temperature and concentration differences. The concentration contribution is especially essential when there are great differences between mixture components of molecular weights.

For laminar boundary layer the solution may be presented as

$$Nu_x = 0,508(Gr_x \cdot Pr)^{1/4} \cdot (1 + \frac{\beta_m}{\beta_t} \cdot \frac{m_{1w} - m_{1\infty}}{t_w - t_\infty} \cdot \frac{1}{\xi})^{1/4} \cdot \left[\frac{Pr}{\eta(20/21\eta + Pr)} \right]^{1/4}$$
$$\cdot \left\{ 1 + Le^{-1} \cdot Du \left[(M_{1w} - M_{1\infty})\xi + S_0 \right] \right\} . \qquad (15)$$

Local mass transfer coefficient d_m is founded from the expression for total mass flux of "active" component (index"1")

$$W_{1w} = j_{1w} + \rho_w V_w \cdot m_{1w} = \rho_w d_m (m_{1w} - m_{1\infty}) \qquad (16)$$

and for Sherwood number

$$Sh_x = \frac{d_m x}{D} = 0,508(Gr_x \cdot Pr)^{1/4} \cdot (1 + \frac{\beta_m}{\beta_t} \cdot \frac{m_{1w} - m_{1\infty}}{t_w - t_\infty} \cdot \frac{1}{\xi})^{1/4}$$
$$\cdot \left[\frac{Pr}{(\eta(20/21\eta + Pr)} \right]^{1/4} \cdot \frac{m_{1w} + m_{1\infty}\xi + S_0}{(1 - m_{1w})(m_{1w} - m_{1\infty})} , \qquad (17)$$

where m_1 – mass concentration of "active" component, β_m and β_t coefficients ofconcentration and thermal expansion correspondingly, $\xi = \delta_T / \delta_m$ –thermal and diffusion boundary layer thickness ratio, Le –Lewis number; η – may be determined from the expression

$$\eta = 1 + \rho_w/\rho_\infty \cdot d_e^{-1} \left[D + 1/(1 - m_{1w}) \right] \cdot \left[(m_{1w} - m_{1\infty})\xi + S_0 \right] +$$
$$+ \frac{\rho_w}{\rho_\infty} Le^{-1} \cdot \frac{C_{p1} - C_{p2}}{C_{p1}} \left[\xi(1 - 1/3\xi) \cdot (m_{1w} - m_{1\infty}) + 2/3 S_0 \right] ,$$

where $Du = (a_T Rc M_c^2 /427 M_1 M_2 C_{pc})(T_w /(T_w - T_\infty) -$ -Dufo number, a_T- coefficient of thermal diffusion, $S_0 = a_T m_{1w}(1 - m_{1w})(T_w - T_\infty)/ T_w$ - Sore number.

For water vapor evaporation and condensation from vapor-air mixture the formulae (15),(17),(18)are grown essentially simplified. Lewis number becomes constant and is equal to 0,84, ξ =1; S_0 =0. Heat transfer which is caused by the specific heat difference of mixture components isn't taken into account.

$$Nu_x = 0,508(Gr_x \cdot Pr)^{1/4} (1 + \frac{\beta_m}{\beta_t} \frac{m_{1w} - m_{1\infty}}{t_w - t_\infty})^{1/4} \left[\frac{Pr}{\eta(20/21 \cdot \eta + Pr)} \right]^{1/4} \cdot$$

$$\cdot \left[1 + Le^{-1} \cdot Du (m_{1w} - m_{1\infty}) \right], \tag{19}$$

$$Sh_x = 0,508(Gr_x \cdot Pr)^{1/4} (1 + \beta_m/\beta_t \cdot \frac{m_{1w} - m_{1\infty}}{t_w - t_\infty})^{1/4} \left[\frac{Pr}{\eta(20/21 \cdot \eta + Pr)} \right]^{1/4} \cdot$$

$$\cdot 1/(1 - m_{1w}), \tag{20}$$

$$\eta = 1 + \rho_w/\rho_\infty \cdot Le^{-1} (m_{1w} - m_{1\infty}) \left[Du + 1/(1 - m_{1w} + 2/3 \right] \cdot \tag{21}$$

For water vapor–air mixture $a_\tau \approx -0.1$. The value of conditional heat transfer coefficient is often used in engineering calculations of heat transfer with phase transformation. It may be determined as

$$q_w^* = q_w + r \cdot W_{1w}, \tag{22}$$

where r– the phase transformation heat. Then for this case Nusselt number is expressed by formula

$$Nu_x^* = 0,508(Gr_x \cdot Pr)^{1/4} (1 + \frac{\beta_m}{\beta_t} \cdot \frac{m_{1w} - m_{1\infty}}{t_w - t_\infty})^{1/4} \left[\frac{Pr}{\eta(20/21 \cdot \eta + Pr)} \right]^{1/4} \cdot$$

$$\cdot \left\{ 1 + Le^{-1} (m_{1w} - m_{1\infty}) \left[Du + K/(1 - m_{1w}) \right] \right\}, \tag{23}$$

where $K = r/C_p (t_w - t_\infty)$.
The comparison of average conditional Nusselt number (Nu^*) with experimental data for air mass portion $0.015 - 0.18$ is presented in Fig.4.

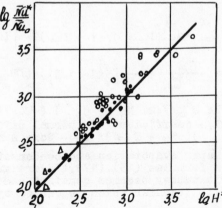

Fig. 4. The average conditional heat transfer coefficient under conditions of water vapor condensation from air-vapor mixture (laminar natural convection):
— -formula (23); experiments: O- Masjukevich (13); ●- Othmer (14), △ -Langen (15).

Approximate solution for binary turbulent boundary layer (effect of thermodiffussion and heat transfer caused difference of component mixture specific heat may be neglected) may be expressed:

$$\frac{Nu_x}{Nu_{x_0}} = \left(\frac{\rho_w}{\rho_\infty}\right)^{1/3} \frac{\left[1+\frac{\beta_m}{\beta_t}\cdot\frac{m_{1\infty}-m_{1w}}{t_\infty-t_w}\cdot\frac{1}{\xi}\right]^{1/3}\left[1+\left(\frac{Pr}{Sc}\right)^{2/3}Du(m_{1w}-m_{1\infty})\cdot\xi^{1/2}\right]}{\left\{1+\frac{2,14}{2,14+Pr^{2/3}}\left(\frac{Pr}{Sc}\right)^{2/3}\cdot\frac{m_{1\bar{w}}\ m_{1\infty}}{1-m_{1w}}\cdot\left[1+Du(1-m_{1w})\xi^{1/2}\right]\right\}^{1/3}} \quad (24)$$

$$Sh_x = Nu_x \frac{(Sc/\ Pr)^{1/3}\cdot\xi^{1/2}}{(1-m_{1w})\left[1+(\ Pr/Sc)^{2/3}\cdot Du\ (m_{1w}-m_{1\infty})\cdot\xi^{1/2}\right]}. \quad (25)$$

Conditional heat transfer coefficient taking into account of phase transformation is determined as

$$\frac{Nu_x^*}{Nu_x} = \frac{1+(\ Pr/Sc)^{2/3}(m_{1w}-m_{1\infty})\cdot\left(\ Du+\frac{K}{1-m_{1w}}\right)\cdot\xi^{1/2}}{1+(\ Pr/Sc)^{2/3}Du\ (m_{1w}-m_{1\infty})\cdot\xi^{1/2}}. \quad (26)$$

In formulae (24)-(26) the value $\xi = \delta_T / \delta_m$ for $Le \leq 1$ is founded from the expression

$$\xi = 0,309\left\{\frac{m_{1w}-m_{1\infty}}{1-m_{1\infty}}Du\ (1-m_w)+1\ \right] + \sqrt{\left\{\frac{m_{1w}-m_{1\infty}}{1-m_{1\infty}}\left[Du(1-m_{1w})+1\right]\right\}^2 + 3,6\left(\frac{1-m_{1w}}{1-m_{1\infty}}Le^{2/3}-0,1\right)}\ \right\}^2. \quad (27)$$

The length of binary boundary layer transient zone may be estimated by formulae (12),(13),

where $\theta_w\ x/D = (m_{1w}-m_{1\infty})\ Sh_x$ \hfill (28)

and $Gr_x^* = g\left[\beta_t(t_w-t_\infty)+\beta_m(m_{1w}-m_{1\infty})\right]x^3/\nu^2$.

4. Impenetrable Horizontal Surface

When the heated surface is turned down boundary layer has steady laminar character with two zones: edge zone and central zone. The boundary layer thickness in the edge zone increases from the edge to the centre up to the value δ_T^* . Then in the central zone this value remains constant. The local heat transfer coefficient changes correspondingly.

Approximate analytical solutions for these two zones give following expressions:

a) for edge zone $Nu_x^H = 0,45 \, (Gr_x \cdot Pr)^{1/5} \cdot (2+Pr^{-1})^{-1/5}$, (29)

b) for central zone $Nu_x^H = 0,0862 \, (Gr_x \cdot Pr)^{1/3}$. (30)

The boundary between two zones is founded from the relation

$$(Gr_x \cdot Pr)_o^H = 2,43 \cdot 10^5 \, {}^5(2+Pr^{-1})^{-3/2}.$$ (31)

For air ($Pr = 0.72$) these formulae are simplified

a) edge zone $Nu_x^H = 0,33 \cdot Gr_x^{1/5}$, (32)

b) central zone $Nu_x^H = 0,0775 \, Gr_x^{1/5}$, (33)

c) boundary between zones $Gr_o^H = 5,4 \cdot 10^4$. (34)

The comparison of solution (32),(33) with experiment is presented in Fig.5.

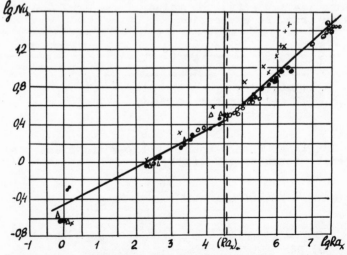

Fig.5. Local heat transfer from heated horizontal surface turned down:— formulae (32),(33); experiments: ●, △ -author,

x -(16), + - (17), ● - (18), 0- (19).
   The average value of heat transfer coefficient for the plate with heated surface turned downwards in air stratum is written as

$$\overline{Nu}_\ell^{-H} = 3,88 + 0,0775 \, Gr_{\ell}^{1/3}. \tag{35}$$

If the heated surface is turned upwards there are two zones of heat transfer too. In the edge zone the boundary layer is laminar, on the periphery of central zone there are irregular vortices, near the centre the rising convective stream appears. In this case the extent of edge zone is more than for surface turned downwards. As for air this length is determined from the the condition

$$Gr_{o}^{6} = 2 \cdot 10^{5}. \tag{36}$$

Heat transfer coefficient for air is calculated by formulae:

   a) edge zone $Nu_{x}^{6} = 0,465 \, Gr_{x}^{1/5}$,   (37)

   b) central zone $Nu_{x}^{6} = 0,12 \cdot Gr_{x}^{1/3}$ (38)

   c) average value for all plates $\overline{Nu}_{\ell}^{6} = 1,92 + 0,12 Gr_{r_\ell}^{1/3}$. (39)

5. Penetrable Horizontal Surface

   For edge zone the results of numerical solutions coincide with approximate solution, the latter being used by means of the integral method. These results agree with experiments satisfactory. For heated surfaces turned up and down there is one formula

$$Nu_{x}/Nu_{x_o} = 1 - 0,615 \, \theta_w \, x \, /\dot{V} \cdot (Gr_{x} /5)^{-1/5}. \tag{40}$$

For heated surface turned down Nu_{x_o} is determined by formula (32), for heated surface turned up Nu_{x_o} is calculated from (37). The boundaries of heat transfer zones are founded from common formula

$$Gr^{*}/ \, Gr_{o} = 1 - 0,25 \, \theta_w \, x /\dot{V} \cdot (Gr_{x} /5)^{-1/5}. \tag{41}$$

The value of Gr_o is determined from the expressions(34) or(36). In central zone heat transfer is calculated from relations:
   a) heated surface turned up
$$Nu_{x}/Nu_{x_o}^{6} = 1 - 1,75\omega + 2,75\omega^{2}, \tag{42}$$

   b) Heated surface turned down
$$Nu_{x}/Nu_{x_o}^{H} = 1 - 4,8\omega + 7,0\omega^{2}, \tag{43}$$

where $\quad \omega = \theta_x \, x \, / \nu \cdot (Gr_x \, / 3)^{-1/3}$.

II. MIXED CONVECTION

The intensity of heat transfer in the case of mixed convection is determined by Gr /Re. It ought to distinguish two cases of mixed convection: in the first place the direction of natural convection coincides with the direction of forced convection (coinciding streams); secondly- the direction of the forced convection is contrary to natural convection flow (opposite streams).

6. Mixed Convection on Impenetrable Vertical Surface

For Pr =0.01 - 100 the analytical solutions are obtained by integral method. These solutions agree with numerical results quite satisfactory. For 0.5 Pr 5 local Nusselt number is founded as

$$
\frac{Nu_x}{Re_x^{0.5}} = \left(\frac{Gr_x}{Re_x^2}\right)^{1/4} \frac{0.344 Pr^{1/2} \cdot \left(\frac{Re_x^2}{Gr_x}\right)^{1/2} \pm 0.508 \left(\frac{Pr}{0.952+Pr}\right)^{1/2}}{\left[1.015 \left(\frac{Re_x^2}{Gr_x}\right)^{1/2} \pm \frac{1}{(0.952 + Pr)^{1/2}}\right]^{1/2}}, \quad (44)
$$

where + attributes to coinciding streams, and — to opposite streams.
It ought to note that for $Gr_x \rightarrow 0$ the equation (44) reduces to expression, which agrees with results of Polhausen's numerical solution for forced convection quite well. For Re = 0 the formula (44) reduces to Squire approximate solution for natural convection.

7. Mixed Convection on Penetrable Vertical Surface

The numerical solution of boundary layer differential equations is used for determination of air filtration influence on the heat transfer in mixed convection conditions. The numerical result approximation is allowed to obtain the local Nusselt number

$$
Nu_x = \left(\frac{Gr_x}{Re_x}\right)^{1/2} \cdot \frac{0.43 \, f_w + 0.42}{(Gr_x /Re_x^2)^{0.12 f_w +0.34}} . \quad (45)
$$

Here $\quad f_w = - 1/3 \, \theta_w \, x \, /\nu \cdot (Gr_x /4)^{-1/4}$.

The formula (45(may be used for coinciding and opposite flows.

REFERENCES

1. Goldstein S., Modern Developments in Fluid Dynamics, 1938.
2. Ostrach S., NACA, TN2365, 1952; NACA, TR1111, 1953.
3. Sparrow E., Gregg I.,Trans.,ASME, S.C.,v80, p.879, 1958.
4. Saunders O., Proc. Roy. Soc.A., v.172, p.948, 1939.
5. Sugawara S., Michigshi S.J. Proc.3-d Japan National Congress for Applied Mechanics, 1953.
6. Schuh H. Göttinger Monographien, Bd .8, 1946.
7. Eigenson L.S. DAH, TXXVI, N5, 1940.
8. Schmidt E., Becremann W.,Forsch. Ing.,-Wes.,BdI,S. 391,1930.
9. Lorenz H. Zeitschribt für Technische Physik, N9,1934.
10. Jakob M. Trans.ASME, s.C, v 70, N1, 1948.
11. Eckert E.,Jacreson T. NACA, Rep. 1015, 1951.
12. Gill W. Int. J.Heat Mass TRansfer, v 8, N8, 1965.
13. Masukesur P.V. Tp. LTUXP, t.XiV, 1956.
14. Othmer D. Ing.Eng. Chem., v 21, pp 577-583, 1929.
15. Langen E. Forsch Geb. Ing. Wes., Bd.2, H10, 1931.
16. Weise R. Forsch. Geg. Ing. Wes., Bd.6, N6, 1935.
17. Жуковский В.С. ЖТФ, т.1, ва 2-3, 1931.
18. Новожилов В.И. ИФЖ, №,6, 1958.
19. Смирнов В.А. Тр МТИЛП, №28, 1963.
20. Kliegel J. Univ.of California, 1959.

HEAT TRANSFER IN BUILDINGS
WITH PANEL HEATING

O. G. MARTYNENKO AND YU. A. SOKOVISHIN

Heat and Mass Transfer Institute
Minsk, USSR

ABSTRACT

Heat transfer of a vertical electrically heated surface is investigated. An analysis is made of the effect of Gr* and Re numbers on heat transfer in buildings at free and mixed convection. Criterial formulas are given for calculation of heat transfer in a wide range of basic parameters. The effect of cocurrent and countercurrent flows is shown.

NOMENCLATURE

x - longitudinal coordinate

y - transverse coordinate

u, v - velocity projections on the x and y axes

l - height of the plate

b - width of the plate

g - gravitational acceleration

β - volume expansion coefficient

μ - dynamic viscosity coefficient

ρ - density of fluid

ν - kinematic viscosity coefficient

λ - thermal conductivity

c_p - specific heat

a - thermal diffusivity

T - temperature

$\vartheta = (T - T_\infty)$ excessive temperature

231

q -   heat flux

δ -   boundary layer thickness

α -   heat transfer coefficient

η, ζ -   similarity variables

ψ -   stream function

F, Θ -   dimensionless stream function and temperature

Q -   volume flow rate

f, φ -   dimensionless velocity and temperature in the integral method

a_i, b_i polynomial coefficients (I8)

$Pr = \dfrac{\nu}{a}$ Prandtl number

$Gr^* = \dfrac{g\beta q_w l^4}{\lambda \nu^2}$ modified Grashof number

$Ra^* = Gr^* Pr$ modified Rayleigh number

$Nu = \dfrac{\alpha l}{\lambda}$ Nusselt number

Subscripts

w -   wall

o -   initial value

x -   local value

∞ -   surrounding medium

T -   stagnation cross-section

Among various methods of central heating a definite place belongs to panel heating, providing an ample possibility for easy and smooth control of heat transfer in buildings, as well as for even distribution of surface temperatures of panels, improvment of sanitary conditions etc. This makes the study of panel heating a matter of practical importance. In electrical panel systems, heat is generated due to electric current passing through a metallic surface (foil, net), and removed from this surface by free or mixed convection, at constant heat flux.

Convective heat transfer in buildings is described by a system of differential equations, including a large number of variables. Analytical solution of such a system is hampered by considerable mathematical difficulties. In this situation, employing of the methods of generalized variables [1,2,3] makes it possible to interprete correctly experimental data and numerical results, and to obtain criterial relationships useful for engineering practice. New dimensionless variables express separate effects, as well as the effects in the aggregate, this allowing determination of the influence of relationships on the process under consideration.

We employ now the dimensional analysis to estimate heat transfer from a vertical surface, l high and b wide. On the surface, the heat flux q_w and the velocity of flow circulation in the building u_∞ are prescribed. In Boussinesque approximation, for the calculation of the buoyancy force, the dependence of heat transfer on dimensional physical values can be given by

$$\alpha = f\left(l, b, \rho, \mu, g\beta, \lambda, c_p, q_w\right) \tag{1}$$

It is assumed that the motion is steady and there is no viscous dissipation.

After ordinary transformations, based on π-theorem and allowing for the vector character of the linear dimension L for the two-dimensional motion (L_x, L_y), we obtain the following functional relations

$$\frac{Nu}{Re^{1/2}} = f_1\left(Pr, \frac{Gr^*}{Re^{5/2}}, \frac{b}{l} \cdot Gr^{*1/5}\right), \tag{2a}$$

$$\frac{Nu}{Gr^{*1/5}} = f_2\left(Pr, \frac{Re}{Gr^{*2/5}}, \frac{b}{l} \cdot Gr^{*1/5}\right). \tag{2b}$$

Note that equations (2) for an infinitely wide plate have been used numerically in [4] for the case of laminar mixed convection over a vertical surface subject to the boundary conditions of the 2nd kind.

For free convection, relation (2b) is reduced to

$$\frac{Nu}{Gr^{*1/5}} = f_3\left(Pr, \frac{b}{l} \cdot Gr^{*1/5}\right). \tag{3}$$

The effectiveness of the method increases, if account is taken of the direction of the linear dimension L (e.g. L_x, L_y). Free convection motion on a cooled surface is directed downwards owing to negative temperature difference. With free and forced convection being opposed, relation (2) can not be obtained because of the flow nature changing essentially behind the separation cross-section. Likewise, for creeping and turbulent motion different relations for calculation of heat transfer can be obtained by the dimensional analysis, allowing for anisotropic character of transfer coefficients. The dimensional analysis does not specify quantitatively the coefficients in criterial relationships, which can be obtained experimentally or calculated.

Let us consider free-convection heat transfer on a vertical surface with even distribution of heating elements. Wall temperature varies, differing slightly from the room temperature, T_∞. Change of density is taken into account only as a buoyancy component. The direction of the coordinate axes is chosen in such a way, that the longitudinal coordinate, x, coincides in direction with the convective flow. Then for the laminar flow, the system of equations in the boundary-layer approximation assumes the following form [6]

$$u \frac{\partial u}{\partial x} + v \frac{\partial u}{\partial y} = g\beta v + \nu \frac{\partial^2 u}{\partial y^2}$$

$$\frac{\partial u}{\partial x} + \frac{\partial v}{\partial y} = 0 \tag{4}$$

$$u \frac{\partial v}{\partial x} + v \frac{\partial v}{\partial y} = a \frac{\partial^2 v}{\partial y^2}$$

with the boundary conditions

$u = 0, \quad v = 0, \quad q_w = -\lambda \left(\frac{\partial v}{\partial y}\right)_0$ at $y = 0$; $u = 0, \quad v = 0$

at $y \rightarrow \infty$. \hfill (5)

The wall heat flux being constant, problem (4),(5) acquires the following similarity solutions for stream function $\psi(x,y)$ and dimensionless temperature

$$\psi(x,y) = 5\nu \left(\frac{Gr_x^*}{5}\right)^{1/5} F(\eta), \quad \Theta = \frac{v}{\frac{q_w x}{\lambda}} \left(\frac{Gr_x^*}{5}\right), \quad \eta = \frac{y}{x} \left(\frac{Gr_x^*}{5}\right)^{1/5}. \tag{6}$$

To determine functions $F(\eta)$ and $\Theta(\eta)$, we obtain the ordinary differential equations

$$F''' + 4FF'' - 3F'^2 + \Theta = 0$$

$$\tag{7}$$

$$\Theta'' + Pr(4F\Theta' - F'\Theta) = 0$$

with the boundary conditions

$F = 0, \quad F' = 0, \quad \Theta' = -1 \quad$ at $\eta = 0$; $\begin{matrix} F' = 0, \\ \Theta = 0 \end{matrix}$ at $\eta \rightarrow \infty$ (8)

Numerical solution of the system (7) with the boundary conditions (8), was carried out by the Runge-Kutta method. The unknown values $F''(0)$ and $\Theta(0)$, were prescribed arbitrarily at the beginning of the interval and then specified by the Newton method. The calculations of the similarity problem were made for air ($Pr \approx 0.7$).

In Fig.1, distribution of dimensionless excessive tempera-

ture $\frac{T - T_\infty}{T_w - T_\infty} = \frac{\Theta(\eta)}{\Theta(0)}$ and velocity $\bar{u} = \frac{ux}{5^{3/2} \nu (Gr_x^*)^{2/5}} = F'(\eta)$.

is presented.

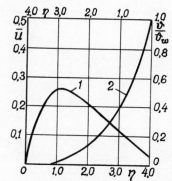

Fig.1. Dimensionless velocity (1) and dimensionless excessive temperature (2)
distribution.

The wall temperature changes along the longitudinal coordinate

$$\vartheta_w = \frac{x\,q_w}{\lambda} \left(\frac{Gr_x^*}{5} \right)^{-1/5} \Theta(0) \qquad (9)$$

as well as along the test plate

$$\frac{\vartheta_w}{(\vartheta_w)_l} = \left(\frac{x}{l} \right)^{1/5} . \qquad (10)$$

The wall temperature equals the ambient temperature on the leading
edge of the plate and increases further in the direction of the
flow.

Heat transfer coefficient is determined by the Newton law

$$\alpha_x = q_w / \vartheta_w ; \qquad Nu_x = \frac{q_w x}{\lambda\,\vartheta_w} , \qquad (11)$$

which leads to criterial relationship (3) for a wide plate

$$\frac{Nu_x}{Gr_x^{*\,1/5}} = \frac{1}{5^{1/5}\Theta(0)} \qquad (12)$$

For Pr=0.7 $\Theta(0) = 1.4985$

$Nu_x = 0.484\,Gr_x^{*\,1/5}$ $Nu_x = 0.519\,Ra_x^{*\,1/5}$.   (12a)

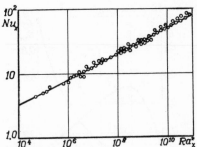

Fig.2. Experimental data on the local air heat
transfer coefficient.

In Fig.2., numerical calculations (solid line) are compared
with experimental results for $10^4 < Ra_x^* < 10$ on local heat transfer [6,8,9] . Deviation from the accurate value (formula (12a))
constitutes ± 6%.

The mean value of the heat transfer coefficient is calculated
by the average temperature at the height of the plate

$$\overline{\vartheta}_w = \frac{1}{l} \int_0^l \vartheta_w \, dx = \frac{5}{6} (\vartheta_w)_l \qquad (13)$$

and for criterial relationship we obtain

$$\frac{\overline{Nu}_l}{Gr_l^{*\,1/5}} = \frac{6}{5^{6/5}\Theta(0)} \, , \qquad \frac{Nu_l}{Gr_l^*} = 0{,}58 \, , \qquad \frac{Nu_l}{Gr_l^*} = 0{,}62 \, . \qquad (14)$$

The applicability of formulas (12)-(13)-(14) is restricted by
a laminar flow regime, observed up to the value
$Ra_l^* < 5 \cdot 10^{10}$ [7,8].

The volume flow rate of air in the boundary layer

$$Q = \int_0^\infty u \, dy = \psi(x, \infty) = 5^{4/5} \nu \, Gr_x^{*\,1/5} F(\infty) = 1{,}912 \, \nu \, Gr_x^{*\,1/5} \quad (15)$$

increases along the plate height $Q \sim x^{4/5}$, as more and more
portions of air are involved into the flow owing to viscosity.
The height of the heat exchanger being known, the velocity of air
circulation in the building can be determined with the help of (15)
It remains to note that for a narrow plate a criteral formula

$$\frac{Nu_x}{Ra_x^{*\,1/5}} = 0{,}519 \left(1 + A \, \frac{l}{6} \, Gr_x^{*\,-1/5} \right) . \qquad (16)$$

has been suggested in [10] , with the coefficient A, which can not be determined because of an insufficient number of experiments.

In the theory of boundary layer, approximate methods of calculation are widely used. In the integral method, integral equations [5] replace differential ones (4). Unknown velocity and temperature profiles are chosen to satisfy the boundary conditions and some general arguments. As a rule, the accuracy of the method is verified by comparing it with the known exact solution or experiments.

For an illustration, the Karman-Pohlhausen method is considered, under the assumption that the boundary layer thicknesses are equal. We integrate the system of equations (4) over the boundary layer thickness

$$\frac{d}{dx} \int_0^\delta u^2 dy = g\beta \int_0^\delta \vartheta \, dy - \nu \left(\frac{\partial u}{\partial y}\right)_0 ,$$

$$\frac{d}{dx} \int_0^\delta u \vartheta \, dy = \frac{a}{\lambda} q_w . \tag{17}$$

It will be assumed that velocity and temperature distribution is expressed in terms of low-order polynomials, conjugation conditions being neglected

$$\frac{u}{u_1(x)} = f(\eta) = \sum_{i=1}^m a_i \eta^i; \qquad \frac{\vartheta}{\vartheta_w} = \varphi(\eta) = 1 - b_1 \eta + \sum_{i=2}^n b_i \eta^i . \tag{18}$$

Substitution of velocity and temperature profiles into equations (17) yields

$$J_1 \frac{d}{dx}\left(u_1^2 \delta\right) = g\beta \vartheta_w \delta J_2 - \frac{\nu u_1}{\delta}$$

$$J_3 \frac{d}{dx}\left(u_1 \delta \vartheta_w\right) = a \frac{q_w}{\lambda} , \tag{19}$$

where $\quad J_1 = \int_0^1 f^2 d\eta , \quad J_2 = \int_0^1 \varphi \, d\eta , \quad J_3 = \int_0^1 f \varphi \, d\eta , \quad \eta = y / \delta(x).$

If the velocity u_1 from the seconf equation of the system (19) is expressed and substituted into the first equation, then the boundary layer thickness δ can be determined in explicit form at arbitrary change of the heat flux

$$\delta = \frac{x^{4/5}}{(Gr_x^* \, Pr^2)^{1/5}} \left(\frac{36_1^3 \, J_1}{5 \, J_2 \, J_3^2}\right)^{1/5} q_w^{-7/15} \left(\int_0^x q_w \, dx\right)^{\frac{1}{3}\left(2 + \frac{J_3}{J_1 \, 6_1} Pr\right)} Y^{-1/5}(x)$$

$$u_1 = \nu \left(\frac{Gr_x^{*2}}{Pr}\right)^{1/5} x^{-8/5} \left(\frac{25 \, J_2^2}{96_1 \, J_1^2 \, J_3}\right)^{1/5} q_w^{-1/5} \left(\int_0^x q_w \, dx\right)^{-\frac{1}{3}\left(1 + 2\frac{J_3 Pr}{6_1 J_1}\right)} Y^{2/5}(x) \quad (20)$$

$$\upsilon_w = \frac{q_w}{\lambda \, 6_1} \delta, \quad Y(x) = \int_0^x q_w^{-1/3} \left(\int_0^x q_w \, dx\right)^{\frac{4}{3} - \frac{5 J_3 Pr}{36_1 J_1}} .$$

In [11] , similar equations are used for the linear distribution of the heat flux at the wall.

For the simplest velocity and temperature profiles [6,7,11]

$$f(\eta) = \eta(1 - \eta)^2, \qquad \varphi(\eta) = (1 - \eta)^2 \qquad (21)$$

the criterial heat transfer equation assumes the form

$$\frac{Nu_x}{Ra_x^{*\,1/5}} = 0.616 \left(\frac{Pr}{0.8 + Pr}\right)^{1/5}; \quad Pr = 0.7; \quad \frac{Nu_x}{Ra_x^{*\,1/5}} = 0.528. \quad (22)$$

The error of formula (22) for Pr=0.7 is approximately 2%.

Explicit relationships for the calculation of heat transfer at arbitrary change of the heat flux at the wall can also be obtained with the help of the thin layer method [12] .

The above calculations, employing the methods of the boundary layer theory, are applicable for an infinite vertical surface, and their accuracy increases with the Gr_x^* number in the laminar flow region. They can also be extended to the region of smaller Gr_x^* numbers through the use of coupled asymptotic expansions in complete equations of motion and heat transfer , with boundary layer equations taken as the first approximation [13] . In case of free convection, the perturbation method can be interpreted as follows. The boundary layer flow near the heated surface makes the major contribution into the solution, while the external flow induced by the boundary layer is secondary and provides for a correction of a higher degree of smallness. The external flow, vice turn, induces the secondary boundary layer, the so-called "boundary layer of the boundary layer", which invokes a corresponding external flow with a still higher order of correction, and so on. Summation of the corrections yields an asymptotic interpretation of the solution for each of the regions. Boundary conditions are chosen with the help of coupling, i.e. correlation of external and internal solutions in the region of their overlap. The joining is effective, provided that the behaviour of the in-

In Fig.3., the dimensionless values against the Gr_x^* number are presented for the air, with subscript 0 referring to the similarity solution of the boundary layer. With the decreasing Gr_x^* number, the rate of flow in the boundary layer and the wall temperature decrease, while dimensionless heat transfer increases. It can be noted that the applicability of the method is restricted in the $Gr_x^* = 0(1)$ region, where viscous flow transits into cellular one.

The first correction to the boundary layer theory for liquid metals on the finite-height plate is calculated in [16].

With air moving in a building, due to external effects as well as to considerable increase of the quantity of air at free motion, heat transfer from the surface becomes a combined one. From relation (2) of the dimensional analysis it follows that in the equations an additional complex, $\dfrac{Re}{Gr^{*\,2/5}}$, appears, to allow for

the relationship between viscous and buoyancy forces. With joint influence of free and forced convection, the laminar regime with formation of a boundary layer on the surface is of greatest practical interest.

Derivation of boundary-layer equations at mixed convection on a vertical surface with a given heat flux at the wall, is presented in [4]. Equations of motion and heat transfer subject to boundary conditions will be of the form

$$u\,\frac{\partial u}{\partial x} + v\,\frac{\partial u}{\partial y} = \nu\,\frac{\partial^2 u}{\partial y^2} \pm g\beta\vartheta, \qquad \frac{\partial u}{\partial x} + \frac{\partial v}{\partial y} = 0$$

$$u\,\frac{\partial \vartheta}{\partial x} + v\,\frac{\partial \vartheta}{\partial y} = a\,\frac{\partial^2 \vartheta}{\partial y^2} \tag{23}$$

$$u = 0, \qquad v = 0 \quad -\lambda\left(\frac{\partial \vartheta}{\partial y}\right)_0 = q_w \text{ at } \quad y = 0; \; u = u_\infty; \vartheta = 0, \text{ at } \; y \to \infty .$$

The double sign in the first equation corresponds to the coinciding oppositely directed free convection.

Similarity solutions of the system (23) at uniform velocity of the external flow and constant heat flux at the wall, are impossible to obtain. That is why it is more convenient for the numerical solution of the problem to make use of the forced convection variables [17].

$$\psi(x,y) = \sqrt{\nu u_\infty x}\; F(\zeta,\xi), \qquad \Theta = \frac{\vartheta}{q_w x}\sqrt{Re_x} ,$$

$$\zeta = \frac{y}{2x}\sqrt{Re_x} , \qquad \xi = \frac{Gr_x^*}{Re_x^{5/2}} . \tag{24}$$

Transformation of equations (23) yields

$$\frac{\partial^3 F}{\partial \zeta^3} + F\,\frac{\partial^2 F}{\partial \zeta^2} \pm 8\xi\Theta = 3\xi\left(\frac{\partial F}{\partial \zeta}\,\frac{\partial^2 F}{\partial \zeta\,\partial \xi} - \frac{\partial F}{\partial \xi}\,\frac{\partial^2 F}{\partial \zeta^2}\right) \tag{25}$$

ternal solution on the external edge of the boundary layer can be
presented analytically. In the vicinity of the leading and rear
edges the method of deformation of coordinates

$$x = X + \varepsilon f(X,Y), \quad y = \varepsilon Y, \quad \varepsilon = Gr_{l}^{*-1/5}.$$

is used, which allows displacement of the peculiarities in the
direction of their real position. The rear edge vicinity is studied
on the basis of the two-layer model of the boundary layer. The
major boundary layer, having the same thickness as the one on the
plate, formes initial conditions for a wake behind the plate. As
to the viscous sublayer, its thickness is one order lower and is
practically determined by complete Navier–Stokes and energy equations. The derivatives of higher order along the longitudinal coordinate provide for a change of the boundary conditions from
adhesion conditions on the plate to symmetry conditions on the
axial line of the wake behind the plate.

    Mathematical interpretation of the method is considerably
bulky and for a vertical isothermal plate is given in [14,15].
We shall consider only its peculiarities due to the change of the
boundary conditions. In the boundary layer region internal coordinates are introduced: the transversal coordinate is multiplied by the boundary-layer thickness, and the longitudinal one
is slightly deformed according to the Lighthill method. The longitudinal coordinate is presented in the form of asymptotic expansion, depending on the boundary-layer thickness, the leading term
of the expansion being the longitudinal coordinate proper. Successive asymptotes, according to which asymptotic expansions are
constructed, are determined from the solution of the coupling
equations in external and internal problems, higher order derivatives being maintained in the internal problem. For every approximation of the internal problem, similarity solutions are developed.
This results in $Gr_{x}^{*-1/5}$ series expansion for the stream function
and temperature. This expansion is amplified with the proper solutions. Because of the deformation of the longitudinal coordinate,
in the solution obtained the boundary layer begins a bit lower
the front edge.

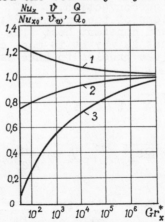

Fig.3. Correction to the boundary layer theory vs Gr_{x}^{*} num-
    ber for dimensionless heat transfer (1),
    wall temperature (2) and flow rate in the boundary
    layer (3).

presents the countrr-current ones.

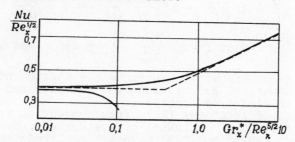

Fig.5. Heat transfer coefficient vs ξ .

Breaking of the lower curve signifies the point of the boundary layer separation. For co-current flows, heat transfer is higher than for free and forced convection taken separately. The greatest deviation from the asymptotic value occurs at $\xi \cong 3$ and is about 20%. Numerical results for other values of Pr numbers are given in [18, 19].

At small ξ , the functions F and Θ are written as a series

$$F(\zeta,\xi) = \sum_{i=0}^{n} \xi^{i} F_{i}(\zeta), \quad \Theta(\zeta,\xi) = \sum_{i=0}^{n} \xi^{i} \Theta_{i}(\zeta). \quad (26)$$

Equations for $F_{i}(\zeta)$ and $\Theta_{i}(\xi)$ are obtained by substituting expansion (26) into equations (25) and by equating the terms with similar powers of ξ . All the equations but the zero one, are presented in a linear form, thus allowing calculation of any approximations of the problem. In [20] , calculation is made of the 1st approximation, and in [21] , three terms of the series (26) for Pr=1 have been evaluated.

In the local similarity method it is assumed that, at low parameter ξ , the derivatives $F(\zeta,\xi)$ and $\Theta(\zeta,\xi)$ in the longitudinal direction can be neglected, the system of equations (25) being transformed into

$$\frac{\partial^3 F}{\partial \zeta^3} + F \frac{\partial^2 F}{\partial \xi^2} \pm 8 \xi \Theta = 0, \quad \frac{1}{Pr} \frac{\partial^2 \Theta}{\partial \zeta^2} + F \frac{\partial \Theta}{\partial \zeta} - \Theta \frac{\partial F}{\partial \xi} = 0. (27)$$

$$F = 0, \quad \frac{\partial F}{\partial \zeta} = 0, \quad \frac{\partial \Theta}{\partial \zeta} = -2 \text{ at } \zeta=0; \quad \frac{\partial F}{\partial \zeta} = 2, \quad \Theta = 0 \quad \text{at } \zeta \to \infty.$$

The value of ξ at every step is assumed constant, and the system of equations is reduced to the ordinary one, simplifying considerably the numerical solution. The accuracy of the method is essentially increased, if the allowance for ξ derivatives is made by the iteration technique. In [22] , calculations have been made up to ξ =2.8 at co-current flows, and heat transfer coefficient in the air 4% too low has been obtained.

The use of the integral methods in the mixed convection problems, because of the absence of similarity, leads to a system of non-linear equations, which must be calculated numerically.

$$\frac{1}{Pr} \frac{\partial^2 \Theta}{\partial \zeta^2} + F \frac{\partial \Theta}{\partial \zeta} - \Theta \frac{\partial F}{\partial \zeta} = 3\xi \left(\frac{\partial F}{\partial \zeta} \frac{\partial \Theta}{\partial \xi} - \frac{\partial \Theta}{\partial \zeta} \frac{\partial F}{\partial \xi} \right)$$

$$F = 0, \quad \frac{\partial F}{\partial \zeta} = 0, \quad \frac{\partial \Theta}{\partial \zeta} = -2 \text{ at } \zeta = 0; \quad \frac{\partial F}{\partial \zeta} = 2, \quad \Theta = 0 \quad \text{at } \zeta \to \infty.$$

The advantage of the above system is the absence of necessity to prescribe in numerical calculations the velocity and temperature profiles in the initial cross-section.

In the system of equations (23), free convection variables (6) can be used, at disturbing influence of the forced convection. However, at $\xi \to \infty$, equations (25) have a logarithmic property, which points at certain difficulties while integrating difference equations. Moreover, variables can not be used when the directions of free and forced convection are opposite.

Numerical calculations of the problem (25) are carried out by the difference method of the second orded of accuracy along both the the coordinates for Pr = 0.7. The results have beeen obtained in the region $0 < \xi < 8$ for co-current flows and up to the stagnation section of the flow (ξ_T =0.09) with opposed free and forced convection.

Velocity and temperature profiles in the boundary layer are

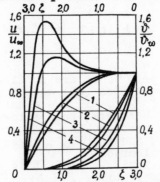

Fig.4. Velocity and temperature profiles :
1. ξ =0; 2 - 0.03; 3 - 1; 4 - 3.

Velocity and temperature profiles in the boundary layer are shown in Fig.4. Notations in the left-hand and lower parts of the figure are related to velocity distribution, while in the right-hand and upper parts, to temperature distribution. The velocity field in the boundary layer limits has a maximum, increasing with the parameter ξ . Temperature distribution is independent of the buoyancy force and its effect on the flow. With the increase of the parameter ξ , the temperature profiles become more complete. The opposite direction of the flow has a minor effect on the temperature profile change in the boundary layer. It is worth noting that the boundary layer thckness increases with decreasing parameter ξ .

The heat transfer coefficient against the parameter ξ is plotted in Fig.5. Dashed lines in logarithmic coordinates refer to asymptotic values of free and forced convection. The upper curve corresponds to the co-current flows, the lower curve

Due to low accuracy, the integral methods are rarely employed in the problems of this kind [4].

Different approaches to the calculation of heat transfer on an electrically heated panel have been analysed and compared above. Further investigation of the problem requires determination of the optimum position of the panel in a building (bearing in mind the comfort of the building) i.e. study of free convection in a rectangular ditch, with different position of the panel and different boundary conditions at the walls. Problems of this kind are explicitly presented in [23,24] and the references quoted.

REFERENCES

1. Sedov, L.I. 1972. Similarity and dimensional methods in mechanics. Nauka. Moscow.

2. Gukhman,A.A. 1973. Introduction to the similarity theory. Vysshaya shkola. Moscow.

3. Huntley, H.E. 1967. Dimensional Analysis. Dover Publications, Inc., New York.

4. Martynenko, O.G., and Sokovishin, Yu.A. 1975. Heat transfer at mixed convection. Nauka i tehnika. Minsk.

5. Brdlik. P.M. 1968. Some problems of heat and mass transfer at gravitational natural convection in an infinite volume. Thesis (Dr. Sci.). NII Stroit. Fiziki, Gosstroy SSSR.

6. Sparrow, E.M., and Gregg, J.L. 1956. Laminar free convection from a vertical plate with uniform heat flux. Trans.ASME, 78:435-440.

7. Gebhart, B. 1973. Natural convection flows and stability. Advances in Heat Transfer, Acad.Press, New York-London. 9:273-348.

8. Rich, B.R. 1953. An investigation of heat transfer from an inclined flat plate in free convection. Trans.ASME. 75:489-499.

9. Darchiya, G.I., and Kusmishvili, G.G. 1972. Free convection heat transfer from vertical and horizontal surfaces at constany heat flux. Bul.sti. Inst.Constr. Bucuresti, 15:61-78.

10. Van Dyke, M. 1975. Perturbation methods in fluid mechanics. The Parabolic Press, Stanford, CA.

11. Oosthuizen, P.H. 1965. An experimental analysis of the heat transfer by laminar free convection from a narrow vertical plate. South Afric.Mech.Eng. 14:153-158.

12. Bobco, R.P. 1959. A closed-form solution for laminar free convection on a vertical plate with prescribed, non-uniform wall heat flux. J.Aero-Space Sci. 26:846-847.

13. Raithty, G.R., and Hollands, K.G.T. 1975. A general method
    for obtaining approximate solutions to laminar and turbulent
    free convection problems. Advances in Heat Transfer, Acad.
    Press, New York-London, 11:265-514.

14. Heber, C.A. 1974. Natural convection around a semi-infinitive
    vertical plate high order effects. Int.J.Heat Mass Transfer,
    17:785-791.

15. Berezovsky, A.A., and Sokovishin, Yu.A. 1975. Method of
    coupled asymptotic expansions in free-convection problems.
    Second All-Union Conference on Modern Heat Convection Problem
    Perm. 19-20.

16. Chang, K.S., Akins, R.G., Bankoff, S.G. 1966. Free convection
    of liquid metals from a uniformly heated verticl plate.
    Ind.Eng.Chem.Fund. 5:26-37.

17. Romanenko, P.N. 1971. Heat and mass transfer and frictions
    at gradient fluid flow. Energiya, Moscow.

18. Oosthuizen, P.H., Hart, R. 1973. A numerical study of laminar
    combined convective flow over flat plate. Trans. ASME, 1:60-6

19. Wilks, G.A. 1974. A separated flow in mixed convection.
    J.Fluid Mech. 62:359-368.

20. Mechigoshi, J., Kikuchi, V. 1969. Buoyancy effects on forced
    convection flow and heat transfer. Mem.Fac.Engng., Kyoto
    Univ. 31:363-380.

21. Wilks, G.A. 1974. The flow of a uniform stream over a semi-
    infinite vertical flat plate with uniform surface heat flux.
    Int.J.Heat Mass Transfer, 17:743-753.

22. Wilks, G. 1973. Combined forced and free convection flow on
    a vertical surface. Int.J.Heat Mass Transfer, 16:1958-1964.

23. Luikov, A.V., and Berkovsky, B.M. 1974. Convection and heat
    waves. Energiya, Moscow.

24. Chu, H.H.S., Churchill, S.W., Patterson, C.V.S. 1976. The
    effect of heater size, location, aspect ratio and boundary
    conditions on two-dimensional, laminar, natural convection
    in rectangular channels. Journ. Heat Transfer, Ser.C.
    98:49-57.

STUDY OF VENTILATION EFFICIENCY IN AN EXPERIMENTAL ROOM: THERMAL COMFORT AND DECONTAMINATION

L. LARET, J. LEBRUN, D. MARRET, AND P. NUSGENS

Thermodynamic Institute
University of Liège (Val-Benoît)
B-4000 Liege, Belgium

ABSTRACT

The present study covers 75 tests performed in a climatic room representing a heated dwelling room. The tests on the comfort involve measurements of the air velocity in the ventilation jet and in the occupied zone. The results of these tests show that it is necessary to increase the supply velocity as the corresponding temperature is lowered, in order to avoid a fall of the jet by buoyancy effect.

In the tests on the ventilation efficiency one observes the "response" of the room to a brutal contamination by a tracer gas.

Contamination tests indicate that the internal air mixing is nearly always "perfect". This can be explained by the effect of natural convection induced by the heating source. For practical cases, one can thus suppose that these conditions of perfect mixing are well obtained.

NOMENCLATURE

A	net area of supply outlet (m^2)
A_r	Archimede number
c	specific heat at constant pressure ($J\ kg^{-1}K^{-1}$)
C	concentration of carbon dioxide in the room at instant τ (%)
C_a	concentration of carbon-dioxide in the ventilating air (%)
C_o	concentration of carbon-dioxide in the room after injection (%)
$\dot{E}"$	buoyancy per source length unit ($N\ m^{-1}s^{-1}$)
g	gravity acceleration ($m\ s^{-2}$)
I	momentum flow (N)
n	actual air change rate (s^{-1})
n'	fictive air change rate (s^{-1})
Q	total emission (W)
Q_c	convective emission (W)

Q'_c linear convective emission (per length unit of heat source (W m^{-1})

t_o outside temperature (°C)

t_s supply air temperature (°C)

t_R resultant temperature (°C)

$\overline{u_s}$ supply velocity (m s^{-1})

V   room volume (m^3)

$\dot{V}'$ linear flow (m^2 s^{-1})

$\dot{V}_s$ supply flow rate (m^3 s^{-1})

x   coordinate in direction of flow jet (m)

z   vertical deviation of the jet (m)

Greck

α   entrainment constant

β   thermal expansion coefficient (K^{-1})

Δt   temperature difference (K)

μ   ratio between thermal and dynamic thickness of the jet

ρ   air density (kg m^{-3})

ρ∞   density outside the jet (kg m^{-3})

τ   time (s)

η   efficiency (%)

INTRODUCTION

In a controlled mechanical ventilation system with double air-flow the supply terms (location of the outlet, orientation, discharge velocity) and exhaust terms (location of the inlet) can be chosen so as to satisfy the comfort requirements and to obtain the best possible "efficiency" for the elimination of air contaminants. The experimental study dealt with hereafter has been performed in a climatic room representing a dwelling room. The study covers 75 tests spread over two series :

- in a first series one measures the air temperatures and velocities inside the room in order to make a good description of the flow pattern and to discover any risk of discomfort due to draught within the occupied zone

- in a second series one observes the response of the room to a brutal initial contamination by a tracer gas (CO_2).

EXPERIMENTAL CONDITIONS

The climatic room [1] [2] [3] has a volume of 3.45 x 4.80 x 2.70 = 44.7 m^3; it consists of one exposed wall called hereafter frontage (3.45 x 2.70 m) with double glazing (2.13 x 1.60 m). Under the nominal thermal conditions (t_R = 22°C, t_o = -3°C), taking into account a minor cooling of the internal walls, the transmission heat losses are about 550 W. To this term must be added the enthalpy flow of the ventilation, which varies with the airflow rate and according to the difference between the supply and exhaust air temperatures.

The room is heated by a single panel radiator under the window.

For the ventilation one considered :

- the three following airflow rates $\dot{V}_s$ = 0.007 , 0.014 and 0.028 $m^3 s^{-1}$
- several supply velocities $\overline{u}_s$ = 1.5 to 6 m s^{-1}
- several supply temperatures $\overline{t}_s$ = 10 [x] to 22°C [xx]
- 5 locations (see Fig. 1)

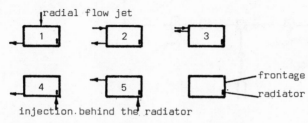

Fig. 1 - Locations of the supply and exhaust openings

The locations 2 and 3 are of course the most economical (air distribution through the false ceiling of the corridor on to the room and air return through the door).

MEASURING TECHNIQUES

- An automatic translation movement apparatus allows for the measurement of the air temperatures and velocities at any point of the room by means of two thermal anemometers [2]. The two magnitudes are alternatively measured every second at fixed points. Each recording is made during periods from 1 to 2 minutes because of the importance of the turbulence.

- For the decontamination tests one injects the tracer gas CO_2 directly into the ventilation circuit. One purges with nitrogen before and after the CO_2 injection in order to obtain a rigorously rectangular impulsion (Fig. 2).

The CO_2 concentrations are measured at nine points inside the room (Fig. 3). A tenth point allows to verify the fresh air concentration (in the laboratory where the air comes from).

These measurement points are connected to an infra-red absorbing gas analyser (ONERA type 80) by means of an electrovalve selector. This system has also a calibration equipment (Fig. 4).

The bulk of the CO_2 injection and measurement equipment is connected to a measurement unit. A digital clock operates the tracer gas injection and the start of the measurement cycles (Fig. 5). These are repeated every 5 minutes (30 seconds per point) during the whole decontamination of the room.

All the measurements are recorded on a punched tape and computed.

(x) 10°C : heat exchanger with efficiency $\eta \simeq$ 50%
(xx) 22°C = t_R luxury solution with re-heater

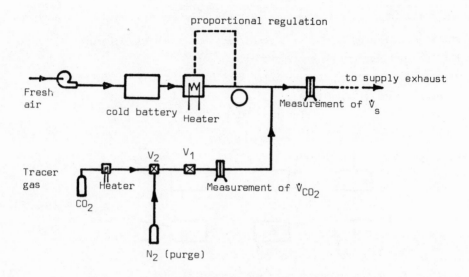

Fig. 2 - Air circuit and tracer gas injection circuit

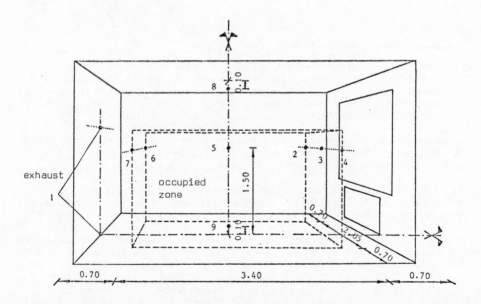

Fig. 3 - Location of the measurement points of the CO_2 concentration in the room

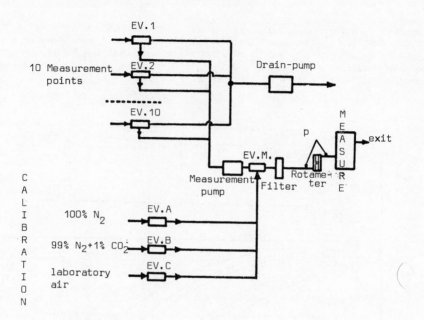

Fig. 4 - CO_2 measurement circuit

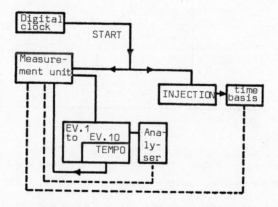

Fig. 5 - Operating diagram of injection and measurement equipment

RESULTS OF THE AIR TEMPERATURE AND VELOCITY MEASUREMENTS : RISKS OF DISCOMFORT

The distributions of the air temperatures are rather uniform in all cases within the occupied zone. The air velocities are also always below 0.1 m s^{-1}, except with the already mentioned most economical supply (locations 2 and 3, see Fig. 1). In this last case, the horizontal supply jet can be deviated downwards due to the buoyancy effect (see examples shown on Fig. 6). From the point of view of thermal comfort, the penetration of the jet into the

occupied zone is unacceptable; it causes excessive mean velocities
(> 0.3 m s^{-1}). The two examples of Fig. 6 are based on same dynamic condi-
tions $\dot{V}_s$ = 0.028 m^3 s^{-1} and $\bar{u}_s$ = 1.5 m s^{-1} except that the supply temperature
has been modified : $t_s \simeq 22°C$ and $t_s \simeq 16°C$ respectively, for a resultant
temperature in the centre of the room t_R = 22°C.

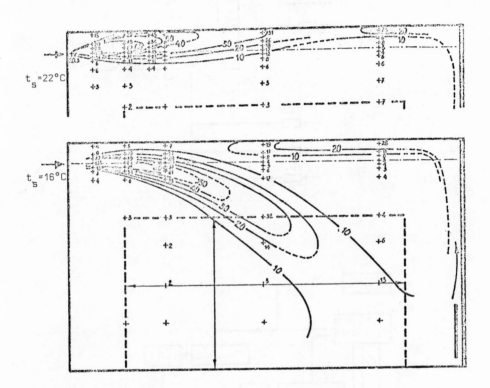

Dynamic supply conditions : $\dot{V}_s$ = 0.028 m^3 s^{-1} and $\bar{u}_s$ = 1.5 m s^{-1}
Inside temperature ; t_R = 22°C

Fig. 6 - Isovelocities in longitudinal symmetry plane of the room

The deviation (z) of the centre line of the jet can be evaluated as a
function of the Archimede number (A_r) and of the horizontal coordinate (x)
(see Fig. 7) by a relation of the type :

$$\frac{z}{\sqrt{A}} = K\, A_r \left| \frac{x}{\sqrt{A}} \right|^3 \qquad\qquad [4]\ [5]\ [6]\ [7]$$

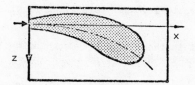

Fig. 7 - Jet deviation z = f(x) due to the buoyancy effect

where A = net area of the supply outlet = $\dfrac{\dot{V}_s}{\overline{u}_s}$

$$A_r = \frac{g \, \beta \, \Delta t \, \sqrt{A}}{(\overline{u}_s)^2} = \frac{g \, \beta \, \Delta t \, \dot{V}_s^{1/2}}{(\overline{u}_s)^{5/2}}$$

with $\Delta t \simeq t_R - t_s$ assimilating t_R with the temperature outside the jet.

Thus in the 2 cases presented in Fig. 6 one has respectively A_r = 0 (isothermal jet) and A_r = 0.012.

If the "Coanda" effect (distance between outlet and ceiling ⩾ length of the outlet) can be neglected and if the net area A differs only slightly from the core area of the outlet, one finds according to Jackman [4] :

    K ∿ 0.04

Other authors [5] [6] [7] propose slightly higher values : K ∿ 0.05 to 0.1, but our experimental results seem to concord better with Jackman's estimation.

In order to avoid any risk of discomfort one must impose a minimum supply velocity depending on the airflow rate and the temperature differential at the supply outlet. Thus for the location of the outlet and the dimensions of the room considered here, one avoid the penetration of the jet into the occupied zone if z ⩽ 0.5 m for x ≃ 4 m which requires that

$$\overline{u}_s \geqslant 0.3 \, \frac{\Delta t^{2/3}}{\dot{V}_s^{1/3}}$$

For instance in the case of Fig. 6, with the supply temperature t_s = 16°C one should have taken $\overline{u}_s \simeq 3.3$ m s^{-1} in order to avoid any risk of discomfort.

RESULTS OF THE TRACER GAS MEASUREMENTS

The climatic room is contaminated by injection of a CO_2 flow rate of $\dot{V}_{CO_2}$ = 2.2 x 10^{-3} m^3 s^{-1} during τ_i = 120 seconds, so as to obtain a theoretical amplitude $(C_o - C_a) \simeq 0.6$ %.

This contamination is in fact very near to a Dirac impulsion for magnitude scale considerated here.

If the mixing were perfect in the room, the decontamination should, at each room point, follow the law of the type :

$$C - C_a = (C_o - C_a) \left[\frac{1 - e^{-n \tau_i}}{n \tau_i} \right] e^{-n \tau} \qquad \text{(see Fig. 8)}$$

where $(C_o - C_a) = \dfrac{V_i}{V}$ with V_i = injected CO_2 volume = $2.2 \times 10^{-3} \times 120$ m^3 = $\underline{0.26 \text{ m}^3}$

and V = room volume = $\underline{44.7 \text{ m}^3}$

$(C_o - C_a) \simeq 0.6$ %

$n = \dfrac{\dot{V}}{V}$ = air change rate

This response is slightly different compared with the one which would be caused by a theoretical impulsion (Dirac funtion) :

$$(C - C_a) = (C_o - C_a) e^{-n \tau}$$

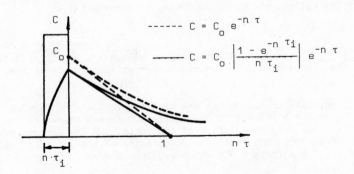

Fig. 8 - Perfect mixing : response to an impulsion of duration τ_i

If the mixing is not perfect, there will appear several decontamination laws in the different measurement points in the room and in principle the experimental results have to be correlated with the equations associating several elementary flow patterns ("perfect piston", "piston with longitudinal dispersion", "short-circuit", etc) [8] [9] [10].

But a first approach (which has proved to be sufficient for this study) consists in interpreting these results as if the mixing were perfect, i.e. in correlating the decontamination at each point with a law of the type :

$$C = C_o e^{-n' \tau} \qquad \text{or} \qquad \boxed{\log_e C = \log_e C_o - n' \tau}$$

i.e. a linear regression where $\log_e C$ and τ are the variables, where $\log_e C_0$ ($C_0 \simeq$ initial concentration) is the intercept and where n' (fictive air change rate) is the slope.

The difference between the fictive air change rate n' and the actual air change rate n allows then to appreciate how far the internal air circulation effectively differs from a perfect mixing. One then must achieve an initial very uniform contamination; for this two small fans have to be used to mix the air in the room only during the tracer gas injection (one has been able to prove that this perturbation had no noticeable repercussion whatsoever). These small fans are also used to determine the actual air change rate n by means of the tracing gas method : in that case, they operate during the whole decontamination so as to achieve as much as possible the assumption of the perfect mixing.

Some examples of typical results are given in table 1 and Fig. 9 : they deal with the location 2 (see Fig. 1).

Table 1 - Examples of tracer gas CO_2 measurement results

Tests nr.	1	2	3	4
Supply conditions : airflow rate ($m^3 \ s^{-1}$)	0.014	0.029	0.015	0.015
velocity ($m \ s^{-1}$)	1.5	1.5	5.3	5.3
temperature (°C)	22	22	9.1	9.8
Differences with the perfect mixing :				
$\frac{n' - n}{n}$ (in %) in the centre 5	+ 19	+ 7	+ 12	-
at the ceiling 8	+ 24	+ 17	+ 10	0
at the floor 9	+ 68	+ 42	+ 16	- 1
Correlation coefficient in the centre r_5 :	.9976	.9992	.9998	.9995
Heterogeneity of the contamination :				
$\frac{C - C_5}{C_5 - C_a}$ (in %) at the moment n τ =	0.11	0.27	0.18	0.19
at point 1	+ 20	- 4	+ 5	- 1
2	- 1	+ 9	+ 3	0
3	- 1	+ 8	- 3	- 5
4	- 4	+ 4	- 5	- 1
6	+ 11	+ 16	+ 2	0
7	+ 9	+ 18	+ 1	+ 1
8	- 15	- 16	- 10	+ 1
9	+ 33	+ 31	+ 5	0
at the moment n τ =	1.04	1.05	1.07	1.14
at point 1	- 10	- 4	+ 3	- 2
2	- 1	+ 2	+ 1	- 4
3	- 5	- 2	- 2	- 5
4	- 7	+ 1	- 6	- 1
6	- 6	- 2	- 1	+ 1
7	- 7	+ 8	+ 5	+ 2
8	- 24	- 22	- 16	+ 3
9	- 9	- 7	+ 1	+ 4

(C_a = outside air concentration)

tests 1 and 2 : isothermal conditions : $t_R = t_0 = 22°C$
tests 3 and 4 : nominal conditions : $t_R = 22°C$
$t_0 = -3°C$
test 4 : small fans operating during the whole decontamination ("perfect mixing").

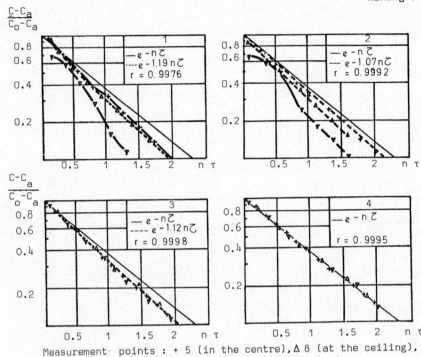

Measurement points : + 5 (in the centre), Δ 8 (at the ceiling), ▽ 9 (at the floor). Test conditions : See table 1.

Fig. 9 - Comparaison of the experimental decontaminations with the perfect mixing assumption

The tests 1 and 2 are performed in strictly <u>isothermal</u> conditions (suppression of the natural convection in the room) : one ascertains a certain heterogeneity of the CO_2 concentration : thus a short time after the contamination ($n \tau \approx 0.1$ to 0.3), the concentration near the floor is comparatively high with the other points of the room :

$$\left(\text{see } \frac{C_q - C_5}{C_5 - C_a} - \text{table 1} \right).$$

As it is precisely on this level that the air exhaust takes place, the result is a ventilation efficiency which is <u>apparently</u> superior to the perfect mixing ($n' > n$).

In fact these are very <u>artificial conditions</u> : as soon as a certain anisothermy is re-introduced in the room, the heterogeneity of the concentration is quickly reduced and one comes near perfect mixing conditions. It therefore seems illusory to search for a more elaborate interior flow pattern (example : compare the actual terms of test nr. 3 with the perfect mixing performed in test nr. 4).

This ascertainment is verified by the whole of the research results (disper-sion : $\frac{n' - n}{n}$ about $\pm$ 10%). Finally, with the exception of certain tests in strictly isothermal conditions and a minor momentum flow, one ascertains always a strong mixing of the air in the room. This mixing is thus not only induced by the ventilation jet, but also by the natural convection jet emit-ted by the heating source.

The room mixing flow $\dot{V}_r$ caused by the ventilation jet cannot be evaluated correctly : the vicinity of the ceiling determines a transversal spread of the jet and one observes an evolution of the axial velocity which differs from the one of the free jet.

One can expect a law of the type :

$$\dot{V}_r \div (\dot{I})^{1/2}$$

with $\dot{I} \approx \rho \, \dot{V}_s \, \bar{u}_s$

and the experience shows that the mixing is already practically perfect ($\left|\frac{n' - n}{n}\right| \leqslant$ 10%) when $\dot{I} \geqslant 0.1$ N.

But this condition is not even imperative taking into consideration the effect of the natural convection.

One knows that the heating body along the frontage can, with a good approxi-mation, be assimilated with a linear natural convection source. The air-flow is sensibly half the one induced by a double source (application of the method of the images to take account of the wall effect) [11] [12].

For a simple linear source, one can prove that this flow rate is :

$$\dot{V}' \approx \frac{(2)^{5/6}}{(\pi)^{1/2}} \, (1 + \mu^2)^{1/6} \, (\alpha)^{2/3} \, (\frac{\dot{E}''}{\rho\infty})^{1/3} \times$$

with $\mu \approx 0.9$

$\alpha \approx 0.16$ [13] [14]

$\dot{E}'' = g \, \beta \, \frac{1}{c} \, \dot{Q}'_c$

$\dot{Q}'_c$ = linear convective emission (per length unit of the source).

The application of this theory to the case studied here is relatively more certain than for the ventilation jet [11] [12]. Taking into account the dimensions of the room and the radiator length ($\sim$ 2 m), one thus has the mixing flow :

$$\dot{V}_r \approx 0.01 \, \dot{Q}_c^{1/3} \quad m^3 \, s^{-1}$$

with $\dot{Q}_c \approx 0.65 \, \dot{Q}$ (convective radiator emission + re-emission of the wall behind the radiator)

$\dot{Q} \approx 550$ W + ventilation enthalpy in nominal conditions.

One has thus $\dot{Q}_c$ > 350 W and consequently $\dot{V}_r$ > 0.07 m^3 s^{-1}, which means
that the continuous mixing flow is always much higher than the ventilation
flow and this suffices to explain the "perfect mixing" ascertained in most
of the decontamination tests.

CONCLUSIONS

When conceiving ventilation systems, one must take into consideration certain
risks of discomfort; in the case of horizontal air supply, one must avoid
that the ventilation jet falls into the occupied zone due to the buoyancy
effect.

This requirement leads to imposing a minimum supply velocity depending on
the temperature differential and the corresponding flow rate.

The decontamination efficiency is practically always the same whatever the
supply and exhaust terms : it corresponds to the perfect mixing, thanks to
the interior natural convection effect. (In fact, one can always not only
rely on the heating source, but also consider the convective emission of the
occupants, the lighting, the possible insolation, etc.).

Efficiency tests in strictly isothermal conditions (and a fortiori on other
patterns which do not take account of the natural convection) would reveal
significant differences in efficiency (especially with low momentum flow),
but these results are without any practical significance.

Finally the tracer gas method is very convenient for measuring the actual
ventilation flows in situ. The interior air can always be mixed artificial-
ly by one or more fans in order to satisfy the assumption of the perfect
mixing.

ACKNOWLEDGEMENTS

This research is part of an IC-IB/IRSIA program "Circuits aérauliques -
Ventilation mécanique contrôlée" run with the cooperation of the Centre
Scientifique et Technique de la Construction (C.S.T.C.).

The results presented could be obtained thanks to this collaboration and
to the active participation of the staff of the C.S.T.C. and the Université
de Liège.

REFERENCES

[1]   LEBRUN, J. - MARRET, D. : Etude du confort thermique et de la con-
                sommation d'énergie dans les conditions d'hiver.
                Rapport de Recherche IC-IB. Mai 1975.

[2]   LARET, L. - LEBRUN, J. - MARRET, D. - NUSGENS, P. - WANNYN, J.P. :
                Confort thermique et consommation d'énergie dans les con-
                ditions d'hiver. Collection des Publications de la Fa-
                culté des Sciences Appliquées de l'Université de Liège,
                n° 56, 1975.

[3]   HANNAY, J. - LARET, L. - MARRET, D. - NUSGENS, P. - WANNYN, J.P. :
                Confort thermique et consommation d'énergie dans les
                conditions d'hiver. C.S.T.C. revue n° 2, june 1976.

[4]   JACKMAN, P.J. : Air movement in rooms with side-wall mounted grille
                design procedure. HVRA Laboratory, Report nr. 65, 1970.

[5]   BATURIN, V.V. : Fundamentals of industrial ventilation.
                Pergamon Press, 1972.

[6]   RECKNAGEL, H. - SPRENGER, E. : Taschenbuch für Heizung, Lüftung und
                Klimatechnik. R. Oldenbourg, München 1970.

[7]   CROOME-GALE, D.J. - ROBERTS, B.M. : Air conditioning and ventilation
                of building. Pergamon Press, 1975.

[8]   LEVENSPIEL, O. - BISCHOFF, K.B. : Patterns of flow in chemical pro-
                cess vessel 3. Advances in Chemical Engineering, vol. 4,
                pp. 95-198, Academic Press, N.Y. 1963.

[9]   DANCKWERTS, P.V. : Continuous flow systems. Chem. Eng. Sc. 2 (1),
                1/1953.

[10]   CHEN, M.S.K. - FAN, L.T. - HWANG, C.L. - LEE, E.S. : Air flow
                models in a confined space. A study in age distribution.
                Building Science, vol. 4, pp. 133-144, Pergamon Press,
                1969.

[11]   DAWS, L.F. : Movement of air stream indoors. Symposium of the
                Society for General Microbiology, Nr. XVII, Airborne
                Microbes, 1967.

[12]   LEBRUN, J. - MARRET, D. : Convection exchanges inside a dwelling
                room in winter. The 1976 Seminar of the International
                Centre for Heat- and Mass-Transfer "Turbulent buoyant
                convection".

[13]   SHAO-LIN LEE - EMMONS, H.W. : A study of natural convection from
                a vertical plane surface. A.S.M.E./C. Transactions,
                February 1968.

[14]   BATCHELOR, G.K. : Heat convection and buoyancy effects in fluids.
                Quarterly Journal Royal Meteorological Society, Nr. 80,
                1954, p. 339.

AIRFLOW IN A ROOM AS INDUCED BY NATURAL CONVECTION STREAMS

H. EUSER, C. J. HOOGENDOORN, AND H. VAN OOIJEN*

Department of Applied Physics
Technical University of Delft
Delft, The Netherlands

ABSTRACT

An experimental study is presented describing the air flow in a room induced by natural convection streams both in a full scale climate chamber and in a 1 to 10 scale-model. The air flows have been investigated. The flow was characterized by the angle under which the cold downdraught of the window and the rising warm air from the radiator placed below it, flows into the room after mixing. The angle appeared to be a function of the ratio of the heat contents of both flows. The measured heat flow for the laminar boundary layers agreed well with literature data, but for the turbulent case there were substantial differences. In particular the measured velocity profile differed from data of Cheesewright. The effects on the angle caused by the form and size of the window sill and the use of glass curtains could be predicted qualitatively from the model experiments.

NOMENCLATURE

a	thermal diffusivity	(m^2/s)
C	constant	
c_p	specific heat at constant pressure	(J/kgK)
g	acceleration of gravity	(m/s^2)
L	length	(m)
M	scale reduction factor	
n	length scale normal to the surface	(m)
Q	total heat flux	(W)
T	temperature	(K)
v	velocity	(m/s)
α	heat transfer coefficient	(W/m^2K)
β	volume expansion coefficient	$(-/K)$
γ	angle	(deg)
ρ	density	(kg/m^3)
μ	viscosity	(Ns/m^2)
λ	thermal conductivity	(W/mK)
ν	kinematic viscosity	(m^2/s)

*Presently at the Unilever Research Laboratory, Vlaardingen, Netherlands

SUBSCRIPTS   DIMENSIONLESS NUMBERS

c   cold boundary layer   Ar = Archimedes number $\dfrac{g\beta L(T-T_o)}{v^2}$
f   full scale
h   hot boundary layer   Nu = Nusselt number $\dfrac{\alpha L}{\lambda}$
l   laminar
m   model   Pr = Prandtl number $\dfrac{v}{a}$
o   reference value
t   turbulent   Ra = Rayleigh number $\dfrac{g\beta L^3(T-T_o)}{va}$
w   wall
$-$ vector   Re = Reynolds number $\dfrac{vL}{v}$

INTRODUCTION

One of the aims of a heating system is to produce optimal thermal confort conditions for the occupants of a room. In order to prevent difficulties
during the installation of the heating system, a definitive choice between the
various types of heating systems has to be made during the design phase of
building. If mechanical ventilation is necessary to provide fresh air, and
if cooling is required during summer, in general an air conditioning system
is choosen. This is mostly the case for large buildings in which a high percentage of glass is used. The hot water radiator system is more frequently
used for individual houses because of its simplicity. In this article we will
concentrate our attention to the latter system.

Because high heat losses take place through single pane windows, radiators are most frequently placed below these windows. Additional radiators may
be placed along cold walls. The cold window pane creates a cold downward air
flow; the radiator generates a warm upward air stream. Combination of the cold
and the hot air streams creates a jet that flows into the room. Under certain
circumstances this will result in thermal discomfort for someone near the
window, depending on the temperature and velocity profile of the jet.

There are several possible ways of predicting the air flow induced by
natural convection streams in a room. A model of the room may be constructed
at full scale, with installed radiators, cold walls and to measure the air
velocities and temperatures.   Though this would be the most reliable
method, the costs are normally too high and the procedure not only takes too
much time but lacks flexibility. A second possibility is to calculate air
velocities and temperatures with a computer model. The basic equations describing the flow - the equations of continuity, motion and energy - have to
be simplified before they can be solved on the computer. This is because a
complete mathematical description of turbulent motion has yet to be developed,
though simplified models based on turbulent kinetic energy and dissipation of
turbulent kinetic energy (k-ϵ model) can provide valuable results [1].
For many practical applications small scale physical models can provide information both more cheaply and more quickly than the afore mentioned methods.
For the reduced scale model similarity principles have to be applied. In practice, this leads to difficulties when the heat transfer through both turbulent
and laminar free convection boundary layers is involved or if the flow by
ventilation is also of importance. It has been the subject of the present
study to find a method to overcome these difficulties.

SCALING

The relation between the original prototype and the reduced scale model,
is given by similarity principles. These are: geometrical similarity, fluid
dynamic similarity and thermal similarity. The required scaling rules can be
derived from the basic equations: the continuity equation (1), the Navier-
Stokes equation (2) and the energy equation (3). If we leave out the time
dependant terms, the terms for mechanical energy and viscous dissipation and
assuming constant fluid properties, these equations can be written as:

$$\nabla \underline{v} = 0 \tag{1}$$

$$\rho(\underline{v}.\nabla). \ \underline{v} - \mu\nabla^2\underline{v} + \rho g\beta(T-T_o) = 0 \tag{2}$$

$$\rho(\underline{v}.\nabla). \ T - \alpha\nabla^2T = 0 \tag{3}$$

In equation (2), buoyancy forces due to temperature induced density gradients are included. We used the Boussinesq approximation, in which all fluid properties are kept constant except for the density in the gravitational term in equation (2). Further the hydrostatic terms in equation (2) have been omitted.

To solve these equations, we need a set of boundary conditions. As we intend to study the case of a closed room (with an exception for the case of ventilation, that will be discussed later) the boundary conditions for velocity and temperature can be written as:

$$\underline{v} = 0 \text{ on the surface} \tag{4}$$

and $\quad\quad\quad$ T or $\dfrac{\partial T}{\partial \underline{n}}$ prescribed on the surface $\tag{5}$

If thermal radiation is involved, the boundary conditions are more complicated.

By making the equations (1) to (3) dimensionless, it can be shown that for complete similarity the Prandtl number, the Reynolds number and the Archimedes number should be identical in the model and the prototype. The Prandtl number is constant because we want to use air both in the model and the prototype. If the length scale L_m in the model is reduced with a factor M; the velocities in the model should be M times as large as in the prototype to keep the Reynolds number constant. In order to keep the Archimedes number constant as well, the temperature differences in the model should be increased with a factor M^3 compared to the differences in the prototype. It is convenient to use a scaling factor of about M = 10. This requires temperature differences in the model to be 1000 times larger than in the prototype, a practical impossibility!

Müllejans [2] has found that for Re>60 the general streamline pattern, which is mainly governed by free turbulence, is independent of the value of the Reynolds and the Prandtl numbers and depends only on buoyancy forces, thus on the Archimedes number. This is because the structure of turbulence at a sufficiently high level of velocity will be similar at different mean velocities and therefore be independent of the Reynolds number. Likewise, the transport of thermal energy by turbulent eddies will dominate the molecular diffusion and will therefore be independent of the Prandtl number.

If we ignore the Reynolds and the Prandtl numbers, we can reduce the velocity in the model to a value at which the flow is still suitably turbulent. This means that with air, and model dimensions $L_m = 0.3$ m, the general stream is turbulent for velocities $v_m > 0.003$ m/s. In this way it is possible to keep the Archimedes number constant even with equal temperature scales in model and prototype. The velocities v_m in the model then will be $M^{\frac{1}{2}}$ times smaller than the velocities v_p in the prototype room.

For laminar flow, model measurements however are improper in such a model. A complication in this respect arises with the natural convection boundary layers along cool or hot walls.

NATURAL CONVECTION BOUNDARY LAYERS

For the laminar free convection vertical boundary layer we have the relation between the average Nusselt number and the Rayleigh number:

$$\overline{Nu} = C_\ell. \ Ra^{\frac{1}{4}} \tag{6}$$

whereas for the case of the turbulent free convection vertical boundary layer
the following relation is used:

$$\overline{Nu} = C_t \cdot Ra^k \tag{7}$$

The relation between the Rayleigh number and the Archimedes number is:

$$Ra = R_e^2 \cdot P_r \cdot A_r \tag{8}$$

For the laminar boundary layer there is a good agreement between different
authors in the literature for the constants. Francke [3] gives a survey of
several authors that find values for C_ℓ between 0.47 and 0.54. Cheesewright
[4] finds a measured value of 0.51 and McAdams [5] gives a value of 0.54 as
an average of several measurements. If we calculate the average Nusselt
number from the theoretical temperature and velocity profile in the boundary
layer as given by Schmidt and Beckmann [6], we find a value of 0.48 for C_ℓ.
A good average therefore will be $C_\ell = 0.51$.

In the case of turbulent boundary layers the differences between the
results of several investigators are rather large. In figure 1 various results
are shown and the values found for C_t and k are tabulated.

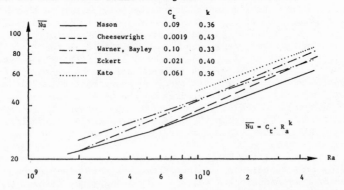

Figure 1. Average Nusselt number for turbulent free convection vertical boundary layers.

Calculations were carried out by Bayley [7], Eckert [8], Kato [9] and Mason
[10]. Cheesewright determined the heat transfer from the measured temperature
gradient in the boundary layer close to the wall. The average Nusselt number
is found by integration of the local Nusselt number. Warner [11] calculated
the average Nusselt number from heat flow measurements, made with a plate that
was divided in small, electrically heated sections. In the turbulent region
($Ra > 2.10^9$)Warner found a smaller local Nusselt value than Cheesewright.
He assumed that this was caused by a decrease of the temperature gradient very
close to the wall within the last 0.5 mm. Cheesewright did no measurements so
close to the wall, so could hot have detected this effect.
The differences for the average Nusselt number are rather high; the value of
the exponent of the Rayleigh number that is found, lies between 0.33 and
0.43. However, the results of Bayley with $C_t = 0.10$ and $k = 0.33$ seem to be
a good average. From this, it follows that for turbulent boundary layers, the
average heat transfer coefficient is nearly independent of the length scale.
For the laminar case, the average heat transfer coefficient is proportional to
$L^{-\frac{1}{4}}$. This means, that scaling of the heat transfer through the boundary layers
is different for laminar and turbulent free convection. Therefore, it is important to know a criterion that tells us whether a free convection boundary
layer is still laminar or already turbulent.

In general, the value of the Rayleigh number is used as a turbulence criterion, but there is a large discrepency between various authors.
Cheesewright has a turbulent situation for Ra>7.10^9 Parczewski [12] finds a turbulent boundary layer for Ra>7.10^7, Davidovici [13] for Ra>2.10^7 and Baturin [14] finds even that for Ra>4.10^5 the boundary layer is already turbulent in an enclosure. It might be that these differences have been caused by differences in free stream turbulence of the surroundings or by a different way of interpretation.

In his paper, Cheesewright gives dimensionless temperature and velocity profiles for the turbulent free convection boundary layer. In figure 2 his results are shown. The temperature profile fits well with the experimental results of Warner, and of Vliet [15], as well as with the numerical calculations of Masons, who used a simple turbulence model. For the velocity profile, the results of Mason are about 25% higher than Cheesewright's experimental results. From the given temperature and velocity profiles it is possible to calculate the average Nusselt number $\overline{Nu}$ from an integral heat-balance over the thickness of the boundary layer at a distance L from the leading edge.

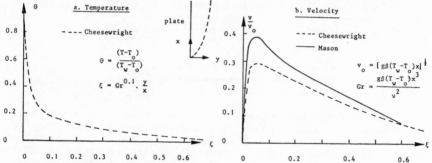

Figure 2. Dimensionless temperature and velocity profile for turbulent free convection vertical boundary layers.

We have calculated the $\overline{Nu}$ values in this way both from Cheesewright's and from Mason's temperature and velocity profiles. The results of these calculations are shown in figure 3 and have been compared with the values as given by the authors themselves. For Mason the two curves should fall close together; the small difference may well be caused by inexact readings taken from the graphs shown in his article. The large difference that is found between the results of Cheesewright and our calculations cannot be explained in this way. We must conclude therefore, that the velocities as measured by Cheesewright in his experiments are too small.

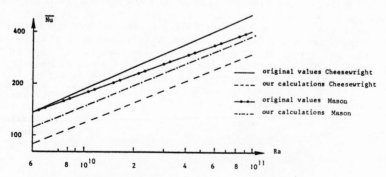

Figure 3. Average Nusselt values calculated with the integral heat balance method.

HEAT TRANSFER

To ensure thermal similarity between the model and the prototype, it is necessary that the relative importance of heat fluxes of different origin remains unchanged. This should be true for the convection heat fluxes and the ventilation heat fluxes. If we use equal temperature differences in model and full scale, and we have a scale reduction factor M, we get for the ratios of the total heat fluxes Q:

$$\frac{Q \text{ by ventilation in the model}}{Q \text{ by ventilation full scale}} = \frac{L_m^2 v_m \Delta T_m}{L_p^2 v_p \Delta T_p} = M^{-5/2} \tag{9}$$

$$\frac{Q \text{ turbulent boundary layer in the model}}{Q \text{ turbulent boundary layer full scale}} = \frac{\alpha_{mt} L_m^2 \Delta T_m}{\alpha_{pt} L_p^2 \Delta T_p} = M^{-2} \tag{10}$$

$$\frac{Q \text{ laminar boundary layer in the model}}{Q \text{ laminar boundary layer full scale}} = \frac{\alpha_{ml} L_m^2 \Delta T_m}{\alpha_{pl} L_p^2 \Delta T_p} = M^{-7/4} \tag{11}$$

$$\frac{Q \text{ laminar boundary layer in the model}}{Q \text{ turbulent boundary layer full scale}} = \frac{\alpha_{ml} L_m^2 \Delta T_m}{\alpha_{pt} L_p^2 \Delta T_p} = C_{1t} M^{-2} L_m \tag{12}$$

If we have turbulent boundary layers both in the prototype and the model, we see from relation (10) that the heat flux densities remain constant as the total heat fluxes reduce by a factor M^2.
With isothermal ventilation, when the ventilation air temperature is equal to the average room temperature, there are no ventilation heat fluxes and scaling may give good results as will be shown later. For nonisothermal ventilation in general it is not possible to scale the ventilation heat fluxes property in a simple way and at the same time to keep the Archimedes number constant. For laminar boundary layers in model and prototype, the heat fluxes scale with a factor $M^{-7/4}$. The most difficult situation in model measurements will be the case where the boundary layers are turbulent in the prototype and laminar in the model. From relation (12) it can be seen, that the ratio of the heat fluxes through the boundary layers is not a constant value but depends on the place in the model. To overcome this problem, we have to remember that we are interested in the heat contents of the boundary layers because without ventilation the thermal buoyancy forces create the air flow pattern in the room. The boundary layers, coming from the cold window and the hot radiator respectively, combine to form a jet that flows into the room. The jet may be characterized by the angle γ between the centre-line of the jet and the horizontal, and by its temperature and velocity profiles. This is schematically shown in figure 4. In general, jets are characterized by inertial forces and buoyancy forces. If buoyancy forces dominate, the jet is called a plume. As this is the case in our experiments, we assume that the angle γ of the resulting jet depends on the total heat fluxes Q_h and Q_c of the two boundary layers and the geometrical situation. We expect a relation of the type:

$$\gamma = f \left(\frac{Q_h}{Q_c}\right) \tag{13}$$

We now are able to overcome the difficulty of thermal scaling of the heat fluxes in the boundary layers for the general case that not all the boundary
layers are turbulent. We simply introduce in the model the typical geometrical
situation of window, radiator and parapet and measure the dependance of the
angle γ on the ratio of the heat contents. The values of Q_h and Q_c of the
boundary layers in the prototype that are required to be able to use relation
(13) may be found by using formula (6) or (7).

EXPERIMENTAL SETUP
        The room type that is most commonly found in houses and offices, is
a room with one outside wall and with the other walls at nearly the same temperatures as the room itself. This type of room has therefore been choosen for
the experiments. The room is shown in figure 5.

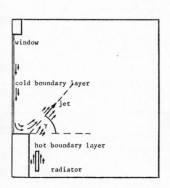

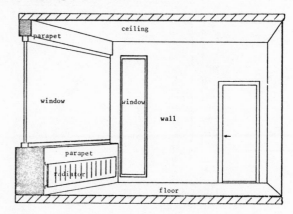

Figure 4. Jet formed by boundary layers.   Figure 5. The experimental full scale test room.

The outside wall contains a window with a height of 1.76 m and a well-insulated
parapet with a height of 0.76 m and a thickness of 0.31 m below it. Above the
glass window there is a parapet to the ceiling with a height of 0.28 m. The
width of the outside wall is 2.40 m, the width of the window is 2.30 m. The
total height of the room was 2.80 m and the total depth 3.20 m.
In   front of the parapet a flat plate hot water radiator with a height of
0.35 m and a width of 2.30 m was placed at a distance of 0.10 m from the
parapet and 0.10 m above the floor. The radiator was supplied with circulating
hot water from a thermostat bath. The radiator was isothermal within 2°C. The
temperature of the window was fixed by means of a stream of cold air on the
outer side. The window was isothermal within 4°C. The temperatures of the
window, the parapet, the walls and radiator were measured with flat surface
thermocouples, the air temperatures in the room with thermocouples with
radiation screens, and the temperatures in the boundary layer were measured
with thin-wire thermocouples with a wire diameter of 0.025 mm. The position in
the boundary layer could be measured within an accuracy of 0.05 mm.
        For the measurement of the general air stream pattern and the local
velocities, the method of flow visualization was used. For this purpose a
small glass window was placed in one of the side walls, close to the outer
wall. A fluorescent lamp with a parabolic reflector was attached to the ceiling.
Lightweight particles were introduced in the air, their movements could be
filmed and photographed with different exposure time. From the results the
stream line pattern and local velocities could be determined. The particles
are formed by heating metaldehyde $(C_2H_4O)_n$, they have a crystal-like structure
and a very low settling speed and are very usefull even at the low velocities
which occur in full scale experiments with free convection.

The velocities in the boundary layers were measured with two different types of
hot-wire anemometers. A constant temperature anemometer (CTA) was used for
velocities above 0.25 m/s. For velovities below 0.30 m/s a special constant
temperature anemometer with a vibrating wire-the low velocity anemometer (LVA)-
was used. With the LVA, the influence of free convection caused by the heated
wire is minimized by the vibration of the probe which causes an extra relative
air velocity.Both instruments were manufactured by DISA. The anemometers
were calibrated in a tube with poiseuille flow. The anomometers could be moved
through the boundary layers with the same instrument used for the thermocouples. In figure 6 the model of the prototype room with a scale reduction
factor M = 10 is shown.

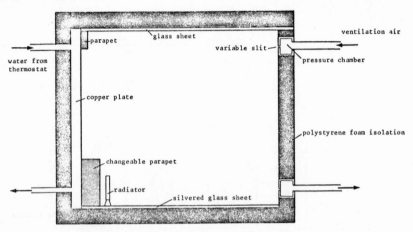

Figure 6. The experimental reduced scale model.

All dimensions were reduced with the same factor 10, except for the width of
the model, which was 0.40 m instead of 0.24 m. As we are concerned with two-
dimensional phenomena, this gives no problems. The window was represented by
a cooled copper plate, that was well-insulated on the outside with 25 mm
polystyrene foam. The parapet was made of polystyrene foam isolation, as
well as the wall opposite of the window. The side walls, the ceiling and
the floor were made of glass sheets to facilitate flow visualization. Behind
the glass sheets, 25 mm polystyrene foam isolation was used. The insulating
sheets could be removed where necessary while photographs were taken. The
model was illuminated with a quartz lamp from outside the model. The light was
reflected out of the model to prevent heating effects. This was achieved by
silvering the bottom side of the glass sheet on the floor so that it acted
as a mirror. For the radiator, a flat plate water heated radiator was made
of copper and kept at a constant temperature with supply from a thermostat bath
Velocities and temperatures were measured in the same way as in the prototype
room. The thermocouples could be introduced through holes in the ceiling;
the anemometer probes through a hole in the wall opposite of the window.
Ventilation could be supplied by two horizontal slots in the wall opposite
the window, one near the ceiling and one near the floor of the room. The slots
had a variable width and a length of 37 mm. Both slots could be used to
introduce the air, behind the slot a pressure chamber was used to ensure
uniform injection velocities.

RESULTS: <u>a</u>. BOUNDARY LAYERS

    The first investigation was to determine the relationship between the Rayleigh number and the laminar or turbulent nature of the boundary layers in our experiments. For this purpose, we have measured the temperature profile in the boundary layer along the glass window in the prototype room at several distances from the top. In figure 7 the results are compared with the theoretical profile for the turbulent boundary layer by Cheesewright. It appeared that the boundary layer was completely turbulent for $Ra > 8.10^9$, as the measured temperature profile fits exactly to the theoretical one. In figure 8 the measured velocity profile in the turbulent boundary layer for $Ra = 6.7 \ 10^9$ is shown.

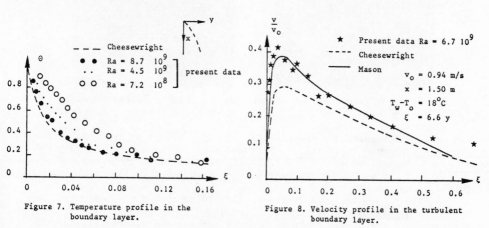

Figure 7. Temperature profile in the boundary layer.

Figure 8. Velocity profile in the turbulent boundary layer.

The measurements fit the theoretical curve of Mason well, but are remarkebly above the values measured by Cheesewright. Outside the boundary layer there was a free-stream velocity of 0.05 m/s. This was induced by the cooperation between the cold window and the hot radiator that during these measurements was placed on the wall opposite, that creates an air circulation in the room In these tests the radiator had been moved to minimize the disturbance of the boundary layer. The free-stream velocity did not affect the heat transfer as can be concluded from measurements of Gryzagoridis [16]. He finds a negligable influence on laminar boundary layers for $Ar > 3$.
We may therefore assume that in the inner part of the boundary layer the velocity profile is not influenced by our free-stream velocity, which conclusion is supported by the close agreement between our maximum velocity and Mason's value. Once again we see that the velocities measured by Cheesewright for the turbulent boundary layer were too low. For the determination of the heat content of the boundary layers, the local Nusselt values have been calculated from the measured temperature profiles and the average Nusselt numbers $\overline{Nu}$ have been found by integration of the local Nusselt values. As these values agree with the results of Bayley, his curve has been used for turbulent boundary layers.

    Further measurements were performed in the model. The velocity profile in the boundary layer at the simulated window in the model was measured at several heights and temperature differences. We found that for $Ra < 1.3 \ 10^7$ the measured velocities fit the theoretical curve of Schmidt and Beckmann for the laminar boundary layer very nicely, we conclude that for this value of Ra the boundary layer is still laminar. As a result of this the heat contents of the laminar boundary layers in the model have been calculated using formula (9) for $Ra < 1.3 \ 10^7$.

b. FLOW IN THE ROOM
    The dependence of the jet angle γ on the ratio of the heat contents
of the boundary layers has been measured for several parapet types. The angles
γ were determined as an average from a set of 10 flow visualization photographs. In figure 9 some typical results are shown.

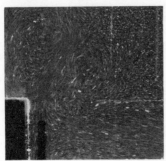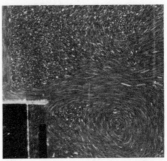

Figure 9. Flow visualization photographs in the 1 : 10 scale model with
        normal window sill (left) and extended window sill (right).

In figure 10 the results are shown for six parapet types, as measured in the
1 : 10 model. Because in practice the radiator has to make good not only the
heat losses through the window, but also the ventilation heat losses and the
losses through the walls, we are particularly interested in the case of values
of the heat content ratio of 2 or higher.
        In figure 10a the parapet system already described has been used. For
a ratio higher than 2, the jet angle is 70° or more. In figure 10b a situation
that occurs very frequently in Dutch homes is shown. The window-sill extends
above the radiator. We see that this reduces the angle γ to 40° for a ratio
of 2. In figure 10c there is no parapet below the window sill, the glass
window completely covers the front wall, though insulation has been placed
behind the radiator. This type of outer wall was frequently used before the
oil crises. For a ratio of 2 the angle γ is now only 20°; even lower than in
figure 10b. A very favourable situation is shown in figure 10d. A glass
curtain has been mounted in the room at the glass window.
The jet goes nearly vertically to the ceiling even for a ratio value of 1 and
at still lower ratio values the angle γ remains positive.
The use of a vertical extension of the window sill, as shown in figure 10e,
lowers the angle γ to 60° for a heat content ratio of 2, a situation worse
than the original one in figure 10a. In figure 10f, the depth of the parapet
has been reduced to half the original value. The corresponding jet angle for
a heat content ratio of 2 has decreased to 60°.
        In the prototype room, the measurements of the jet angle γ have been
carried out for two situations. In figure 10a the results for the original
room as described before are shown as a dotted line, whereas in figure 10b
the situation with the extended window sill in practice is shown. From both
measurements we see that the general form of the curve is the same for model
and prototype, but the values of the jet angle in the prototype room are
10° to 20° lower than the values as measured in the model. This difference is
mainly caused by a difference in the parapet temperature behind the radiator.
For the model measurements a black painted radiator has been used. This
resulted in relatively high surface temperatures of the parapet behind the
radiator. With a radiator temperature of 80°C, the surface temperature was
about 30°C. With a polished copper radiator this value decreased to 17°C
whilst the jet angle was about 20° lower. It follows that the jet angle is
very sensitive to small details and modelling has to be performed very
thoroughly in order to get reliable results.

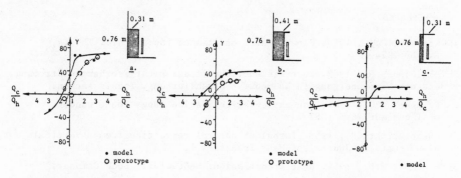

a. Standard parapet.   b. Extended window-sill.   c. Without parapet.

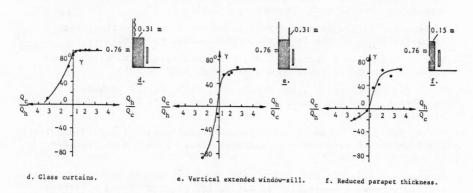

d. Glass curtains.   e. Vertical extended window-sill.   f. Reduced parapet thickness.

Figure 10. Dependence of the jet angle γ on the heat content ratio of the boundary layers for six parapet types.

SUMMARY AND CONCLUSIONS

For thermal comfort in a room when close to a window, the combination of the cold downdraught from the window and the hot air stream coming from the radiator are of great importance. In particular the jet resulting from the mixing of these two air streams can create local thermal discomfort. It is shown that it is not simply possible to use reduced scale measurements for the determination of the temperature and velocity pattern in general, because the boundary layer heat fluxes that create the bouyancy forces do not scale correctly. However, the angle γ between the axis of the resulting jet and the horizontal can be predicted rather well from model measurements provided that the heat content ratio of the boundary layers from window and radiator in the model is choosen to be the same as the ratio in the full scale experiment. The value of the heat content ratio has been calculated from literature values.

The use of glass curtains, a well insulated parapet behind the radiator, a window sill that does not extend above the radiator and a deep parapet is recommmanded in order to achieve a high value for the jet angle even for situations where the heat content of the boundary layer coming from the radiator does not exceed the heat content of the cold downdraught from the window.

REFERENCES

[1] Nielsen, P.V., 1976. Flow in air conditioned rooms. Ph.D. Thesis, Lyngby 1974.

[2] Müllejans, H., 1963. Uber die Ahnlichheit der nicht-isothermen Strömung und den Wärmeübergang in Räumen met Strahllüftung. Thesis Aachen.

[3] Francke, G.G., 1964. Konvektie-verschijnselen langs koude wanden. De Ingenieur 76 : 1087.

[4] Cheesewright, R., 1968. Turbulent natural convection from a vertical plane surface. Journal of Heat Transfer: 1.

[5] McAdams, W.H., 1954. Heat Transmission. McGraw-Hill Book Company, London.

[6] Schmidt, E., und Beckmann, W., 1930. Das Temperatur-und Geschwindigkeitsfeld von einer Wärme abgebenden senkrechten Platte bei natürlicher Konvektion. Techn. Mech. Thermodyn. 1 : 341.

[7] Bayley, F.J.,1955. An analysis of turbulent free convection heat transfer. Proc. of the Institute of Mech. Eng. 169 : 361.

[8] Eckert, E.R.G., and Jackson, T.W., 1951. Analysis of turbulent free convection boundary layer on a flat plate. NACA TR 1015.

[9] Kato, H., Nishiwaki, N., and Hirata, M., 1968. On the turbulent heat transfer by free convection from a vertical plate. Int. J. Heat Mass Transfer 11 : 1117.

[10] Mason, H.B., and Seban, R.A., 1974. Numerical prediction for turbulent free convection from vertical surfaces. Int. J. Heat Mass Transfer 17 : 1329.

[11] Warner, C.Y., and Arpaci, V.S., 1968. An experimental investigation of turbulent natural convection in air at low pressure along a vertical heated flat plate. Int. J. Heat Mass Transfer 11 : 397.

[12] Parczewski, K.I., and Renzi, P.N., 1960. Scale model studies of temperature distributions in internally heated enclosures. ASHRAE Journal : 60.

[13] Davidovici, P., 1964. Etudes de températures intérieures sur maquettes dans la ventilation naturelle et l'air pulsé. Industries Thermiques 10 : 513.

[14] Baturin, V.V., 1972. Fundamentals of industrial ventilation. Pergamon Press, Oxford.

[15] Vliet, G.C., and Liu, C.K., 1969. An experimental study of turbulent natural convection boundary layers. Journal of Heat Transfer : 517.

[16] Gryzagoridis, J., 1975. Combined free and forced convection from a isothermal vertical plate. Int. J. Heat Mass Transfer 18 : 911.

PREDICTIONS OF BUOYANCY INFLUENCED FLOW IN VENTILATED INDUSTRIAL HALLS

MATS LARSSON

Dept. of Applied Thermo and Fluid Dynamics
Chalmers University of Technology
Göteborg, Sweden

ABSTRACT

   Numerical predictions of buoyancy influenced recirculating flow in
a ventilated room are presented and comparison is made with experimental data.
The flow is turbulent with steady, two-dimensional meanflow. The Archimedesnumbers
are chosen to be representative for industrial halls. This means that the effects
of buoyancy are strong enough to create a velocity field which shows very little
resemblance to the isothermal velocity field in the same space.
   The calculations are made with the k-ε model of turbulence and the Boussinesque
approximation is used for the buoyance influence. The boundary value treatment is
discussed and a special wallfunction for heated or cooled vertical walls is
presented.

NOMENCLATURE

c, c_{1T}, c_2, c_μ, σ_κ, σ_ϵ, σ_T	constants in the turbulence model
a, a_E, b, c_E	constants in the velocity profiles
Ar_{in}	Archimedes number based on inlet height and maximal temperature difference in the room
G_{MN}	generation term in the M-equation, $N = S$ Shear stress generation $N = B$ Buoyancy generation
g	acceleration due to gravity
k	turbulent kinetic energy
Re_{in}	Reynolds' number based on inlet values
T	mean temperature
T_0	average temperature of the room (K)
U_{in}	inlet velocity
U_i	velocity vector
u_{max}	parameter in the Eckert-Jackson velocity profile

271

V U_2

u_* friction velocity

x_i cartesian coordinates (i = 1, 2, 3)

x x_1

x_+ $\dfrac{u_* x}{\nu}$

δ   parameter in the Eckert-Jackson velocity profile

ΔT   temperature difference between the inlet jet
                                      and the inlet wall

ε   dissipation of turbulent kinetic energy

κ   von Kàràman's constant

μ   viscosity

ν   kinematic viscosity

ν_t eddy viscosity (kinematic)

ν_e $\nu_t + \nu$

τ_w wall shear stress

φ   general dependent variable (k or T)

Ψ   stream function

ω   vorticity of the mean velocity field

INTRODUCTION

   Design of ventilation systems for large halls is a difficult task especially i:
detailed knowledge of the resulting temperature and velocity distribution is
important. The most common technic at present is to perform experimental investigations. As these are very expensive to carry out in the real hall, the experiments mostly have to be made in a model on a reduced scale. Generally the flow
is affected by buoyancy and the scaling laws of combined convection will determir
the relation between size and temperatures in the model. The result will often be
a temperature field which is very difficult to obtain with usual experimental
equipment.
   This paper presents results from a project with purpose to replace some of the
experimental work with numerical predictions.
   Predictions have been made for a model of an indsutrial hall, see fig 1, fo
which experimental data are available. Three different thermal cases are considere

1. Isothermal flow

2. The temperature of the inlet air is higher than the room average.

3. The temperature of the inlet air is lower than the room average.

In cases 2 and 3 the inlet wall and the ceiling are kept at a constant temperature and the floor and the outlet wall are insulated. The thermal loads are chosen to give strong effects of buoyancy comparable to those in industrial halls. The isothermal case is reported in [1] and will therefore only be briefly discussed here.

Similar calculations have been performed by Nielsen [2] and by Hjertager and Magnussen [3].

The present contribution is the extension to cases where the buoyancy has a considerable influence on the whole flowfield and futhermore the design of a method for boundary value treatment on heated or cooled vertical walls.

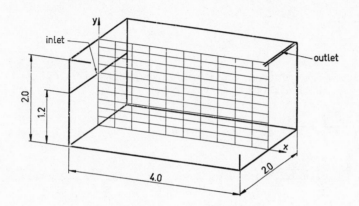

Figure 1. A model of an industrial hall. Distances in meters.

BASIC DIFFERENTIAL EQUATIONS

The flow in ventilated halls is normally turbulent and recirculating. Consequently the computations have to be based on the complete Navier-Stokes' equations, the continuity equation and the energy equation. For the buoyancy influence the Boussinesque approximation is used. In the present cases the mean-flow is assumed stationary and twodimensional which makes it advantageous to transform the basic flow equations into equations for meanvorticity and stream-function. The set of equations is closed by applying the k-ε-model of turbulence, described by Launder and Spalding [4]. The equations to be solved, expressed in tensornotation, then read:

$$\frac{\partial}{\partial x_j} (U_j \omega) = \frac{\partial}{\partial x_j} [\frac{\partial}{\partial x_j} (\nu_e \cdot \omega)] + G_{\omega B} + G_{\omega S} \tag{1}$$

$$\frac{\partial^2 \Psi}{\partial x_j \partial x_j} = - \omega \tag{2}$$

$$\frac{\partial}{\partial x_j} (U_j T) = \frac{\partial}{\partial x_j} (\frac{\nu_e}{\sigma_T} \cdot \frac{\partial T}{\partial x_j}) \tag{3}$$

$$\frac{\partial}{\partial x_j} (U_j k) = \frac{\partial}{\partial x_j} (\frac{\nu_e}{\sigma_k} \frac{\partial k}{\partial x_j}) + G_{kS} + G_{kB} \tag{4}$$

$$\frac{\partial}{\partial x_j} (U_j \varepsilon) = \frac{\partial}{\partial x_j} (\frac{\nu_e}{\sigma_\varepsilon} \frac{\partial \varepsilon}{\partial x_j}) + \frac{\varepsilon}{k} (c_1 G_{kS} + c_{1T} \cdot G_{kB} - c_2 \varepsilon) \tag{5}$$

$$\nu_t = c_\mu \cdot \frac{k^2}{\varepsilon} \tag{6}$$

$$G_{\omega B} = \frac{g}{T_o} \frac{\partial T}{\partial x_1}$$

$G_{\omega S}$ is composed of terms containing second derivatives of ν_t. It can be shown to be negligible in this case.

$$G_{kS} = \nu_t (\frac{\partial U_i}{\partial x_j} + \frac{\partial U_j}{\partial x_j}) \frac{\partial U_i}{\partial x_j}$$

$$G_{kB} = - \frac{g}{T_o} \frac{\nu_t}{\sigma_T} \cdot \frac{\partial T}{\partial x_2}$$

VALUES OF CONSTANTS

c_1	c_{1T}	c_2	σ_T	σ_k	σ_ε	c_μ
1.43	0.0	1.92	0.5*	1.0	1.3	0.09

The choice of c_{1T} = 0.0 is based on conclusions of Hossain and Rodi [5].

The differential equations are put into finite difference form and the solution is obtained by Gauss-Seidel iteration as described by Gosman et al [6]. The order of evaluation of gridpoints goes from inlet to outlet wall, starting alternately at the bottom or at the top horizontal level. At each gridpoint new values of the variables are calculated sucessively in this order ω, Ψ, k, ε, T. The convergence is checked by observation of the local maximum change and the change in the average field value between consecutive iteration cycles.

BOUNDARY VALUE TREATMENT

As the basic equations are elliptic, boundary values have to be prescribed on a closed boundary surrounding the field of interest. The real boundary values of the five dependent variables are easily found, but problems arise near solid wall

*Near solid walls σ_T = 0.9 is used.

or at the inlet where the variables have steep gradients and change nonlinearly.
As the finite difference presuppose a linear variation between gridpoints the
distance between them have to be very small. This would increase the number of
gridpoints to an uneconomical level. Instead the number of gridpoints needed to
describe the flow remote from the walls and the inlet jet is used. The near wall
flow and the inlet jet are treated separately using analytical results from
boundary layer theory.

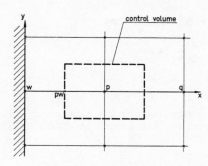

Figure 2.

Solid walls

Isothermal flow

For streamfunction the real wall values can be used direct. For the dissipation
of turbulent kinetic energy, ε, all points p (Fig 2) are used as boundary and
the value of ε at p is determined from

$$\varepsilon_p = \frac{k_p^{1.5}}{\kappa y_p}$$

Even for the mean vorticity, ω, the real wall value, τ_w/μ, probably is a good
choice in spite of the fact that ω is highly nonlinear near the wall. This can
be seen from a control volume approach to the finite difference formulation which
shows that the transport from the wall side into the control volume around p in
Fig 2 is dominated by molecular diffusion and turbulent transport. The term of
equation (1) describing this and its numerical approximation have the following
form:

$$[\frac{\partial}{\partial x} (\nu_e \, \omega)]_{pw} = \frac{(\nu_e \cdot \omega)_p - (\nu_e \cdot \omega)_w}{x_p}$$

From this it follows

$$\omega_w = (\frac{\nu_e}{\nu} \cdot \omega)_p - x_p \cdot [\frac{\partial}{\partial x} (\frac{\nu_e}{\nu} \cdot \omega)]_{pw} \qquad (7)$$

If the derivative on the right hand side of (7) which is approximatly the derivative
of the shear stress can be determined, the value of ω_w can be calculated. As the flow
near the main part of the solid walls in the present case is similar to ordinary

turbulent boundary layer flow thoretical results can be used for this purpose. If it is assumed that the points p and pw are situated in the constant stress layer the derivate in (7) is obviously zero. Using a logarithmic velocity-distribution for the determination of ω_p and the Prandtl mixinglength assumption for the eddy viscosity the result is:

$$\omega_w = \frac{u_*^2}{\nu} = \frac{\tau_w}{\mu}$$

The friction velocity u_* is determined from

$$|\Psi_p - \Psi_w| = u_* [\int_0^{x_{+1}} x_+ dx + \int_{x_{+1}}^{x_{+p}} ax_+^b dx] ,$$

$$x_{+1} = 11.8$$
$$a = 8.3$$
$$b = 1/7$$

For the turbulent kinetic energy, k, and the temperature, T, slipvalues are used. The dominating transport term and its numerical approximation have the following form:

$$(\frac{\nu_e}{\sigma_\phi})_{pw} \cdot (\frac{\partial \phi}{\partial x})_{pw} = (\frac{\nu_e}{\sigma_\phi})_{pw} \cdot \frac{\phi_p - \phi_w}{x_p}$$

which gives

$$\phi_w = \phi_p - x_p (\frac{\partial \phi}{\partial x})_{pw}$$

The right hand derivative for $\phi = k$ is derived from Wolfshteins wallfunctions [7] and for $\phi = T$ from a 1/7-power profile.

Nonisothermal flow

The same relations as for isothermal flow are used except for the mean vorticity at vertical walls, where buoyancy effects are dominant. If the mean velocity is higher between the gridpoints p and w in Fig 2 than between q and p the velocity-distribution near the wall is described by the relation proposed by Eckert and Jackson [8] for turbulent free convection on a vertical wall. The profile relation is

$$V(x) = a_E u_{max} \cdot (\frac{x}{\delta})^b (1 - \frac{x}{\delta})^{c_E} ,$$

$$a_E = 1.86$$
$$b = 1/7$$
$$c_E = 4.0$$

The maximum velocity, u_{max}, and the thickness parameter, δ, are unknown. They are determined from the following relations:

$$\omega_p = (\frac{\partial V}{\partial x})_p$$

$$\Psi_p - \Psi_w = \int_0^{x_p} V(x) dx$$

To be treatable the result from the integration has to be approximated with a linear function. When the two unknown parameters have been determined, the profile relation can be used together with a mixing length assumption to calculate ω_w from (7).

The wall functions of Wolfshtein, which are used to determine boundary values for k even in the nonisothermal cases, do not include buoyancy effects. This should produce errors especially in the predictions of the distributions of turbulent quantities along horizontal boundaries where steep vertical temperature gradients exist. Tests have shown that although errors are produced changes in these boundary values have no significant effect on the meanflow in the rest of the room.

Inlet

The width of the inlet slot is only 6[†]mm, so it is impossible to cover the inlet jet with a sufficient dence net of gridpoints. Instead a box of 100 x 40 mm is formed around the inlet and analytical solutions of the jet flow inside this box are used to produce boundary values for the numerical procedure for the rest of the field.

Outlet

Derivatives of all variables are assumed to be zero in the flow direction.

EXPERIMENTAL INVESTIGATION

An experimental investigation has been performed in the model of an industrial hall shown in Fig 1. The experimental technic is reported in [1] and will there-fore only be briefly described here.

Measurements have been made in 11 x 11 points in the room. The velocities have been measured with a 'hot sphere' anemometer especially designed for low air velocities. Its time constant is about 2 seconds, which excludes measuring tur-bulent quantities. Futhermore it is fairly insensitive to the velocity direction. Therefore the directions have been studied visually by adding sublimed metaldehyd to the inlet air. As this is a rather uncertain method, only the measured absolute velocities are used for the following quantitative comparison with predictions.

The mean temperatures have been measured with thermocouples.

Three different thermal cases have been considered, all with $Re_{in} = 1180$:

1. Isothermal flow $Ar_{in} = 0.0$ $U_{in} = 3.0$ m/s $\Delta T = 0.0$

2. Warm inflow $Ar_{in} = 7.8 \cdot 10^{-4}$ $U_{in} = 3.0$ m/s $\Delta T = 35^\circ C$

3. Cold inflow $Ar_{in} = -4.9 \cdot 10^{-4}$ $U_{in} = 3.6$ m/s $\Delta T = -38^\circ C$

Case 1 and 2 are found very stable and the results are probably fairly accurate. The flow in case 3 showed instability in the upper half of the room which made the measurements difficult. The flow near the floor were found more stable and has been used in the following comparison with the predictions.

† For the cold inflow case the width of the inlet slot is 5 mm.

DISCUSSION OF THE PREDICTIONS AND COMPARISON WITH EXPERIMENTS

General remarks

Fig 3, 4 and 5 show predicted velocity vector fields and fig 6, 7 and 8 show a comparison between predicted and experimental velocity contours. All three predicted velocity fields show very good resemblance to the visualized experimental fields. For the recirculation points on the ceiling in the isothermal and the warm inflow cases, fig 6 and 7, the predicted positions are slightly too far from the inlet. This is probably partly caused by the boundary values on the 'inlet box'. Inside the box the jet is supposed to develop symmetrically in the horizontal direction. The Coanda effect, which tends to force the jet towards the nearest wall, is thus prevented from acting near the inlet. This also affects the cold inflow case. Experiments show that near the inlet the jet is forced upwards, obviously by the Coanda effect, and a short distance down stream, the buoyancy takes over and forces the cold jet towards the floor. Predictions with a horizontally directed jet inside the 'inlet box' show no Coanda effect, so the recirculating zone under the inlet jet becomes too small. To compensate for this the inlet jet must be slightly directed upwards. For the present results a deviation from the horizontal plane of $4,5^{\circ}$ is used. In general the positions of the recirculating zones are well predicted. For the warm inflow case fig 7 shows very good agreement between measured and predicted absolute velocities. For the isothermal, fig 6, and the cold inflow case, fig 8, the velocities and the recirculating masses are slightly underestimated. The largest discrepancy is found in the cold inflow case, fig 8, along the outlet half of the floor. This could possibly be explained by the experimental values in the recirculating zone of the outlet half of the room. This zone extends vertically from the floor to the ceiling and was therefore directly influenced by the earlier mentioned instabilities along the ceiling.

Buoyancy influence on the turbulence

The term G_{kB} in Eqn (4), describing buoyancy generation couples the turbulence equations to the temperature distribution. This term has negative effects on the convergency and tests have therefore been performed to examine the influence of G_{kB} on the final solution. This also reveals the different characters of the two nonisothermal flows considered.

The warm inflow case is stably stratified with a thin, almost horizontal inversion layer about 80 cm above the floor. The temperature gradient in the layer is about 10 K/m. This reduces the vertical momentum transport through the layer and the air movements under the layer become very small. The velocity contours in fig 7 show that the velocity in this zone everywhere is less than 4% of the inlet velocity. As expected tests have shown G_{kB} to have a decisive influence on the final solution shown in fig 4 and 7.

The cold inflow case is very well mixed, the temperature is almost constant in the main body of the air. As a result there is only a very weak buoyancy influence on the turbulence. The term G_{kB} has no significant influence on the mean flow and its influence on the turbulence quantities is limited to a thin layer near the ceiling. To improve the convergency G_{kB} was eliminated from the equations used for this case.

Temperatures

For the warm inflow case the inlet temperature is 50°C. The inlet wall and the ceiling are kept at 15°C. The predicted average temperature in the room is 28°C while experiments give 25°C. The predicted temperature gradient in the inversion layer agrees well with the experimental value of 10 K/m.

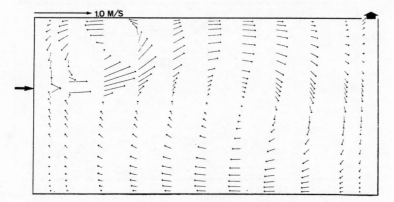

Figure 3. Predicted velocity vector field for the isothermal case.

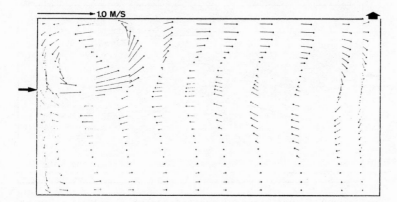

Figure 4. Predicted velocity vector field for the warm inflow case.

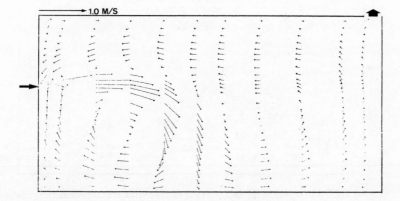

Figure 5. Predicted velocity vector field for the cold inflow case.

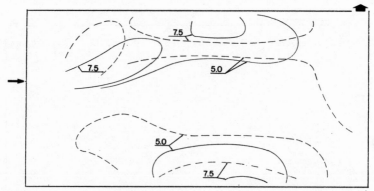

Figure 6. Comparison of predicted and measured velocity contours for the isothermal case. Numbers denote percentage of the inlet velocity.
——————, predictions: – – – – – , measurements.

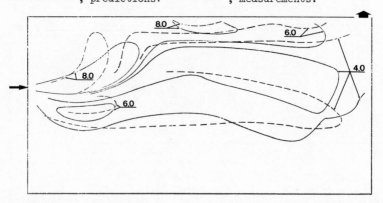

Figure 7. Comparison of predicted and measured velocity contours for the warm
inflow case. Numbers denote percentage of the inlet velocity.
——————, predictions: – – – – – – , measurements.

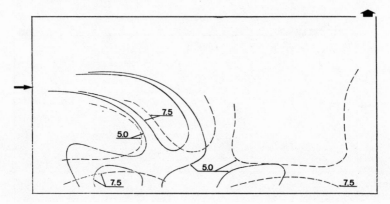

Figure 8. Comparison of predicted and measured velocity contours for the cold
inflow case. Numbers denote percentage of the inlet velocity.
——————, predictions: – – – – – , measurements.

For the cold inflow case the inlet temperature is $2^\circ C$ and the isothermal boundaries are kept at $40^\circ C$. The predicted average temperature is $21^\circ C$, which deviates less than $1^\circ C$ from the measured value.

CONCLUSIONS AND FURTHER WORK

The comparisons with experiments have shown that the isothermal case and the warm inflow case are well predicted by the numerical method used. For the cold inflow case the main features are qualitatively well predicted, but it is difficult to draw conclusions from a detailed comparison because instabilities disturbed the experimental investigations. For this case a drawback is the need for directing the inletjet slightly upwards to compensate for the lack of Coanda effect. At present no satisfactory method to calculate the angle to the horizontal plane is known to the author. The further work will deal with improvments of the method and also with a project together with ventilation industries to test the method on a suitable real case.

ACKNOWLEDGEMENT

This work has been supported by the Swedish Board for Technical Development and the Swedish Council for Building Research.

REFERENCES

1.   Holmberg, R., Larsson, M., Sundkvist, S.G. 1975. Calculation of velocity distribution in a ventilated room. VVS, No5, pp 59 66. (In Swedish.)

2.   Nielsen, P.V. 1974. Flow in air conditioned rooms. Thesis for Ph.D. at the Technical University of Denmark, Copenhagen, Denmark.

3.   Hjertager, B.H., Magnussen, B.F. 1976.Numerical prediction of three-dimensional turbulent buoyant flow in a ventilated room. 1976 international seminar on Turbulent Buoyant Convection, ICHMT, Dubrovnic, Yugoslavia, aug 30 - sept 4, pp 429 - 441.

4.   Launder, B.E., Spalding, D.B. 1972. Mathematical models of turbulence. Academic Press, London.

5.   Hossain, M.S., Rodi, N. 1976. Influence of buoyancy on the turbulence intensities in horizontal and vertical jets. 1976 International Seminar on Turbulent Buoyant Convection, ICHMT, Dubrovnik, Jugoslavia, aug 30 - sept 4, pp 39 - 51.

6.   Gosman, A.D., Pun, W.M., Runchal, A.K., Spalding, D.B, Wolfshtein, M. 1969. Heat and Mass Transfer in recirculating Flows. Academic Press, London.

7.   Wolfshtein, M. 1969. The velocity and temperature distribution in one-dimensional flow with turbulence augmentation and pressure gradient. Int. J. Heat Mass Transfer, vol 12, pp 301 - 318.

8.   Eckert, E.R.G., Jackson, T.W. 1951. Analysis of turbulent free-convection boundary layer on a flat plate, NACA TR 1015.

THREE-DIMENSIONAL FREE JETS AND WALL JETS: APPLICATIONS TO HEATING AND VENTILATION

PASQUALE M. SFORZA

Polytechnic Institute of New York
Aerodynamics Laboratories
Farmingdale, New York, USA 11735

ABSTRACT

An overview of the state-of-the-art of basic research in three-dimensional turbulent jet diffusion as it applies to engineering problems in heating and ventilation is presented. Two cases are treated in some detail: three-dimensional free jets and wall jets. The characteristics of such flows such as velocity decay, transverse growth, mean velocity fields, thermal effects, etc. are described. Emphasis is placed upon experimental results and empirical approaches for the prediction of such flow fields in practical applications.

NOMENCLATURE

d   orifice minor axis dimension

d_e effective momentum diameter = $(8\int_o^\infty \rho U^2 r dr)^{\frac{1}{2}}/\rho^{\frac{1}{2}} U_{0c}$

d_{et} effective thermal diameter = $(8\int_o^\infty \rho c_p UT r dr)^{\frac{1}{2}}/(\rho c_p UT)^{\frac{1}{2}}_{0c}$

e   "eccentricity" or fineness ratio of orifice = d/ℓ

ℓ orifice major axis dimension

m   total mass flux = $2\pi\int_o^\infty \rho U r dr$

M   total momentum flux = $2\pi\int_o^\infty \rho U^2 r dr$

r   radial coordinate

Re   Reynolds Number = Ud/ν

T   mean temperature

U   mean velocity component in the X direction

ν kinematic viscosity

ρ density

τ_w wall shearing stress

Subscripts

a   denotes ambient atmospheric conditions

i   denotes conditions at the initial X station

m   denotes a maximum condition

0   denotes centerline conditions

0c   denotes centerline conditions at jet exit

½   denotes conditions at $U=\frac{1}{2}U_0$ or $(T-T_a)/(T_{0c}-T_a)=\frac{1}{2}(T_0-T_a)/(T_{0c}-Ta)$

INTRODUCTION

The characteristics of jet diffusion are of great importance
for determining the environment of the interior of buildings. Despite the vast literature produced in the study of jets over the
past fifty years, remarkably little basic research has been reported
on configurations other than round or quasi-two-dimensional jets.
For heating and ventilating purposes, design constraints often require the use of non-axisymmetric duct exits. Therefore, fundamental knowledge of three-dimensional jet flow fields is of great value
for such engineering purposes.

Over the past decade, the author and his students have been engaged in basic three-dimensional jet mixing research, both theoretical and experimental in nature. Included among the many different
cases that have been studied are two that are of interest to those
engaged in the investigation of heat and mass transfer in buildings:
jets exhausting into an essentially free space (free jets) and those
exhausting near a solid surface (wall jets). Much of this work appears in Refs. 1-8, but due to space limitations in journals a great
deal is still unpublished.

The purpose and scope of this work is to illustrate and explain
the bulk properties of three-dimensional turbulent jet flows, based
primarily upon our continuous studies at the Polytechnic's Aerodynamics Laboratories. Emphasis will be placed upon those experimenta
results and empirical formulations which bear on the prediction of
such flows for practical engineering purposes.

THREE-DIMENSIONAL FREE JETS

In the interests of completeness, the general features of three
dimensional turbulent jets, as outlined by Sforza et al [1], are reviewed. Such jets are found to be characterized by the presence of
three distinct regions in terms of the axis velocity decay (Fig. 1).
These regions may be classified as follows:

1) Potential core (PC) region: Here the mixing initiated at
the jet boundaries has not yet permeated the entire flow field, thus
leaving an interior region characterized by a uniform axis velocity
approximately equal to the jet exit velocity.

2) Characteristic decay (CD) region: Here the axis velocity
decay is dependent upon orifice configuration, and velocity profiles
in the plane of the minor axis of the orifice are found to be "similar" (i.e. affinely related), whereas those in the plane of the majo

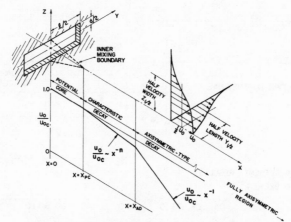

Fig. 1   Schematic diagram of a typical three-dimensional free jet flow field illustrating characteristic features.

axis are "nonsimilar". Hence this region is said to be "characteristic" of the orifice geometry and in it $U_0 \sim x^{-n}$, where n depends on
orifice geometry.

3) Axisymmetric decay (AD) region: The axis decay in this
region is axisymmetric in nature, i.e. $U_0 \sim x^{-1}$, and the entire flow
is found to be approaching axisymmetry, thus becoming oblivious of
the initial orifice geometry. Velocity profiles in this region are
found to be similar in both symmetry planes. Far downstream, the
jet is observed to become fully axisymmetric.

It is important to note that the bulk characteristics of the
jet in the PC and CD regions (near field) are dependent on orifice
geometry, and that the flow in the AD region is essentially independent of orifice geometry. The basis for these conclusions is
treated in detail in [1] and [2], and is briefly described in the
present work.

Axis Velocity Decay

The exponent in the power law variation of the centerline velocity is shown as a function of fineness ratio $e=d/\ell$ for various
bilaterally symmetric jets in Fig. 2. It is evident that there is
a geometry which leads to a "slowest" decay in the CD region, and
this occurs for e $\sim$0.1. The centerline velocity decay for various
three-dimensional jets is shown in Fig. 3 as a function of the non-
dimensional streamwise distance X/d_e. The quantity $d_e=(8\int_0^\infty U^2 r dr)^{\frac{1}{2}}/U_{0c}$
is an effective exit diameter for a given jet orifice. A good ap-
proximation for d_e is the diameter of a circular jet of equal area
and exhaust velocity. The total momentum flux of the jet,
$2\pi\int_0^\infty \rho U^2 r dr$, is a constant for a free jet flow and it sets the
asymptotic behavior. Thus the ultimate "throw" of the jet depends
upon the initial exit momentum flux integral and one can best compare
jets by nondimensionalizing streamwise distance by the effective

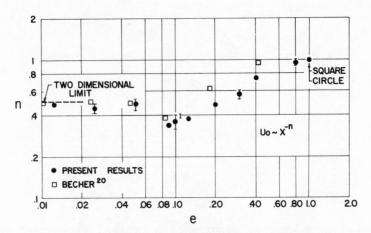

Fig. 2   Variation of the exponent in the power law
         for axis velocity decay in the CD region
         of three-dimensional jets.

momentum diameter of the jets in question.   In Fig. 3 it is clear

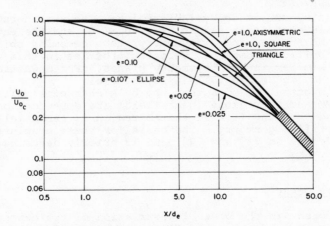

Fig. 3   Axis velocity decay for various three-di-
         mensional free jets.

that all the different jets display the same far field behavior although the intermediate field is dependent upon orifice geometry.
This is indicative of the results one would obtain from a comparison
of jets of various geometry but equal area exhausting at the same
velocity; the more slender jets slow down soonest but all reach the
same (low) velocity level at the same (far) downstream point, within the accuracy of the initial integrated momentum flux.

The extent of the PC and CD regions is dependent upon the orifice geometry, because the formation of these regions is influenced
by the lateral extent of mixing originating at the boundaries of the
orifice (see Fig. 1).   A plot of these boundaries, as determined from
axis decay plots is shown in Fig. 4, including results from other
investigations.   Note that X_{PC} and X_{AD} correlate with the initial

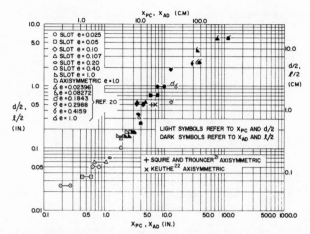

Fig. 4 Correlation of X_{PC} and X_{AD} with orifice geometry for various three-dimensional jets.

half-width and half-length of the orifice in a fairly orderly fashion. These results suggest a relation for the extent of the PC and CD regions given by: $X_{PC}/d = 6.6$ and $X_{AD}/d = 4/e$. From these experimentally derived relations and the results for the power law exponent in Fig. 2 it is possible to construct, for engineering purposes, a reasonably accurate description of the axis velocity field for bilaterally symmetric free jets.

Half-Width Boundaries

The half-width boundaries for "slender" jets (i.e. for $e \leq 0.1$) may be normalized by introducing new variables: $\hat{Y}_1 = (Y_1/d)(e/e_r)$ and $\hat{Z}_1 = (Z_1/d)(e/e_r)$, where e_r is the eccentricity d/ℓ of a reference slender jet (in this report $e_r = 0.1$). In these new coordinates the half-width boundaries of different eccentricity jets are found to fall, to a reasonable approximation, onto one pair of curves, as can be seen in Fig. 5. The results serve to indicate that for slender three-dimensional jets the major axis half-width decreases initially, whereas the minor axis half-width grows; at some intermediate station they cross over, grow similarly but at different rates, and finally tend to approach each other far downstream, where the jet tends toward axisymmetry.

It may be noted that a similar type of coordinate transformation may be used for "bluff" jets ($e \sim 1.0$), where the half-width growth is, for all practical purposes, the same in both the Y and Z directions. This is accomplished by introducing $\hat{Y}_1 = (Y_1/d)(h/h_r)$ and $\hat{Z}_1 = (Z_1/d)(h/h_r)$, where h_r is the hydraulic diameter of the reference bluff jet, which in this report is the axisymmetric one. This transformation leads to good agreement between all bluff jets but due to space limitations no results are included here; see instead, [2].

For jets that fit neither extreme case there are no hard and fast rules for simple engineering calculations other than what has already been suggested. However, it should be reasonable to extend

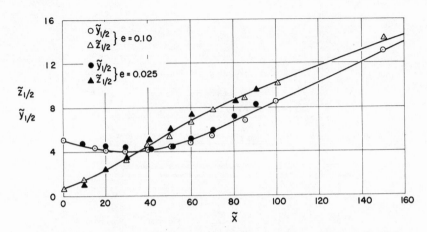

Fig. 5   Correlation of half-width data for two different three-dimensional jets. Here the
notation $(\tilde{\ }) = (\tilde{\ })/d(e_r/e)$ and e_r is chosen
as 0.1.

the slender jet ideas up to about e=0.4 and the bluff jet ideas
down to that value.

Velocity Profile Similarity

It is found that the velocity profiles in the AD region of the
flow are similar, i.e. $U/U_0 = f(s)$, in both symmetry planes. Here
the similarity variable $s=\tilde{Y}/Y_{\frac{1}{2}}=Z/Z_{\frac{1}{2}}$; it is further found that the
function f(s) is well represented by the result obtained for cir-
cular jets with constant eddy viscosity as shown in [5], i.e.
$f=(1+0.414\ s^2)^{-2}$.
In the CD region, the profiles in the minor axis (i.e. the Z)
direction the profiles are found to be similar and the above remarks
apply. However, in the major axis direction the profiles are not
similar and no simple calculations are suggested. For most practical purposes diffusive-like profiles may be sketched in from knowledge of the various boundary conditions without too great a sacri-
fice in accuracy. The more detailed calculations of such flows
still waits upon progress; for recent developments see [9].

Mass Entrainment

Results of experiments on the mass entrainment characteristics
of three-dimensional and axisymmetric jets are presented in Fig. 6
along with the results of previous investigations of circular jets
alone, [10] and [11]. These data are all for Reynolds numbers
$Re=U_0 d/\nu>2.5x10^4$; the studies of [10] indicate that below this
value the mass entrainment becomes Reynolds number dependent, for
axisymmetric jets. This was found to be true in the cases studied
here and the trends noted in [10] were repeated.
Also included in Fig. 6 are the results for the total momentum
flux integral $M=2\pi \int_0^\infty \rho U^2 rdr$, evaluated from the data at the various

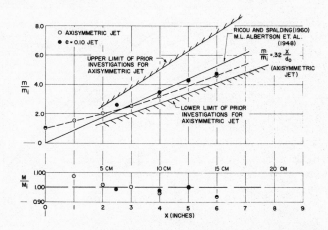

Fig. 6 Mass entrainment and momentum conservation as a function of axial position.

axial stations corresponding to those at which the mass entrainment
data were taken. For all jets considered, the total momentum flux
integral remained reasonably constant (within about 4%) throughout
the entire flow field except in the very near and very far fields
where the accuracy drops off to about 6%.

The mass entrainment rate of the e=0.1 jet is found to approach that of the axisymmetric jet in the far field but the three-
dimensional jet's rate is higher. Such enhanced mass entrainment
behavior of jets with eccentricities e≈0.1 has been observed by
other researchers, e.g. [12] and [13].

Thermal Effects

When the three-dimensional jet is heated the basic character
of the flow field as described with the aid of Fig. 1 is unchanged.
The axial temperature variation $(T_0-T_a)/(T_{0c}-T_a)$ vs. X/d_{et} is found
to have the three distinct regions first illustrated for the unheated velocity field, i.e. PC, CD, and AD regions, as is shown in
Fig. 7. The axial distance has been nondimensionalized with respect
to an effective thermal length $d_{et}=(8\int_0^\infty \rho C_p UTrdr)^{\frac{1}{2}}/(\rho C_p UT)_0^{\frac{1}{2}}$. This
corresponds to normalizing the jet flows to equivalent initial total
enthalpy flux at the exit station; for heated free jets this quanti-
ty, $2\pi\int_0^\infty \rho C_p UTrdr$, is a constant throughout the flow. Note that in
the far field all the jets display essentially the same, axisymmetric
behavior.

The velocity field in the heated case is very similar to that in
the unheated case and the two can be fairly well correlated by intro-
ducing a new coordinate $\theta=(X/d)(\rho_{0c}/\rho_a)^{\frac{1}{2}}$, which accounts for the density changes brought about by heating, [14]. Thus most of the infor-
mation discussed previously for the unheated jet velocity field may
be extended to the heated case by use of the new independent variable, θ.

It must be noted that the PC, CD, and AD regions are different
in extent for temperature and velocity in the heated case: typically
these regions occur sooner in X for the former. This is seen from

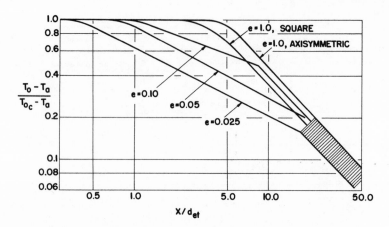

Fig. 7   Centerline behavior of the temperature
         field of three-dimensional free jets.

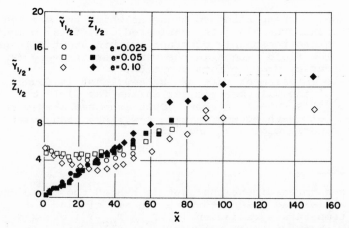

Fig. 8   Correlation of the thermal half-widths of
         three-dimensional free jets. Here the
         notation $(\tilde{\ }) = (\tilde{\ })/d(e_r/e)$ and $e_r = 0.1$.

the normalized plots of the thermal half-width boundaries shown in
Fig. 8, which illustrates that the thermal half-width boundary
crossover point occurs much sooner than that for velocity.

The profile function f(s) used previously for unheated jet
velocity profiles will serve adequately in the heated jet case; the
difference, of course, is that the half-widths are not the same, so
the actual profiles, in the physical plane, will also be different.

THREE-DIMENSIONAL WALL JETS

Nonaxisymmetric jets issuing into a quiescent atmosphere,

tangent to, and at the surface of a wall are termed three-dimensional
wall jets. Such wall jets may also be characterized by three distinct regions in terms of the streamwise decay of the maximum velocity, [3]. This feature is analogous to that encountered in the description of three-dimensional free jets. The regions are illustrated
in Fig. 9, where it can be seen that the PC and CD regions are de-

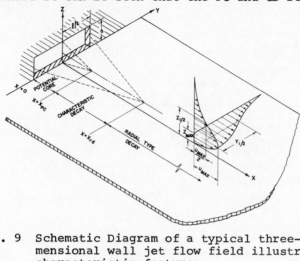

Fig. 9   Schematic Diagram of a typical three-di-
         mensional wall jet flow field illustrating
         characteristic features.

fined as in the case of the three-dimensional free jet. The far
downstream region, however, is called the Radial-Type (RD) decay
because the maximum velocity there decays like that of a radial
wall jet, i.e. an axisymmetric jet which impinges normally upon a
wall and spreads out radially. In the RD region, as in the AD
region of a free jet, the flow becomes increasingly oblivious of
the orifice geometry.
     The velocity decay along the mainstream direction in a three-
dimensional wall jet is very much like that of the three-dimensional
free jet, as can be seen in Fig. 10. Note that the maximum velocity

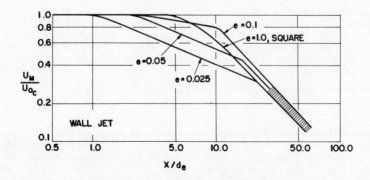

Fig. 10   Axis decay of maximum velocity in various
          three-dimensional wall jet flows

decay in the CD region is orifice geometry dependent while in the
RD region the velocity $U_m \sim X^{-1.14}$, the value typical of radial wall
jets. The variation of the exponent in the power law behavior of
the maximum velocity decay in the CD region of wall jets is shown
in Fig. 11, including results from other investigations concerned
with two-dimensional and radial wall jet behavior. It is con-

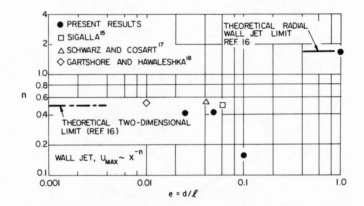

Fig. 11   Variation of the exponent in the power
         law for maximum velocity decay in the
         CD region of three-dimensional wall
         jets.

cluded [3] that the difference between the free and wall jet decay
laws are primarily due to the alteration of the mass entrainment
capabilities caused by the presence of the wall and secondarily
due to the influence of surface friction.

Half-Width Boundaries

    The half-width boundaries of three-dimensional wall jets, de-
fined in Fig. 9, are not as readily correlated as are those of
comparable free jets. This can be seen in Figs. 12 and 13, where
these boundaries are depicted, in the physical plane, for the
various cases tested. From these diagram two important conclusions
can be deduced. First, the $Z_{\frac{1}{2}}$ (normal half-width) growths in the
streamwise direction are essentially the same for all cases tested.
Second, far downstream the $Y_{\frac{1}{2}}$ (transverse half-width) growth becomes oblivious of orifice geometry, and this growth rate becomes
identical for all the flows.

Surface Friction

    Centerline skin friction data was obtained for all jets tested,
while spanwise data was taken primarily for the e=0.1 case. This
data is shown in Fig. 14 where $C_f = \tau_w / \frac{1}{2} \rho U_m^2$ is plotted against
$Re_m = U_m Z_{\frac{1}{2}} / \nu$ along with a correlation for two-dimensional wall jet
cases[19], i.e. $C_f = 0.048 Re_m^{-0.24}$. The wall shear is characterized
by a similarity distribution in the spanwise direction given by

$\tau_w/\tau_{w0} = (1+0.414Y^2/Y_{\frac{1}{2}}^2)^{-2}$ where the ratio is that of local wall shear stress to centerline wall shear stress. The variable $Y_{\frac{1}{2}}$ is, for the

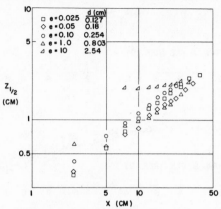

Fig. 12   Growth of $Z_{\frac{1}{2}}$ with X. for three-dimensional wall jets.

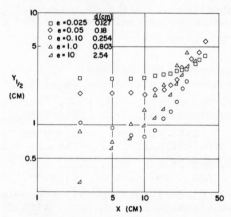

Fig. 13   Growth of $Y_{\frac{1}{2}}$ with X for three-dimensional wall jets.

greatest accuracy, the half-width of the shear stress distribution, but acceptable results are obtained if one uses the velocity half-width distribution instead. It seems reasonable to expect that the results for skin friction can be extended to heat transfer by means of the Reynolds analogy for turbulent flow [23]. Note that the quantity Z_m denotes the height above the surface at which $U=U_m$. Along the centerline, the following behavior is noted: RD region, all jets studied in [3], $Z_m/d_e = 0.35\pm 0.05$, and CD region, $e \lesssim 0.05$, $Z_m/d_e = 0.0135X/d_e$. The intermediate case of the CD region of the $e=0.1$ wall jet yielded $Z_m/d_e = 0.027X/d_e$, twice the value for the more slender orifices.

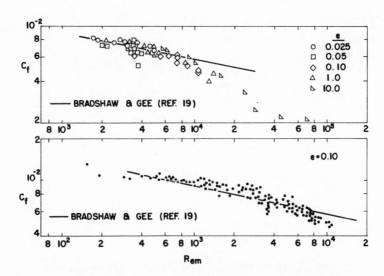

Fig. 14   Skin friction coefficient variation along
          centerline of wall jets. Lower figure
          illustrates off-axis behavior for e=0.1.

REFERENCES

1. Sforza, P.M., Steiger, M.H., and Trentacoste, N.P. 1966.
    Studies in Three-Dimensional Viscous Jets. AIAA Jl. 4:800-806.

2. Trentacoste, N.P. and Sforza, P.M. 1967. Further Experimental
    Results for Three-Dimensional Free Jets. AIAA Jl. 5:885-891.

3. Sforza, P.M. and Herbst, G. 1970. A Study of Three-Dimensional
    Incompressible Turbulent Wall Jets. AIAA Jl. 8:276-283.

4. Trentacoste, N.P. and Sforza, P.M. 1968. Some Remarks on Three-
    Dimensional Wakes and Jets. AIAA Jl. 6:2454-2456.

5. Sforza, P.M. 1969. A Quasi-Axisymmetric Approximation for Tur-
    bulent Three-Dimensional Wakes and Jets. AIAA Jl. 7:1380-1383.

6. Trentacoste, N.P. and Sforza, P.M. 1970. Studies in Homogeneous
    and Nonhomogeneous Free Turbulent Shear Flows. AIAA 8th Annual
    Aerospace Sciences Meeting Paper No. 70-125.

7. Sforza, P.M. and Mons, R.F. 1977. Mass, Momentum, and Heat
    Transport in Turbulent Free Jets. Int. Jl. Heat and Mass Trans.
    (accepted for publication).

8. Sforza, P.M. and Stasi, W.J. 1977. Heated Three-Dimensional Tur-
    bulent Jets. (submitted to the 1977 ASME Winter Annual Meeting).

9. McGuirk, J.J. and Rodi, W. 1977. The Calculation of Three-Di-
    mensional Turbulent Free Jets. Symposium on Turbulent Shear
    Flows, Pennsylvania State University, April 18-20.

10. Ricou, F.P. and Spalding, D.B. 1961. Measurements of Entrain-
    ment by Axisymmetric Turbulent Jets. Jl. of Fluid Mech. 11:21-32

11. Albertson, M.L. et al 1948. Diffusion in Submerged Jets. Amer.
    Soc. Civil Eng. December:1571-1596.

12. Tuve, G.L. et al 1942. Entrainment and Jet Pump Action of Air Streams. Jl. Heat and Vent. 48:241-266.

13. Manganiello, E.J. and Bogatsky, D. 1944. Natl. Advisory Comm. Aero. NACA WRE-224.

14. Laufer, J. 1969. Turbulent Shear Flows of Variable Density. AIAA Jl. 7:706-712.

15. Sigalla, A. 1958. Measurements of Skin Friction in a Plane Turbulent Wall Jet. Jl. Roy, Aero. Soc. 62:873.

16. Glauert, M.B. 1956. The Wall Jet. Jl. Fluid Mech. 1:625.

17. Schwarz, W.H. and Cosart, W.P. 1961. The Two-Dimensional Turbulent Wall Jet. Jl. Fluid Mech. 10:481-495.

18. Gartshore, I.S. and Hawaleshka, O. 1964. The Design of a Two-Dimensional Blowing Slot. McGill Univ. ME Report 64-5, June.

19. Bradshaw, P. and Gee, M.T. 1962. Turbulent Wall Jets With and Without an External Stream. Aero. Res. Council R&M No. 3252.

20. Becher, P. 1949. Calculation of Jets and Inlets in Ventilation. I. Kommission Hos. Jul. Gjellerups Forlag. Copenhagen.

21. Squire, H.B. and Trouncer, J. 1944. Round Jets in a General Stream. Aero. Res. Council Tech. Rept. R&M No. 1974.

22. Keuthe, A.M. 1935. Investigation of the Turbulent Mixing Regions Formed by Jets. Jl. App. Mech. 3:A87-A95.

23. Chapman, A.J. 1974. Heat Transfer. Macmillan Co. New York.

CALCULATION OF ROTATING AIRFLOWS IN SPACES BY CONFORMAL MAPPING

R. D. CROMMELIN

Environmental Hygiene TNO
Delft WIJK 8, The Netherlands

SUMMARY

A method has been developed to calculate 2-dimensional rotating air flows
in rooms by conformal mapping. The agreement with the velocities measured
in a test room and a scale model is in general good except at places where
the influence of the air jet is strong. The velocities depend upon the air
inlet velocity, the place of the air supply opening and the roughness of
the walls.

1. Introduction

Air flow in closed spaces is a subject of special interest in research
of the indoor environment. But it is believed that knowledge of flows in
closed spaces is also of interest in other fields of research.
Air flows can be investigated by:

- experimental research in test rooms,
- experimental research in scale models,
- theoretical models

If reliable theoretical models are available air flows in spaces can be
predicted with far less effort than by experiments. Therefore many theoretical models have been developed solving the fundamental equations of
transport of momentum and energy by numerical methods.
In closed spaces like rooms the air usually behaves as a rotating mass.
Therefore in this research project an attempt has been made to give a
mathematical formulation of a rotating flow within a boundary which
obviously is not a circle but a rectangular if the flow is approximated as
2-dimensional. Such a method will give a wider meaning to the concept of
rotation and will require far less computer capacity than a numerical
model. The reason is that no difference scheme with a large number of
points is required and no iteration procedure of a large amount of data
is needed. In fact this method proved suitable for a minicomputer having
a storage capacity of 8 Kb. In order to investigate the applicability
and shortcomings of this method experiments in a test room and in a
scale model of this room were carried out.

2. General Concept

The air mass in a room can be subject to two kinds of forces:

- supply of air (air jet momentum),
- heating or cooling equipment (buoyancy forces).

The direction of these forces is usually not towards the center of
gravitation and therefore the air mass is subject to an angular momentum
causing a rotation of the air mass. At the walls skin friction will occur
and in the corners of the room secondary vortices will cause extra loss
of momentum. This gives an angular momentum opposite the angular momentum
of the driving force, i.e. the air jet or the buoyancy force.
By fundamental law these two angular momentums should be equal which
makes possible the calculation of an overall resistance coefficient. For
a systematic approach some simplifications must be made:

a. The rotation of the air mass is 2-dimensional, i.e. takes place in a
   vertical cross section of the room determined by the length and the
   height.

b. The rotation is described by only one vortex covering the whole rectangular cross section.

Ad a. Experiments have been done with air blowing in from a slit over the
full width of the room. From literature data [1] it is known that at the
ratios between length, width and height of the test room and of the model
used in our experiments the approximation of 2-dimensional flow is justified.

Ad b. At the ratio between the length and the height in our experiments
this assumption is justified as has been verified by smoke tests in the
scale model. In the corners small secondary vortices are present. If the
length is more than about 2.5 times the height the flow breaks up in 2
vortices [2].
For the calculation of a rotating flow first the rectangular boundary has
been transformed into a circle. This was possible by a succession of conformal mappings of one complex plane into another. The flow is first considered as being surrounded by a rotating circle. The fluid then rotates
as a solid mass and the stream lines are concentric circles within the
boundary circle. By reverse transformations the boundary circle is now
transformed into a rectangular and the concentric inner circles become
closed curves within the rectangular.
We now suppose that these closed curves are still the stream lines of the
flow and that the mass flow between two stream lines is invariant for
transformations. This means that velocities are relatively high at places
where the stream lines are close together and relatively low if the
stream lines are far from each other as is the case near the corners.
This is analogous to the way potential flow problems are solved but in
our case there is no potential flow.
In the experiments there was supply and exhaust of air which is not taken
into account in this calculation. But the rotating air mass was much
larger than the mass of supply air and therefore this calculation method
was used.
It is clear that in this way the velocities can only be calculated from
the center of the room until the outer edge of the boundary layers. In
the relatively thin boundary layers the velocities decrease to zero at
the walls but here we are more interested in the friction forces than in
velocities.

3. Theory

3.a. Stream lines

First the transformation of the rectangular to the circle will be described.

In the complex plane z_1 the rectangular has the sides with dimensions a and b, see fig. 1.

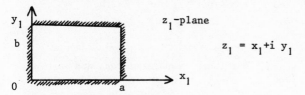

Fig.1. Rectangular in the z_1-plane

The first transformation is: $z_2 = \frac{\pi}{b} z_1$ (1)

The second transformation is: $z_3 = \cosh z_2$ (2)

Now the rectangular has been transformed into a half ellips, see fig.2.

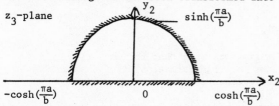

Fig.2. Half ellips in the z_3-plane

The length of the horizontal half axis is $\cosh(\frac{\pi a}{b})$ and of the vertical half axis $\sinh(\frac{\pi a}{b})$. If a>b the difference between these values is always below 0.25% and therefore these values are assumed to be equal. Under the condition a>b the rectangular has then nearly been transformed into a half circle, so it is always necessary that the largest side is taken as horizontal.

The third transformation is:

$$z_4 = z_3 + \frac{\cosh^2(\frac{\pi a}{b})}{z_3} \qquad (3)$$

This transforms the half circle into the lower half plane $y_4 < 0$, see fig.3.

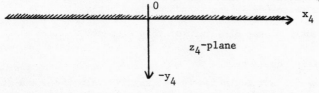

Fig.3. Half plane of z_4.

In the rectangular each point can be taken as the center of rotation. This point z_{1c} in the z_1-plane is mapped into a point z_{4c} in the z_4-plane by equation (1)-(3). The point z_{4c} is replaced to the point $-i$, so the next transformation is:

$$z_5 = \frac{z_4 - x_{4c}}{-y_{4c}} \qquad (4)$$

This includes a horizontal shift over a distance x_{4c} and a contraction with a factor y_{4c}.

The z_5 half plane is transformed into a circle with radius 1 by the following equation:

$$z_6 = \frac{z_5 + i}{z_5 - i} \tag{5}$$

In the appendix the equations in x and y for these transformations and the reverse transformations are given.

3.b. Velocities

In the z_6-plane a rotating flow like of a solid disk with angular velocity ω is assumed. For the absolute value and the direction of the velocity vector in the z_6-plane the following equations hold:

$$\left| v_6 \right| = \omega \sqrt{x_6^2 + y_6^2} \tag{6}$$

$$\phi_6 = - \text{arctg} \left(\frac{x_6}{y_6} \right) \tag{7}$$

In the previous paragraph the assumption was made that the mass flow between two stream lines is invariant for transformation. If the mass flow between two stream lines close together is $d\psi$ then

$$d\psi = \left| v_i \right| \cdot \left| dz_i \right| = \left| v_{i+1} \right| \cdot \left| dz_{i+1} \right|$$

ψ is called the stream function which is the general solution of the continuity equation.

In the z_1-plane the absolute value and the direction of the velocity then are:

$$\left| v_1 \right| = \left| v_6 \right| \cdot \left| \frac{dz_6}{dz_1} \right| = \left| v_6 \right| \cdot \prod_{i=1}^{5} \left| \frac{dz_{i+1}}{dz_i} \right| \tag{8}$$

$$\phi_1 = \phi_6 - \arg\left(\frac{dz_6}{dz_1} \right) = \phi_6 - \sum_{i=1}^{5} \arg\left(\frac{dz_{i+1}}{dz_i} \right) \tag{9}$$

These equations are a result of the characteristic properties of conformal transformations that in a point an infinitesimal small line element is multiplied with a factor $\left(\frac{dz_{i+1}}{dz_i} \right)$ and turned over an angle $\arg\left(\frac{dz_{i+1}}{dz_i} \right)$ independently of the direction of this element.

3.c. Resistance coefficient

One of the input data for a calculation is the angular velocity ω and it is clear that there must be a relationship between its value and the resistance which the air flow meets at the walls and the corners by skin friction and secondary vortices.

The resistance which the flow meets can be expressed by an overall resistance coefficient.

As an example the case of an isothermal 2-dimensional jet from a vertical wall into the room is considered.

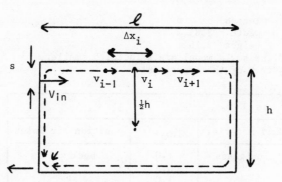

Fig.4. Isothermal jet from the back wall into a room.

The angular momentum caused by the jet per unit of width of the room is:

$$M_1 = \tfrac{1}{2}\rho V_{in}^2 \cdot s \cdot \tfrac{1}{2}h \tag{10}$$

The air outlet causes no angular momentum because the air flows to the outlet opening from all directions and behaves there as a potential flow. Over a length Δx_i the resistance force by skin friction is $C_f \cdot \tfrac{1}{2}\rho v_i^2 \cdot \Delta x_i$.

This results in an angular momentum:

$$M_2 = C_f \cdot \tfrac{1}{2}\rho \left\{ \tfrac{1}{2}h \cdot 2\overset{\ell}{\Sigma}\Delta x_i \cdot v_i^2 + \tfrac{1}{2}\ell \cdot 2\overset{h}{\Sigma}\Delta x_i \cdot v_i^2 \right\} \tag{11}$$

The first term includes a summation along the floor and ceilings and the second term includes summation over the two vertical walls. In order to determine M_2 it is thus necessary to measure the velocity in a large number of points near the walls and just outside the boundary layers. It must be remarked that in equation (11) the secondary vortices which will be present in the two corners opposite the air inlet and air outlet are not taken into account. The velocities are very low there however and the contribution of the terms $v_i^2 \Delta x_i$ are therefore negligable.

4. Experiments

A number of experiments in a climate room and a scale model 1:5 of this room have been carried out to test the validity of the calculation method. The experiments were isothermal, i.e. no heat sources were present. Air was blown in through a slit of 9 cm in the climate room and 1.8 cm in the model over the full width to give a two dimensional flow. The exhaust opening was a slit with the same dimensions. The place of the inlet and exhaust openings are indicated in fig. 4 but in the climate room also experiments have been done with air blown in from the floor in the corner opposite the exhaust opening. The jet was then directed along the vertical wall. These two ways of air supply are representative for actual air conditioning systems. A smooth and a more rough material for the walls was used to investigate the influence on the velocities. In the climate room in a number of experiments a plate of 70 cm width was placed at an angle of 45° in the corners opposite the inlet and exhaust opening to eliminate the influence of secondary vortices for the most part.

The dimensions of the climate room are:

                    length   5   m
                    width   3.85 m
                    height   3.15 m
The dimensions of the scale model are a factor 5 smaller.
Tabel 1 gives a survey of the experiments.

Scale model			
Exp.nr.	wall material	V_{in}, m/s	position air inlet
1	plexiglass	1.0	back wall
2	"	0.5	"
3	board	1.0	"
4	"	0.5	"
Climate room			
5	board	0.5	back wall
6	"	1.0	"
7	hard board	0.5	"
8	"	1.0	"
9	"	1.0	floor
10	"	0.5	"
11 x)	"	0.5	"
12 x)	"	1.0	"
13 x)	"	1.0	back wall
14 x)	"	0.5	"

Table 1. Survey of the experiments

x) plates in the corners opposite the inlet and exhaust openings

The air velocities were measured with an anemometer suitable for a velocity range of 0.06- 2 m/s and insensible for the direction of flow.
In the climate room measurements were done in 5 vertical planes in the
length of the room.by automatically steered equipments. In this way
the velocities were measured at 40 levels and 20 distances from the
back wall i.e. in 800 points per plane. In the scale model the measurements were carried out in 4 vertical planes in 16x24 points, i.e.
384 points per plane.

5. Results

5.a. Velocity profiles

In the z_6-plane the radius of the circle to be transformed into the
rectangular (5x3.15 m) is taken as 1 m. Forthe inner circles values
of the radius of 0.1 m, increasing with 0.1 to 0.8 m, then increasing with 0.05 to 0.95 were selected. At each inner circle 36
points determined by ϕ values of 0°, increasing with 10° to 350°,
were selected for transformation. The corresponding points in the
z_1- plane (rectangular) were obtained using equation (5b) to (1b)
(see Appendix). The absolute value and direction of the velocity in
each point was calculated using equation (8) and (9). For these
calculations suitable values of x_{1c}, y_{1c} and ω were required to give
the best agreement with measurements.

Experimental values of the velocity (only absolute values) were obtained by linear interpolation between the velocities measured in the 4 points around the considered point of the transformed circle. The following interpolation formula has been used:

$$v = \frac{1}{4}\Big[v(i+1, j+1)+v(i,j+1)+v(i+1,j)+v(i,j)+$$

$$+ \frac{2x-x_i-x_{i+1}}{x_{i+1}-x_i}\{v(i+1,j+1)+v(i+1,j)-v(i,j+1)-v(i,j)\}+$$

$$+ \frac{2y-y_j-y_{j+1}}{y_{j+1}-y_j}\{v(i+1,j+1)+v(i,j+1)-v(i+1,j)-v(i,j)\}\Big] \qquad (12)$$

Fig. 5 shows some transformed circles and velocities calculated above the center of rotation ($\phi = 270^\circ$) which has been chosen here at the intersection of the diagonals. But fig. 6 shows that the center of rotation should be chosen higher than the half height to fit the experimental velocity profile. This is due to the influence of the jet along the ceiling caused by the air supply in the back wall at the upper corner. Very low velocities as predicted by the calculations are not observed experimentally. This is due to the anemometer which cannot indicate very low velocities and due to slow fluctuations which can be expected at low velocities.
Fig. 7-9 show some examples of theoretical and experimental velocity profiles along transformed circles near the walls (R = 0.8 and 0.85) in the climate room and the model. For convenience the dimensions of the model are taken as equal to the climate room dimensions (5x3.15 m^2). In fact these dimensions and the values of ω are a factor 5 smaller.
The direction of the flow is from 360° to 0° in fig. 7-9.
In fig. 5 the direction of rotation is clockwise in case of air supply from the back wall (left) and counter clockwise in case of air supply from the floor along the front wall (right side).
Especially in fig. 7 the influence of the jet is visible by the sharp rise of the experimental velocity curve between 340° and 310°. At higher values of R this increase becomes even sharper and this is not explained by the theoretical curve at whatever value of x_{1c} and y_{1c}. It means that the calculation method is not suitable for the jet near the inlet opening but for this other theories are available.
Table 2 shows the values of ω, x_{1c} and y_{1c} which give the best agreement between theoretical and experimental velocity profiles along the transformed circles. The values are averaged over the 4 or 5 vertical planes. For the experiments in the climate room the same values could generally be used for all planes but for the model experiments, especially if V_{in} = 1 m/s, different values for each plane had to be used. The maximum deviation from the average value was then 15%.

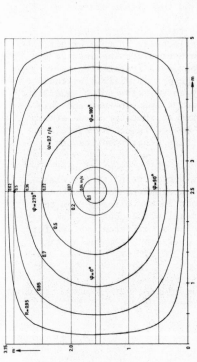

Fig. 5. Transformed circles and calculated velocities along the vertical axis ($\varphi = 270°$)

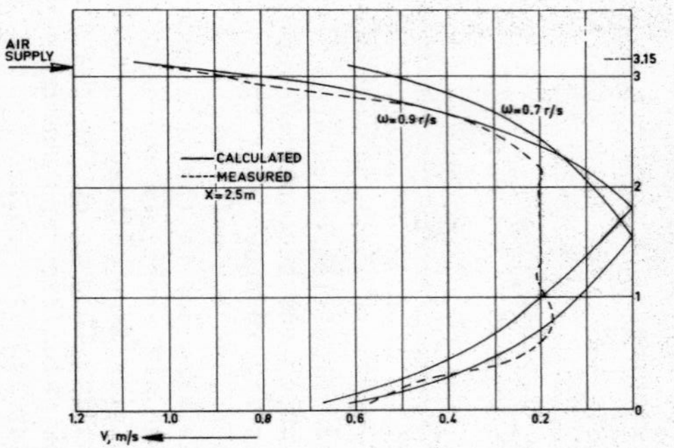

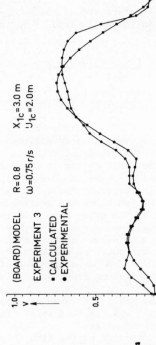

Fig. 6. Theoretical and experimental velocity profiles along the vertical axis ($\varphi = 90°$ and $270°$)

(PLEXIGLASS) R=0.85 X_{1c} = 2.5 m
EXPERIMENT 1 ω=0.94 r/s y_{1c} = 1.9 m

■ CALCULATED
● EXPERIMENTAL

Fig. 7. Velocity profile in a plane in the plexiglass model along a transformed circle (R = 0.85)

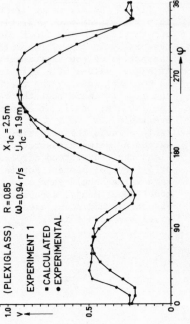

(BOARD) MODEL R=0.8 X_{1c}=3.0 m
EXPERIMENT 3 ω=0.75 r/s y_{1c}=2.0 m

■ CALCULATED
● EXPERIMENTAL

Fig. 8. Velocity profile in the board model along a transformed circle (R = 0.8)

304

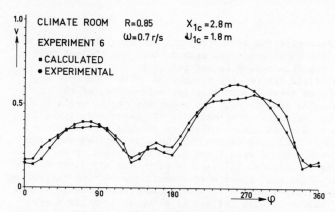

Fig. 9. Velocity profile in the
climate room along a
transformed circle.

5.b. Resistance coefficient

The angular momentum of the driving force (jet) on the air mass
was calculated using equation (10). The terms between the brackets
in the right hand side can also be calculated for each vertical
plane and then averaged over the 4 or 5 planes for each experiment.
Equating of the two formulae gives c_f.
Because the flow in the room and the model is turbulent it may be
expected that ω is proportional to V_{in}. The value of c_f however is
determined by the squared values of V_{in} and the local velocities
near the walls just outside the boundary layers. Therefore it is
worthwile to calculate also $\dfrac{\omega \sqrt{c_f}}{V_{in}}$ and see if this value is more

or less constant for the experiments. Table 2 shows these values,
together with ω, c_f, x_{1a} and y_{1c}.

Experiment nr.	V_{in} m/s	ω, rad/s (average)	c_f	$\dfrac{\omega \sqrt{c_f}}{V_{in}}$, rad/m	x_{1c}, m	y_{1c}, m
1	1	0.957	0.0161	0.121	2.5-3	1.7-2
2	0.5	0.417	0.0223	0.124	2.3-3	1.8-1.9
3	1	0.719	0.0232	0.110	2.8-3	1.9-2
4	0.5	0.325	0.0310	0.114	2.5-2.8	1.8-2
5	0.5	0.35	0.0386	0.138	2.8	1.8
6	1	0.704	0.0325	0.127	2.8	1.8
7	0.5	0.375	0.0341	0.138	2.8	1.8
8	1	0.750	0.0289	0.128	2.8	1.8
9	1	0.450	0.118	0.155	3.1	1.95
10	0.5	0.250	0.124	0.176	3.1	1.95
11	0.5	0.256	0.0709	0.136	3.1	1.95
12	1	0.472	0.0936	0.144	3.1	1.95
13	1	0.720	0.0313	0.127	2.8	1.8
14	0.5	0.325	0.0423	0.134	2.8	1.8

Table 2. Values of ω, c_f, x_{1c} and y_{1c} in the climate room and the model

5.c. Discussion

Fig. 7-9 indicate that the velocities can be calculated rather
accurately by the above described method. Only for the jet near
the inlet opening this method is not suitable.

The influence of the jet becomes clear from table 2 because x_{1c}
and y_{1c} have to be taken higher than the half values of the rectangular sides. The influence of the roughness of the walls is
demonstrated by the model experiments (1-4). Experiment 1 and 2
show higher ω values and lower resistance coefficients than experiment 3 and 4. In a less degree this is also noticed when comparing experiment 5 and 6 with 7 and 8. In most cases c_f decreases
with increasing velocities. This agrees with boundary layer theory
which predicts lower wall shear stress coefficients with increasing
Reynolds numbers. Experiment 9-12 show considerably lower ω values
and higher c_f values because the air is blown in from the floor
and follows a wall of 3.15 m instead of 5 m. Plates in the corners
give a small increase of ω but a strong decrease of c_f in this
case (exp. 11, 12). When blowing in air from the back wall the
effect of plates in the corners is opposite but much smaller.
Experiments 9 and 10 show much higher values of $\frac{\omega \sqrt{c_f}}{V_{in}}$ than the

other experiments. But for the experiments in the climate room with
air inlet in the back wall and the model experiments this value
can be considered as constant. One finds:

- model experiments (1-4): $\frac{\omega \sqrt{c_f}}{V_{in}} = 0.117$, max. deviation 6%.

- climate room experiments, air inlet in back wall
(exp. 5-8, 13, 14):
$$\frac{\omega \sqrt{c_f}}{V_{in}} = 0.132, \text{ max. deviation } 4.5\%$$

For air inlet from the floor such empirical relationships cannot
yet be established. Much more experimental work is needed for that.
But also the above given relationships need more experimental
support and should still be considered as preliminary.

6. Conclusions

6.1. The calculation method described in this paper looks promising for
predicting velocities in a room, except near the air inlet opening
if the air behaves as a 2-dimensional rotating mass.

6.2. For the calculations reliable data for the coordinates of the
center of rotation and the resistance coefficient are required.
Much experimental work has still to be done for this.

6.3. The angular velocity can be calculated from the inlet velocity
and the resistance coefficient by empirical relationships as
given above.
These relationships are determined mainly by the dimensions, the
position of the air inlet opening and the wall roughness.

7. Nomenclature

a	longest side of the rectangular	m
b	shortest side of the rectangular	m
c_f	resistance coefficient (equation 11)	
h	height of the room	m
1	length of the room	m
M	angular momentum per m width	N
R	radius of a circle	m
s	seize of a slit	m
v_{in}	inlet velocity	m/s
v	velocity	m/s
x,y	coordinates	m
x_{1c}, y_{1c}	coordinates of the center of rotation in the rectangular	m
z	complex number ($z = x + i\,y$)	
ρ	density of air	kg/m^3
ω	angular velocity	rad/s

indices

1, 2, ..., 6	sequence of transformation
i, j	sequence of points in x and y direction

8. References

[1] Nielsen, P.V.   Flow in air conditioned rooms.
                        Ph.D. Thesis, Techn. Univ. of Denmark,
                        August 1976.

[2] Regenscheit, B.   Modellversuche zur Erforschung der
                        Raumströmungen in belüfteten Räumen.
                        Staub 24, Januar 1964.

Appendix. Equations of transformation

$$z_2 = \frac{\pi}{b} z_1 \text{ gives: } x_2 = \frac{\pi}{b} x_1 \text{ and } y_2 = \frac{\pi}{b} y_1 \tag{1a}$$

$$z_3 = \cosh z_2 \text{ gives: } x_3 = \cosh x_2 \cdot \cos y_2 \text{ and } y_3 = \sinh x_2 \sin y_2 \tag{2a}$$

$$z_4 = z_3 + \frac{\cosh^2(\frac{\pi a}{b})}{z_3} \text{ gives: } x_4 = x_3 \left(1 + \frac{\cosh^2(\frac{\pi a}{b})}{x_3^2 + y_3^2}\right) \text{ and}$$

$$y_4 = y_3\left(1 - \frac{\cosh^2(\frac{\pi a}{b})}{x_3^2 + y_3^2}\right) \quad \text{and} \tag{3a}$$

$$z_5 = -\frac{z_4 - x_{4c}}{y_{4c}} \text{ gives: } x_5 = -\frac{x_4 - x_{4c}}{y_{4c}} \text{ and } y_5 = -\frac{y_4}{y_{4c}} \tag{4a}$$

$$z_6 = \frac{z_5 + i}{z_5 - i} \text{ gives: } x_6 = \frac{x_5^2 + (y_5^2 - 1)}{x_5^2 + (y_5 - 1)^2} \text{ and } y_6 = \frac{2x_5}{x_5^2 + (y_5 - 1)^2} \tag{5a}$$

The reverse transformations are:

$$x_5 = \frac{2y_6}{(x_6 - 1)^2 + y_6^2} \text{ and } y_5 = \frac{(x_6^2 - 1) + y_6^2}{(x_6 - 1)^2 + y_6^2} \tag{5b}$$

$$x_4 = -x_5 \cdot x_{4c} + x_{4c} \text{ and } y_4 = -y_5 \cdot y_{4c} \tag{4b}$$

$$\text{If } x_{41} = \left\{ 0.5(x_4^2 - y_4^2 - 4\cosh^2(\frac{\pi a}{b})) + (0.25(x_4^2 - y_4^2 - 4\cosh^2(\frac{\pi a}{b}))^2 \right.$$

$$\left. + x_4^2 y_4^2)^{\frac{1}{2}} \right\}^{\frac{1}{2}}, \text{ then}$$

$$\left.\begin{array}{l} x_3 = 0.5 \ (x_4 - x_{41}) \text{ if } x_4 > 0 \\[4pt] x_3 = 0.5(x_4 + x_{41}) \text{ if } x_4 < 0 \\[4pt] y_3 = \frac{x_3 y_4}{2x_3 - x_4} \end{array}\right\} \tag{3b}$$

$$\text{If } x_{31} = \left\{ -0.5(x_3^2 - y_3^2 - 1) + (0.25(x_3^2 - y_3^2 - 1)^2 + x_3^2 y_3^2)^{\frac{1}{2}} \right\}^{\frac{1}{2}} \quad \text{and}$$

$$x_{32} = \frac{x_3 y_3}{x_{31}}, \text{ then :}$$

$$\left.\begin{array}{l} x_2 = 0.5 \ \ln((x_3 + x_{32})^2 + (y_3 + x_{31})^2) \\[4pt] y_2 = \text{arctg} \ (\frac{y_3 + x_{31}}{x_3 + x_{32}}) \text{ if } x_3 > 0 \\[4pt] y_2 = \text{arctg} \ (\frac{y_3 + x_{31}}{x_3 + x_{32}}) + \pi \text{ if } x_3 < 0 \end{array}\right\} \tag{2b}$$

$$x_1 = \frac{b}{\pi} x_1 \text{ and } y_1 = \frac{b}{\pi} y_1 \tag{1b}$$

POROSITY AND THE
VENTILATION OF
ANIMAL HOUSING

D. R. PATTIE

School of Engineering
University of Guelph
Guelph, Ontario, Canada N1G 2W1

ABSTRACT

In areas where winter temperatures are often below -25°C, it is found that the insulated mechanically-ventilated building will not give satisfactory results. In these conditions satisfactory results are obtained with porous type buildings, the result of many years of research, observation, and experiment.

LATENT HEAT OF RESPIRATION

Animals produce heat and water vapour: the ratio of latent heat of respired moisture to total heat produced, may be as high as 40%, as in swine, and only the difference, or what is known as the sensible heat, is usually available to warm the building. The continuous removal of water vapour is the prime requirement of the ventilation system for a building housing animals. When this is accomplished it is usually found that air distribution and odour level are satisfactory.

Ventilation is usually achieved by air movement induced by electrically-driven fans. Fresh air moves in, replacing the stale air which is exhausted. Fans are controlled by thermostat and sometimes by humidistat. In this way water vapour is expelled from the building. Unfortunately, a considerable loss of heat is incurred in this type of ventilation, and even with very good insulation, it is found that as the outside temperature drops a point is reached where it is no longer possible to maintain the relative humidity at an acceptably low level of 75% or so, and at the same time be able to maintain a desirable level of temperature within the building. Increased ventilation rate results in a lowering of temperature with little effect on relative humidity, while a reduction of ventilation rate will result in an increase of temperature and an increase in relative humidity which leads to condensation of moisture on the coldest surfaces. Interestingly, condensation of moisture is accompanied by a release of heat within the building which assists in maintaining a comfortable temperature, however, it is associated with a rise in odour level and staleness of air. The temperature at which the onset of condensation occurs varies with the type of livestock housed and with many other factors. As an indication, it may be expected that difficulties are encountered in housing swine at temperatures below -10°C, in housing cattle at temperatures below -20°C, and in housing chickens at temperatures below -40°C.

The importance of porosity in ventilation became evident when it was found that a number of newly constructed dairy buildings would not operate satisfactorily in an area of Northern Ontario, where the temperature drops

309

below −40°C. At the same time a survey of a number of buildings showed that some of the older type buildings were operating quite well, evidently on account of an incidental degree of porosity. It was evident that some degree of porosity, due to cracks between boards, for example, and some materials of construction could lead to performance superior to that obtainable with the insulated, vapour-tight type of building.

Experimentally, a dairy barn was converted to a porous type of construction, with very satisfactory results. Also, laboratory work was carried on, the focus being on measurement of simultaneous heat and moisture transmission through porous materials. The aim was to measure the heat loss per unit mass of moisture transmitted through porous buildings, ventilated to some extent by diffusion and by natural convection, and to show that this could be less than that occuring with the conventional vapour-tight building and ventilation system.

Generally the results have shown that the greatest heat loss is incurred with the conventional ventilation system and that the smallest heat loss is the reslut of employing devices which retain some part of the latent heat of the respired moisture within the building: that it is possible to devise systems capable of retaining part of the latent heat of respired moisture is, perhaps, by far the most important finding made.

The system which was evolved for dairy barns in Northern Ontario, is that shown in Fig. 1. Cold air enters from the attic space, which should

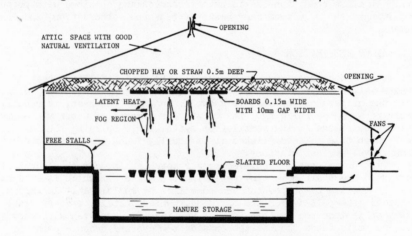

FIG. 1 POROUS CEILING VENTILATION SYSTEM DEVELOPED
FOR DAIRY BARNS IN NORTHERN ONTARIO

have good natural ventilation to avoid problems with condensation of moisture, and filters down through the porous material. The depth of the porous material does not appear to be critical. Chopped hay or straw have been employed successfully, although, eventually, a more permanent material will likely be found. The gaps between the boards in the ceiling regulate and distribute the flow of air. The portion of the ceiling treated in this manner has varied from 20% to 100% of the ceiling area. The cold air coming into the space occupied by the animals strikes the warmer and more humid air causing a fog. The fog region may extend down to the concrete slats covering the manure storage pit, depending upon how cold the weather is. The colder the weather, the more pronounced is the fog region, and the more likely is the fog to persist in the air passing down through the slatted floor. The air is expelled from the manure pit by fans of which at least one should operate continuously.

The fog region is where the respired water vapour is condensed and its latent heat released to the interior environment. The amount of the fog which persists and passes through the slatted floor is important as there is a tendency for the liquid droplets of the fog to re-evaporate, withdrawing again the necessary latent heat from the interior environment. The greater the quantity of moisture which leaves in the liquid state, the greater is the amount of the latent heat that has been retained in the interior environment. The intensity of the fog increases as outdoor temperature drops resulting in a self-compensating effect on the interior temperature, in that as the outdoor temperature becomes lower, more latent heat is available. It would appear evident that a lower ceiling height would be an advantage to the operation of this system, but as yet there is no practical evidence whether this is the case or not.

In some cases porous ceiling type of housing has been shown to be able to operate satisfactorily without fans throughout a whole winter, the ventilation being by diffusion and convection of air through the ceiling. The feature of not being entirely reliant on electrical power is most attractive in providing against power failure. Satisfactory operation over a lengthy period, however, depends upon securing the correct degree of porosity which at the present time is difficult to specify: difficulties would also be experienced in circumstances where the number of animals varied, for example, so that forced convection by the use of fans leads to a more practical system.

The fact that it is possible to retain part of the latent heat of the moisture to heat the air space was first observed under laboratory conditions, when experimenting with a somewhat different system. Tests were being made on a 2.5 m cube model, as shown in Fig. 2. Electrical heaters supplied

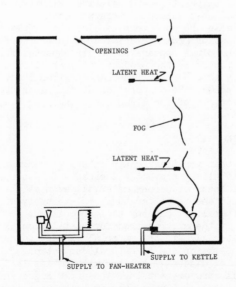

FIG. 2   2.5m CUBE WITH HEATER AND KETTLE

sensible heat and an electric kettle supplied water vapour. Experiments
were being made on the effects of natural convection through holes in the
ceiling. It had been suspected that respired air being less dense on
account of its higher temperature and on account of its higher water vapour
content, would rise to the ceiling and escape if suitable exits were
available. In this way moisture would be expelled without having been
mixed throughout the air space, with the possibility of eliminating moisture
with a smaller loss of heat. Calculations from data obtained showed that
a good part of the latent heat of the water could be retained to warm the
interior. A similar test employing a surface type humidifier evaporating
water at the same temperature as that of the interior, showed that no latent
heat could be retained even when the relative humidity of the interior was
made very high: no fog was being formed and none of the water was being
expelled in the liquid state: the water, once having been evaporated,
remained in the vapour state. However, when a heater was installed to heat
the water which passed over the humidifier element, thus raising the
temperature of evaporation, calculations from the data showed that part of
the latent heat of the moisture was retained, but only in tests made at
high relative humidity. That is, a fog was formed where the vapour from the
humidifier contacted the interior air, the latent heat being released to the
interior environment. A process of re-evaporation then occurs when the heat
necessary for evaporation is withdrawn from the air. When the relative
humidity is high enough the re-evaporation is slow enough that some of the
moisture passes through an opening in the ceiling, in the liquid state,
thus part of the latent heat is retained. In these tests the transfers
of latent heat are evident by the fact that the temperature in the lower
region can be greater than the temperature in the higher region, that is,
an inversion of temperature occurs. Normally, temperature at the ceiling
will be 2C to 5C greater than the temperature at the floor.

In respiration of animals water is evaporated at the body temperature
of about 39C. The fog produced by respiring into a cold atmosphere and its
persistance depending on temperature and relative humidity, are well known
to everyone from personal observation of this phenomena.

On a much larger scale the phenomena is also observable in very cold
weather, in the plume issuing from the stack of a gas-fired furnace. The
visible white part of the plume appears to be disconnected from the top of
the stack, where the moisture in the products of combustion has not yet
been cooled sufficiently to condense. The condensation is evident by the
dense whiteness of the plume. Re-evaporation of the water droplets can be
observed by the disappearance of the plume occurring over a considerable
distance.

FILTRATION OF VENTILATING AIR

It has been shown that ventilating air entering downwards through a
porous ceiling will effectively prevent the escape of heat upwards through
the ceiling. A 3 cm thickness of porous material, even at the very low flow
rates involved, less than one metre per hour, is extremely effective. It
could be stated as a generality that any wall or ceiling, insulated or not,
will conduct less heat if the air required for ventilation is permitted to
filtrate through the material. If the material is very porous, there is a
possibility of over-ventilation due to natural convection. There is some
evidence that even a thin diaphragm ceiling with no insulating material may
permit only a very small loss of heat into the attic space when holes have
been drilled through the material in a regular pattern to permit air to be
drawn in from the attic space. This effect was noted in a large building.
It was not found possible, in a limited range of tests, to demonstrate the
magnitude of effect on a model that was evident in the large building,

probably because the air over the ceiling of the model was not reduced to a sufficiently stilled condition. The air motion above a thin perforated ceiling certainly influences the heat transmission, and filtration from a still-air region would appear to be the most effective in reducing heat loss.

Surprisingly, reversing the air flow, that is, blowing air upwards through a porous ceiling, has a very similar effect in greatly reducing the rate of conductive heat transmission through the ceiling. The upward movement of air is not so practical in animal housing, and is also somewhat less effective, as the hotter air which tends to rise to the ceiling is expelled sooner and does not have the same effect in warming the building. In this connection, it has been found that the air velocities due to convective movement of the air are many times greater than the average velocity of air being filtered through the porous ceiling. However, if slots or holes are employed, then the air does enter as a perceptible jet.

In hot summer conditions, the upward movement of air through a porous ceiling is found to be quite effective in reducing the heat load from a very hot attic. An explanation of these effects of filtration of air may be seen in Fig. 3.

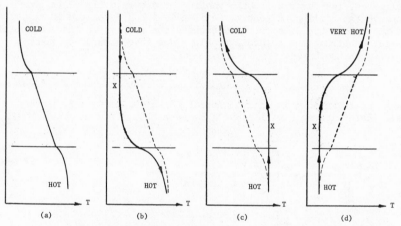

FIG. 3 TEMPERATURE DISTRIBUTION CURVES THROUGH POROUS CEILING

    (a) WITH NO FILTRATION OF AIR
    (b) WITH DOWNWARD FILTRATION OF AIR
    (c) WITH UPWARD FILTRATION OF AIR
    (d) UPWARD FILTRATION OF AIR INTO VERY HOT REGION

IN (b) (c) AND (d) FILTRATION OF AIR PRODUCES A REGION MARKED X WHICH IS PRACTICALLY AT CONSTANT TEMPERATURE INDICATING THAT THE RATE OF HEAT TRANSMISSION BY CONDUCTION IS VERY MUCH REDUCED FROM THE NORMAL AMOUNT. THE BROKEN LINE INDICATES TEMPERATURE DISTRIBUTION WITH NO FILTRATION OF AIR.

The effect of air filtration upon conductive heat transmission has apparently gone unnoticed, but, to be practical, while it is an advantage to employ this effect, the result of the reduction in conductive heat loss is small, when a substantial thickness of insulation is employed.

VENTILATION BY DIFFUSION

Although it was recognized in early times that ventilation could be
achieved by the diffusion of gases and vapours through the material of walls
and ceilings, recent practice has been to install a vapour barrier ostensibly
to prevent the migration of water vapour from the interior into the material
of the wall or ceiling. If a low enough temperature were reached, within
the wall, condensation or icing would occur, decreasing the effectiveness of
thermal insulation and causing damage. It is now recognized that damage
due to frost is due to moisture from a flow of moist air wich is transported
through the wall at structural faults, and that the quantity of moisture
which would actually diffuse by molecular diffusion, in a whole season,
would be much too small to cause any damage. Damage due to condensation has
occurred where a material which was effectively a vapour barrier has been
placed on the outside of a building of poor construction or in a poor state
of repair. Flat roofs and also sloping roofs with insulation on the underside are notoriously troublesome due to the fact that it is difficult to
provide a completely effective vapour barrier under the insulation. In
colder climates, ice forms under the roofing, and eventually melts and
leaks through the ceiling, appearing as if a leak had developed in the roof.
A ventilated space should be provided between the insulation and the roofing,
in flat roof construction, to avoid problems due to condensation and
freezing of moisture.

It can be shown that in animal housing a vapour barrier is not
essential. Referring to Fig. 4 it will be seen that for a material of

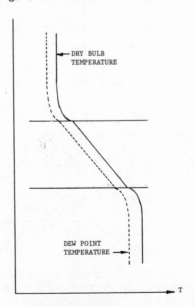

FIG. 4   CONDENSATION WILL NOT OCCUR
AT ANY POINT WITHIN A
UNIFORMLY PERMEABLE CEILING
OR WALL.

uniform permeability, and in this case, in the absence of any filtration of air, it is unlikely that condensation will occur at any point within the material when it does not occur at either surface: similar reasoning can be applied when filtration of air is present. The escape of moisture by diffusion is an advantage, as no loss of sensible heat is involved, however, to obtain a high rate of diffusion, the material would have to be quite thin, when its thermal insulation value would be small. In ventilation by diffusion the ability to transmit moisture is much more important than thermal insulation value, however, no material was found with the characteristics to perform satisfactorily in severe climates by the action of molecular diffusion alone; satisfactory materials may be obtained for moderate climates.

In one test, two men were enclosed inside a 2.5 m cube constructed from wood fibre board 12 mm thick. Moisture and carbon dioxide diffuse outwards maintaining low concentrations inside, and oxygen diffuses inwards so that depletion of oxygen does not occur. The oxygen content decreased 0.1% and the carbon dioxide content increased by 0.08% above the level in the outer space, these changes occurring the first 30 minutes of the test, the levels remaining constant thereafter to the end of the three hour test: the inside conditions were a comfortable 25°C at 40% relative humidity, whereas the outside conditions were 0°C at 80% relative humidity: the ventilation is entirely by diffusion. It is quite possible that this principle might provide a simple inexpensive shelter against dust, overcoming the ventilation and filtration problems simultaneously.

It has been demonstrated with this permeable wall of wood filter board that if a piece of metal foil is glued onto the inner surface, condensation will appear on this impermeable surface when no condensation appears on the adjacent surface of the permeable material: this shows that permeable materials permit higher relative humidity without showing wetness.

OTHER ASPECTS OF POROUS CEILINGS

Organic materials which have the property of transmitting liquid water at relatively high rates may well be preferable to inorganic materials which may not have this characteristic. Moisture which condenses on the inner surface, gives up latent heat and increases somewhat the temperature at the inner surface. The liquid water is conducted through the material and is re-evaporated at the outer surface. At the outer surface, latent heat is absorbed and the temperature at the outer surface lowered. This leads to a higher rate of heat transfer through the material, but is due to the transfer of latent heat from the inner to the outer surface. The beneficial effect is that the inner surface temperature is raised resulting in a reduced rate of loss of sensible heat. This principle is sketched in Fig. 5.

Measurements were made to determine whether moisture will diffuse against the direction of filtration of air through porous material. It was found that the rate of diffusion is rapidly reduced as the filtration rate is increased and that at the usual rates required for ventilation, it may be taken that the rate of moisture transmission by diffusion is reduced to zero.

The incorporation of porosity in a building is associated with a marked reduction in odour level, which may be due to the fact that there is less wetness, or it may possibly be due to odorous molecules diffusing through the porous material at higher rates than other molecules, or it may be due to the fact that the concentration gradient is proportionately higher for the odourous molecules, than it is for water molecules, which are likely to be present in sizeable concentration in the atmospheric air.

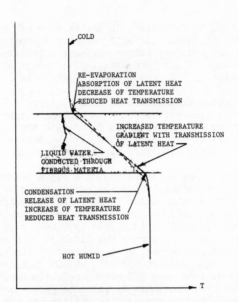

FIG. 5 FIBROUS ORGANIC MATERIAL TRANSMITTING
WATER IN THE LIQUID STATE RESULTING IN
A REDUCTION OF CONDUCTIVE HEAT LOSS.

THE BROKEN LINE INDICATES TEMPERATURE
DISTRIBUTION FOR DRY AIR.

THERE IS NO FILTRATION OF AIR IN
EITHER CASE.

A number of barns have been constructed with the vapour barrier
deliberately omitted and no deleterious effect has been observed. No condensation or frost has been observed within the material of porous ceilings:
however, at high rates of vapour transmission and with the aid of convective
flow upwards through the ceiling, condensation may form on the topmost
protruding fibres, and eventually the drops will become large enough to fall
or run back into the material. In cold enough conditions a snow-like layer
forms on top of a porous ceiling, in these circumstances.

CONCLUSION

It will be evident that very significant advances can be made in
ventilation, by quite simple means.
The outstanding conclusion of this work is that, in animal housing,
contrary to the accepted view, it is possible to retain part of the latent
heat of respiration, and employ it in heating the building. The heat
transmission characteristics of porous insulation are completely changed
when air filtrates through them: the rate of conductive heat transmission
is reduced almost to zero by very small rates of filtration such as
required in ventilation, and this is the case with air flow in either
direction. Permeance in the materials of construction may have definite
advantages: the ability of a building material to diffuse moisture may be
more important than its value as thermal insulation.

Although these developments have taken place in areas where winter conditions are very severe, satisfactory results having been achieved whereas otherwise housing conditions for animals were untenable, much of the development will be applicable in areas of more moderate climate. Adoption of the porous ceiling principle is taking place in severely cold areas of Canada and more general adoption can be expected: it is possible also that the porous ceiling principles could be employed in the ventilation of buildings having high density occupancy by humans, with resultant energy saving.

BIBLIOGRAPHY

CONDUCTION HEAT TRANSFER,
    Schneider, P.J., Addison Wesley Publishing Company, Chapter 9,
    pp. 218-226, 1955.

MATHEMATICS OF DIFFUSION,
    Crank, J., Oxford, 1956.

HEAT EXCHANGE OF A POROUS PLATE IN A GAS STREAM,
    Grodzkov, N.N. and Ye. P. Valuin,
    A.I.D. Report, T-64-24, sponsored by Academy of Construction
    and Architecture, USSR

HEAT TRANSMISSION OF POROUS MATERIALS IN VENTILATION,
    Pattie, D.R., Transactions of the American Society of Agricultural
    Engineering, Vol. 9, No. 3, pp. 409, 410 and 416, 1966.

PRINCIPLES OF ANIMAL ENVIRONMENT,
    Esmay, Merle L., The AVI Publishing Company, Inc. 1969.

HEAT TRANSMISSION FROM A RADIANT HEAT SOURCE WITH FILTRATION OF AIR
    THROUGH A POROUS SLAB.
    Kosiba, I., and D.R. Pattie,
    Proceedings of the Third Canadian Congress of Applied Mechanics,
    Calgary, 1971.

ENVIRONMENTAL CONTROL FOR ANIMALS AND PLANTS,
    ASHRAE Handbook of Fundamentals, Chapter 8, pp. 151-166, 1972.

VENTILATION BY DIFFUSION AND FILTRATION,
    Pattie, D.R., and F. Lederer,
    Proceedings of the Fourth Canadian Congress of Applied Mechanics,
    Montreal, 1973.

VENTILATION OF DAIRY BARNS WITH POROUS CEILING INLET SYSTEMS - PART I.
    Turnbull, J.E. and C.G. Hickman,
    Canadian Agricultural Engineering, Vol. 16, pp. 91-95, 1974.

VENTILATION OF DAIRY BARNS WITH POROUS CEILING INLET SYSTEMS - PART II.
    Turnbull, J.E. and J.P.F. Davisse,
    Canadian Agricultural Engineering, Vol. 17, pp. 59-62, 1975.

UNIVERSITY OF GUELPH, MSc THESES,
   Airflow Patterns with Various Types of Air Inlets Employed in
   Poultry House Ventilation Systems. W.R. Milne, 1962.

   Heat Transmission of Porous Insulation with Infiltration and
   Exfiltration of Air, Gerald A. Callen, 1967.

   Diffusion of Water Vapour Through a Porous Ceiling with Infiltration
   and Exfiltration of Air, Fridrick Lederer, 1969.

   Heat Transmission with Filtration of Air in a Porous Plate
   Exposed to Radiant Heat, Ivan Kosiba, 1970.

   Radiant Heat Load with Filtration of Air, George B. Holder

ACKNOWLEDGEMENTS

   Laboratory facilities were provided by the School of Engineering
and the Department of Horticulture, of the Ontario Agriculture College,
at the University of Guelph. Field studies were made jointly with
Ontario Ministry of Agriculture and Food, and with the Engineering Research
Service, Canada Agriculture. The research was supported, from its
commencement in 1962, by Ontario Ministry of Agriculture and Food, by the
National Research Council Canada since 1967, and by Agriculture Canada
since 1974. The author gratefully acknowledges the willing support of this
project over such a lengthy period. The author also acknowledges the
input of the students whose theses are listed in the Bibliography.

MATHEMATICAL MODELLING AND EVALUATION OF ENERGY REQUIREMENTS IN HEATING AND COOLING OF BUILDINGS

FUNDAMENTALS OF BUILDING HEAT TRANSFER

TAMAMI KUSUDA

Thermal Engineering Section
Building Environment Division
National Bureau of Standards
Washington, D.C., USA 20234

ABSTRACT

Basic problems and unique features of building heat transfer are described in relation to the heating and cooling load calculation, which is a starting point for building energy consumption analysis and equipment sizing. Detailed discussion is given of the relationship between heat loss (heat gain) and heating load (cooling load). Also outlined is a discussion of the multi-space heat transfer problem in which the air and heat exchange equations among adjacent spaces in a building are solved simultaneously with the radiant heat exchange equations for the surfaces of each room.

Key Words: Air-leakage; dynamic heat transfer; energy analysis; heat and
             cooling loads; heat loss and heat gain; multi-room problems.

NOMENCLATURE

A   Matrix element as defined in the text

B   Matrix element as defined in the text

C_p Specific heat of air

DB   Outdoor dry-bulb temperature

F   Exterior surface convective heat transfer coefficient

G   Air flow between adjacent spaces

H   Interior surface convective heat transfer coefficient

H_k Interior surface radiative heat transfer coefficient

I   Solar radiation indicent upon an exterior surface

QG   Strength of heat source/sink

r   Radiant heat incident upon each interior surface

QS   Net long-wave radiant heat to an exterior surface

S   Interior surface area

TA or T_A Space air temperature

TI   Interior surface temperature

TO or T_0 Exterior surface temperature

TQ   Temperature of heat source/sink

TS   Supply air temperature to a space

X_n, Y_n, Z_n $n = 1 \ldots N$ thermal response factor

α solar absorptance of exterior surface

τ solar transmittance of window

SUBSCRIPTS

t   time

SUPERSCRIPTS

i, j   denotes i-th and j-th spaces

k, l   denotes k-th and l-th surfaces

INTRODUCTION

The purpose of this paper is to summarize state-of-the-art information on various aspects of building heat transfer, and to discuss two selected subjects in detail; namely, the multi-space heat transfer problem and the relationship between heat loss and heating load.

Building heat transfer calculations are performed for different applications such as:
  (a) Heat loss and heat gain through exterior envelope--
- conduction through exterior envelope,
- conduction heat transfer through basement walls and slab-on-grade floor (. . . to semi-infinite region),
- short wavelength (or solar heat) transmission, absorption, and reflection for fenestration,
- air leakage through exterior envelopes as well as the interior partition walls, ceilings and floors,
- thermal storage in exterior masses of buildings.

  (b) Interior environmental analyses--
- radiant heat exchange among interior surfaces and heat sink/sources,
- convective heat exchange between room air and room interior surfaces,
- room air convection . . . inter- and intra-room convective motion,
- convective and radiative heat transfer of internal heat sources such as heaters, coolers and occupants,
- thermal storage in interior masses.

(c) Material or building element-related problems--
  o   cold-bridge effect,
  o   convection within porous insulation,
  o   moisture condensation due to simultaneous flow of air,
       moisture, and heat.
Basic to all of these applications is the fact that heat transfer processes for
buildings are usually ill defined, time dependent, multi-dimensional, and in
many cases non-linear. Thus all the solutions available today for any one of
these applications are based upon numerous simplifying assumptions. In many
cases refined and advanced solutions may be available, but they tend to be overly
complex and have very little practical value. In addition, until recently,
at least in the United States, most applications of advanced heat-transfer
analysis have traditionally been explored by the aerospace and nuclear power
industries. For this reason, many of the challenging problems in building
heat transfer have been left unsolved.

CONDUCTION HEAT TRANSFER

    Conduction heat transfer problems relevant to buildings include:
    (a) Exterior wall conduction--
      o   transient heat transfer responding to climatic effects, such as
           temperature fluctuation, solar radiation, wind and precipitation;
           thermal storage . . . damping and lag effect; and cold-bridge
           effect (two-dimensional and non-linear heat flow path).
    (b) Interior mass conduction--
      o   heat storage in partition walls, floor/ceiling sandwich.
    (c) Conversion from heat gain/loss to cooling and heating load.
    (d) Ground heat loss from slab-on-grade floor and basement walls.

    Most of the conduction problems are multi-dimensional and transient,
requiring numerical (either finite difference and/or finite element analyses)
or analog computer-simulation calculation. For one-dimensional multi-layer
problems, numerous analytical solutions are available, for both steady-state
and transient conditions. These solutions are frequently used for heat gain
and loss calculation for exterior envelopes and heat storage in interior
structures. Most notable contributions in recent years for the multi-layer
problems are the thermal-time-constant concept of Raychaudhuri [1]*, the
admittance solution of Laudon [2], the frequency response solution of Muncey [3]
and the response factor solution of Mitalas and Arseneault [4]. These solutions
permit the accurate evaluation of heat conduction through building walls, roofs
and floors so long as the heat flow is normal to the surface. Missing from
these methods and awaiting development are similar solutions for multi-dimensional problems related to building corners, floor slab-on-grade, and basement
walls.

CONVECTION HEAT TRANSFER

    Convective heat transfer problems relevant to buildings include:
    (a) Heat transfer at the exterior surface considering both wind and
         surface roughness characteristics,
    (b) Convection in and through the cavity walls,
    (c) Convection between the window glass panes,
    (d) Inter- and intra-space air motion due to temperature gradient,

---

* Numbers in brackets indicate references at end of text.

(e)   Convection heat transfer due to air leakage through exterior walls.

(f)   Convection heat transfer within the porous insulating structure.

The surface heat transfer coefficients are affected by the nature of the air boundary layer which is strongly influenced by the surface geometry, temperature gradient, and the flow outside the boundary layer region. Unfortunately most of the textbook solutions and analytical expression available are based on very simplistic boundary conditions. They should be considered as only approximate solutions to the real building problems. The air flow around the building, along the internal surfaces, and in the wall cavities is very complex, erratic and not conducive to exact analytical solutions. Moreover, the actual building surface geometry is seldom well defined.

Many inexperienced building heat transfer analysts have mistaken notions that the exterior surface or interior surface heat transfer coefficients as found in the heat transfer textbook or in the ASHRAE Handbook of Fundamentals [5] are absolutely accurate. This notion was conveyed to the author recently in one of the thermographic heat-flux measurement meetings, at which an attempt was made to convert the surface temperature measured by a sophisticated infrared scanning apparatus to a heat flux value by applying the ASHRAE Handbook value for the surface heat transfer coefficient. In reality, the surface heat transfer coefficient could differ from the published value by as much as 100%, depending upon the local wind gust and the irregularity of the surface geometry. The fact is that, except those reported by Ito and Kimura [6], the exterior surface heat transfer coefficients for actual buildings have never been studied experimentally. The same is true for the interior surface heat transfer coefficients, except for the original work of Wilkes and Peterson [7].

A number of studies have been reported in recent years in which numerical solutions of the Navier Stokes fluid dynamics equation and the energy equation were obtained for two-dimensional air flow within a confined space for the laminar-flow regime characteristics of room convection, and into the turbulent-flow regime (the Grashoff number of 10^{12}). In addition, steady turbulent-flow solutions to the modified Navier-Stokes equations were obtained by Nielsen [9]. The computer time and computer memory size required to deal with realistic three-dimensional flow problems in a room are formidable. Bankvall [10] and Tien [11] are studying flow equations within porous media, and Tien is extending the solution to partially filled insulation cavities, as well as to the permeable boundary surface problems, to simulate air leakage in insulated walls.

RADIANT HEAT EXCHANGE

Radiation heat transfer is very important in building application in the following areas:
(a)   Short-wavelength radiation:
   o   solar heat absorption on opaque exterior surfaces,
   o   solar heat transmission through transparent surfaces,
   o   solar heat absorption by interior building surfaces,
   o   absorption and reflection of solar heat by window glass.
(b)   Long-wavelength radiation:
   o   heat emission by the exterior surfaces to the sky,
   o   heat exchange among interior surfaces,
   o   heat exchange between interior surfaces and occupants,
   o   heat exchange between the lighting fixture and interior surfaces.

While extensive work has been reported in the area of solar energy exchange with exterior surfaces, basic irradiation data are still insufficient with respect to the diffuse sky radiation component, particularly for vertical surfaces and for cloudy sky conditions.

Perhaps the most difficult and tedious problem in dealing with solar heat exchange analysis for building applications is the analysis of direct or beam radiation that is transmitted through fenestration, absorbed, reflected and reemitted by the interior surfaces. Because of the complex geometry of the time dependent shade and sunlight patterns, exact solutions to simulate realistic solar heat exchange in a room are virtually impossible. It is usually assumed that the interior surface is gray (non-spectral) and diffuse (no specular reflection), and the solar heat is diffused once it enters through the window and interior shading devices.

The long-wavelength radiant heat emission to the sky from the building exterior surface has not been well explored except for the clear sky condition, although some work is going on at CSTB in France [12].

HEATING AND COOLING LOADS CALCULATIONS

Because of the increased urgency for energy conservation in the design of buildings, and because the essential part of the energy conservation design is the accurate determination of building heating and cooling requirements, a great emphasis has been placed upon load calculations in recent years.

It might be mentioned that at least in the United States past practices for heating and cooling load calculations were based upon steady-state heat transfer performed at design conditions that represent extremely cold and hot days. This is because the heating and cooling loads in the United States were calculated solely for the purpose of selecting and sizing the heating and cooling equipment to provide comfort for the extremes of climatic conditions. Since these extreme design conditions rarely represent day-by-day average operating conditions, the heating and cooling equipment designed or selected is usually oversized and operates at part-load conditions that result in poor efficiency.

On the other hand, the European and Japanese engineers have been predicting indoor temperature as a function of the ever-changing outdoor climate conditions. Their interest has been primarily to determine the need for air-conditioning by estimating the number of hours in the summer when the indoor temperature could exceed the comfort requirement.

Since most North American homes and commercial buildings are already air-conditioned and centrally heated, major emphasis in the United States is to be able to predict energy consumption by this heating and cooling equipment. It has been well accepted in the United States that the most accurate way of estimating the annual energy consumption of a given building is to simulate the heat transfer performance of the building and the performance of its heating, ventilating and air conditioning systems on an hourly basis throughout the year, thus requiring 8760 calculations. During the past decade, a number of sophisticated computer programs have been developed to do the hourly simulation [13]. The American Society for Heating, Refrigerating and Air-Conditioning Engineers has, for example, developed recommended procedures [14, 15] to simulate the building heat transfer, and system and equipment performances.

A difficult portion of this hourly simulation approach is the coupling between the building heating and cooling requirement and the available capabilities of the building heating and cooling systems -- especially for the building with many zones, each requiring different heating and cooling with

respect to time-dependent use schedule and occupancy patterns. The coupling
calculation becomes especially difficult when the heating and cooling capacity
provided by the heat distribution system and by the building central plant
cannot match with the space heating and cooling requirements.

Thus when a mismatch between the calculated requirement and the available
system capacity occurs, the space temperature drifts from the set-point for
which the original load calculation was performed. The amount of this drift
is dependent upon the thermal storage characteristics of the space under consideration, the operating characteristics of the heat distribution system,
and the part-load or the overload characteristics of the central heating and
cooling plant. To the best of this author's knowledge, no existing energy
analysis computer program can handle this problem in an exact manner.

Simultaneous calculations of multi-space heating and cooling requirements,
together with the heating and cooling capability of central HVAC systems, under
dynamic conditions, (where the climate as well as operating parameters are
constantly changing) are extremely difficult. Thus a common American practice
is to calculate the space heating and cooling requirement at a given set of
required space temperature conditions. The coupling between the load and the
system capacity is evaluated by the use of the weighting factor concept. The
weighting factors are transfer functions generated by Mitalas [4] for three
typical rooms representing light, medium and heavy construction, by solving
detailed room surface heat balance equations.

1. Heat Balance Equation at Exterior Surfaces

The building exterior surfaces receive solar radiation, exchange long-
wavelength radiation with the surroundings including the sky, exchange heat
with the outside air by the convection process, and conduct heat into the
solid structure. Equation (1) represents a complete heat balance on the
exterior surface with the response factor being used for the heat conduction
term.[4]

$$q_t^{(i,k)} = \sum_{n=0}^{\infty} Y_n^{(i,k)} \cdot TI_{t-n}^{(i,k)} - \sum_{n=0}^{\infty} Z_n^{(i,k)} \cdot TO_{t-n}^{(i,k)}$$

$$= F^{(i,k)} \cdot (TO_t^{(i,k)} - DB_t) + QS_t^{(i,k)} - \alpha \cdot I_t \tag{1}$$

2. Heat Balance Equation at Interior Surfaces (Inter-Space Heat Exchange)

Equation (2) shows all the components involved in the heat balance of an
interior surface; namely, the long-wavelength radiant heat exchange with the
rest of the surfaces, convective heat exchange with the air, incident solar
radiation (through the windows), long-wavelength radiation from lighting and
equipment, and heat stored into the solid material, which is expressed in
response-factor terms.

$$q_t^{(i,k)} = \sum_{n=o}^{\infty} X_n^{(i,k)} \, TI_{t-n}^{(i,k)} - \sum_{n=o}^{\infty} Y_n^{(i,k)} \, TO_{t-n}^{(i,k)}$$

$$= H^{(i,k)} \, (TA_t^{(i)} - TI_t^{(i,k)}) + \sum_{j=1}^{N_i} H_j^{(i,k)} \, (TI_t^{(i,j)} - TI_k^{(i,k)}) + r_t^{(i,k)}$$

$$(2)$$

By rearranging the terms, equation (2) becomes

$$\sum_{j=1}^{N_i} A_j^{(i,k)} \, TI_t^{(i,j)} + AI^{(i,k)} TA_t^{(i)} + AO^{(i,k)} \, TO_t^{(i,k)} = B^{(i,k)} \qquad k = 1, N_i$$

$$(3)$$

where N_i = total number of heat transfer surfaces in the i-th room

$$B^{(i,k)} = -\sum_{n=1}^{\infty} X_n^{(i,k)} \, TI_{t-n}^{(i,k)} + \sum_{n=1}^{\infty} Y_n^{(i,k)} \, TO_{t-n}^{(i,k)} + r_t^{(i,k)}$$

$$A_j^{(i,k)} = X_o^{(i,k)} + H^{(i,k)} + \sum_{j=1}^{N_i} H_j^{(i,k)}$$

$$A_K^{(i,k)} = -H_K^{(i,k)}$$

$$A_0^{(i,k)} = Y_o^{(i,k)}$$
and
$$AI^{(i,k)} = H^{(i,k)}$$

Equations similar to (3) should be prepared for all the interior surfaces in a given space (room). On the other hand, the overall heat balance equation of the space air should include the convection heat exchange with the surfaces, heat sources, air leaked into the space, and heat given off and supplied by the heat sources/sinks (including space conditioning devices), as follows:

$$\sum_{k=1}^{N_i} H_\bullet^{(i,k)} \, (TI_t^{(i,k)} - TA_t^{(i)}) \, S^{(i,k)} + \sum_{k=1}^{N_i} GC_p^{(i,k)} \, (AT_t^{(i,k)} - TA_t^{(i)})$$

$$+ QG^{(i)} = 0$$

$$(4)$$

By letting
$$M_i = N_i + 1$$

$$A_K^{(i \cdot M_i)} = H^{(i,k)} S^{(i,k)}$$

$$AI^{(i,M_i)} = \sum_{k=1}^{N_i} H^{(i,k)} S^{(i,k)} - \sum_{k=1}^{N_i} G^{i,k} \cdot C_p$$

$$AO_k^{(i,M_i)} = G^{(i,k)} C_p$$

$$B^{(i, M_i)} = -QG^{(i)},$$

The space heat balance equation (4) then becomes:

$$\sum_{k=1}^{N_i} A_K^{(i,M_i)} TI_t^{i,k} + AI^{(i,M_i)} \cdot TA_t^{(i)} + \sum_{k=1}^{N_i} AO_K^{(i,M_i)} \cdot AT_t^{(i,K)} = B^{(i,M_i)}$$

$$(5)$$

A set of $(N_i + 1)$ simultaneous equations comprising N_i equations of type (3) and equations (5) must be solved simultaneously for a given space to yield the interior surface temperature and space air temperature.

3. Inter-Space (Inter-Room) Heat Exchange

When the space under consideration in the previous section is adjacent to other spaces, all of the inner surface temperatures and air temperatures for all of the adjacent spaces are being affected by one another. Figure 1 illustrates the heat transfer process through a boundary wall between the i-th space and j-th space, indicating conductive and convective heat exchange between the two spaces. A simplified and yet relatively typical four-space problem, as depicted in figure 2, is used herein to illustrate the complexity of the multi-room problem.

In order to simplify the mathematical manipulation, figure 3 is provided to show the interaction of four spaces by thermal coupling through adjacent interior surfaces. Figure 4 shows the matrix form of the complete heat balance equation to be solved for this four-space problem using matrix notations provided in equations (2) and (3). In this matrix presentation air leakage between the adjacent spaces is presented. Air leakage is, however, in reality a function of the temperature difference and pressure difference between the space due to wind effect. Tamura and Sanders [16] developed and later Fothergill [17] expanded a comprehensive computer program to calculate air flow between building spaces as affected by the external wind pressure as well as by the thermal stack effect. The program was primarily developed for the study of smoke migration pattern analysis under a fire research program. It does not incorporate the thermal storage characteristics of the building internal thermal mass; however, there is a real need for an efficient computation procedure by which the multi-space air and heat transfer processes can be analyzed in a comprehensive manner.

The above discussions indicate that a comprehensive analysis of multi-room heat transfer problems requires a large-size matrix manipulation, which requires a large and fast computer. The complexity of the analysis is basically due to the inclusion of the interior surface radiation heat exchange terms. Considerable simplification is possible, if it can be assumed that all the room interior surface temperatures are the same as the room air temperatures. A brief discussion of this subject is provided in the Appendix.

4. Coupling with the Central System Capacity

Heat source/sink term QG indicated in equation (4) could represent the
heating and cooling capacity of the supply air system. If the supply air
quantity and its temperature are expressed as GS_t and TC_t respectively, the
term QG takes on the value $GS_t \cdot C_p \cdot (TC_t - TA_t)$.

In this manner the space air temperature can be determined as a balance
between the space heating and cooling requirement and the heating and cooling
capacity of the supply air system. The values of GS and TC may vary, however,
according to the mode and type of the building air distribution system. For
example, a constant air volume system would provide GS constant and regulate
TC, whereas a variable volume system would regulate GS with TC held constant.

Difficulty arises, however, in simulating the regulation or control of GS ·
and TC in the computational process. The heat exchanges among the space thermostat, space air and room surfaces enter the heat balance equations, thereby
increasing the complexity of the problem. In addition, due to the time response
characteristic of the thermostat, the simulation of the control system requires
minute-by-minute calculations, rather than the hour-by-hour calculations.

SUMMARY

Basic problems and difficulties inherent in the comprehensive analysis of
building heat transfer are due to many parameters which are ill defined and
some of which are time dependent and multi-dimensional.

One of the crucial areas for building heat transfer analysis is the hour-
by-hour simulation of the balancing process between the heating and cooling
requirements of many spaces and the central heating and cooling capacity of the
building. Analytical solutions available for any of these problems are based
upon simplistic assumptions.

Detailed discussions of radiant heat exchange effects for building heat
loss calculations are included in the Appendix.

REFERENCES

1. Raychaudhuri, B.C., Transient Thermal Response of Enclosures; The Integrated
   Thermal Time Constant, International J. Heat Mass Transfer, 8, 1439 (1965).

2. Laudon, A. G., Summertime Temperatures in Buildings, Building Research
   Studies, 1968. Building Research Establishment, Building Research Station,
   Garston, Watford, WD2 7JR.

3. Muncey, R. W., The Thermal Response of a Building to Sudden Changes of
   Temperatures or Heat Flow, Australian J. Sci. 14, 123, 1963.

4. Mitalas, G. P., and J. G. Arsenault, "Fortran IV Program to Calculate Z-
   Transfer Functions for the Calculation of Transient Heat Transfer Through
   Walls and Roofs," Use of Computers for Environmental Engineering Related to
   Buildings, NBS-BSS 39, 1971, pp. 663-668.

5. American Society for Heating, Refrigerating and Air Conditioning Engineers,
   1977.

6. Ito, N., and K. Kimura, "Field Experiment Study on the Convective Heat
   Transfer Coefficient of Exterior Surfaces of a Building," ASHRAE Trans.
   Vol. 78-1 (1970).

7. Wilkes, G.B., and C. M. F. Petersen, "Radiation and Convection from Surfaces
    in Various Positions," ASHVE Trans. Vol. 44, 1, 1938, pp. 513.

8. Fromm, J. E., "A Numerical Method for Computing the Non-Linear, Time
    Dependent Buoyant Circulation of Air in Rooms," Use of Computers for
    Environmental Engineering Related to Buildings, NBS-BSS 39, (1971) pp. 451-467

9. Nielsen, P. V., "Flow in Air Conditioned Rooms," Ph.D. Thesis, Technical
    University of Denmark, Nordberg, 1974.

10. Bankvall, C. G., "Natural Convective Heat Transfer in Insulated Structures,"
    Lund Institute of Technology, (Sweden) Report 38, 1972.

11. Tien, C. L., and P. J. Burns, "Convection in a Vertical Slot Filled with
    Porous Insulation," International Heat and Mass Transfer, Dubrovnik, Yugo-
    slavia, 1977.

12. Bertolo, L., "Instrumentation for Measuring Heat Transfer to the Sky," an
    unpublished CSTB document, CSTB, France, circa 1975.

13. Crall, C., "Bibliography on Available Computer Programs in the General
    Area of Heating, Refrigerating, Air Conditioning and Ventilating,"
    ASHRAE Research Project Report GPR-153, October 1975.

14. Task Group on Energy Requirements, "Procedure to Determine Heating and
    Cooling Loads for Computerized Energy Calculations for Buildings,"
    special ASHRAE Bulletin, 1974.

15. Stoecker, W. C., "Procedure to Simulate HVAC System and Equipment for
    Computerized Energy Calculations," Special ASHRAE Bulletin, 1974.

16. Tamura, G., and D. Sanders, "A Fortran IV Program to Simulate Air Movement
    in Multi-Story Buildings," DBR-NRC Computer Program No. 35, March 1973.

17. Fothergill, J. W., et al, " Development of an Air Movement Simulation
    Program," NBS Contract Report, Integrated Systems Inc., Rockville, Maryland,
    1976.

APPENDIX

HEATING LOAD VS HEAT LOSS FOR CENTRAL HEATING SYSTEMS

When a building is centrally heated, the heating of a given space is provided by the hot supply air to the system. The hot air entering the space is
usually assumed instantly mixed with the space air to raise the space temperature.
Thus the manner in which the warm space air loses its heat to the cold exterior
surfaces present in the space dictates the heating requirement. In many calculations, however, it is tacitly assumed that the space air temperature is equal
to the space surface temperatures and the heating load is calculated by a simple
formula such as:

$$Q = U_o(T_A - T_o) \cdot S_o$$

where U_o = overall heat transfer coefficient of exterior surface

T_A = room air temperature

T_o = outdoor temperature

S_o = exterior surface areas.

This simple relationship is incompatible with reality, where the space
surface temperature is seldom equal to the space air temperature.

To examine the difference in results between the common assumption and
the exact solution, a simple room of one single-glazed window was studied.
In this room the walls are assumed massless and have the same overall and surface heat transfer coefficients and the same interior surface temperatures, and
are exposed to the same outdoor air temperature as shown in figure A-1. Ignoring
the solar heat absorbed by the glass, the heat exchange equation for the inner
surface of the window for the model will be:

$$H_G \cdot (T_{SG} - T_A) + H_{GW}(T_{SG} - T_{SW}) + U_{GT}(T_{SG} - T_o) = 0 \qquad (A-1)$$

where H_G = convective heat transfer coefficient over the interior surface of
the window.

H_{GW} = radiative heat transfer coefficient of the window to the interior
surface of the opaque wall.

U_{GT} = thermal conductance between the inner surface of window and outside
ambient

T_{SG} = interior surface temperature of the window

T_{SW} = interior surface temperature of all the walls

T_A = room air temperature

T_o = outside ambient temperature.

A similar equation can be constructed for the wall portion, again ignoring
the heat storage effect, as follows:

$$H_W \cdot (T_{SW} - T_A) + H_{GW} \cdot (T_{SW} - T_{SG}) + U_{WT}(T_{SW} - T_o) = I \frac{A_G}{A_W} \tau \qquad (A-2)$$

where H_W = convective heat transfer coefficient over the interior surface of
the wall

H_{WG} = radiative heat transfer coefficient of the interior surface of the
wall to the window

U_{WT} = thermal conductance between the interior surfaces of the wall and the outside ambient

I = solar radiation incident upon the window

τ = solar transmittance through the window

A_G = window area

A_W = wall area.

The values of U_{GT} and U_{WT} can be calculated by knowing the overall heat transfer coefficient for window U_G and that for wall U_W as follows:

$$\frac{1}{U_{GT}} = \frac{1}{U_G} - \frac{1}{H_G + H_{GW}}$$

$$\frac{1}{U_{WT}} = \frac{1}{U_W} - \frac{1}{H_W + H_{WG}}$$

(A-3)

On the other hand, the radiative heat transfer coefficiencts H_{GW} and H_{WG} are related to one another as follows:

$$H_{GW} \cdot A_G = H_{WG} \cdot A_W \qquad (A-4)$$

A convectional way of calculating the heating load by the use of U value is

$$Q = (U_G \cdot A_G + U_W \cdot A_W)(T_A - T_o) - I \cdot A_G \cdot \tau \qquad (A-5)$$

An exact way of calculating the heating load is to say that the heating load is the heat lost by the room air to the surrounding surfaces, or

$$Q' = H_W A_W (T_A - T_{SW}) + H_G A_G (T_A - T_{SG}). \qquad (A-6)$$

In order to use the latter equation the values of T_{SW} and T_{SG} must be determined by solving equations A-1 and A-2 simultaneously. Several simple calculations are performed as follows:

It is assumed that:

A_G = 10,32, and 100 ft^2

A_W = 900 ft^2

T_A = 70° F

T_o = 0° F and 40° F

U_W = 0.01, 0.05, 0.10 and 0.50 Btu/hr ft^2 °F

I = 0, 100 and 200 Btu/ft^2 hr

H_{RG} = 0.918 Btu/hr ft^2 °F

$H_G = H_W$ = 0.542 Btu/hr ft^2 °F

Figure 2 depicts the value of R = Q'/Q as affected by the value of U_W for the case of I = 0. On the other hand, Table A-1 shows the case of solar heat gains I = 100 and 200 being included for the calculations.

These results show that the error due to the assumption of the same air and exterior surface temperatures is rather small as long as the heat loss through the wall portion is small and the window area is less than 10% of the total room surface area. The conclusion is valid even in the case where the solar heat gain is included.

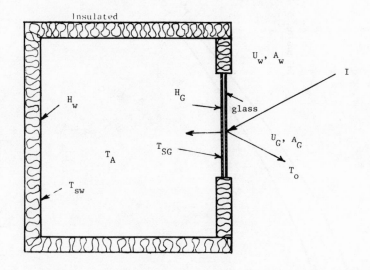

$$Q = (U_G - A_G + U_w \cdot A_w)(T_A - T_o) - I \cdot \tau$$

$$Q' = A_G \cdot H_G \cdot (T_{SG} - T_A) + A_w \cdot H_w (T_{sw} - T_A)$$

$$R = \frac{Q'}{Q}$$

Figure A-1. A simplified room with all the opaque envelopes having the same thermal characteristics.

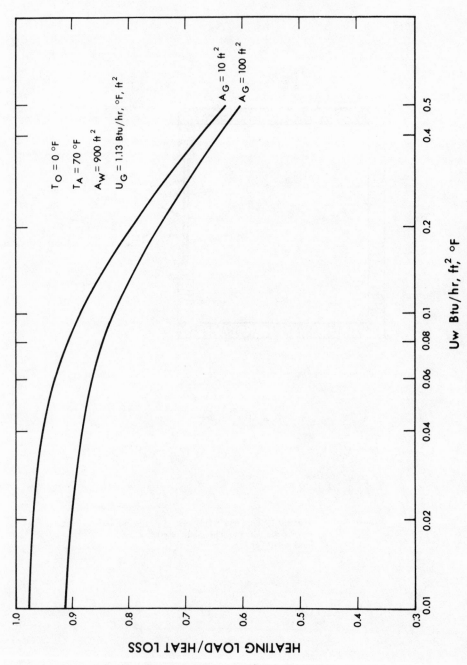

Figure A-2. Relationship between the heating load Q' and heat loss Q.

334

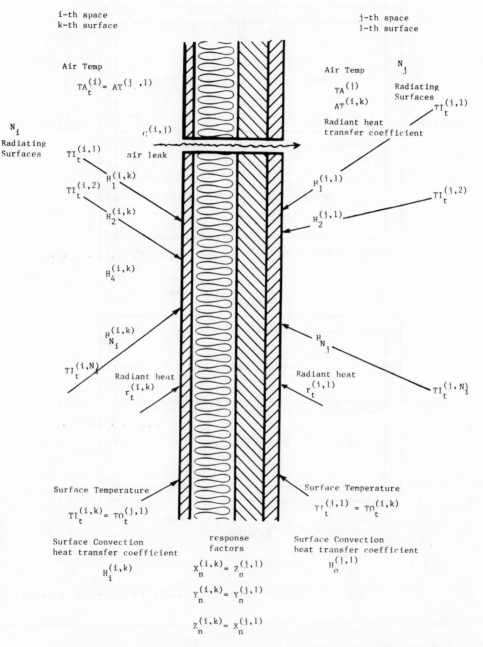

Figure 1. Heat transfer of a partition wall between the i-th and j-th space.

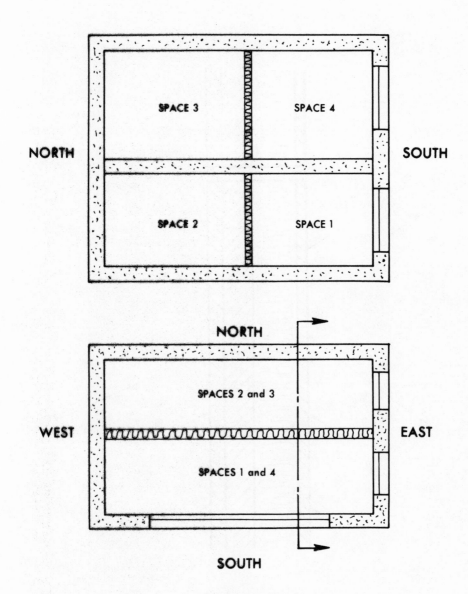

Figure 2. Plan of a Sample 4-space model.

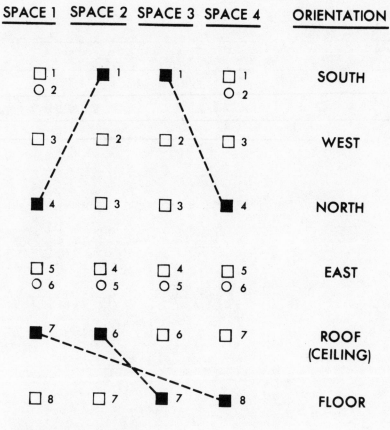

SPACE 1	SPACE 2	SPACE 3	SPACE 4	ORIENTATION
□ 1 O 2	■ 1	■ 1	□ 1 O 2	SOUTH
□ 3	□ 2	□ 2	□ 3	WEST
■ 4	□ 3	□ 3	■ 4	NORTH
□ 5 O 6	□ 4 O 5	□ 4 O 5	□ 5 O 6	EAST
■ 7	■ 6	□ 6	□ 7	ROOF (CEILING)
□ 8	□ 7	■ 7	■ 8	FLOOR

□ OPAQUE WALL ....EXTERIOR

■ OPAQUE WALL ....INTERIOR

O TRANSPARENT WALL (WINDOW)

△ OPEN PASSAGE

Figure 3. Interior-exterior wall relationship of
a 4-space model as depicted in Figure 2.

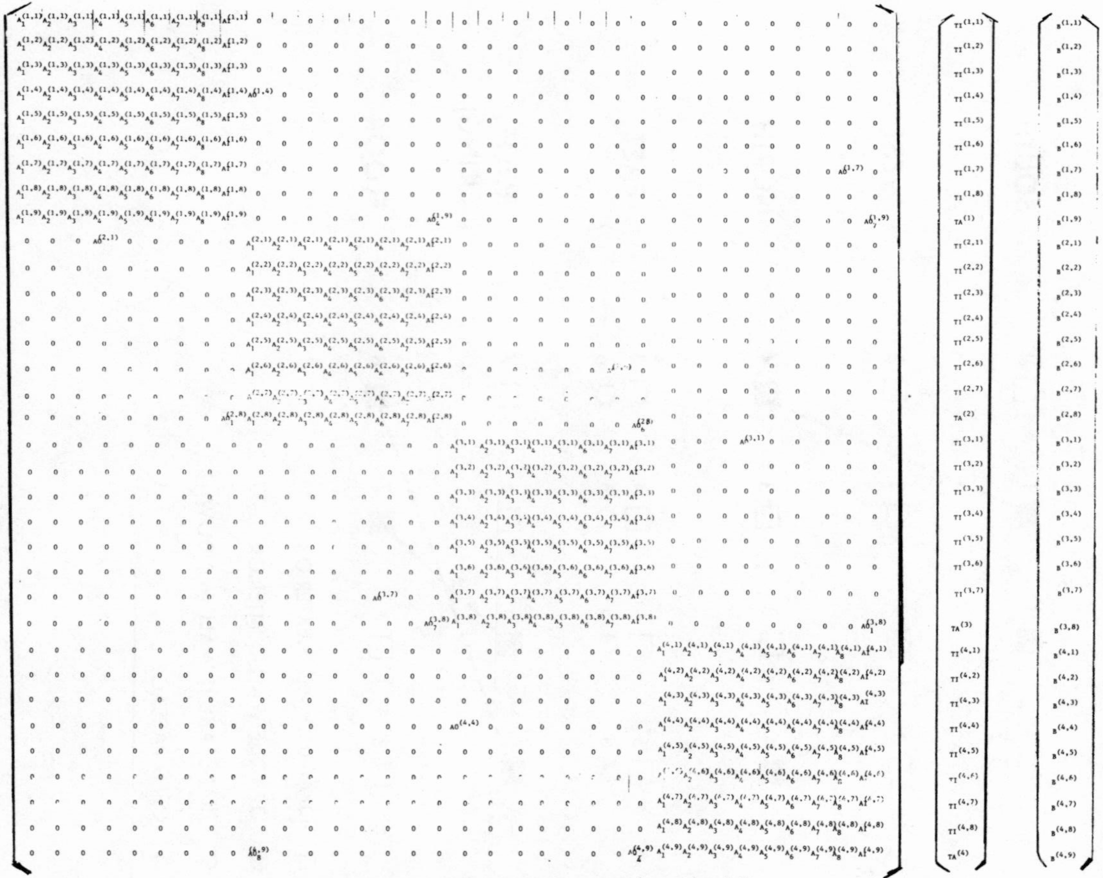

Figure 4. Heat balance equation of a 4-space model.

HEAT TRANSFER CALCULATION
FOR INDUSTRIAL ENTERPRISES

V. P. MOTULEVICH

Moscow Power Engineering Institute
Moscow, USSR

Industrial enterprises require temperature, humidity and aerodynamic conditions to be maintained at prescribed limits.

This is primarily necessary to set up comfortable conditions
with resulting the best physical state and the highest labour
productivity of the stuff available inside the enterprises.

Again, setting up the reqired temperature-humidity conditions
is frequently caused by technological requirements of such industries as radio engineering, nuclear, electronic, instrument-
making, chemical, food and textile, etc.

Optimal conditions are mainly provided by a temperature level
of air and surfaces faced into the room.

Air temperature into the rooms plays an important part in
convective heat transfer both from human beings and technological equipment.

Temperature of the surfaces determines a level of radiation
thermal fluxes. It is also important in developing natural convection.

Besides, it seems to be important if human beings or technological equipment may appear in the region of the boundary layer or turbulent wake.

It is a physiological thermoregulation system that dictates
the required temperature of the human being body at level of
36.6°C. A person feels comfortable if temperature conditions
into the room do not result in overstress of the thermoregulation system.

Depending on labour intensity heat output of grown up person
organism amounts to 140W (rest or little job), to 175W (easy job),
to 290W (average amount of job), and more than 290W (hard job).
With intensive loads of short duration heat output may reach
1000W [I] .

On average with normal loads about half amount of heat is released by radiation, a quarter by convection and almost the same
by evaporation. When increasing intensity of job a share of
amount of heat by evaporation sharply rises.

Heat release output and mechanism of heat removal from different components of technological equipment imposing stringent
requirements upon temperature conditions are quite different.

Operating staff and technological equipment requiring the
specified temperature conditions occupy usually only a part of

the space, so comfortable* conditions should be set up just at
those zones.

Along with air temperature (t_a) and temperature of radiation
surfaces (t_R) which have qualitatively the same effect on the
comfortable conditions, so called room temperature (t_r) is also
introduced

$$t_r = \frac{1}{2}\left(t_a + t_R\right)$$

(I)

Then, depending on intensity of job room temperature at the
middle part of the working zone should be [I]: at rest from 21
to 23°C, with easy job from 19 to 21°C, with moderate job from
16 to 19°C, and with hard job from 14 to 16°C.

Naturally these recommendations as well as definition of the
job intensity level are of conditional nature.

When setting up comfortable conditions for operation of the
staff it is important to know, apart from general room temperature, under what conditions are some parts of a person's body,
and primarily his head and feet [2].

Head is extremely sensitive to radiation heat transfer, which
requires heavy demands to be placed on temperature of the surface
from which such heat transfer occurs (t_R):

heated surface:

$$t_{Rh} \leqslant 19,2 + \frac{8,7}{g}$$

(2)

cooled surface:

$$t_{Rc} \geqslant 23 - \frac{5}{g}$$

(3)

where g is coefficient of radiation from an elementary area
of persons' head towards radiating or absorbing surface [3].

Floor temperature is permitted to be within the range from
22 to 34°C. It should not be below air temperature more than by
2 to 2.5°C.

Thus, setting up comfortable conditions necessitates calculation of distribution of air temperature in the rooms and temperature of the surfaces faced into the hall.

Here conditions of heat and mass transfer through the construction surfaces, intensity and nature of heat sources available
inside the room, their geometry and specific features of internal aerodynamics are assumed to be known.

Solution to the problem posed involves much difficulties. But
the aim may be achieved using balance relations between the sour-

---

* Here by term "comfortable" are meant conditions which are
  optimal from thermal point of view both for the operating
  staff and for technological equipment (they may be different
  at different zones).

ces and removal of heat as well as equations relating intensity of heat and mass transfer of some components of the rooms and their associated objects between each other and to environmental air.

In doing so, one may derive a closed system of equations which, though very complex, may be solved by present-day computers [6].

$$\sum_j C_0 \varepsilon_{i-j} \beta_{i-j} (\mathcal{T}_i - \mathcal{T}_j) \mathcal{G}_{i-j} F_i + d_i (\mathcal{T}_i - t_a) F_i + K_i' (\mathcal{T}_i - t_{env\,i}) F_i + \Theta_i = 0 \quad (4)$$

$$\sum_j C_0 \varepsilon_{n-j} \beta_{n-j} (\mathcal{T}_n - \mathcal{T}_j) \mathcal{G}_{n-j} \Delta F_n + d_n (\mathcal{T}_n - t_n) \Delta F_n + K_n' (\mathcal{T}_n - t_{env\,n}) \Delta F_n + \Theta_n = 0 \quad (5)$$

$$G_{n-1} C_p (t_{n-1} - t_n) + \Delta G C_p (t_a - t_n) + d_n (\mathcal{T}_n - t_n) \Delta F_n = 0 \qquad (6)$$

$$\sum_i d_i (\mathcal{T}_i - t_a) + G_f C_p (t_f - t_a) + G_i C_p (t_a - t_f) + \Theta_r = 0 \qquad (7)$$

In this system of equations eq.(4) is balance of heat for i-element of the surface. The number of such equations is equal to that of the surfaces considered.

Here the first term characterizes radiation heat flux between i and j all the elements of the room. It is the first reflection that is considered here since calculation of multiple reflections would make the problem too involved but would not affect the accuracy of the results obtained.

The second term describes convection heat transfer of i-element to the environment whose temperatures are equal to $\mathcal{T}_i$ and t_a, respectively.

The third term characterizes heat fluxes through the construction surface. In this case heat output coefficient K_i relates to the unit of i surface and environmental air temperature is accepted as t_{env} .

Finally, Θ_i concerns with other heat sources which may occur on the surface.

The number of such equations should be equal to that of elements into which the wall is divided.

Equation (5) is similar to eq.(4) and differs from the latter

by the fact that it describes heat balance of n element of the
surface subjected to the air jet moving along the surface. The
temperature of n element is assumed to be t_n.

The number of such equations should be equal to that of elements into which the jet is divided.

Equation (6) expresses heat balance of the jet itself. Here
the first term characterizes the difference between enthalpy
fluxes at output and input of the element considered, the second
deals with change in enthalpy due to adding the environmental
air (ΔG) and the last term describes heat transfer to the
blown element of the surface.

The number of equations of type (5) is equal to that of
type (6).

If the supplied air jet is free then it may be solved either
by equations of type (6) without the last term or by generalized
relationships described in paper [7].

Finally, the last equation (7) is the general heat balance in
the rooms where "f" symbol relates to the jet cross section which
is conditionally assumed to be the last and G_l is air flow rate
of inlet ventilation.

In equations (4) and (5) C_0 is constant in equation of Step-
hen-Boltzmann, ε_{i-j} is effective blackness for the given pair
of surfaces exchanging radiation energy. It depends on blackness
of each of these surfaces, their relative positions and sizes.

So, for infinitely large parallel plates one may express:

$$\varepsilon_{i-j} = \left[\frac{1}{\varepsilon_i} + \frac{1}{\varepsilon_j} - 1 \right]^{-1} \tag{8}$$

For the case when a small convex surface with F_i area is
covered with another having F_j area it is written:

$$\varepsilon_{i-j} = \left[\frac{1}{\varepsilon_i} + \frac{F_i}{F_j} \left(\frac{1}{\varepsilon_j} - 1 \right) \right]^{-1} \tag{9}$$

If sizes of the surfaces are small as compared to distance
between them then back radiation exchange may not be considered:

$$\varepsilon_{i-j} = \varepsilon_i \varepsilon_j \tag{10}$$

In general case effective blackness is determined in a rather
complicated manner. But for the industrial halls it will be
within the range determined by relationships (8) and (10) with
the difference not exceeding 5%.

Radiation coefficient $\mathscr{G}_{1-2}$ is a share of radiation flux
incident to surface 2 against the total flux radiated by surface 1.

This quantity known from radiation heat transfer theory depends on geometric features of bodies exchanging heat energy,
and for three surface arrangements most typical of industrial
enterprises it may be determined from diagrams shown in Fig.1,2
and 3. The latter case is of interest for calculating intensity

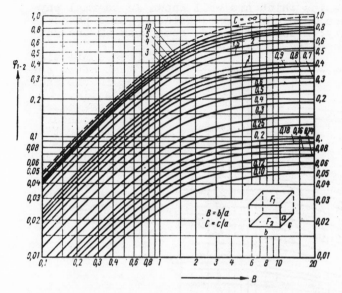

Fig.1. Coefficient of radiation from surface tu surface located in a parallel plane.

of radiation through an open hole connecting operating room with the zone of elevated temperature. In such cases radiation from the zone is assumed to be equal to that of black body.

This diagram enables to consider the weakening effect of radiation of the side walls of the hole. Coefficient b_{1-2} entering into equations (4) and (5) is of minor importance helping in representing the final relation in a form most suitable for calculations. It is determined as follows:

$$\beta_{1-2}\left(\tau_1 - \tau_2\right) = \left(\frac{T_1}{100}\right)^4 - \left(\frac{T_2}{100}\right)^4 \tag{11}$$

Within the range of temperatures of interest for industrial enterprises it may be approximated by a linear function of an average arithmetic temperature (τ_{av}) of the surfaces exchanging heat energy:

$$\beta_{1-2} \simeq 0{,}81 + 0{,}01\,\tau_{av} \tag{12}$$

Expressions for convection heat transfer coefficients α_i entering into the system of equations (4) + (7) are derived either theoretically or studied experimentally depending on specific conditions to which the given element of the surface is subjected. Appropriate predicted relations obtained by different authors are given in [6].

Air temperatures inside the rooms are most commonly leveled off due to mixing processes.

However, the temperature distribution may be extremely non-uniform in specific cases typical of industrial enterprises with zones of extensive heat release and with nonisothermic jets supplying for thermal regulation.

So, there may occur a zone of heat air over a heat source and at the region where the jet is not mixed its temperature will correspond to the initial one.

Heat transfer of some elements of the surface may be calcula-

ted based on relationships which are well known in thermal physics and valid for idealized conditions excluding side phenomena
affecting its intensity: a lack of space, roughness of the surface, radiation heat sources, etc.

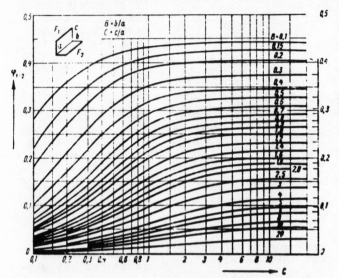

Free convection
is most often rea-
lized in the indus-
trial halls.

In these cases a
laminar boundary
layer is first de-
veloped on the in-
ner surfaces, and
then a turbulent
one occurs. Heat
transfer intensity
is determined by
the product of num-
bers of Grasgoff
and Prandtl, and
for Pr = 0.709:

Fig.2. Coefficient of radiation from surface
to surface located in a normal plane.

$$Nu_x = 0,356\ Gr_x{}^{1/4} \tag{13}$$

from which at $t_a \approx 20°C$ an average value of α seems to
be equal to:

$$\alpha_{av} = 1,38\left(\frac{\Delta t}{\ell}\right)^{1/4} \tag{14}$$

where ℓ is height of the room.

If according to paper [9] it is assumed that transition of the
laminar layer to the turbulent one takes place with $Gr = 10^9$,
then the height at which the transition occurs is determined by
the following formula:

$$h_{cr} = 1,89\ \Delta t^{-1/3} \tag{15}$$

_____
* In all formulas MKS system is used.

and an average value of heat transfer coefficient at this region is

$$\alpha_{av} = 1,17 \, \Delta t^{1/3} \tag{16}$$

In developing the turbulent boundary layer heat transfer is expressed as follows:

$$Nu_x = 0,135 \left(Gr_x Pr \right)^{1/3} \tag{17}$$

which provides a constant value of over the surface length:

$$\alpha_{av} = 1,67 \sqrt[3]{\Delta t} \tag{18}$$

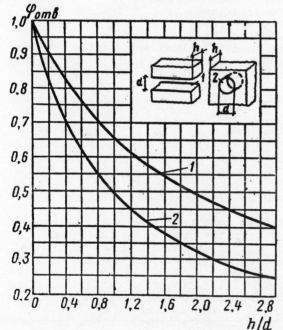

If the surface is horizontal then due to the complex behavior of the air flow the coefficient in formula (18) will change: for the heated surface faced upwards and for the cooled surface faced downwards it increases and if vice versa it decreases by 33%.

It should be noted that in case of horizontal surfaces due to some difficulties in supplying air heat transfer intensity decreases with increase in size of the surfaces.

Owing to general mobility of air in the halls along with natural convection there may also occur both forced and mixed convection.

Fig.3. Coefficient of radiation of outlet surface with consideration for reflection of side surfaces in slot (1) and cylindrical (2) holes.

In the first case with developing the laminar boundary layer we have:

$$Nu_x = 0,297 \, Re_x^{0,5} \tag{19}$$

or

$$\alpha_{av} = 3,93 \left(\frac{2r}{X} \right)^{0,5} \tag{20}$$

- 8 -

With the turbulent layer, respectively:

$$Nu_x = 0,032 \ Re_x^{0,8} \tag{21}$$

$$\alpha_{av} = 5,95 \ \frac{v^{0,8}}{x^{0,2}} \tag{22}$$

Strictly speaking, the presented formulas are held true at the constant temperature of the surface. If it changes then, in principle, one may use different semiempiric methods taking account of this fact (for example, see [10]).

However, based on studies conducted in [9] one may state that, as applied to calculation of heat transfer in the halls, we shall not make a great error if prehistory of the flux will be neglected in developing the turbulent boundary layer and the presented criterion relations will be considered locally valid.

In industrial halls there may often occur such conditions when free convection is combined with the forced one.

Heat transfer under such conditions has been little studied.

For example, in paper [11] heat transfer coefficient is suggested to be determined as follows:

$$Nu = 0,46 \ Re_{eff}^{0,5} \tag{23}$$

where

$$Re_{eff} = Re + \sqrt{Gr/2} \tag{24}$$

Validity of the given method requires to be additionally verified.

It should be noted that heat transfer of some elements of the halls often differs greatly from the process in ideal conditions.

This is affected by the space being enclosed and finite as well as by availability of surfaces heated up to different temperatures, etc.

A typical example is found in results of investigation of natural convection in a special chamber in which arrangement and sizes of some elements were close to real ones [12] .

In particular, this paper studied heat transfer under natural convection for a narrow heated band extended from floor till ceiling.

Three typical zones of the flow were found: laminar flow in the vicinity of the floor, turbulent in the middle part of the surface and stagnation zone near the ceiling.

The process development is greatly affected by the general intensification of air movement and in particular by occurrence of the flow normal to the wall near the floor.

This flow intensifies transition of the laminar flow to the turbulent one and results in increase in the heat transfer coefficient as compared to free surface: at the outer boundary of the laminar section it increases up to 25% with increase in its average value by 9%.

At the turbulent layer zone heat transfer first somewhat de-

creases (~ by 5%) as compared to the free surface and then it
increases by 25%.

The particular feature of the turbulent zone is nonexistence
of automodelling: heat transfer coefficient increases in direction of movement.

Near the ceiling there occure the zone substagnated movement
(Δh) whose value may be determined from the formula:

$$\Delta h = 1,35 - \frac{1450 \, h}{(Gr_h \cdot P_r)^{1/3}} \qquad (25)$$

where h is height of the room.

Average intensity of heat transfer at the stagnation zone is
determined by the expression:

$$Nu_{h-\Delta h} = 0,02 \, (Gr_{h-\Delta h} \, P_r)^{0.46} \frac{h}{\Delta h} \, \ell n \frac{h}{h-\Delta h} \qquad (26)$$

Still more complicated situation occurs in case of free convection on the heated element surrounded by the cold surfaces
from three sides and confined by the floor on the underside:
a heating element under window, hot surface of the furnace door
and so on.

Due to substagnating effect of the emerging air fluxes heat
transfer intensity decreases. According to data of [6] heat
transfer at the lower and upper parts of such an element may decrease, as compared to the free surface, by 44% and 85%, respectively.

The presented data are typical of different halls, though
they are obtained for the narrow range of change in geometric
characteristics of the elements.

Besides, they point to the fact that one should be somewhat
cautions when using relations obtained in idealized conditions
for calculations.

It should be also pointed to possible penetrations of air
through the surfaces enclosing the halls. It is known that both
with the laminar [13] and with the turbulent flow [14] a rather
small flow rate through these surfaces may result in a significant decrease or increase in the heat transfer intensity with
injection or with suction, respectively.

For control of heat transfer in industrial enterprises an air
jet is often generated along the enclosing surfaces. The temperature of the jet may be different from its average value in the
rooms.

Under some conditions when the range of the jet is not sufficient and the flow is deviated from the surface by the gravitational forces the jet may be broken away and appear in the working zone, which is usually undesirable.

Paper [7] deals with conditions under which the cold near-
wall jet is broken away from the ceiling. This paper also shows

that the air flow rate per unit of the jet width at the cross
section X is equal to

$$G_x = \rho \mathcal{U}_o \, (ax)^{0,5} \tag{27}$$

where $\mathcal{U}_o$ is initial velocity of the jet and a is width of
the inlet slot.

Heat transfer of the flat jet may be determined by the formula [15] :

$$Nu_x = 0{,}104 \, Re_x^{0,8} \left(\frac{a}{X}\right)^{0,4} \tag{28}$$

from which it follows that the heat transfer coefficient along
the flow decreases, so that

$$\alpha_{av} = 2{,}5 \, \alpha_x$$

To protect the enclosing surfaces from the cold fluxes directing downwards sometimes the heated air flux is generated upwards.

These jets come into contact and as a result of interaction
they direct inside the room.

The air temperature of the jets is often unfavourable for the
staff, so that it is desirable that the mixed jets should be deviated upwards.

According to studies conducted in paper [16] this may be
achieved by fulfilling the following inequality:

$$0{,}09 \, \frac{h-x_e}{x_e} \, Gr_{h-x_e}^{-0,1} \left[1 + 9 \left(\frac{x_e}{a} - 8\right)\right]^{-1} \leqslant 1 \tag{29}$$

where h is height of the room; x_e is range of the jet determined by the relationship

$$\frac{x_e}{h} = \frac{13{,}7 \, \mathcal{U}_o^2 \, a}{\left\{21{,}2 \, \frac{g}{c_p T_a} \left[0{,}6 q_w h - c_p G_o \, (t_o - t_a)\right]\right\}^{2/3}} \tag{30}$$

In equation (30) the notations q_w and t_o denote the heat
flux from the jet to the wall and initial temperature of air in
the jet, respectively.

The heat transfer intensity in the downwards flux is determined by the above relations for natural convection and in the
upwards flux by the following formula:

$$Nu_{ax} = 0{,}118 \, Re_a^{0,8} \left(\frac{X}{a}\right)^{-0,625} \tag{31}$$

and at the stagnation zone of the contact of two jets where the heat transfer intensity sharply decreases it is determined by the approximate relation:

$$Nu_{a\ av} = 2.5 \cdot 10^{-3}\ Re_a^{q_{\gamma}}\ Gr_a^{q_{33}} \tag{32}$$

Let the heat sources (heat removal) denoted as the last term of equation (7) be: heating elements (θ_h), people (θ_m) machine-tools and electric motors (θ_{mach}), special equipment (furnaces, boilers, tanks, etc.) (θ_{eq}), material supplied to the enterprises (θ_{mat}), and electric light (θ_{il}) (Fig.4).

Let us consider some of them in more detail 4 .

From the point of view of the heat and mass transfer process a human being is a very complicated object whose heat release to the room may be apparent radiating-convective or latent due to moisture evaporated from the surface of the body and contained in the breathed out air.

The relation between these constituents is greatly affected by thermal protecting characteristics of cloth, and improvement of these characteristics results in increase in the latent heat release.

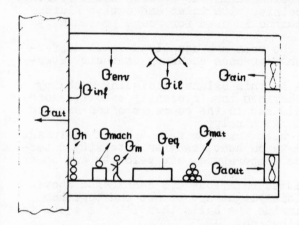

Fig.4. Diagram of heat fluxes in industrial halls.

However, the whole heat release practically does not depend on the kind of cloth and is mainly determined by job intensity. Data concerning the whole heat release were presented above.

As for the apparent heat which is very important for calculation of ventilation it may be determined from the formula [5]:

$$\theta_m = \beta_i \beta_d \left(2.5 + 10.3 \sqrt{v_a}\right)\left(35 - t_r\right) \tag{33}$$

where β_i is factor of job intensity which is equal to 1.0; 1.07; 1.15 for easy, moderate and hard job, respectively; β_d is coefficient taking account of thermal protecting characteristics of cloth which is equal to 1.0; 0.65 and 0.4 for light, ordinary and coldproof cloth, respectively, v_a is air velocity.

Heat releases from electric motors and their associated equipment depend on the rate of their utilization, simultaneous or alternate work and other factors:

$$\theta_{mach} = N K_p K_u K_s \left(1 - \mathcal{C} + K_t \mathcal{C}\right) \qquad (34)$$

where N is nominal power of electric motors; K_p is factor of
its utilization (0.7 + 0.9); K_u is factor of loading (0.5 + 0.8);
K_s is factor of simultaneous work of electric motors (0.5 + 1);
$\mathcal{C}$ is efficiency of motor; K_t is coefficient taking account
of heat fluxes to cooling emulsion, water or other fluid taken
off from the room (0.1 + 1).

Heat flux from special equipment θ_{eq} (tanks, furnaces and so
on) is entirely governed by particular technology of the process.

It may be basically determined from the balance relation,
since the amount of energy released in the process and difference
of the enthalpy fluxes associated with inlet and outlet fluids
are known. Sometimes this method is used for rough preliminary
predictions.

However, heat losses in the room are usually a small difference of great numbers, which decreases accuracy of their determination.

Besides, some components in this balance including enthalpy
of the outlet fluid are determined insufficiently exactly. Moreover, the amount of heat releases in the rooms is often assumed
to be known for calculation of enthalpy.

Therefore calculation of heat transfer from special equipment
is done on elements using data on heat transfer intensity determined by its nature, surface temperature, air velocity and other
factors.

The corresponding calculated relations are considered above.

If the sources of artificial light are preset and together
with fittings are located inside the halls then θ_{ie} is equal to
their total output of illumination.

If the light sources are located out of the rooms then only
radiative heat of visible and ultraphotic radiation comes inside
the rooms, the share of the radiative heat ($\mathcal{C}_{ie}$) for luminescent tubes is of 0.55 and for filament-type tubes reaches up
to 0.85.

In design calculations output of lighting apparatus is determined by the required illuminance of working sites (E):

$$\theta_{ie} = E F q_{ie} \mathcal{C}_{ie} \qquad (35)$$

where F is area of the illuminated surface; q is specific
heat release per unit of illuminance. This coefficient changes
within the range from 0.05 to 0.13 for luminescent tubes and
from 0.13 to 0.25 for filament-type tubes.

If the technological process utilizes material supplied to the
halls with temperature different from that of the room (for example, from the furnace and from outside in cold season) then the
material is also additional source of heat whose integral intensity is determined by the expression:

$$Q'_{mat} = G_{mat} \left(i_i - i_r \right) \tag{36}$$

where G_{mat} is mass of material and i_i and i_r are initial and final value of its enthalpy with consideration for phase transformations if the latter take place.

As the temperature of material and that of environment are levelled, the heat transfer intensity decreases.

It is often necessary to determine the intensity in the limited initial period of time Δt . In this case it (Q''_{mat}) may be calculated from the formula:

$$Q''_{mat} = Q'_{mat} \cdot B \tag{37}$$

where B is a share of excess enthalpy which for the given period of time depends on thermophysical properties of the material, its geometry and coefficient of the external heat transfer.

For approximate estimates one may use relationship between B and Fourier criterion (Fig.5):

$$F_0 = \frac{\Delta t}{c \, G_{mat} \left(\dfrac{G_{mat}}{\rho \lambda F^2} + \dfrac{1}{\alpha_w F} \right)} \tag{38}$$

where c , ρ and λ are thermal capacity, density and thermal conductivity of the material; F is area of its heat releasing surface and α_w is external coefficient of heat transfer.

To calculate heat intensity through the outer surfaces of the premises which is conditionally expressed through the third terms of equations (4) and (5) it is necessary to know the coefficient of heat transfer K, outside air temperature t_{env} which is different from its true temperature with consideration for radiation and also to take into account accumulating properties of enclosing surfaces for those cases when temperature of the environment is approximately determined by the expression:

$$t_{env \; eff} = t_{env} + \frac{\rho q_{r \, ad}}{\alpha_o} \tag{39}$$

where ρ is coefficient of absorption of solar radiation by the outer enclosing surfaces and α_o coefficient of convective heat transfer of these surfaces.

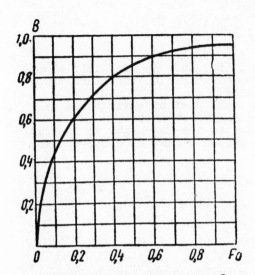

Fig.5. Relationship between B and Fourier criterion.

Due to restriction of its volume the present paper deals only with cases of stationary conditions.

Estimates show that if non-stationary heat fluxes do not exceed in magnitude 25% of the total ones then the solutions obtained have a small error.

Usually this condition is fulfilled in the closed industrial halls with constant technological heat releases.

If the variable constituent reaches from 25 to 60% of the total one then the approximate analysis of non-stationary process may be used when heat sources are not divided into convective and radiative and the law of heat release is assumed to be harmonic.

If the share of variable heat releases exceeds 60% of the total ones and calculation of output of air conditioning system would not produce an error exceeding 15% then complex calculation is required with consideration for harmonic and discontinuous heat supply and specific features of convective and radiative heat transfer.

At present similar methods are well developed (see, for example, [17]).

In the system of equations (4) – (7) usually preset are flow rate and temperature of air supplying to the room, environmental temperature (including also conditional), sources of the internal heat releases. Also conditions of heat transfer allowing to determine the appropriate coefficients of heat transfer are assumed to be known.

In this case unknown quantities are temperatures of the surface elements and air at some zones of the halls.

Sometimes also other combinations of unknown and preset quantities may be found.

However, in all cases a starting system of equations is closed.

Since coefficients of heat transfer, in turn, depend on unknown temperatures in some cases the problem is solved by method of successive approximations. But the process is easily solved and does not involve additional difficulties.

Conclusions

1. Distribution of temperatures and heat transfer between some elements of industrial halls are described by a closed system of equations whose solution may be done by present-day computers.

2. Accuracy of the solutions obtained depends primarily on validity of the relations used for determining heat transfer between some elements.

3. Theoretical and empirical relationships describing heat transfer intensity in classical idealized conditions may be great-

ly different from those which are valid inside the rooms due to the occurrence of different disturbing factors.

4. To verify and sometimes to establish appropriate relationships it is necessary to carry out additional theoretical and experimental studies both using simulations and under conditions close to natural ones.

5. It is desirable that data obtained on existing installations should be widely used since expenditures for conducting **the experiment itself are rather small compared to the cost of the premises where experiments are carried out.**

REFERENCES

1. Kamenev P.N., Skanavi A.N., Bogoslovsky V.N., Egiazarov A.G., Scheglov V.P., Otoplenie i ventiliatsia, I, M., Stroiizdat, 1975.
2. Goromosov M.S., Tsyper N.A., Vodosnabzhenie i sanitarnaya tekhnika, I, 1957.
3. Shorin S.N., Teploperedacha, M., Vysshya shkola, 1964.
4. Bogoslovsky V.N., Novozhilov V.I., Simakov B.D., Titov V.I., Otoplenie i ventilyatsya, 2, M., Stroiizdat, 1976.
5. Sanitarnie normi proektirovaniya promishlennykh predpriyaty, M., Striizdat, 1972.
6. Bogoslovsky V.N., Stroitelnaya teplophyzika, M., Vysshaya shkola, 1970.
7. Shepelev I.A., Sb. trudov NIIST, 23, Stroiizdat, 1967.
8. Shorin S.N., Sb. Sovremennie voprosi otopleniya i ventilyatsii, Stroiizdat, 1949.
9. Eckert E., Drake R., Introduction to the Transfer of Heat and Mass, 1959.
10. Kutateladze S.S., Leontiev A.I., Turbulentny teplo-massoobmen, M., Energia, 1972.
11. Krisher O., Nauchnie osnovi tekhniki sushki, IL, 1961.
12. Rat D., Dissertation, MISI, 1969.
13. Motulevich V.P., I, Ph. J., v.6, 4, 1963.
14. Motulevich V.P., Forced convection (with Injection). Survey Paper, IV Int. Heat Transf. Conf. Versailles, Sept. 1970.
15. Mayers, Shouer, Yustins, Trans. ASME, N 3, 1963, Heat Transfer.
16. Pukemo N.M., Burtsev V.I., Sb. Nauchnikh rabot IOT VTsSPS, N 5, Profizdat, 1965.
17. Sotnikov A.G., Systemi konditsionirovaniya vozdukha s kolichestvennim regulirovaniem, Leningrad, Stroiizdat, 1976.

THE CONVOLUTION PRINCIPLE AS APPLIED TO BUILDING HEAT TRANSFER PROBLEMS, INCLUDING FUNDAMENTALS OF EFFICIENT USE*

METIN LOKMANHEKIM

Lawrence Berkeley Laboratory
University of California
Berkeley, California, USA 94720

The major "Energy Utilization Analysis of Buildings" computer programs utilize the Convolution Principle for the following purposes:

(1)   to compute heat gain (loss is considered as negative heat gain) through a wall/roof;

(2)   to compute exterior suface temperature of a wall/roof;

(3)   to compute the time delay between heat gain to a space and resulting loads on the heating, ventilating and air conditioning system.

To understand the use of Convolution Principle, as an example, heat gain into the building through a wall/roof is explained as follows.

The value of heat gain Q into the building through a wall/roof depends on the present value, and the past history of the temperature difference ΔT between the inside air and the outside surface of the wall/roof. In other words, the graph of the schedule of Q vs. time t depends on the graph of the schedule of ΔT vs. t as shown in Fig. 1.

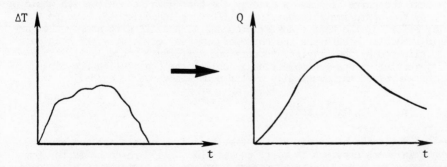

Fig. 1. Dependence of heat gain schedule on temperature difference schedule.

---

*Work supported by the U.S. Department of Energy

If it were necessary to compute Q for each hour, on the basis of the hourly history of ΔT, the differential equation of heat conduction would have to be repeatedly solved. This time-consuming operation, however, can be simplified so that Q need be determined as a function of t for only one temperature difference schedule. When a unit height isosceles triangle temperature pulse is used as one temperature schedule, the values of Q at successive equal-time intervals elicited by this unit height isosceles triangles, are called the response factors r_0, r_1,... of the wall/roof construction as shown in Fig. 2.

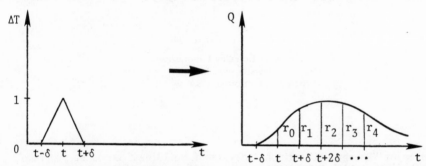

Fig. 2   Heat gain schedule for a unit height isosceles triangle temperature pulse showing response factors.

Any arbitrary schedule of ΔT (Fig. 3a) may be approximated by a schedule of ΔT', whose values agree with those of ΔT at integral multiples of the time interval δ. This schedule of approximate temperature differences ΔT', may be resolved into a series of isosceles triangle pulses (ΔT_1, ΔT_2, ΔT_3, ΔT_4, and ΔT_5 in Fig 3c) which, when added together, give exactly ΔT'. Each of these component pulses has a base width, or duration of 2δ, a peak occurring at each integral multiple of δ, and a height equal to the value of ΔT' at the time of the pulse's peak. Each pulse alone would elicit its own heat gain schedule as shown in Figs. 3d, 3e, 3f, 3g and 3h. The heat gain schedules elicited by the individual pulses are all the same except for two differences. Their heights are proportional to the heights of the pulses which elicit them, and each is moved to the right on the time axis as far as the pulse which elicited it.

The values of the individual responses, Q_1 ... Q_5, may be added at each value of time, to give the curve of sums (Fig. 3i). The superposition theorem asserts that the curve of sums is exactly the heat gain schedule which would be elicited by the approximate temperature difference schedule ΔT'. Due to the smoothing effect of the heat transfer process, ΔT and ΔT' give nearly the same heat gain schedule. Therefore the curve of sums is very nearly the heat gain schedule elicited by the original temperature difference schedule ΔT. This method of resolution and recombination is called the Convolution Principle, and can be expressed mathematically by the equation

$$Q_t = \sum_{i=0}^{n} r_i \Delta T_{t-i} , \qquad (1)$$

whereas Q_t equals the heat gain at the time t; ΔT_{t-1} equals the temperature difference in hours previous to t; r_i equals the $t-1_i$th response factor for the wall/roof; and n equals the number of hours of the temperature difference history which significantly effects Q_t. It should be noted that, the response

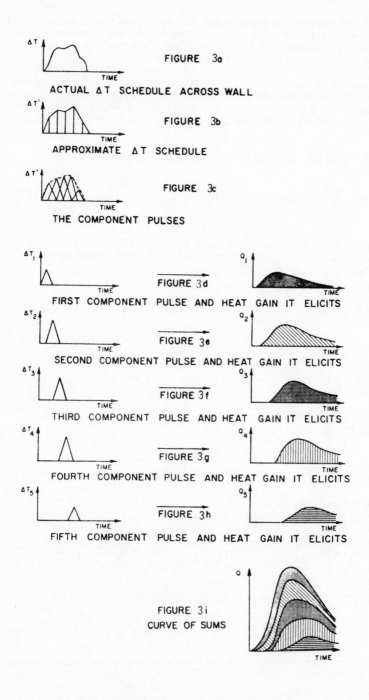

FIGURE 3a
ACTUAL ΔT SCHEDULE ACROSS WALL

FIGURE 3b
APPROXIMATE ΔT SCHEDULE

FIGURE 3c
THE COMPONENT PULSES

FIGURE 3d
FIRST COMPONENT PULSE AND HEAT GAIN IT ELICITS

FIGURE 3e
SECOND COMPONENT PULSE AND HEAT GAIN IT ELICITS

FIGURE 3f
THIRD COMPONENT PULSE AND HEAT GAIN IT ELICITS

FIGURE 3g
FOURTH COMPONENT PULSE AND HEAT GAIN IT ELICITS

FIGURE 3h
FIFTH COMPONENT PULSE AND HEAT GAIN IT ELICITS

FIGURE 3i
CURVE OF SUMS

Fig. 3 The convolution principle.

factors are the only information about the wall/roof which appears in Eq. (1). Thus, the response factors characterize completely the thermal properties of the structure of the wall/roof and, alone describe how the structure absorbs and releases heat over a prolonged period of time.

Utilization of Eq. (1), as is, requires extensive computation time at each simulated hour. However, use of a subtle modification of it which allows n to equal infinity, if it is necessary, saves tremendous amounts of computation time. The explanation of the fundamentals of this efficient technique is given as follows.

At time $t-1$, Eq. (1) can be written explicitly as

$$Q_{t-1} = r_o \, \Delta T_{t-1} + r_1 \, \Delta T_{t-2} + \cdots + r_m \, \Delta T_{t-m-1} + r_{m+1} \, \Delta T_{t-m-2} + r_{m+2} \, \Delta T_{t-m-3} + \cdots$$

$$(2)$$

If response factors reach a Common Ratio at the mth hour after a unit height isosceles triangle pulse is applied, then the Common Ratio R can be expressed as

$$R = \frac{r_{m+2}}{r_{m+1}} = \frac{r_{m+3}}{r_{m+2}} = \cdots \tag{3}$$

Using Eq. (3), Eq. (2) can be written as

$$Q_{t-1} = r_o \, \Delta T_{t-1} + r_1 \, \Delta T_{t-2} + \cdots + r_m \, \Delta T_{t-m-1} + r_{m+1}(\Delta T_{t-m-2} + R \, \Delta T_{t-m-3} + \cdots)$$

$$(4)$$

Similarly, at time t, Eq. (1) can be written as

$$Q_t = r_o \, \Delta T_t + r_1 \, \Delta T_{t-1} + \cdots + r_m \, \Delta T_{t-m} + r_{m+1} \, (\Delta T_{t-m-1} + R \, \Delta T_{t-m-2} + \cdots) \tag{5}$$

Multiplying both sides of Eq. (4) by R, subtracting the result from Eq. (5), rearranging the terms and solving for Q_t gives

$$Q_t - RQ_{t-1} = r_o \, \Delta T_t + r_1 \, \Delta T_{t-1} + \cdots + r_m \, \Delta T_{t-m}$$

$$+ r_{m+1}(\Delta T_{t-m-1} + R \, \Delta T_{t-m-2} + \cdots)$$

$$- R \, r_o \, \Delta T_{t-1} - R \, r_1 \, \Delta T_{t-2} - \cdots - R \, r_m \, \Delta T_{t-m-1}$$

$$- r_{m+1} \, (R \, \Delta t_{t-m-2} + R^2 \, \Delta T_{t-m-3} + \cdots)$$

$$= r_o \, \Delta T_t + (r_1 - R \, r_o) \, \Delta T_{t-1} + \cdots + (r_{m+1} - R \, r_m) \Delta T_{t-m-1}.$$

Let $\quad r'_0 = r_0$

$$r'_1 = r_1 - R\,r_0$$

$$\vdots$$

$$r'_{m+1} = r_{m+1} - R\,r_m \;.$$

Then

And

$$Q_t - R\,Q_{t-1} = r'_0\,\Delta T_t + r'_1\,\Delta T_{t-1} + \cdots + r'_{m+1}\,\Delta T_{t-m-1},$$

$$Q_t = R\,Q_{t-1} + \sum_{i=0}^{m} r'_m\,\Delta T_{t-i} \;. \tag{6}$$

Careful examination of Eq. (6) shows that once Q_{t-1} is calculated, it can be stored and later used in the calculation of Q_t along with a few multiplications and additions saving tremendous amounts of computation time compared to repetitive use of Eq. (1) for each simulated hour.

BIBLIOGRAPHY

1. Procedures for Determining Heating and Cooling Loads for Computerized Energy Calculations - Algorithms for Building Heat Transfer Subroutines, ASHRAE, 1975.

2. Metin Lokmanhekim, et al., Computer Program for Analysis of Energy Utilization in Postal Facilities, published by GPO: Vol. 1—User's Manual—GPO Pub. No. 889-667; Vol. 2— Engineering Manual—GPO Pub. No. 889-666; Vol. 3—Operation Manual GPO Pub. No. 889-665; Vol. 4— Fortran.

3. Metin Lokmanhekim, Calculation of Cooling and Heating Loads with Proposed ASHRAE Procedure and Energy Utilization Analysis of Buildings, University of Hawaii, Department of Mechanical Engineering, 1971-72 Academic Year National Science Foundation Distinguished Visiting Scholar Seminar, Dec. 10-13. 1971.

4. Metin Lokmanhekim, Calculation of Energy Requirements with the Proposed ASHRAE Algorithms for U.S. Postal Buildings, presented at First Symposium on Use of Computers for Environmental Engineering Related to Buildings, Nov. 30-Dec. 2, 1970.

PERSPECTIVE METHODS
OF NUMERICAL SOLUTION
OF FREE-CONVECTION
PROBLEMS FOR BUILDINGS

B. M. BERKOVSKY AND E. F. NOGOTOV

Heat and Mass Transfer Institute
Minsk, USSR

ABSTRACT

Two finite-difference schemes designed for solution of problems on free convection in enclosed regions are considered. Both schemes allow calculations over a wide range of process parameters, give accurate display of local and integral heat transfer characteristics and possess high speed of convergence. The methods of automatic choice of the optimum time step size providing the highest speed of convergence in solution of unsteady-state problems are described.

— — —   — — —

Successful design, building and maintenance of various premises and apartments requires the knowledge of thermal regimes of these constructions at different ambient conditions. The objective of this work is to emphasize the possibilities which appear in connection with the development of numerical methods of calculating the problems of this kind. Consideration is given to the methods adjusted for modern high-speed computers.

Heat transfer and temperature distribution in apartments and premises are exactly described by the Boussinesq equations /1-3/ which in Cartesian coordinates (note, that the majority of practical problems allow two-dimensional statement) may be written as

$$\frac{\partial \Theta}{\partial t} = \frac{\partial}{\partial x}\left(\frac{1}{Pr}\frac{\partial \Theta}{\partial x} - u\Theta\right) + \frac{\partial}{\partial y}\left(\frac{1}{Pr}\frac{\partial \Theta}{\partial y} - v\Theta\right) + Q, \quad (1)$$

$$\frac{\partial \omega}{\partial t} = \frac{\partial}{\partial x}\left(\frac{\partial \omega}{\partial x} - u\omega\right) + \frac{\partial}{\partial y}\left(\frac{\partial \omega}{\partial y} - v\omega\right) + Gr\frac{\partial \Theta}{\partial x}, \quad (2)$$

$$-\omega = \frac{\partial^2 \psi}{\partial x^2} + \frac{\partial^2 \psi}{\partial y^2}, \quad u = \frac{\partial \psi}{\partial y}, \quad v = -\frac{\partial \psi}{\partial x}. \quad (3)-(5)$$

Equations (1)-(5) determine five unknown functions, that is temperature Θ, vorticity ω, stream function ψ and velocity vector projections u and v on the coordinate axes x and y, respectively. They include two dimensionless complexes: Prandtl number $Pr = \nu / \varkappa$ and Grashof number $Gr = g\beta\Delta T L^3 / \nu^2$, where g is the acceleration of gravity; β the temperature expansion coefficient, $\varkappa$ the thermal diffusivity. The function Q characterizes heat sources. It may depend on spatial coordinates, time and temperature.

For greater generality, equations (1)-(5) are written in dimensionless form. The characteristic dimension L, characteristic temperature difference ΔT, kinematic viscosity $\sqrt{}$ and $\sqrt{}/L$ and $L^2/\sqrt{}$ are used as the scales of distance, temperature, stream function, velocity and time.

System (1)-(5) is solved at definite initial and boundary conditions determined by concrete statement of the problem being solved. On solid impermeable boundaries, for example, the adhision conditions

$$\psi = \frac{\partial \psi}{\partial n} = 0 \qquad (u = v = 0). \tag{6}$$

are usually prescribed. For temperature, the conditions of the first, second and third kind as well as the conjugation conditions may be given.

The solution of this system determines temperature, distribution and velocity of convective flows inside the domain.

Below two difference algorithms are given which are, to the authors' opinion, most promising for numerical solution of the problem on free convection in enclosed regions.

Both finite-difference schemes are developed especially for the problems of the above kind [4-5]. Their indubitable advantage is in that they are monotonic, conservative and have the second order of accuracy relatively the steps of the difference scheme. That is why they allow calculations over a wide range of process parameters, give accurate presentation of the conservation laws, inherent in real physical processes, on the grid and yield the results of satisfactory accuracy for relatively coarse grids (h $\sim$ 1/20). As compared to other schemes, they give more accurate display of local heat transfer characteristics. In addition, both algorithms possess sufficiently high speed of convergence and are relatively simple to be realized on computers.

For solution of steady-state problems(herein, the time derivatives of the first two equations are equal to zero) most effective is the difference scheme of the form

$$\Theta^{s+1}(k,l) = [AXT \cdot \Theta^{s+1}(k-1,l) + BXT \cdot \Theta^s(k+1,l) + AYT \cdot \Theta^{s+1}(k,l-1) +$$
$$+ BYT \cdot \Theta^s(k,l+1) + h^2 Q(k,l)]/(AXT + BXT + AYT + BYT), \tag{7}$$

$$\omega^{s+1}(k,l) = (1-\gamma_\omega)\omega^s(k,l) + \gamma_\omega [AXW \cdot \omega^{s+1}(k-1,l) + BXW \cdot \omega^s(k+1,l) +$$
$$+ AYW \cdot \omega^{s+1}(k,l-1) + BYW \cdot \omega^s(k,l+1) +$$
$$+ 0.25\, Grh(\Theta^s(k+1,l) - \Theta^{s+1}(k-1,l))]/(AXW + BXW + AYW + BYW), \tag{8}$$

$$\psi^{s+1}(k,l) = (1-\gamma_\psi)\psi^s(k,l) + \gamma_\psi[\psi^{s+1}(k-1,l) + \psi^s(k+1,l) +$$
$$+ \psi^{s+1}(k,l-1) + \psi^s(k,l+1) + h^2 \omega^{s+1}(k,l)]/4, \tag{9}$$

where
$$AXT = \left[Pr\left(1+0.5\, Pr\, h\, |u(k-\tfrac{1}{2}\, l)|\right)\right]^{-1} + h\, u^+(k-\tfrac{1}{2},l),$$

$$BXT=\left[Pr\left(1+0.5Prh\mid u\left(k+\tfrac{1}{2},l\right)\mid\right)\right]^{-1}-hu^{-}\left(k+\tfrac{1}{2},l\right),$$

$$AYT=\left[Pr\left(1+0.5Prh\mid v\left(k,\ l-\tfrac{1}{2}\right)\mid\right)\right]^{-1}+hv^{+}\left(k,\ l-\tfrac{1}{2}\right),$$

$$BYT=\left[Pr\left(1+0.5Prh\mid v\left(k,\ l+\tfrac{1}{2}\right)\mid\right)\right]^{-1}-hv^{-}\left(k,l+\tfrac{1}{2}\right),$$

$$AXW=\left[1+0.5\ h\mid u\left(k-\tfrac{1}{2},l\right)\mid\right]^{-1}+hu^{+}\left(k-\tfrac{1}{2},l\right),$$

$$BXW=\left[1+0.5\ h\mid u\left(k+\tfrac{1}{2},l\right)\mid\right]^{-1}-hu^{-}\left(k+\tfrac{1}{2},l\right),$$

$$AYW=\left[1+0.5\ h\mid v\left(k,l-\tfrac{1}{2}\right)\mid\right]^{-1}+hv^{+}\left(k,l-\tfrac{1}{2}\right),$$

$$BYW=\left[1+0.5\ h\mid v\left(k,l+\tfrac{1}{2}\right)\mid\right]^{-1}-hv^{-}\left(k,l+\tfrac{1}{2}\right),$$

$$u^{\pm}(k,l)=\left[u(k,l)\pm\mid u(k,l)\mid\right]/2,\quad v^{\pm}(k,l)=\left[v(k,l)\pm\mid v(k,l)\mid\right]/2$$

The velocity projections u and v are calculated by the formulae

$$u\left(k\pm\tfrac{1}{2},l\right)=\left[\phi(k\pm1,l+1)+\phi(k,l+1)-\phi(k\pm1,l-1)-\phi(k,l-1)\right]/(4h)$$

$$v\left(k,l\pm\tfrac{1}{2}\right)=\left[\phi(k-1,l\pm1)+\phi(k-1,l)-\phi(k+1,l\pm1)-\phi(k+1,l)\right]/(4h) \quad (10)$$

Finite-difference scheme (7)-(9) is written as the Zeidel iteration process with the relaxation parameters used for the vorticity and stream function equations. It is not reasonable to apply relaxation for the temperature equation, as the calculation experience shows that this does not result in noticeable acceleration of the iteration convergence. Iteration process (7)-(9) converges at all h .

The boundary conditions for the functions Θ and ψ are also expressed in a difference form, and no mathematical difficulties arise here. Approximate formulae for the calculation of the boundary conditions for the vorticity ω can be obtained, for example, from equation (3) assuming its validity at the boundary and allowing for the boundary conditions for the function ψ . However, because of their local nature, the conditions of this kind deteriorate noticeably the stability of the computing procedure. We think that another approach to the solution of the problem of boundary conditions for the vorticity is more promising. It does not use the function ω at the boundary at all. This method has been used in /6/, its high effectiveness being verified by multiple computations.

This method implies that for the calculation of the function ω at all of the internal nodes of the domain, the nodes being at a distance h from the boundary, formula

$$\omega^{s+1}(k,l)=(1-\gamma_{\omega})\,\omega^{s}(k,l)+\gamma_{\omega}\left[4\psi^{s+1}(k,l)-\psi^{s+1}(k-1,l)-\right.$$
$$\left.-\psi^{s+1}(k+1,l)-\psi^{s+1}(k,l-1)-\psi^{s+1}(k,l+1)\right]/h^{2}, \quad (11)$$

is used instead of (8), and the values of the stream function ψ at these nodes are corrected to satisfy the condition for $\dfrac{\partial\psi}{\partial n}$

Thus, for example, if conditions (6) are prescribed at the boundary $x=0$, then it is assumed that

$$\psi(1,l)=0,\qquad \psi(2,l)=0.25\,\psi(3,l). \quad (12)$$

Here the derivative $\dfrac{\partial \psi}{\partial x}$ is approximated by the expression

$$\frac{\partial \psi}{\partial x}\bigg|_{(1,l)} = -\frac{3\psi(1,l)-4\psi(2,l)+\psi(3,l)}{2h} + 0\left(h^2\right).$$

(13)

The computing algorithm for the solution of system (7)-(9) is constructed in such a fashion that at each grid point the values of $\Theta^{S+1}(k,l)$, $\omega^{S+1}(k,l)$ and $\psi^{S+1}(k,l)$ are computed simultaneously. The iteration process ceases once the condition

$$\max_{k,l}\left|\frac{\omega^{S+1}(k,l)-\omega^S(k,l)}{\omega^{S+1}(k,l)}\right| \leqslant \varepsilon \qquad (\omega^{S+1}(k,l) \neq 0).$$

(14)

is satisfied. It is senseless to verify similar conditions for the functions Θ and ψ, as their speed of convergence is higher than that of the function ω.

The relaxation parameters are chosen to ensure the highest speed of iteration convergence.

The numerical modelling of unsteady-state thermoconvective processes makes more serious demands to the finite-difference techniques. Since the solution, as a rule, should be obtained for large time interval, it is also important that the numerical algorithm allowed computations with larger time step sizes.

The suggested scheme makes it possible to calculate with $\tau \sim h$, the optimum value of τ, which provides the highest speed of convergence, being chosen automatically in the course of computation.

The computing algorithm is based on the method of variable directions and provides for consistent solution to the equations of temperature, vorticity and stream function. The system of difference equations to approximate initial system (1)-(5) is written as follows

$$AXT\cdot\bar{\Theta}(k-1,l)-(CXT+\varrho)\bar{\Theta}(k,l)+BXT\cdot\bar{\Theta}(k+1,l)=-FXT,$$ (15)

$$AYT\cdot\Theta^{n+1}(k,l-1)-(CYT+\varrho)\Theta^{n+1}(k,l)+BYT\cdot\Theta^{n+1}(k,l+1)=-FYT,$$ (16)

$$AXW\cdot\bar{\omega}(k-1,l)-(CXW+\varrho)\bar{\omega}(k,l)+BXW\cdot\bar{\omega}(k+1,l)=-FXW,$$ (17)

$$AYW\cdot\omega^{n+1}(k,l-1)-(CYW+\varrho)\omega^{n+1}(k,l)+BYW\cdot\omega^{n+1}(k,l+1)=-FYW,$$ (18)

$$\bar{\psi}(k-1,l)-(2+\sigma)\bar{\psi}(k,l)+\bar{\psi}(k+1,l)=-FXP,$$ (19)

$$\psi^{S+1}(k,l+1)-(2+\sigma)\psi^{S+1}(k,l)+\psi^{S+1}(k-1,l)=-FYP,$$ (20)

where $\quad \varrho = 0.5\tau/h^2, \quad \sigma = 0.5\tau'/h^2,$

α (1) and β (1) are evaluated from the conditions at the left-hand boundary. The coefficients α (k), β (k) being determined, the condition at the right-hand boundary is used to estimate $\bar{\Theta}$ (K , l) , $\bar{\Theta}$ (K-1, l) , $\bar{\Theta}$ (K-2, l) , ... $\bar{\Theta}$ (2, l) by the formula

$$\bar{\Theta}(k,l) = \alpha(k)\,\bar{\Theta}(k+1,l) + \beta(k) \qquad (23)$$

By substituting all indices l , find all $\bar{\Theta}$ (k, l).

Difference equation (16) is solved in the same way. The index k is fixed and the computation is made of the coefficients

$$\alpha(l) = \text{BYT}\left[\text{CYT} + \varphi - \text{AYT}\cdot\alpha(l-1)\right]^{-1},$$

$$\beta(l) = \left[\text{AYT}\cdot\beta(l-1) + \text{FYT}\right]\cdot\alpha(l)\big/\text{BYT}, \, l=1,2,3,\ldots L(24)$$

α (1), β (1) are estimated from the condition at y = 0. Then, using the condition at the other boundary, the values of Θ^{n+1} (k, l) are determined by the formula

$$\Theta^{n+1}(k,l) = \alpha(l)\cdot\Theta^{n+1}(k,l+1) + \beta(l) \qquad (25)$$

Further, the same procedure is used to evaluate the functions ω^{n+1} (k, l) and ψ^{n+1} (k, l) . The vorticity ω is determined only inside the reduced domain, whose boundary is at a distance h from the main boundary, while ω at the boundary of the reduced domain is evaluated by formula (11). To calculate ψ^{n+1} (k, l) from equation (19)-(20), the iteration process is constructed. Iterations cease once the condition

$$\max_{k,l}\,\left|\frac{\psi^{S+1}(k,l) - \psi^{S}(k,l)}{\psi^{S+1}(k,l)}\right| \leqslant \varepsilon_{\psi} \qquad (\psi^{S+1}(k,l) \neq 0).$$

$$(26)$$

is fulfilled, and it is assumed that ψ^{n+1} (k, l) = ψ^{S+1} (k, l). Here S is the number of iterations. At the near-boundary points the values of ψ^{S+1} (k, l) are corrected in accordance with the condition for $\dfrac{\partial\psi}{\partial n}$.

For the specification of Θ^{n+1} (k, l) , ω^{n+1} (k, l) and ψ^{n+1} (k, l) , an external iteration cycle is developed. On the first iteration in formula (21) the values of ψ^{n+1} (k, l) are taken instead of the unknown ψ^{n} (k, l) . In the following iterations, when calculating the velocity components, boundary-value vorticities and source $Gr\,\dfrac{\partial\Theta}{\partial n}$, the computed Θ^{n+1} (k, l) , ω^{n+1} (k, l) and ψ^{n+1} (k, l) are used. The departure from the external iteration cycle is determined by the condition

$$CXT = AXT + BXT + h\left[u^{+}\left(k+\tfrac{1}{2}, l\right) - u^{-}\left(k-\tfrac{1}{2}, l\right)\right],$$

$$CYT = AYT + BYT + h\left[v^{+}\left(k, l+\tfrac{1}{2}\right) - v^{-}\left(k, l-\tfrac{1}{2}\right)\right],$$

$$CXW = AXW + BXW + h\left[u^{+}\left(k+\tfrac{1}{2}, l\right) - u^{-}\left(k-\tfrac{1}{2}, l\right)\right],$$

$$CYW = AYW + BYW + h\left[v^{+}\left(k, l+\tfrac{1}{2}\right) - v^{-}\left(k, l-\tfrac{1}{2}\right)\right],$$

$$FXT = AYT \cdot \Theta^{n}(k, l-1) - (CYT - \varrho)\Theta^{n}(k, l) + BYT \cdot \Theta^{n}(k, l+1) + h^{2}Q(k, l),$$

$$FYT = AXT \cdot \bar{\Theta}(k-1, l) - (CXT - \varrho)\bar{\Theta}(k, l) + BXT \cdot \bar{\Theta}(k+1, l) + h^{2}Q(k, l),$$

$$FXW = AYW \cdot \omega^{n}(k, l-1) - (CYW - \varrho)\omega^{n}(k, l) + BYW \cdot \omega^{n}(k, l+1) +$$

$$+ 0.25\, Grh\left[\Theta^{n}(k+1, l) - \Theta^{n}(k-1, l) + \Theta^{n+1}(k+1, l) - \Theta^{n+1}(k-1, l)\right],$$

$$FYW = AXW \cdot \bar{\omega}(k-1, l) - (CXW - \varrho)\bar{\omega}(k, l) + BXW \cdot \bar{\omega}(k+1, l) +$$

$$+ 0.25\, Grh\left[\Theta^{n}(k+1, l) - \Theta^{n}(k-1, l) + \Theta^{n+1}(k+1, l) - \Theta^{n+1}(k-1, l)\right],$$

$$FXP = \varphi^{s}(k, l-1) - (2-\sigma)\varphi^{s}(k, l) + \varphi^{s}(k, l+1) + h^{2}\omega^{n+1}(k, l),$$

$$FYP = \bar{\varphi}(k-1, l) - (2-\sigma)\bar{\varphi}(k, l) + \bar{\varphi}(k+1, l) + h^{2}\omega^{n+1}(k, l).$$

Here the same notations as in procedure (7)–(9) are used. The velocity components are evaluated by the averaged formulae of the form

$$u\left(k\pm\tfrac{1}{2}, l\right) = \left[\varphi^{n}(k\pm 1, l+1) + \varphi^{n}(k, l+1) - \varphi^{n}(k\pm 1, l-1) - \varphi^{n}(k, l-1) +\right.$$

$$\left. + \varphi^{n+1}(k\pm 1, l+1) + \varphi^{n+1}(k, l+1) - \varphi^{n+1}(k\pm 1, l-1) - \varphi^{n+1}(k, l-1)\right]/(8h),$$

$$v\left(k, l\pm\tfrac{1}{2}\right) = \left[\varphi^{n}(k-1, l\pm 1) + \varphi^{n}(k-1, l) - \varphi^{n}(k+1, l\pm 1) - \varphi^{n}(k+1, l) +\right.$$

$$\left. + \varphi^{n+1}(k-1, l\pm 1) + \varphi^{n+1}(k-1, l) - \varphi^{n+1}(k+1, l\pm 1) - \varphi^{n+1}(k+1, l)\right]/(8h).$$

$$(21)$$

Difference equations (15)–(20) are solved by the two-step method of elimination /7/. Equation (15) is to be solved first. To do this, index l is fixed and the values of α (k) and β (k) (k=1,2,3 ... K) are calculated using the recurrent formulae

$$\alpha(k) = BXT \left[CXT + \varrho - AXT \cdot \alpha(k-1)\right]^{-1},$$

$$\beta(k) = \left[AXT \cdot \beta(k-1) + FXT\right] \cdot \alpha(k)/BXT.$$

$$(22)$$

$$\max_{k,l} \left| \frac{\omega^{n+1,S1+1}(k,l) - \omega^{n+1,S1}(k,l)}{\omega^{n+1,S1+1}(k,l)} \right| \leqslant \varepsilon_\omega \qquad \left(\omega^{n+1,S1+1}(k,l) \neq 0\right).$$

(27)

where $S1$ is the number of the external iteration.

The conditions for the stability of the two-step method of elimination are ensured by the validity of the unequality

$$\frac{2}{\tau} > \max_{k,l} \left(\frac{u\left(k-\frac{1}{2},l\right) - u\left(k+\frac{1}{2},l\right)}{h}, \frac{v\left(k,l-\frac{1}{2}\right) - v\left(k,l+\frac{1}{2}\right)}{h} \right)$$

(28)

For the decrease of the computing time, the choice of τ depending on the number of external iterations performed on the previous time layer is rather effective. If the number of $S1^n$ - iterations of the n -th layer proves to be greater than S_0 , then on the (n+1)th layer the step is chosen to be equal to $\tau^{n+1} = 0.7 \tau^n$. If $S1^n < S_0$, the step increases, $\tau^{n+1} = 1.2 \tau^n$. At $S1^{n+1}$ greater than S_m , the step decreases by the factor of two, and the computation is repeated. The latter ensures the computing stability, as the sharp increase of the number of iterations procedes the instability. The computing experience shows that at $S_0 = 4$ and $S_m = 8$ the program chooses the maximum possible step within the intervals with smooth change of the solution. The value τ is also controlled by condition (28). If this is not fulfilled, then the step τ decreases by 30 per cent.

Both difference schemes have been multiply applied for the computation of free-convection problems, and our computations have verified their high effectiveness.

REFERENCES

1. Landau L.L., and Lifshits E.M. 1954. Solid Mechanics, Moscow.

2. Luikov A.V., and Berkovsky B.M. 1974. Convection and Thermal Waves, Moscow.

3. Berkovsky B.M., and Nogotov E.F. 1976. Difference Methods of Heat Transfer Problems Investigation, Minsk.

4. Berkovsky B.M., and Polevikov V.K. 1973. The Prandtl number effect of the free-convection structure and heat transfer. J.Engng Physics 24: 842-849.

5.   Nogotov E.F., and Sinitsyn A.K. 1976. On numerical investi-
      gation of unsteady-state convection problems. J.Engng
      Physics 31: 1113-1119.

6.   Polezhaev V.I., and Gryaznov V.L. 1974. Method of calculating
      the boundary conditions for the Navier-Stokes equations in
      the vortex, stream function variables. Dokl. AN SSSR 219:
      301-304.

7.   Godunov S.K., and Ryabenky V.S. 1962. Introduction into
      the difference scheme theory, Moscow.

A UNIFIED APPROACH TO MASS AND HEAT TRANSFER CALCULATIONS IN ENCLOSURES

M. G. DAVIES

Department of Building Engineering
The University of Liverpool
Liverpool L69 3BX, England

ABSTRACT

A procedure frequently used in the U.K. for the design of heating and cooling systems and estimating the response of the fabric of a building, is based on a combination of air and radiant temperature, called environmental temperature, rather than air temperature alone. The paper sketches its key features and indicates how it can be extended to include the effect of step changes in thermal excitation, and to include moisture movement in an enclosure.

INTRODUCTION

The environmental temperature (e.t.) procedure is a design procedure to assist the heating services engineer and architect to estimate the temperature field in an enclosure. It assumes that he will a) have details of its construction including orientation, b) know the patterns of behaviour of the occupants insofar as they effect the control of the heating services, control of ventilation and imposition of metabolic and other heat loads, c) have information on the probability of occurrence of values of ambient air temperature, insolation and perhaps wind speed for the site.

The procedure was developed by Mr. E. Danter and his co-workers at the U.K. Building Research Station, Watford, during the 1960's. It came about because of the high incidence of summer overheating in certain lightweight buildings erected in the post war years. Two leading papers in its description are refs. 1 and 2.

The procedure was adopted by the Institution of Heating and Ventilating Engineers (now the Chartered Institution of Building Services) as its preferred method of design and it is described from the designers point of view in the I.H.V.E. Guide Book A, (ref. 3).

The present author has more recently been concerned with the ideas underlying the procedure and has extended its scope, so as to estimate the transient response of an enclosure and also to evaluate the consequences of introducing water vapour into the enclosure. This work is discussed in detail in a series of articles, most of which are as yet unpublished (refs. A1-A10).

In this article, some of the main features of the e.t. procedure will be set out (Section 2), a way in which they may be arrived at will be presented, (Section 3), some theory regarding the response of solid walls to sinusoidal and step excitation will be given (Sections 4 & 5) and finally a method of including moisture movement will be discussed (Section 6). The article thus attempts to provide a bird's eye view of the capability of the procedure, based on refs. A1-A10 and other papers.

2. THE ENVIRONMENTAL TEMPERATURE PROCEDURE

Heating calculations in an enclosure have usually been carried out in terms of air temperature. In the e.t. procedure temperature and heat flows are referred instead to "environmental temperature", which is a combination of air temperature, the temperature of the cool surface of the enclosure and the temperature of any heat source present. It thus takes better account of the important role of radiation as a mechanism for the transfer of heat in an enclosure.

It attempts to steer a middle course between the over simplification associated with use of air temperature alone, and the prohibitively complicated calculations which take fuller account of radiant exchanges. It is a design procedure which does not claim to be exact but which limits errors in heat flows as calculated by more exact methods, to errors of less than 5% (ref. 2). The labour associated with the calculation is quite light. The set of ideas behind the procedure however is extensive and would not appear to be fully understood by all architects and engineers.

The main features of the e.t. procedure follow:

a) Environmental temperature, T_{ei} is related to air temperature T_{ai} and the space averaged radiant temperature T_r as

$$T_{ei} = \frac{\frac{6}{5} Eh_r}{h_c + \frac{6}{5}Eh_r} T_r + \frac{h_c}{h_c + \frac{6}{5}Eh_r} T_{ai} \tag{1}$$

T_r includes the effect of any hot body source present. With the conventional values h_c, the convection heat transfer coefficient of 3.0 W/m^2K and Eh_r the radiation heat transfer coefficient of 0.9 x 5.7 W/m^2K,

$$T_{ei} \simeq \frac{1}{3}T_{ai} + \frac{2}{3} T_r \tag{2}$$

b) Heat may be input to an enclosure in a variety of forms. If input from a local convector heater, it is all input at T_{ai}. The energy from the sun is all received at solid surfaces whose mean temperature is around T_r,(if hot bodies are absent.) A radiator, lamp or occupant emits partly convectively and partly by radiation, some heat being received at T_{ai} and some at T_r.

Now if the heat from a hot body source is in the ratio h_c : $\frac{6}{5} Eh_r$ or about $\frac{1}{3}$:$\frac{2}{3}$ convection to radiation, the entire heat input can be considered to be input at T_{ei}. As appears later, it is convenient to input all heat of whatever origin at T_{ai}. If the source actually delivers its input at T_{ai}, for example, its effect is the same as a reduced input at T_{ei}.

c) A steady state heat loss can be taken to be generated by the difference between T_{ei} and ambient air temperature T_{ao}. This takes place by ventilation, and via one or more paths through the fabric of the enclosure. These paths are in parallel and this feature permits the advantageous ease of computation.

d) The conductance between T_{ei} and an area A' at T_s of the walls of the enclosure is expressed as $A'(h_c + \frac{6}{5} Eh_r)$

For convenience write $A'h_s = A'(h_c + \frac{6}{5} Eh_r)$ (3)

The further conductance between T_s and T_{ao} will be denoted by $A'h_o$, where h_o includes the outer film and any transmittance due to the fabric. The two can be combined to form an overall conductance:

$$A'U = ((A'h_s)^{-1} + (A'h_o)^{-1})^{-1}$$ (4)

e) The conductance between T_{ei} and the inside air temperature is expressed as $\Sigma Ah_s \, h_c / (\frac{6}{5} Eh_r)$

where ΣA denotes the total internal surface area, $6A$ for a cubic enclosure.

The ventilation loss between T_{ai} and T_{ao} is represented by the conductance nVs where n is the number of volume air changes in unit time, V is the volume of the enclosure and s is the volumetric specific heat of air, often taken as $1200J/m^3K$. These quantities too can be combined.

$$C_v = ((\Sigma Ah_s \, h_c / (\frac{6}{5} Eh_r))^{-1} + (nVs)^{-1})^{-1}$$ (5)

f) It follows that a hot body steadily inputting $\frac{1}{3}Q$ convectively and $\frac{2}{3}Q$ radiantly to an enclosure, from which heat is lost only by ventilation and through A', will generate an inside e.t. of

$$T_{ei} = T_{ao} + Q/(A'U + C_v)$$ (6)

g) If the heat source has a sinusoidal component of amplitude δQ, the amplitude of T_{ei} can be estimated in a similar way. A discussion of heat storage however is needed before this can be obtained.

h) It should be mentioned that ref. 3 uses $h_c + Eh_r$ rather than $h_c + \frac{6}{5}Eh_r$ and usually quotes it in its reciprocal form: $(h_c + Eh_r)^{-1} = 0.123m^2K/W$. The conductance $\Sigma Ah_s h_c / (\frac{6}{5} Eh_r)$ is usually quoted as $4.8\Sigma A$ W/K.

3. A DERIVATION OF THE PROCEDURE

The temperature varies continuously over the air and surfaces of an enclosure. Any thermal model assumes that the temperature is uniform over some limited area, (or alternatively represents a local distribution by its mean value.)

The e.t. procedure offers a simple and approximate prescription for handling the mutual exchanges between the air (at a uniform temperature) and however many local uniform temperature surfaces are assumed. In the case of a particular enclosure in which just three temperatures are specified: (a) the basic equations describing the convective and radiant exchanges can be written down very simply, and (b) these equations transform exactly to give the e.t. procedure. The details are given in ref.A6.

For consider an enclosure consisting of a cube of total internal area $6A$. It has one external wall whose internal surface is at T_s, and five internal walls whose surfaces are all at T_r'. The air is at T_{ai}. There are three internal exchanges of heat, arranged in delta form: radiation from

T_r' to T_s, $C_1 = AEh_r$; convection from T_{ai} to T_s, $C_3 = Ah_c$, and convection from T_r' to T_{ai}, $C_4 = 5Ah_c$.

These conductances transform to an equivalent set (K_s, K_r and K_{ai}) in star form:

$$K_s C_4 = K_r C_3 = K_{ai} C_1 = C_4 C_3 + C_3 C_1 + C_1 C_4 \tag{7}$$

so $K_s = A(h_c + \frac{6}{5}Eh_r)$, $K_r = 5A(h_c + \frac{6}{5}Eh_r)$ and $K_{ai} = 6A(h_c + \frac{6}{5}Eh_r)h_c/(\frac{6}{5}Eh_r)$ (8)

The star configuration offers a further temperature compounded of T_s, T_r' and T_{ai}:

$$T_{ec} = \frac{K_s T_s + K_r T_r' + K_{ai} T_{ai}}{K_s + K_r + K_{ai}} = \frac{\frac{1}{5}Eh_r T_r + Eh_r T_r' + h_c T_{ai}}{\frac{6}{5}Eh_r + h_c} = \frac{\frac{6}{5}Eh_r T_r + h_c T_{ai}}{\frac{6}{5}Eh_r + h_c} \tag{9}$$

where $T_{rc} = \frac{1}{6}T_s + \frac{5}{6}T_r'$, and represents the mean surface temperature.

Thus T_{ec} is an equivalent temperature representing the cool parts of the enclosure, and the conductances from it to the walls and to the air are those noted in 2d and 2e.

Now suppose that a hot body source at T_{hb} emitting Q Watts is placed within the enclosure.   It is taken to be a pure heat source, emitting Q Watts irrespective of the temperature of the enclosure.   Suppose that the conductances associated with its heat transfer to T_s, etc., are k_s, k_r and k_{ai}.

$$Q = k_s(T_{hb} - T_s) + k_r(T_{hb} - T_r') + k_{ai}(T_{hb} - T_{ai}) \tag{10}$$

$$= (k_s + k_r + k_{ai})T_{hb} - (k_s T_s + k_r T_r' + k_{ai} T_{ai}) \tag{11}$$

If, but only if,

$$k_s/K_s = k_r/K_r = k_{ai}/K_{ai} \quad (= \alpha \text{ say}) \tag{12}$$

the heat transfer from the hot body has the same consequences as the heat transfer from a source at T_{eh} emitting Q Watts through the superposed conductances $K_s + k_s$, $K_r + k_r$, $K_{ai} + k_{ai}$, where

$$T_{eh} = (T_{ec} + \alpha T_{hb})/(1 + \alpha) \tag{13}$$

If the source is now supposed to shrink to very small dimensions, the k's tend to zero, α tends to zero and αT_{hb} tends to $(Q/6A)$ $(\frac{6}{5} Eh_r)/h_s^2$.   Thus T_{eh} tends to $T_{ec} + (Q/6A)(\frac{6}{5}Eh_r)/h_s$, a   fixed elevation above T_{ec}.

If now the external conductances beyond T_s, T_r' and T_{ai}, namely Ah_o, zero and nVs, are introduced and combined with their respective internal conductances using eqs. 4 and 5, heat continuity requires that

$$Q = (T_{eh} - T_{ao})(AU + C_v) \tag{14}$$

Comparison with eq. 6 shows that T_{ei} and T_{eh} are the same.   Thus the operation defined by eq. 6 yields a temperature, at any rate in these very restricted circumstances, which exceeds the mean equivalent temperature T_{ec} of the enclosure by a fixed amount depending on the strength of the heat source and

and the heat exchange characteristics of the room.

$$\text{Since} \quad T_{ei} = T_{eh} = \frac{\frac{6}{5}Eh_r(T_{rc} + Q/(6Ah_s))}{h_s} + \frac{h_c T_{ai}}{h_s} \tag{15}$$

it follows from eq.1 that the average radiant temperature T_r is the sum of T_{rc} from the cool enclosure, together with $Q/(6Ah_s)$ due to the hot source.

The restriction includes the provision that $k_s:k_r:k_{ai}$, and therefore in the limit, the heat flows to T_s, T_r' and T_{ai}, should be in the ratio $1:5: 6h_c/\frac{6}{5} Eh_r$ or $1:5:2.92$. This could be satisfied by a small source placed at the centre of a cubic room with suitable convective and emissive properties. The combined radiative to convective output is nearly $2/3:1/3$.

One would suppose that T_r', the overall averaged radiant temperature is comparatively insensitive to the position of the source within the room. Danter has considered the question of non cubic rooms, and shows that the factor $\frac{6}{5}$ is a satisfactory mean value with which to multiply Eh_r for rooms of various shapes.

If the source inputs in a ratio significantly different from the $\frac{2}{3}:\frac{1}{3}$ ratio, the several inputs can be scaled to an equivalent input at T_{ei}; (see ref. A6, section 6).

Only for the three temperature enclosure is the delta to star transformation possible. No single quantity T_{ec} for a four or more temperature enclosure exists with which T_{hb} can be combined. However, it would appear a reasonably approximation to work in terms of a star temperature T_{ec}' with the conductances noted in section 2, (A'h_s to the surface A' and $\Sigma Ah_s h_c/\frac{6}{5}Eh_r$ to the air). Danter (ref. 2) presents figures which show that heat flows so computed are within 5%, (and usually better) of their exactly computed values.

If the surface at T_r' permitted a steady state heat loss with a conductance of $5Ah_o'$ between T_r' and T_{ao}, then

$$T_{ei} = Q/(AU + 5AU' + C_v) + T_{ao} \tag{16}$$

$$\text{where} \quad U' = (h_s^{-1} + h_o'^{-1})^{-1} \tag{17}$$

The following two sections deal briefly with the situation where the T_r' wall has heat storage properties.

4. SINUSOIDAL EXCITATION

If the surface of an infinitely thick slab of material (conductivity k, density ρ and specific heat c) undergoes a sinusoidal temperature variation of amplitude δT_o, and period P, the ratio of the sinusoidal amplitude of variation of heat flux δq, and δT_o is

$$a = \delta q/\delta T_o = \sqrt{2\pi k\rho c/P} \cdot \exp(i\pi/4) \tag{18}$$

a is complex. The term $\exp(i\pi/4)$ implies that the maximum of δq falls $\pi/4$ radians, $45°$, 1/8 of a cycle, or 3 hours in 24 hours before the maximum of δT_o.

If the slab has a finite thickness X and is insulated on its rear surface, the surface admittance, y_s is

$$y_s = \sqrt{\frac{2\pi k\rho c}{P}} \sqrt{\frac{\cosh 2\tau - \cos 2\tau}{\cosh 2\tau + \cos 2\tau}} \exp(i(\pi/4 + \arctan \frac{\sin 2\tau}{\sinh 2\tau})) \qquad (19)$$

where $\tau = \sqrt{\pi\rho c/Pk}$. The thick admittance a, (units W/m^2K) and τ (dimensionless) completely determine the transfer properties of the slab undergoing sinusoidal excitation.

If a wall section is composed of several layers, its surface admittance can be computed as discussed in refs. A1 and A2.

In the environmental temperature procedure, transfer quantities are referred where appropriate to environmental rather than surface and air temperatures. Accordingly an admittance based on e.t. can be written down

$$Y = (h_s^{-1} + y_s^{-1})^{-1} \qquad (20)$$

Y is a laborious quantity to compute from first principles. Tabulated values are provided in ref.3, Table A9, and more extensively in ref.4.

If a sinusoidally varying component of heat of amplitude δQ can be considered input at T_{ei}, the corresponding amplitude of variation of T_{ei} is then

$$\delta T_{ei} = \delta Q/(AU + 5AY + C_v) \qquad (21)$$

The largest contributor to δQ is solar gain. Ref.3 provides a wealth of information on solar radiation, effect of orientation of windows, and the effect of sunshading devices, so as to permit estimates of steady state and swings in temperature. The peak temperature reached is of course the sum $T_{ei} + \delta T_{ei}$.

Ref.A3 indicates from first principles how the magnitudes and phases of T_s, T_r', and T_{ai} can be calculated. In principle these quantities might be estimated via e.t. but ref. 3 does not provide details of how this is to be done.

5. STEP EXCITATION

Another basic form of time varying excitation, which of course involves heat storage, consists in imposing a sudden step in the form of excitation. Consider the infinite slab again. Suppose it is initially at T_j throughout and that its exposed surface is in contact with air, heat transfer coefficient h. Suppose that at time zero, the air temperature falls to zero and stays there. The surface temperature T_r' starts to fall. Its rate of fall depends upon time, non dimensionalised as $h^2t/k\rho c$.

If the slab has a thickness X and is insulated, T_r' initially falls at a rate independent of X. After a time given approximately by another non dimensionalised time, $kt/\rho cX^2$ of about 0.3, T_r' falls relatively faster. This solution is discussed in ref. A4, where a table is given relating T_r' to time and slab thickness. Charts providing similar information are widely available in heat transfer texts. (From the designers point of view, they are laborious to compute.)

In ref. A5 two developments are discussed: (a) the single heat loss mechanism h is generalised to a network of heat loss mechanisms, convection, radiation, ventilation and fabric conduction (without storage), with the intermediate temperature nodes T_s and T_{ai}; (b) it is shown how these temperatures may be computed when the slab is initially at zero throughout and at zero time step heat inputs an imposed at T_s, T_r', and T_{ai}.

In ref. A7, this process is extended to computing the variation in T_{ei} following a step in heat input.

Given tables or charts of transient response characteristics, computation of this variation is simple.

The response time of T_r' is defined to be the time t_r taken for T_r' to accomplish 63.2% of its total change.   It is shown in ref. A4 that if the ratio (conductance from T_r' to T_{ao})/(conductance from T_r' to the back surface, i.e. 5Ak/X in the current example) is less than about 0.4, the response time is given by (thermal capacity of the enclosure)/(conductance from T_r' to T_{ao}).

t_r cannot be observed directly, for the weather changes too rapidly.   A similar quantity can be found by evaluating the autocorrelation function of a time series of values of T_r'.   Details are in ref. A5.

6.   MOISTURE MOVEMENT

The steady state flow of water vapour introduced into an enclosure may be lost by ventilation, by diffusion through the fabric and by condensation on, for example, the window.

These processes can be included with the heat flows in an enclosure in an approximate way by making two transformations.

(i)   The mass flows are expressed as latent heat flows by multiplication by the latent heat.

(ii) The generating vapour pressures are expressed by their corresponding dewpoint temperatures.

Thus a system of latent heat and dewpoint temperature sources can be included in the e.t. formulation, together with the corresponding system of conductances representing ventilation, diffusion and condensation/evaporation. This development to the e.t. procedure is described in ref. A10.

Results are presented there showing how relative humidity, environmental temperature and the rate of condensation or evaporation at a window surface will vary with the amount of moisture input to the enclosure.   Parameters controlling this variation are the amount of heat input to the enclosure, the ventilation rate, insulation, ambient temperature and ambient humidity.

The calculations are somewhat more laborious from the designers point of view than the estimation of steady, periodically varying and transiently varying temperatures.

7.   DISCUSSION

The environmental temperature procedure as given in ref.3 provides the building designer with a ready means of estimating the mean temperature and extent of periodic variation in temperature when the constructional details (dimensions, materials, sunshading devices, orientation), occupational factors (heating, ventilation rate) and meteorological factors (air temperatures and sunshine) are known.   The set of personal articles cited are intended to provide a summary of basic ideas on heat and mass transfer, how they feature in a simple form of enclosure, and how two further forms of excitation can be included within the formulation of the environmental temperature procedure. Fig.1 shows their interrelation.

REFERENCES

1.   Loudon A.G., Summertime temperatures in buildings without air conditioning, Journal of the Heating and Ventilating Engineers, 37, 280-292, 1970.

2.   Danter E., Heat exchanges in a room and the definition of room temperature, Building Services Engineer, 41, 231-243, 1974.

3.   Institute of Heating and Ventilating Engineers Guide, Book A, I.H.V.E., London 1971.

4.   Milbank N.O and Harrington-Lynn J., Thermal response and the admittance procedure, Building Services Engineer, 42, 38-51, 1974.

A1   The thermal admittance of layered walls. Build.Sci. 8, 208-220, 1973.

A2   The response of a layered construction to sinusoidally varying thermal excitation.   Int. note.

A3   Daily temperature swings in an enclosure with heavy internal walls. Int. note.

A4   The structure of the transient cooling of a slab.   Int. note.

A5   The transient response of an enclosure with heavy internal walls. Int. note.

A6   On the basis of the environmental temperature procedure.   Int. note.

A7   The estimation of environmental temperature following a step change in heat input or ambient temperature.   Int. note.

A8   Computing the rate of superficial and interstitial condensation. Build. Sci. 8, 97-104, 1973.

A9   Estimation of loss of water vapour from an enclosure.   Build.Sci.10, 185-188, 1975.

A10 The environmental temperature procedure and moisture movement in an enclosure.   Int. note.

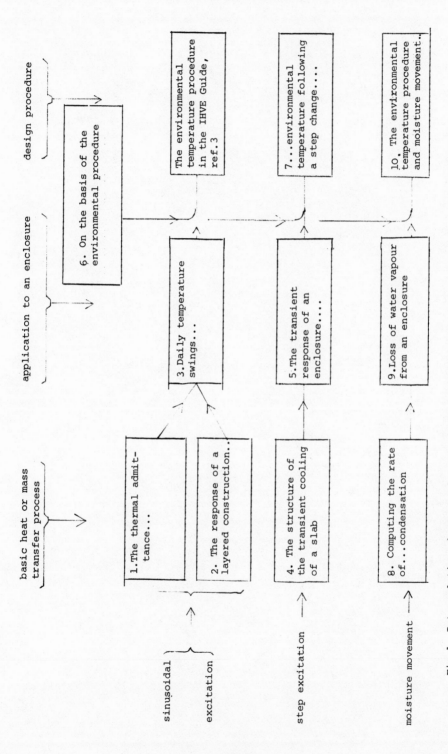

basic heat or mass transfer process

application to an enclosure

design procedure

sinusoidal excitation

1.The thermal admit-tance...

2. The response of a layered construction..

3.Daily temperature swings...

6. On the basis of the environmental procedure

The environmental temperature procedure in the IHVE Guide, ref.3

step excitation

4. The structure of the transient cooling of a slab

5.The transient response of an enclosure....

7...environmental temperature following a step change....

moisture movement

8. Computing the rate of...condensation

9.Loss of water vapour from an enclosure

10. The environmental temperature procedure and moisture movement..

Fig.1 Interrelations between the references and the environmental temperature procedure

A DIGITAL COMPUTER PROGRAM FOR THE CALCULATION OF YEARLY ROOM ENERGY DEMANDS AND TEMPERATURE EXCEEDING RATES, USING HOURLY WEATHER DATA

H. J. NICOLAAS, K. TH. KNORR, AND P. EUSER

Institute of Applied Physics TNO–TH
Delft, The Netherlands 2208

SUMMARY

A description is given of a computer program for the calculation of the indoor temperatures, temperature exceeding rates, the design heating and
cooling loads, and the yearly heating and cooling demands in the rooms of
a building. The program is based on a thermal response factor method. The
response factors are determined by means of a program in which the equations
for the heat transfer processes in the room are solved by an implicit
differential method.
Comparable results obtained by other methods are given.

1 Introduction

Investigations and design calculations with the purpose to evaluate the
practical possibilities to reduce the energy consumption in buildings
involved the need of accurate methods to calculate the yearly heating and
cooling demands in the rooms of a building. For these calculations we
developed a computer program, which demands sufficient low computer time for
practical applications. The basic computer model makes it possible to apply
the program for a wide range of wall constructions and fenestrians.
The convection and radiation heat transfer in the room are considered separately. The heat transfer in glazing and sunshading is simulated by a resistance network which is provided with heat sources according to the absorbed
solar radiation flows. The operation of the sunshading equipment can be controlled on the incident solar radiation. The operation of the lighting is
controlled on the required illumination in the room.
A procedure is developed to break off the response functions in dependence on
the values of the response function and the exitation function, the restterms
taken into account.
Calculation results will be given to compare the applied response factor
method with a hybrid method and with a basic implicit differential method
used for the calculation of the response factors.

2 Description and formulation of the heat transfer processes in rooms under non-steady conditions

Detailed descriptions of the heat transfer processes are given in literature.
We refer to [1] and confuse ourselves here to the formulation of the
equations to be solved.

2.1 Heat transfer by conduction

Non-steady heat conduction in a homogeneous medium is described by the partial differential equation of Fourier. When $T(x,t)$ is the temperature at place x and time t then:

$$\lambda \frac{\partial^2 T}{\partial x^2} = \rho.c. \frac{\partial T}{\partial t} \tag{1}$$

where λ is the thermal conductivity and $\rho.c$ the volumetric heat capacity. The boundary and initial conditions are:

$$T(x,0) = f(x) \tag{2}$$

$$a_0 \left. \frac{\partial T}{\partial x} \right|_{x=0} + a_1.T(x_0,t) = g(t) \tag{3}$$

where a_0 and a_1 are constants.

The heat flux by conduction in the material is given by:

$$\phi_c = \lambda \frac{\partial T}{\partial x} \tag{4}$$

and the gradient of the accumulation flux by:

$$\frac{\partial \phi_c}{\partial x} = \rho.c \frac{\partial T}{\partial t} \tag{5}$$

When the medium is not homogeneous but consists of different layers, then eq (1) is valid for each layer and additional conditions occur on account of contineous heat flow at the internal boundaries. For the numerical evaluation of the partial differential equations of type eq (1) in case of composite wall constructions is chosen for an implicit method with place and time dis- cretisation on account of its total stability. A homogeneous layer is suppo- sed to be devided in m segments. Let the temperature in segment k at time $n.\Delta t$ be $T_{k,n}$ then:

$$\frac{\partial T}{\partial t} = \frac{T_{k,n} - T_{k,n-1}}{\Delta t} \tag{6}$$

$$\frac{\partial^2 T}{\partial x^2} = \frac{T_{k-1,n} - 2T_{k,n} + T_{k+1,n}}{(\Delta x)^2} \tag{7}$$

Substitution of eqs. (6) and (7) into (1) gives the following implicit difference equation:

$$\frac{\lambda}{\Delta x} T_{k-1,n} - (\frac{2\lambda}{\Delta x} + \frac{\rho.c.\Delta x}{\Delta t}) T_{k,n} + \frac{\lambda}{\Delta x} T_{k+1,n} =$$

$$\frac{\rho.c.\Delta x}{\Delta t} T_{k,n-1} \tag{8}$$

for k=1,2,3,....m, with $T_{0,n}$ and $T_{m+1,n}$ as boundary conditions, and with $T_{0,0}$, $T_{1,0}$, $T_{2,0}$...$T_{m,0}$ as initial conditions.

With eq (6) it is possible to set up a computation model (fig. 1) where $R_1=R_2= \frac{\Delta x_1}{F.\lambda_1}$ and $R_3 = \frac{\Delta t}{\rho.c.F.\Delta x_1}$ with F being the area of the segment.

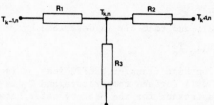

<u>Figure 1</u> - Base scheme for the used heat conduction difference equation.

Eq (8) may also be written as a recurrent matrix equation:
$$A \underline{z}(n) + B \underline{u}(n) = C \underline{z}(n-1) \qquad (9)$$

where $\underline{z}(n)$ is a vector with temperatures $T_{1,n}; T_{2,n}; \ldots; T_{m,n}$

$\underline{u}(n)$ is a vector with boundary conditions $T_{0,n}$ and $T_{m+1,n}$

A, B and C are coefficient matrices of equation (8)

$\underline{z}(0)$ is the initial condition.

The solution of eq (9) is obtained by applying matrix inversion:
$$\underline{z}(n) = A^{-1}\left[C\, \underline{z}(n-1) - B\, \underline{u}(n)\right] \qquad (10)$$

By repeated substitution eq (10) passes to:
$$\underline{z}(n) = A^{-1}\, C\, \underline{z}(0) - \sum_{1=1}^{n} (A^{-1}\, C)^{1-1}\, A^{-1}\, \underline{u}(1) \qquad (11)$$

The solution $\underline{z}(n)$ only depends on the initial condition $\underline{z}(0)$ and the boundary conditions $\underline{u}(\overline{1})$.

2.2 Radiation heat transfer

The radiation heat transfer between two grey surfaces is described by the Stefan Boltzmann law:
$$\emptyset_{r_{12}} = 4.\sigma.F_{\varepsilon}.A_1.F_{12}\,(T_1^{\,4} - T_2^{\,4}) \qquad (12)$$

where $\emptyset_{r_{12}}$ is the radiation heat flux from surface 1 to surface 2, σ is the Stefan Boltzmann constant, F_ε is an emissivity factor dependent on emissivities ε_1 and ε_2 of both surfaces and on the geometrical arrangement, A_1 is the area of surface 1, F_{12} is the configuration or geometric factor defined as the radiation fraction leaving surface 1 which falls on surface 2, T_1 and T_2 are the absolute temperatures of surfaces 1 and 2.
If $(T_1-T_2) = \Delta T$ is small and $\overline{T} = (T_1+T_2)/2$, eq (6) can be approximated by:
$$\emptyset_{r_{12}} = 4.\sigma.F_{\varepsilon}.A_1.F_{12}.\overline{T}^3.\Delta T \qquad (13)$$

The heat resistance for radiation R_r for two surfaces 1 and 2 is then:
$$R_r = \frac{T}{\emptyset_{r_{12}}} = \frac{1}{4.\sigma.F_{\varepsilon}.A_1.F_{12}.\overline{T}^3} \qquad (14)$$

In most cases of radiation transfer in rooms the temperature range is sufficiently small. By approximation R_r will be constant. Moreover for most room wall surfaces $F_{\varepsilon} \approx 1$, so that then $R_r = \text{constant}/A_1.F_{12}$.

The geometric factors F for perpendicular and parellel planes between the wall surfaces of a room is calculated applying well-known formulaes. Radiation at the blades of venetian blinds is a special case. For both heat radiation and the reflection, absorption and transmission of solar radiation several geometric factors were to be calculated [2].

2.3 Convection heat transfer

The heat flow between wall and room air is described by:

$$\emptyset_c = \alpha_c \cdot A \cdot (T_w - T_a) \tag{15}$$

where α_c is the convection heat transfer coefficient, A is the area of the wall surface, T_w is the wall surface temperature and T_a is the room air temperature. The heat resistance for the heat transfer between the wall surfaces and the room air may be presented by:

$$R_c = \frac{1}{\alpha_c \cdot A} \tag{16}$$

2.4 Ventilation heat transfer

The heat flow by ventilation can be described as:

$$\emptyset_v = \rho \cdot c \cdot \dot{m}_v \cdot \Delta T \quad , \quad \Delta T = T_{a1} - T_{a2} \tag{17}$$

where $\dot{m}_v$ is the volume flux of the air, T_{a1} is the temperature of the air supply and T_{a2} is the temperature of the return air.

3 The room model

The applied room model is based on the following simplifying assumptions:
a) All walls are homogeneous or build up by a few homogeneous layers.
b) The heat transfer in the wall is considered in one direction, namely perpendicular to the wall.
c) The glazing and the shadings are considered as a separation wall without heat capacity. The heat transfer through these walls is described with resistances for convection and radiation.
d) The air in the room has one uniform temperature.
e) The time is discretizised in steps of one hour.
If the model is schematized in this way it is possible to compose a calculation model based on the described equations. The computation model of the applied room is schematically shown in figure 2, page 383. The resistances drawn in this figure represent the convection resistances between the surfaces and the air temperatures. The radiation resistances have been left out for clearness' sake. The blocks represent the capacitive walls and layers. The room model can simulate various room configurations. The number of outer walls may be one up to four. Each wall may consist of four layers (fig. 3, page 383). The ceiling may be a part of a ceiling floor construction or a part of a roof construction, both construction types consisting of six layers. Each outer wall may have a non-transparent and a transparent part.
At the glazing and the sunshading devices the direct solar radiation and the diffuse radiation fluxes are partly reflected, absorbed and transmitted. An example of one of the possible glazing and sunshading systems is shown in figure 4, page 383). To calculate the various absorbed fractions, the properties of the glass planes and sunshading elements, the basic calculation method is taken from Parmelee [3,4].
Besides the properties of the separate glass planes etc. also the absorption, the transmission and the reflection factors of the combined glazing and sunshading systems are calculated. The direct solar radiation, the diffuse sky radiation and the diffuse radiation from the ground are considered separately. The solar radiation data are derived from meteo data tapes of the Royal Dutch Meteorological Institute (K.N.M.I). These hourly data, which are related to the radiation on a horizontal plane, were converted to the values corresponding to vertical planes in dependence on the latitude [5].

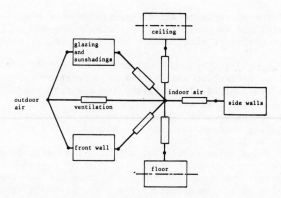

Figure 2 - The thermal model of the unit room (radiation resistances are omitted for clearness' sake)

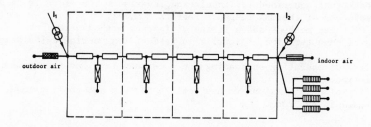

Figure 3 - The computation scheme for the front wall

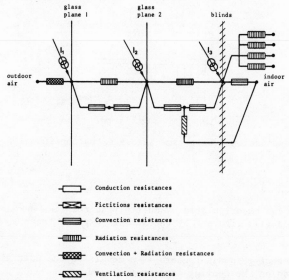

Figure 4 - The computation scheme for double glass with indoor blinds

383

The calculation of the absorptance and transmittance have been carried out for
the following glazing and sunshading systems: single and double glass, without
blinds, with indoor blinds or with outdoor blinds and double glass with blinds
in between

Glazings with sun absorbing planes are calculated in another way. For each
plane the absorption, reflection and transmission fraction for a mean incidental angle of 45^0 is applied. If necessary the quantities are determined
experimentally with a spectrometer.

Glazing with a reflecting coating, which are as a rule slightly tinted, are
calculated while taking into account the dependence of the reflection and
transmission on the angle of incidence.

The part of the incident solar radiation, which is transmitted through the
glazing and sunshading system which enters the room is spread out over the
various wall surfaces. The values of the absorptance and the transmittance
of the different glazing and sunshading systems are given in appendix I,
figure 5, page 390.

4 The applied response factor method

The room model just described is lineair and invariant. Therefore it is appropriate to make use of thermal response factors $\lfloor 6,7 \rfloor$.

Advantages of these methods above finite difference methods are:

a) decrease of computer time, specially in case of long period calculations
   (a year e.g.).
b) possibility of operating sunshadings on a certain level of the incident
   sun radiation.
c) possibility of using maximum values for the heating and cooling loads
   (capacities) in the room.

4.1 Principle of the method

The response of the room model to a unit time-series exitation function
$(1,0,0,0 \ldots)$ is called the unit response function. The time-series representation of this response function is the set of response factors (fig. 6).

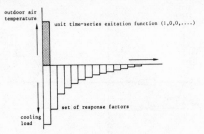

Figure 6 - The response on a unit time-series exitation function

$$R = \sum_{j=0}^{\infty} r_j \qquad\qquad (18)$$

The cooling load can be calculated by multiplying the time-series of the
exitation function (I), figure 7, by the set of response factors, R,

$$CL = R.I \qquad\qquad (19)$$

where R is the set of response factors as a result of the unit time-series
exitation function on the cooling load in the room.

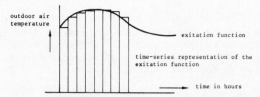

Figure 7 - The exitation function

The general form of the cooling load as a result of the different time-series exitation functions is:

$$cl_n = \sum_{j=0}^{\infty} r_{k,j} * i_{k,n-j} \tag{20}$$

where $k = 1,2,\ldots,1$ is the number of a exitation function and the set of response factors belonging to it, such as outdoor air temperature, sun radiation, etc.

To calculate room air temperatures in partly conditioned rooms it is necessary to calculate the set of response factors ($\sum_{j=0}^{\infty} r_{a,j}$) of the indoor air temperature on the cooling load in the room.
Then the cooling load q_n is:

$$q_n = cl_n + \sum_{j=0}^{\infty} r_{a,j} * i_{a,n-j} \tag{21}$$

The room air temperature at time n is:

$$i_{a,n} = (q_n - cl_n - \sum_{j=1}^{\infty} r_{a,j} * i_{a,n-j})/r_{a,0} \tag{22}$$

4.2 The introduced restterm function

The unit response function calculations are hourly carried out for 144 steps. The rest of the function is approximated by an e-function $a.e^{-bt}$. The constants a and b can be calculated by taking two points in the set of response factors, e.g. 120 and 144 (figure 8).

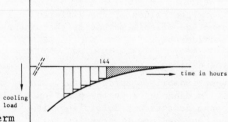

Figure 8 - The restterm

A total set of response factors can be represented by:

$$R = \sum_{j=0}^{144} r_{k,j} + \int_{145}^{\infty} a.e^{-bt}.dt \tag{23}$$

This integration is made for every set of response factors.

4.3 Weight factors

To reduce the length of the sets of response factors there are introduced
weight factors. They are related to the product of the maximum values of the
exitation functions and the restterms of the sets of response factors e.g.:
maximum value of sunradiation on the roof 1000 W/m^2, with a set of response
factors:

$$R_1 = \sum_{j=0}^{144} r_{1,j} + \int_{145}^{\infty} a_1 e^{-b_1 t} dt.$$

maximum value of sunradiation on the front wall 800 W/m^2, with a set of
response factors:

$$R_2 = \sum_{j=0}^{144} r_{2,j} + \int_{145}^{\infty} a_2 e^{-b_2 t} dt.$$

The products then are:

$$\int_{145}^{\infty} a_1 e^{-b_1 t} dt \times 1000 = A_{144} \quad \text{and} \quad \int_{145}^{\infty} a_2 e^{-b_2 t} dt \times 800 = B_{144}$$

For the biggest sum is always taken into account 144 steps (in this case e.g.
A_{144}).

Now is calculated a number m so, that:

$$(\sum_{m}^{144} r_{2,j} + \int_{145}^{\infty} a_2 e^{-b_2 t} dt) \times 800 = B_m \simeq A_{144}$$

The error which has been made by cut-off the set of response factors is
weighted to the magnitude of the different exitation functions.
The cooling load (cl) resulting from the radiation falling on the front wall
(i_2) at time n is:

$$cl_n = \sum_{j=0}^{m} r_{2,j} \times i_{2,n-j} + B_m \times (\sum_{k=0}^{24} i_{2,n-k-m})/24 \tag{24}$$

where $(\sum_{k=0}^{24} i_{2,n-k-m})/24$ is the average of the sun radiation in a day m steps

back.
Every 24 hours the new average values of the different exitation functions
are calculated. The advantage of this method is that the number of calculations is decreased considerably.

4.4 Operating sunshading devices

In case of operating venetian blinds up/down, two sets of response factors
are calculated (R_u, R_d). In dependence on the position of the blinds the
different sets of response factors are multiplied with "1" or "0" (s or $\bar{s}$).
The general term in the cooling load time-series is:

$$cl_n = \sum_{j=0}^{\infty} r_{u,j} \times i_{u,n-j} \times s_j + \sum_{j=0}^{\infty} r_{d,j} \times i_{d,n-j} \times \bar{s}_j \tag{25}$$

5 Possibilities of the program

The program is not only applicable for 'long period' calculations but also
for design calculations. Design outdoor air temperature, solar radiation data,
heat loads of lighting and occupants with separate radiation and convection
fractions can be stated. The program calculates hourly the room air temperatures and the heating and cooling demands for the conditioned periods of
the day.

In case of 'long period' calculations, meteo data tapes of the K.N.M.I (De
Bilt, The Netherlands) are applied. Then it is possible to calculate:
i) The heating and cooling demands in a room summarised for each week and a
    whole year. As an example a plotted graph is shown in appendix II
    (fig. 9, page 388).
ii) The number of hours during which the air temperature exceeds certain
    temperature levels. The maximum room air temperatures in a summer of the
    period under consideration are also been presented (fig. 10, page 388).
Further possibilities of the program are:
a) Applying a maximum capacity of the heating and cooling units.
b) The lighting can be operated in dependence on the illumination on a certain
   place in the room. This illumination is hourly calculated as a function of
   the sun radiation flux, the lighttransmittance of glazing and sunshading
   and the glass area.
c) Operating the sunshading in dependence on the outer solar radiation flux.
d) Natural ventilation flow can be stated as a constant value or as a lineair
   function of windvelocity $v.(c_1 + c_2 * v)$.
e) Mechanical ventilation air flow during a part of the day and with a certain
   temperature difference in respect to the outdoor air or with a constant
   temperature value can be stated.

6 Comparison with other programs

Results of the program with the applied response factor method are compared
with two other programs.
i) The program based on the finite difference method, by means of which the
    response factors are determined, is used to check the applied restterm
    approximation. We found differences in temperatures and heating and coo-
    ling demands less than 1%.
ii) A hybrid method is used for comparing the absolute values of the room air
    temperatures and the heating and cooling loads. With this method the
    equations representing the heat transfer in the room are solved with an
    electric analog model [8]. A digital computer arranges the in- and
    output of this model. Said method was earlier checked by means of expe-
    rimental results. Comparing calculations were carried out for several
    room configurations. The maximum difference in the yearly heating or
    cooling demands was about 6%. A result is given in figure 11, page 388.

7 Computing time

The computing time for a heating and cooling demand year calculation in case
of operated sunshading and lighting is about 275 sec. on a CDC cyber Cyber 72.

8 Further developments

These are:
i) Extending the program in order to calculate the heating and cooling de-
    mands for the different zones of a building.
ii) Coupling this extended program with programs for air conditioning instal-
    lations in order to calculate real total energy consumptions.

Acknowledgement

The prescribed program is developed in cooperation with Installatie Techniek
Bredero B.V. (Utrecht), Rijksgebouwendienst (The Hague), Van Swaay Installa-
ties B.V. (Zoetermeer) and Advies- en Constructiebureau TEBODIN B.V. (The
Hague), which sponsored the project.

THE TEMPERATURE LEVEL EXCEEDINGS BETWEEN 7 O'CLOCK AND 17 O'CLOCK IN A WORKWEEK OF 5 DAYS.
STARTING DATE: APRIL 28, 1977.

WEEKNUMBER	TEMPERATURE LEVEL EXCEEDINGS IN HOURS 24°C	28°C	MAX. TEMPERATURE °C	LIGHTING ON HOURS
5	0	0	23.8	40
6	3	0	25.9	33
7	9	5	28.6	30
8	15	2	28,3	36
9	31	9	30.9	42
10	14	5	29.7	42
11	34	15	31.0	34
12	12	0	26.8	38
13	19	4	30.7	34
14	3	0	25.2	41
15	2	0	25.0	42
16	41	33	31.5	36
17	18	4	29.1	43
18	26	0	27.6	38
19	19	4	29.4	38
20	5	0	24.9	40
21	9	4	30.2	35
22	24	18	32.6	29
23	11	0	27.2	30
24	4	0	25.5	34
25	8	1	28.1	37
26	0	0	21.5	31
total 22 weeks	307	104		803

Figure 10 - The maximum room air temperatures and the number of hours during which the air temperature exceeds certain temperature levels for a certain room in the summer of 1964.

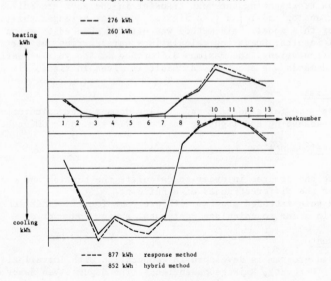

Four weekly heating/cooling demand calculation 1964/1965

Figure 11 - Comparing the response method with a hybrid method for a certain room for heating and cooling demands 1964-1965.

388

REFERENCES

[1] Proc. First Symposium on the Use of Computers for Environmental Engineering related to Buildings, Washington 1970.

[2] Oegema, S.W.T.M.; Euser, P, An accurate computing model for the analysis of the non-steady thermal behaviour of office buildings, published in [1].

[3] Parmelee, G.V., Aubele, W.W., Trans. ASHVE 58,377 (1952).

[4] Parmelee, G.V., Aubele, W.W., Huebscher, R.G., Trans. ASHVE 54, 165 (1948).

[5] Hoogendoorn, C.J., Den Ouden, C., Calculations based on hourly meteorological data of the possibility of using solar energy for heating of buildings in The Netherlands. Proc. International solar energy congress and exposition 1975, section 40/5.

[6] Stephenson, D.G., Mitalas, G.P., Cooling load calculations by thermal response factor method. ASHRAE Trans. 1967, no. 2018, p. III 1.1-1.7.

[7] Kusuda, T., Thermal response factors for multi-layer structures of various heat conduction systems. ASHRAE Trans. 1969, no. 2108, p. 246-270.

[8] Euser, P., Studies over energieverbruik en energiebesparing bij ruimteverwarming TNO-project 1 (1973) p. 298-302.

SYSTEM		A_I	A_{II}	A_{III}	D	a_1	a_2	a_3	a_4	a_5	a_6	a_7	a_8	a_9	
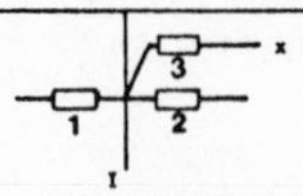 outside inside	SINGLE CLEAR GLASS	0,12	----	----	0,77	23	3	6	----	----	---	---	----	---	
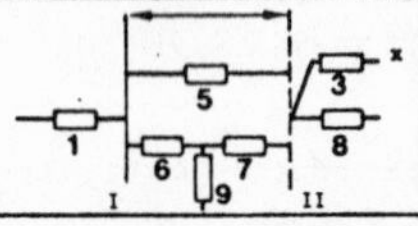	SINGLE CLEAR GLASS WITH INDOOR BLINDS	0,17	0,35	----	0,15	23	----	6	----	6	5	7	7	32	
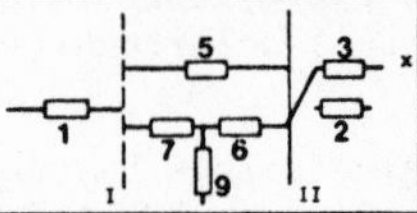	SINGLE CLEAR GLASS WITH OUTDOOR BLINDS	0,38	0,03	----	0,13	23	3	6	----	6	8	12	----	200	
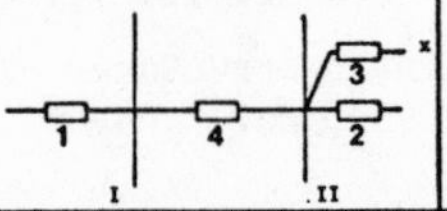	DOUBLE CLEAR GLASS	0,13	0,10	----	0,62	23	3	6	7	----	---	----	----	---	
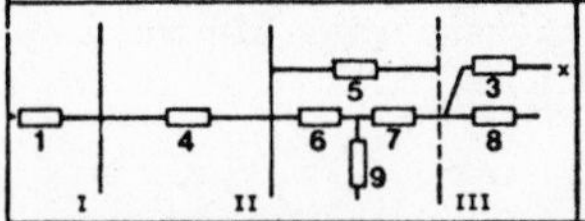	DOUBLE CLEAR GLASS WITH INDOOR BLINDS	0,16	0,13	0,29	0,13	23	----	6	7	6	5	7	7	32	
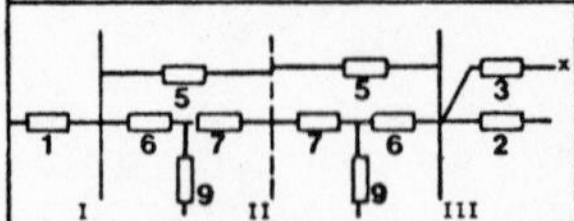	DOUBLE CLEAR GLASS WITH BLINDS IN BETWEEN	0,16	0,37	0,02	0,12	23	3	6	----	6	8	10	---	0	
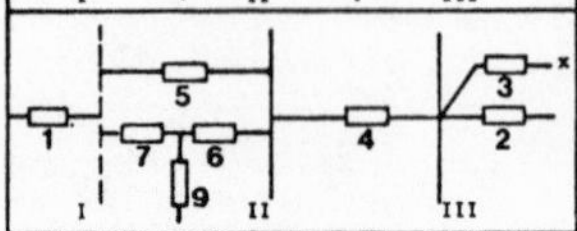	DOUBLE CLEAR GLASS WITH OUTDOOR BLINDS	0,38	0,03	0,01	0,11	23	3	6	7	6	8	12	---	200	
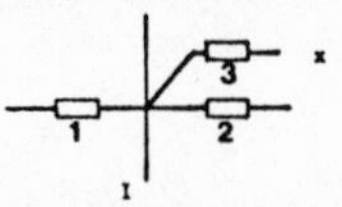	SINGLE ABSORBING GLASS	0,50	----	----	0,42	23	3	6	----	---	---	---	----	---	
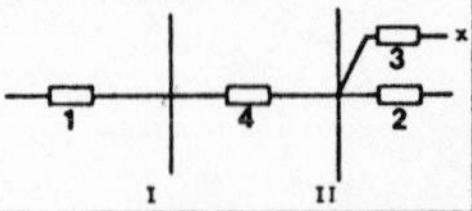	CLEAR GLASS INDOOR SIDE ABSORBING GLASS OUTDOOR SIDE	0,53	0,05	----	0,34	23	3	6	7	---	----	---	----	---	
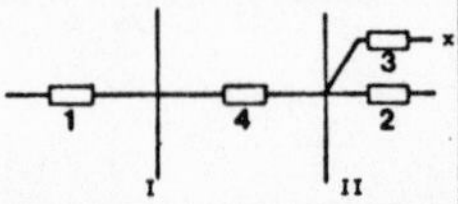	REFLECTING DOUBLE GLASS STOPRAY 38/28	0,37	0,03	----	0,24	23	3	6	3	---	----	---	----	---	

Slat angle 45⁰

A = Absorptance   a = Heat transfer coefficients

D = Transmittance   x = To each of the inside wall surfaces

ventilation heat transfer coefficients are based on real practical conditions as to air velocity and air flow

Figure 5 - The properties of glazing and sunshading systems

390

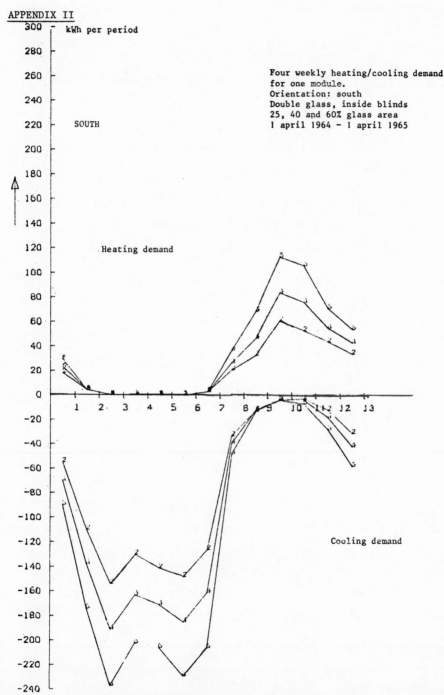

Figure 9 - The heating and cooling demands in a room with three percentages of glass area in the front wall for hourly weather data of the year 1964-1965 (De Bilt, The Netherlands)

FUNCTIONAL MODELLING OF ROOM TEMPERATURE RESPONSES AND THE EFFECTS OF THERMOSTAT CHARACTERISTICS ON ROOM TEMPERATURE CONTROL

K. M. LETHERMAN

Department of Building, University of Manchester
Institute of Science and Technology
Manchester M60 1QD, England

1.   SUMMARY

A digital simulation was performed on a number of simple rooms to compute the response of the air temperature and the surface temperature to a step input of heat. The step responses were thus obtained in numerical form. A suitable form of Laplace transfer function was obtained empirically to give an accurate functional representation of the responses. The numerical and functional representations of the responses are compared.

The control characteristics of an on-off room thermostat are then combined with each room response function and the controlled cycling characteristics of the combination of thermostat and room are computed. The effects of the sensing properties of the thermostat and of the dynamic responses of the thermostat and heating plant are examined and compared.

NOMENCLATURE

C	coefficient	(^{o}C)
D	pure time delay	(sec)
K	thermal diffusivity	(m^2/sec)
L	section thickness	(m)
s	Laplace operator	(sec^{-1})
t	time	(sec)
T	room time constant	(sec)
TT	thermostat time constant	(sec)
TP	plant time constant	(sec)
Y	convective response factor	(-)
θ	temperature	(^{o}C)
W	angular frequency	(radian/sec)

Subscripts

A	air
P	plant
S	surface
T	thermostat
X	environment

393

2.1 Heat transfer within the room

Conventionally, it is assumed that the surface temperature of each wall of a
room is uniform over its whole area and that the air temperature is uniform
over the volume of the room. Similarly, a constant air velocity is assumed to
exist over each surface so that a constant coefficient of heat transfer by
convection can be taken. Also, by a well-known approximation a constant value
of the coefficient of heat transfer by radiation can be taken, provided that
the temperature difference between surfaces is not too large.

With these assumptions it is possible to assign constant thermal resistances
to the various heat-flow paths: between one wall and another, and between a
wall and room air. For an ordinary six-sided room, this results in 15 radiation resistances between the surfaces and 6 convection resistances between air
and surfaces: 21 heat flow paths in all. By a technique originally due to
Danter (1974) this network can be replaced by an approximately equivalent 'star'
network of only 7 paths. The error involved is almost always less than 5% and
is usually less than 1%. This simplified network was originally developed for
steady state conditions, but it has also been used by Bloomfield (1975) for
intermittent heating studies, and by Loudon (1968) for solar overheating
studies. Because of its simplicity and relative accuracy Danters network was
used in these studies.

2.2 Heat transfer within the structure

There are two main problems associated with modelling of wall thermal characteristics and response:

(1) the fact that corners and edges cause the lines of heat flow in the steady
    state to be non-parallel, and

(2) the fact that there is thermal capacity as well as thermal resistance
    distributed throughout the thickness of the fabric. The diffusion of heat
    through the solid structure causes great complexity in any mathematical
    solution for the response.

An approximate solution to the first difficulty is available from the work of
Herpol (1972), who states that, for a cube with homogeneous walls of internal
side A and wall thickness H, the formulae for plane walls can be used without
significant error, if the dimensions are taken as $(A + H/2)$.

The general area of the second problem, that of prediction of the temperatures
produced by conduction heat flow, is one to which finite-difference methods of
computation have been applied for a long time. The procedure involves 'lumping'
of the conduction path, i.e. rearranging the capacitance in the wall so that,
instead of being continuously distributed, it is collected into a number of
discrete sections. The partial differential equation which applies to the
distributed system is then replaced by a set of ordinary differential equations
which can be solved relatively easily. The question of the relationship
between modelling accuracy and the size of the discrete sections has been
investigated by Paschkis and Heisler (1944, 1946) and by Letherman (1977).
From the latter study it can be concluded that the value of the dimensionless
number WL^2/K should be 3.0 or less.

The section thickness for a thermostatic cycle period of 1000 seconds is thus
about 10 mm. The magnitude error involved in using discrete sections will
then be not more than 3% and the phase error not more than 5°.

3. Functional modelling of room step responses

The methods outlined above were applied to the modelling of a variety of simple
rooms. For simplicity the rooms were taken to be windowless cubes with homogeneous slab walls, roof and floor. The exposure of all surfaces was taken to
be similar so that the external surface resistance was the same for all
surfaces. The modelling network was thus reduced to the simplest possible case,
as shown in Fig. 1. A digital computer program (Electric Circuit Analysis
Program) was available at the University of Manchester Regional Computer Centre
for calculating the response of networks and this was used to find the step
responses of the room models.

The mode of heat supply was purely convective, and a constant air change rate
of 2.0 air changes per hour was used. The materials used and their thermal
properties are listed in Table 1. The internal dimensions of the rooms were
3 x 3 x 3 m. The wall thicknesses varied from 5 mm to 250 mm.

MATERIAL	THERMAL CONDUCTIVITY W/mK	SPECIFIC HEAT CAPACITY J/kgK	DENSITY kg/m^3	THERMAL DIFFUSIVITY m^2/s
Dense Concrete	1.4	840	2100	7.94×10^{-7}
Brick	0.84	800	1700	6.18×10^{-7}
Lightweight Concrete	0.19	1000	600	3.17×10^{-7}
Foamed Plastic	0.033	1380	25	9.57×10^{-7}
Air		1019	1.19	-

TABLE 1 - Thermal Properties

The responses, since they were obtained numerically, exist as individual
values of temperature at given time values: they are points, not continuous
curves. It is then necessary to express these responses by a set of empirical
equations. These will give continuous functions, and if the functions are
chosen to be of a certain form, then Laplace transform techniques will allow
these original basic responses to be easily modified to incorporate the effect
of adding a time constant in the heating system and another in the thermostat.

A function was adopted which takes the sum of three simple exponential terms.
As a function of time this is

$$\theta A(t) = \sum_{n=1}^{3} Cn \left[1 - \exp(-t/Tn) \right] \qquad (1)$$

and in Laplace transformed form

$$\theta A(s) = 1/s \sum_{n=1}^{3} Cn/(1+sTn) \qquad (2)$$

where the 1/s denotes a unit step forcing function and the Cn, Tn are to be
determined. These constants were found numerically, by taking six pairs of
values of θA and t for each response, and substituting into six simultaneous
equations of a form similar to (1) above. The set of equations were then
solved iteratively by a method described by Letherman (1976). Once the coefficients in the modelling functions had been determined, the computed step
responses of the rooms were now represented by continuous functions. Figures 2
and 3 show comparisons of the calculated response values and the equivalent

modelling functions for the air and surface temperature responses respectively. It can be seen from Figures 2 and 3 that good agreement between the individual computed values and the continuous functions was obtained over a wide range of time.

4. Thermostat response characteristics

4.1 Introduction

One of the functions of a room thermostat is to act as a sensor of room temperature. It is well known that sensors can be designed to be particularly responsive either to the air temperature or to the radiant temperature of their surroundings. The same thing must apply to room thermostats and, in order to predict their control characteristics, it is necessary to know to what extent they are sensitive to radiation. Similarly, as far as dynamic response is concerned, the sensor output should ideally be able to respond immediately to any change in surrounding conditions. Because the thermostat has other functions than simply temperature sensing, it is not possible to provide ideal design for high speed of response.

4.2 Steady-state characteristics

The point must first be made that there is hardly ever such a thing as a single unique 'room temperature'. The heat flowing within a room volume and through the walls causes, even in the steady state, differences in air temperature from point to point throughout the room. Also, because of convective heat transfer coefficients, the wall surface temperatures will generally be below the adjacent air temperature (for convective heating). This may be so not only for an external wall, but also for an internal wall if the temperature on the other side is lower than on the 'room' side, and under transient conditions also. The usual assumptions are that the air temperature is uniform throughout the room volume, and that the temperature of each surface is uniform over its whole area. On these assumptions, the steady temperature θT reached by any non-generating body within the room will be some linear combination of air and surface temperatures:

$$\theta T = Y . \quad \theta A + (1 - Y) . \quad \theta S \qquad (0 < Y < 1) \qquad (3)$$

where Y is dependent on such factors as the emissivity properties of the body and walls, the velocity of air over the body, and (strictly) on the position of the body within the room. This is so since if different walls have differing surface temperatures the radiant temperature is a function of position within the room. For the simple rooms considered in this study, all six surfaces are at the same temperature and are assumed to be of constant high emissivity. The temperature reached by a sensor within the casing of a thermostat also depends upon whether the casing is enclosed or ventilated. For completely enclosed thermostats the sensor 'sees' only the inner surface of the case. For ventilated thermostats the sensor is also influenced by the temperature of air passing through the case.

This question has been investigated by Baxter and Longworth (1974), who tested a range of commercially available room thermostats. In a series of tests, Y was found to range from 0.59 to 0.91. Their results may be divided into two groups:

(a) those relating to thermostats with pierced casing, giving a 'ventilated' effect, and

(b) those relating to thermostats with cases which are not pierced, giving an 'enclosed' effect.

For these two groups the parameter Y was generally about 0.85 to 0.91 and 0.60 to 0.70 respectively , with some few results falling between these ranges. Approximate mean values can be taken as 0.9 and 0.67. It thus appears that a thermostat can conveniently be classified either as 'ventilated', so that nine-tenths of the response is contributed by the air dry-bulb temperature, giving

$$\theta T = 0.9 \ \theta A \ + \ 0.1 \ \ \theta S \tag{4}$$

or as 'enclosed', with two-thirds of the response due to air temperature:

$$\theta T = 0.67 \ \theta A \ + \ 0.33 \ \ \theta S \tag{5}$$

The steady state temperatures reached by the two types of thermostat can be illustrated by dividing up the air-surface temperature difference into thirty units:

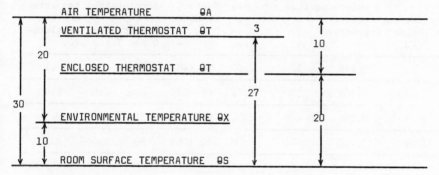

4.3 Accelerator heaters

One of the major factors in the dynamics of thermostatic switching control is the time-lag existing within the room because of the physical distance between the heater and the sensor. This can result in oscillations in controlled temperature which are of large amplitude and low frequency. In order to reduce this lag effect, it is usual to incorporate a small heater within the casing of the thermostat, called the 'accelerator'. This heater is switched on and off with the main room heater, being operated by the same set of contacts. The effect is to introduce a 'minor loop' of feedback within the main control loop. This extra negative feedback is a common feature of practical controllers. It is exemplified by the motor feedback potentiometer in electrical proportional control and by the feedback bellows or diaphragm in pneumatic controllers.

The accelerator heaters used in room thermostats may be arranged in two altern-ative ways; either as very low-value resistors (approximately one ohm) in series with the load, or as rather high value resistors (about 0.1 M ohm) in parallel with the load. The series type are usually adjustable to provide allowance for various load currents. The parallel arrangement is probably more reliable and is more common than the series arrangement. For a given supply voltage it needs no adjustment for load current.

4.4 Dynamic characteristics

Dynamic response measurements were carried out on a number of commercially available room thermostats to determine the characteristics of their dynamic

responses. Each thermostat was mounted in a cabinet similar to that specified
in British Standard Specification No. 3955 and the responses to three different
types of step input were recorded. The three types of input were:

1. change in surrounding air temperature

2. change in radiant temperature

3. rated voltage applied to the accelerator heater

The responses were recorded by a fine-wire thermocouple bonded to the temperature sensor of the thermostat. In each case it was found that the step
response could be represented fairly accurately by a pure time delay with a
simple lag term. The transfer function was thus of the form $\exp(-Ds)/(1+Ts)$.

The experimental results are presented in Table 2.

Thermo-stat type	Thermal resistance of casing $^{o}C/W$	Accelerator heater power W	Air tempera-ture response		Radiant response		Accelerator input response		Final temp.rise due to accelera-tor ^{o}C
			D sec	T sec	D sec	T sec	D sec	T sec	
A	11.9	0.178	30	1200	40	1400	60	900	2.1
B	17.4	0.109	35	700	40	820	35	590	1.9
C	19.0	0.311	40	1000	45	1150	35	920	5.9
D	11.4	0.185	40	900	50	1000	30	390	2.1

TABLE 2 - Thermostat responses

4.5 Step response of the complete system including thermostat and heating plant dynamics

Equation (2) of section 3 can now be modified to include thermostat and heating
plant dynamics, as follows.

The air temperature step response transform now becomes:

$$\theta A(s) = \frac{\exp(-Ds)}{s(1+TTs)(1+TPs)} \sum_{n=1}^{3} Cn/(1+sTn) \qquad (6)$$

multiplying out and taking one of the three similar terms gives:

$$F(s) = Cn/s \, (1+TTs) \, (1+TPs) \, (1+Tns) \qquad (7)$$

The inversion of this has been shown by Letherman (1976) to be:

$$f(t) = Cn \left[1 - \frac{TT^2\exp(-t/tt)}{(TT-TP)(TT-Tn)} - \frac{TP^2\exp(-t/TP)}{(TP-TT)(TP-Tn)} - \frac{Tn^2\exp(-t/Tn)}{(Tn-TT)(Tn-TP)} \right] \qquad (8)$$

and this function is one term in the total step response. There are six such
terms, three for the air temperature and three for the surface temperature.

5. Calculation of the cycling characteristics of the controlled system

There are two relatively well-known methods for calculating the response of a
non-linear controlled system. The describing function method (Gille et al.

1959) makes the assumption that the cycling oscillations are steady sinusoidal
and is therefore an approximate method. The method of J.S. Tsypkin (1955) does
not make this assumption and so is in principle more exact, but both methods
give information only about the cycle period and amplitude. A third method,
due to Lepage (1950) makes no assumptions about the waveshape, and is capable
of giving information about all aspects of the response and its build-up from
zero. This may be termed the step response method since it uses the sum of a
sequence of positive and negative step responses, each initiated at the switching instant of the thermostat contacts. Its graphical application is illustrated in Fig. 4, where q(t) represents the unit step response function and the
oscillations due to a thermostat with a total hysteresis h and a heater of
power M are shown. It is convenient to apply this method with the aid of a
digital computer program. The properties of the thermostat and of the controlled system, the room, can be specified independently, and hence their separate
effects can be examined.

6. Results and conclusions

In the combination of the room, heating plant and thermostat there are up to
two dozen separate variables, the effects of which could be investigated. A
multi-dimensional search could be performed for a combination of values which
give an optimum performance. However, this procedure would be very laborious
and expensive in computer time. It seems more efficient to proceed by examining the effect of variations in one parameter at a time, starting from some
central typical case. The case is defined as follows:

ROOM: dimensions 3 x 3 x 3 m, ventilation rate 2 air changes/hr.
                   All surfaces 100 mm solid slab lightweight concrete.
                   External temp. constant at 0°C.

HEATING PLANT: sized to give 20°C environmental temperature at maximum steady
                   output. 100% convective heating, plant time constant 600
                   seconds.

THERMOSTAT: convective response factor Y = 0.9, time constant = 1000 sec,
                   hysteresis 2°C, accelerator temperature rise 2ºC, accelerator
                   time constant 600 seconds, set point variable 1°C to 19°C.

The results for the range of set points 1-19°C are shown in Fig. 5 and are
generally typical of the behaviour of real rooms under thermostatic control:
for a set point at the centre of the control range the performance is optimal,
with zero offset and minimum values of period and amplitude of the controlled
temperature cycle. The period and amplitude are 2150 seconds and 2.4°C. When
the set point departs from its central value, the control behaviour becomes
worse, eventually approaching an infinite periodic time. This behaviour
compares well with that reported by Roots (1969).

6.1 The effect of the accelerator heater

If the accelerator heater effect is removed and the set point is maintained at
10°C, the period and amplitude of steady oscillations are now found to be 2770
sec and 3.71°C, which represent increases of 29% and 55% respectively.

6.2 The effect of the thermostat sensing characteristics

Returning to the original case but using a thermostat convective response
factor Y of 0.67 instead of 0.9 to model a thermostat with an enclosed case
instead of one with a ventilated case, this causes the cycling characteristics

to change to 2300 sec period and $2.20^{\circ}C$ amplitude. The slightly longer period
and smaller amplitude might be expected since the wall surface temperature is
less rapidly responsive than the air and oscillates over a narrower range.

The effect of thermostat hysteresis can easily be investigated, and the
responses were calculated for hysteresis values from $0.1^{\circ}C$ to $5.0^{\circ}C$. The
results are presented in Table 3.

Total Hysteresis $^{\circ}C$	Cycle Period sec	Cycle Amplitude $^{\circ}C$
0.1	150	0.075
0.2	270	0.169
0.5	960	0.49
1.0	1560	1.19
2.0	2150	2.4
5.0	3300	5.4

TABLE 3 - Effect of hysteresis band width

6.3 The effect of the thermal properties of the structure

The controlled cycling characteristics of the room specified as the original
case were computed for the four different wall materials given in Table 1. In
each case, the heat input was calculated to maintain a $20^{\circ}C$ rise in the
environmental temperature in the steady state. The set point was $10^{\circ}C$ throughout. The results are set out in Table 4.

These results are interesting in that the lighter materials might be expected
to give fast cycling and large amplitude of oscillation, whereas the effect
seems to be the opposite of this. The range of thermal properties of the
materials is about 50:1, but the cycling characteristics have a range of only
about 2:1.

Results were also obtained for a range of wall thicknesses of a single material.
These were 5 mm to 250 mm of lightweight concrete. These results are also
given in Table 4.

Wall material	Wall thickness mm	Heat input W	Cycle period sec	Cycle amplitude (p/p) $^{\circ}C$
Dense concrete	100	5340	1670	3.50
Brickwork	100	4520	1740	3.24
Lightweight concrete	100	3140	2150	2.40
Foamed plastic	100	740	3170	2.49
Lightweight concrete	5	6400	920	3.34
"	10	5680	1240	2.93
"	50	3650	1980	2.72
"	100	3140	2150	2.40
"	250	2610	2320	2.31

TABLE 4 - The effects of wall material and thickness

Each step in the table indicates approximately a doubling of thickness, so that the increment of thickness is increasing rapidly. As the wall thickness increases, the cycle period increases rapidly at first, but then the increase becomes much slower. Once the thickness has reached about 50 mm, further increases produce little change in the cycling.

6.4 General conclusions

The conclusions may be summarised as follows:

1. For typical rooms, the minimum amplitude of the controlled cycle occurs when the set point is at the centre of the control range, and is approximately equal to the hysteresis band of the thermostat.

2. For a given thermostat, dense structural materials tend to give shorter cycling period and smaller amplitudes than lightweight materials.

3. The controlled cycling characteristics are determined mainly by the inner-most 50-100 mm of slab walls.

References

1. Baxter A J and A L Longworth - The thermal response of room thermostats, Heating and Ventilating Engineer, Sept 1974 p 103-6.

2. Bloomfield D P - The effect of intermittent heating on surface temperatures, Building Science, 10 p 111-25, 1975.

3. British Standard Specification 3955, section 2F, 1967, Room thermostats.

4. Danter E - Heat exchanges in a room, Building Services Engineer, 41 p 232-45, 1974.

5. Gille J C et al. - Feedback control systems, McGraw Hill, New York, 1959.

6. Herpol G - Cursus van Aanwendung der Brandstoffen. 1st Deel Warmte-over-dracht. Cursus gedoceerd aan de Rijksuniversiteit te Gent. (Reference 1 of : E Tavernier - Characteristics for the thermal behaviour of rooms in summer, HVRA translation No. 229 Chal. Clim, May 1972 71 437 35-66).

7. Lepage C, CEMV (30) Paris 1950 (quoted in Gille 1959, p 394).

8. Letherman K M - A rational criterion for accuracy of modelling of heat conduction in plane walls. Building and Environment, 1977 (to be published).

9. Letherman K M - Thermostatic control of room temperature, Ph.D. thesis, Department of Building, University of Manchester Institute of Science and Technology, 1976.

10. Loudon A G - Summertime temperatures in buildings, Proc. IHVE/BRS symposium on thermal environment, London 1968.

11. Paschkis V and Heisler M P - Accuracy of measurements in lumped RC circuits, Elect. Eng. 63 p 165, 1944.

12. Paschkis V and Heisler M P - Accuracy of lumping in an electric circuit, J. Appl. Phys. 17, 4 p 246-54, 1946.

13. Roots W K - Fundamentals of temperature control, Academic Press, London 1969.

14. Tsypkin J S - Teorija relejnykh sistem automaticheskovo regulirovanija, Gostekhizdat, Moscow 1955 (quoted in Gille et al. p 394).

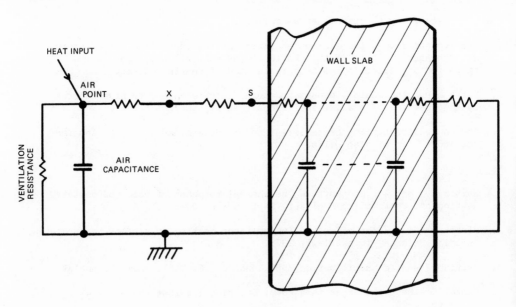

FIGURE 1: THE SIMULATION NETWORK

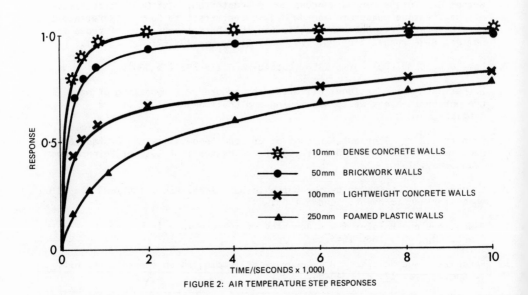

FIGURE 2: AIR TEMPERATURE STEP RESPONSES

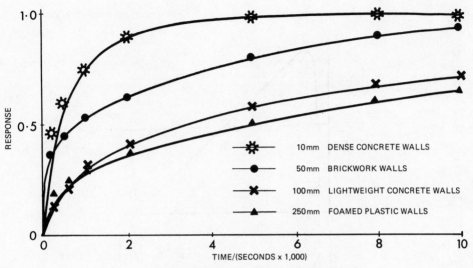

FIGURE 3: SURFACE TEMPERATURE STEP RESPONSES

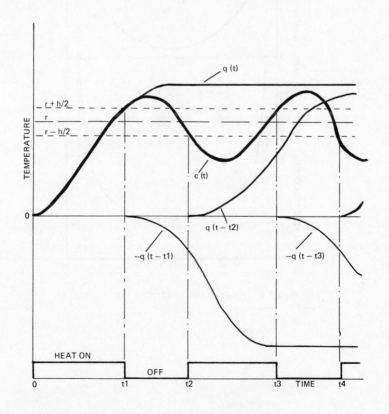

FIGURE 4: CALCULATION OF CONTROLLED CYCLE RESPONSE

403

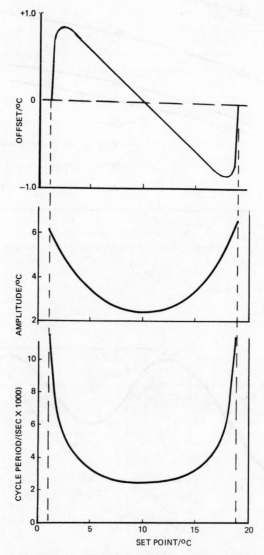

FIGURE 5: CYCLE CHARACTERISTICS AT VARIOUS SET POINTS

A METHOD FOR CALCULATING THE THERMAL LOAD ON AIR-CONDITIONED BUILDINGS

KAMAL-ELDIN HASSAN

Arab Development Institute
Tripoli, Libya

GALAL M. ZAKI

Al-Fateh University
Tripoli, Libya

ABSTRACT

A mathematical method to predict the thermal effect of a building component is presented. Dimensionless groups are used; they give generality to the solution obtained in the sense that it could be used for various building materials, localities,...etc. The method separates the effect of outside air temperature from that of solar radiation, a feature that adds to the flexibility and generality of a solution. It also gives, as the steady-state part of the solution, the average load on the air conditioning unit and, therefore, an indication of the total energy required per, say, day. Further, the solution gives general mathematical expressions that could be used directly for any special case that may be encountered, without need for re-solving the problem anew.

NOMENCLATURE

F   dimensionless function defined by Eq. 29

$\bar{H}$ transfer function

h   heat transfer coefficient ($W/ m^2 K$)

I   intensity of solar radiation (W/ m^2)

i $\sqrt{-1}$, imaginary unit

k   thermal conductivity ($W/ m K$)

L   wall thickness (m)

N   dimensionless number defined from Eq. 8

n   integer, 1, 2, 3, ...

P   real part of function F

p   dimensionless variable defined by Eq. 28

Q   imaginary part of function F

q   heat flux density (W/ m^2)

R   ratio defined from Eq. 8

r   modulus of function F

s   Laplace's variable

T   dimensionless time defined by Eq. 6

t   temperature (K)

w   water equivalent of room contents per unit wall area (J/ m^2 K)

X   dimensionless distance defined by Eq. 6

x   distance (m)

α   thermal diffusivity (m^2/ s)

γ   dimensionless variable defined by Eq. 37

ε   emissivity (dimensionless)

θ   reduced temperature defined by Eq. 7 (dimensionless)

τ   time (s)

∅   argument of function F

ψ   time lag (rad)

Superscripts

‾   Laplace transform

Subscripts

a   outside air   i   inside surface

amp amplitude   n   n th. term

c   cosine   o   outside surface

cy   cycle   r   room air

I   radiation   s   sine

INTRODUCTION

Buildings are essentially constructed as shelters from adverse weather conditions. Their thermal behavior should be, therefore, a very important factor in their design. An accurate method that would predict their behavior is essential for proper design. Indeed, there are a few methods in the literature for this prediction, both analytic and simulative [1 to 8] [1].

In the present work, the formulation follows closely that used earlier by one of the authors [4]. However, the formulation here is more general and, therefore, could be solved for either passive or positive control (or any required combination of both) of the building temperature. Further, the problem is stated, and solved, in terms of dimensionless groups. This renders the solution more general. Indeed, a single solution could be used for a broad range of conditions: materials, latitudes,...etc.

Further, in this approach, the effect of outside air temperature and of solar radiation are separated, and the solution gives their separate effects. Because the effect of the outside air temperature is the same for all orientations, this adds to the flexibility and generality of the solution. The problem treated here is that of controlled indoor temperature; i.e., for a refrigerated or air conditioned space.

The method of solution used here is purely mathematical. The dimensionless periodic driving functions are analysed to their harmonics, where steady-state terms also appear. The solutions of the steady-state parts are straightforward, and give the average load on the air conditioning unit; hence indicate directly the total energy needed per cycle (day). The solution for a single general harmonic is obtained by the "frequency response" method used by the authors in another kind of problem [9]. This general harmonic solution could be used directly for other special situations that may not fall under a general case.

With the effects of the steady-state terms, and of each harmonic determined, the final solution for a given general situation is obtained by simple superposition.

FORMULATION

The problem is formulated here in the same manner used before [4]. Conduction through a large wall is represented by the general equation of conduction in one dimension. For an isotropic wall as shown in Fig. 1, it is

$$\alpha \frac{\partial^2 t}{\partial x^2} = \frac{\partial t}{\partial \tau} \tag{1}$$

The condition of periodicity gives

---

[1] Numbers in brackets designate References at end of paper. Those cited here are not the only ones in the literature, some of them [1, 2, 8] list other pertinent works

$$t(x,\tau) = t(x, \tau + n\, \tau_{cy}) \tag{2}$$

where $n = 1, 2, 3, \ldots$, and $\tau_{cy} = 24$ h is the period. For the exposed wall surface

$$-k\left.\frac{\partial t}{\partial x}\right|_{x=0} = \epsilon I + h_o (t_a - t_o) \tag{3}$$

In a similar manner, for the inner surface of the wall

$$-k\left.\frac{\partial t}{\partial x}\right|_{x=L} = h_r (t_i - t_r) \tag{4}$$

The above relations constitute the boundary value problem, the solution of which gives the temperature distribution in the wall in space and time.

The heat convected to the room as given by Eq. 4 is used to change the temperature of the room contents, and to supply any other internal heat load $q = q(\tau)$, such as that of an air conditioning unit. Therefore

$$h_r (t_i - t_r) = w\frac{dt_r}{d\tau} + q \tag{5}$$

In this relation, the room temperature t_r is assumed constant throughout the room. The water equivalent w and thermal load q are both prorated to a unit surface area of the wall.

Dimensionless Groups

As done before [4], the following dimensionless independent variables are defined

$$X = x/L \qquad\qquad T = \tau/\tau_{cy} \tag{6}$$

Also, a reduced temperature θ is defined by

$$\theta = (t - t_{a,min})/(t_{a,max} - t_{a,min}) \tag{7}$$

Dimensionless parameters are chosen to separate the effects of the various dimensional parameters as follows

$$N_B = h_r L/k$$

$$N_I = \epsilon I_{max}/h_r (t_{a,max} - t_{a,min})$$

$$N_k = \alpha h_r^2 \tau_{cy}/k^2 \tag{8}$$

$$N_w = w/(h_r \tau_{cy})$$

$$N_q = q/h_r (t_{a,max} - t_{a,min})$$

$$N_h = h_o / h_r$$
$$R_I = I / I_{max}$$
(8)

Dimensionless Form

Using the above groups, Eqs. 1 to 5 become, respectively

$$\frac{\partial^2 \theta}{\partial X^2} = \frac{N_B^2}{N_k} \frac{\partial \theta}{\partial T}$$
(9)

$$\theta(X,T) = \theta(X,T+n) \qquad n = 1, 2, 3, \ldots$$
(10)

$$\left. \frac{\partial \theta}{\partial X} \right|_{X=0} = N_B N_h (\theta_o - \theta_a) - N_B N_I R_I$$
(11)

$$\left. \frac{\partial \theta}{\partial X} \right|_{X=1} = N_B (\theta_r - \theta_i)$$
(12)

$$\theta_i - \theta_r = N_W \frac{d\theta_r}{dT} + N_q$$
(13)

The case of uncontrolled room temperature without internal heat load (N_q =0) was solved before by finite differences. In the present work, the case of an air conditioned space is treated. In this case θ_r is a constant and, therefore, $d\theta_r / dT = 0$. Equation 13 reduces to

$$\theta_i - \theta_r = N_q$$
(14)

SOLUTION

The problem, as formulated above, is to be solved for $N_q = N_q(T)$, a periodic function. It should be noted that, in these equations, θ_a and R_I are periodic functions of time only and independent of position.

In the method presented here, θ_a and R_I are analysed to their various harmonics by practical Fourier analysis [10], i.e.,

$$\theta_a = \theta_{a,0} + \sum_{n=1}^{m} \left[\theta_{a,c,n} \cos(2n\pi T) + \theta_{a,s,n} \sin(2n\pi T) \right]$$
(15)

and

$$R_I = R_{I,0} + \sum_{n=1}^{m} \left[R_{I,c,n} \cos(2n\pi T) + R_{I,s,n} \sin(2n\pi T) \right]$$
(16)

Each of Eqs. 15 and 16 contains a constant term, subscripted 0, and a series of harmonics. The constant term, from Fourier analysis, is the factor for $n = 0$ and represents the integrated mean of the function, whereas the harmonics represent fluctuations about this mean. The solution for the constant term is, naturally, a steady-state solution. It gives for Eq. 15 a steady conduction flux density that is due to the difference between the mean outside air temperature and the room temperature. For Eq. 16, the constant term gives the steady conduction flux density due to the mean solar irradiance if falling steadily on the wall.

A solution for a general harmonic n gives the reduced flux density $N_{q,n}$ for this harmonic, itself a harmonic function that gives fluctuations about the steady mean. The final solution for the original periodic function could then be obtained by superposition of the steady parts and the harmonics.

Steady-State Part

For this part, the temperature is independent of time; Eqs. 9 to 12 reduce to, respectively

$$\frac{d^2\theta}{dX^2} = 0 \qquad \text{or} \qquad \frac{d\theta}{dX} = \text{const.} \tag{17}$$

$$\theta = \theta(X)$$

$$-\frac{d\theta}{dX} = \theta_0 - \theta_1 = N_B N_h (\theta_{a,0} - \theta_0) + N_B N_I R_I \tag{18}$$

$$= N_B (\theta_1 - \theta_r) \tag{19}$$

Solving for θ_0 and θ_1 in the last two equations and substituting into Eq. 14 gives

$$N_{q,0} = \frac{N_h (\theta_{a,0} - \theta_r) + N_I R_{I,0}}{1 + N_h (1 + N_B)} \tag{20}$$

The first term on the right hand side gives the effect of the outside air temperature; the second term gives the effect of solar radiation. Indeed, Eq. 20 gives the mean load on the air conditioning unit during the day.

Harmonics Part

Noting that a sine function is a cosine function with a phase shift, the method outlined above is implemented here by replacing θ_a and R_I in Eq. 11 by their general harmonic form, i.e.

$$\theta_a = \theta_{a,n} \cos(2n\pi T) \tag{21}$$

$$R_I = R_{I,n} \cos(2n\pi T) \tag{22}$$

In the "frequency response" method [11], the transfer function as obtained from the Laplace transform of the problem would be used. The transfer function is used to give the amplitude of the established-state solution and its phase shift from the driving function as follows.

Laplace Transformation: Equations 9, 11, 12 and 14 when transformed in the usual manner give, respectively

$$\frac{\partial^2 \bar{\theta}}{\partial X^2} - \frac{N_B^2}{N_k} s \bar{\theta} = 0 \tag{23}$$

$$\frac{\partial \bar{\theta}}{\partial X}\bigg|_{X=0} = N_B N_h (\bar{\theta}|_{X=0} - \bar{\theta}_a) - N_B N_I \bar{R}_I \tag{24}$$

$$-\frac{\partial \bar{\theta}}{\partial X}\bigg|_{X=1} = N_B (\bar{\theta}|_{X=1} - \bar{\theta}_r) \tag{25}$$

$$\bar{N}_{q,n} = \bar{\theta}|_{X=1} - \bar{\theta}_r \tag{26}$$

The solution of the boundary value problem given by Eqs. 23 and 25 is

$$\bar{\theta} = \Big\{ N_B (p \cosh pX + N_B N_h \sinh pX) \bar{\theta}_r$$
$$+ N_B N_h \big[p \cosh p(1-X) + N_B \sinh p(1-X) \big] \bar{\theta}_a$$
$$+ N_B N_I \big[p \cosh p(1-X) + N_B \sinh p(1-X) \big] \bar{R}_I \Big\} / \bar{F}(s) \tag{27}$$

In this expression

$$p = p(s) = N_B \sqrt{s/N_k} \tag{28}$$

$$\bar{F}(s) = (p^2 + N_B^2 N_h) \sinh p + p N_B (1 + N_h) \cosh p \tag{29}$$

The transformed reduced flux density could now be obtained from Eq. 26 using Eq. 27 and 29, it gives

$$\bar{N}_{q,n} = \frac{p(s)}{\bar{F}(s)} \big[N_B N_h \bar{\theta}_a + N_B N_I \bar{R}_I$$
$$- (p \sinh p + N_B N_h \cosh p) \bar{\theta}_r \big] \tag{30}$$

Transfer Function: A transfer function $\bar{H}(s)$ is the ratio between the Laplace transforms of the dependent and driving functions. In the present case there are two driving functions as given by Eqs. 21 and 22. Their effects appear as the first two terms of Eq. 30. The third term gives the effect of the room air temperature, a constant in the present problem, and was handled in the steady-state part

The transfer function for the nth harmonic of the radiative part is obtained as follows

$$\overline{H}_{I,n}(s) = \overline{N}_{q,I,n} / \overline{R}_I = N_B N_I p(s) / \overline{F}(s) \tag{31}$$

For the outside air temperature part

$$\overline{H}_{\theta,n}(s) = \overline{N}_{q,\theta,n} / \overline{\theta}_a = N_B N_h p(s) / \overline{F}(s) \tag{32}$$

Established-State Solution: In the frequency response method [11] , the amplitude of the dependent function is obtained by multiplying the amplitude of the excitation function by the modulus of the corresponding transfer function in which the Laplace transform variable s is replaced, in this case, by $2n\pi i$. Hence, for the radiative part for example, the amplitude of the nth harmonic is given by

$$N_{q,I,n,amp} = R_{I,n} |\overline{H}_{I,n}(2n\pi i)| \tag{33}$$

The phase shift ψ is also obtained as the argument of the same transfer function, i.e.

$$\psi_{I,n} = arg\ H_{I,n}(2n\pi i) \tag{34}$$

To implement this method for the nth harmonic of the radiative part, the following process is used.

$$\overline{H}_{I,n}(2n\pi i) = N_B N_I p(2n\pi i) / \overline{F}(2n\pi i) \tag{35}$$

From the theory of complex variables, Eq. 28 gives

$$p(2n\pi i) = \gamma_n(1 + i) = \sqrt{2}\ \gamma_n\ e^{i\pi/4} \tag{36}$$

where

$$\gamma_n = N_B \sqrt{n\pi / N_k} \tag{37}$$

Using the relations

$$\sinh(y + iz) = \cos z\ \sinh y + i \sin z\ \cosh y$$

$$\cosh(y + iz) = \cos z\ \cosh y + i \sin z\ \sinh y$$

the function $\overline{F}(2n\pi i)$ is determined as

$$\overline{F}(2n\pi i) = P_n + i\ Q_n = r_n\ e^{i\varnothing_n} \tag{38}$$

where

$$P_n = N_B^2 N_h \cos \gamma_n\ \sinh \gamma_n - 2 \gamma_n^2 \sin \gamma_n\ \cosh \gamma_n$$
$$+ N_B (1 + N_h)\ \gamma_n (\cos \gamma_n \cosh \gamma_n - \sin \gamma_n \sinh \gamma_n) \tag{39}$$

$$Q_n = N_B^2 N_h \sin \gamma_n \cosh \gamma_n + 2 \gamma_n^2 \cos \gamma_n \sinh \gamma_n$$
$$+ N_B (1 + N_h) \gamma_n (\cos \gamma_n \cosh \gamma_n + \sin \gamma_n \sinh \gamma_n) \qquad (40)$$

$$r_n = \sqrt{P_n^2 + Q_n^2} \qquad (41)$$

$$\phi_n = \arctan (Q_n / P_n) \qquad (42)$$

Substituting from Eqs. 36 and 41 into Eq. 35

$$\bar{H}_{I,n}(2n\pi i) = \sqrt{2} N_B N_I \gamma_n \exp i \left(\frac{\pi}{4} - \phi_n\right) / r_n \qquad (43)$$

Using this relation in Eqs. 33 and 34 gives

$$N_{q,I,n,amp} = \sqrt{2} N_B N_I R_{I,n} \gamma_n / r_n \qquad (44)$$

and $\qquad \psi_n = \frac{\pi}{4} - \phi_n \qquad (45)$

Therefore, the contribution of the nth harmonic driving function of Eq. 22 to the heat flux density to the room is given by

$$N_{q,I,n} = \sqrt{2} N_B N_I \gamma_n R_{I,n} \cos(2n\pi T + \phi_n - \frac{\pi}{4}) / r_n$$

The sine part of Eq. 16 has the same effect as the cosine part above. The contribution of all the radiative harmonics to the heat flux density is obtained by superposition as

$$N_{q,I} = \sqrt{2} N_B N_I \sum_{n=1}^{m} (\gamma_n / r_n) \left[R_{I,c,n} \cos(2n\pi T + \phi_n - \frac{\pi}{4}) \right.$$
$$\left. + R_{I,s,n} \sin(2n\pi T + \phi_n - \frac{\pi}{4}) \right] \qquad (46)$$

In a similar manner, the contribution of all the "outside air temperature harmonics" is given by

$$N_{q,\theta} = \sqrt{2} N_B N_h \sum_{n=1}^{m} (\gamma_n / r_n) \left[\theta_{a,c,n} \cos(2n\pi T + \phi_n - \frac{\pi}{4}) \right.$$
$$\left. + \theta_{a,s,n} \sin(2n\pi T + \phi_n - \frac{\pi}{4}) \right] \qquad (47)$$

Total Flux Density: The total flux density is obtained by adding the steady and established periodic parts of Eqs. 20, 46 and 47

$$N_q = \frac{N_h (\theta_{a,0} - \theta_r) + N_I R_{I,0}}{1 + N_h (1 + N_B)}$$

$$+ \sqrt{2} \, N_B \, N_I \sum_{n=1}^{m} (\gamma_n / r_n) \left[R_{I,c,n} \cos(2n\pi T + \phi - \frac{\pi}{4}) \right.$$

$$\left. + R_{I,c,n} \sin(2n\pi T + \phi_n - \frac{\pi}{4}) \right]$$

$$+ \sqrt{2} \, N_B \, N_h \sum_{n=1}^{m} (\gamma_n / r_n) \left[\theta_{a,c,n} \cos(2n\pi T + \phi_n - \frac{\pi}{4}) \right.$$

$$\left. + \theta_{a,s,n} \sin(2n\pi T + \phi_n - \frac{\pi}{4}) \right] \qquad (48)$$

NUMERICAL EXAMPLE

In the following, the results obtained by the previous analysis for a roof or a southern wall are presented. For this purpose, the dimensionless functions θ_a and R_I should be known. A study of meteorological data for the summer months in a number of cities in Egypt and Sudan showed that θ_a follows closely the curve shown in Fig. 2 [3]. Fig. 3 gives R_I for a roof or a southern wall in a haze-free atmosphere. These curves represent, quite closely, the data published by ASHRAE [1]. Admittedly, these curves merit a wider study than has been possible. Efforts should be made to standardize these normalized curves of θ_a and R_I ; values for the latter should be determined for different orientations.

Using hourly values from the above mentioned curves, each was represented by 24 terms of Fourier series, including the "zero term". Using the calculated Fourier coefficients, hourly values of N_q were computed; the results are shwon in Fig. 4 for $N_h = 1$ and 2, and for $N_B = 3$, 6 and 9 . As could be seen, an increase in N_B decreases the maximum flux density and retards it. On the other hand, an increase in N_h decreases the maximum flux density without affecting its time.

CONCLUSIONS

A more general method for the determination of thermal behavior of buildings is presented. Generality is obtained through the use of dimensionless groups, and the normalization of meteorological data. Generality is also attained by analising the periodic meteorological functions to their harmonics, and giving solution for a general harmonic. This solution is obtained mathematically by the "frequency response" method, is general, and could be used directly for any situation. The final solution for a given general case could be obtained by superposition of the various harmonics, and of a mean steady-state solution.

An example is solved for the case of a south facade which is practically equivalent, except for the magnitude of solar irradiance, to the case of a roof.

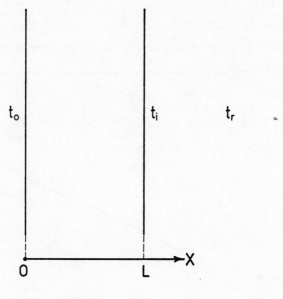

Fig. 1

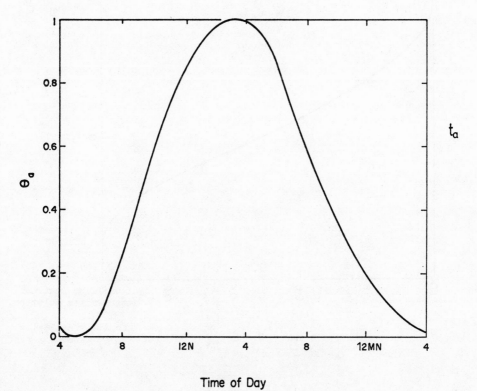

Time of Day

Fig. 2

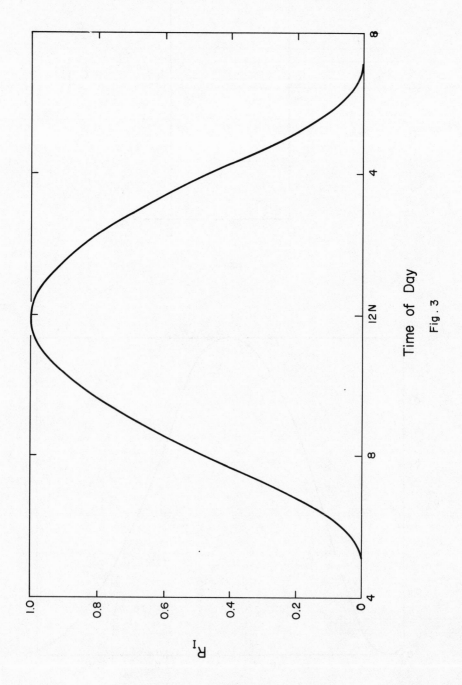

Time of Day

Fig . 3

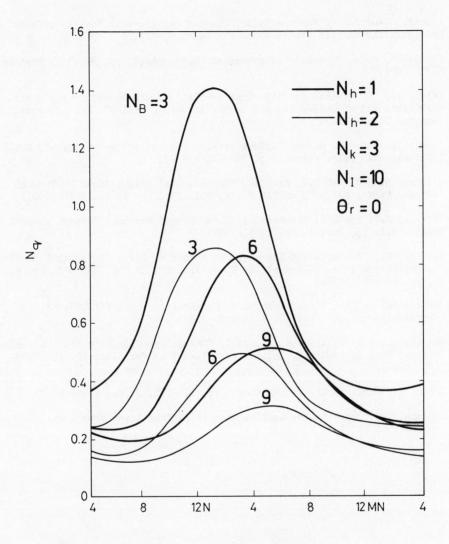

Time of Day

Fig. 4

REFERENCES

1 - ASHRAE, "Handbook of Fundamentals", Chapter 22, American Society of Heating, Refrigerating and Air Conditioning Engineers (1972).

2 - Threlkeld, J.L., "Thermal Environmental Engineering", pp. 340-355, Prentice-Hall (1970).

3 - Saleh, A., "Grundsätze fur die wärmetechnish und wirstschaftlich günstige ausbildung von gebäudewänden und dächen in warmen Länden", Dr. Dissertation, Hannover T.H. (1962).

4 - Hassan, K. and G.B. Hanna, "Effect of Walls on Indoor Temperatures", BUILD International, July/ August, pp. 220-226 (1972).

5 - Stephenson, D.G. and G.P. Mitalas, "Cooling Load Calculations by Thermal Response Factor Method", ASHRAE Trans. Vol. 73, Part 1, No. 2018 (1967).

6 - Mitalas, G.P. and D.G. Stephenson, "Room Thermal Response Factors", ASHRAE Trans., Vol. 73, Part 1, No. 2019 (1967).

7 - Gautier, E., "Comportement Thermique de Locaux en Été - Étude par l'Homologie Hydrodynamique", Revue Generale de Thermique, No. 88, pp. 331-345, and No. 89, pp. 469-486 (1969).

8 - Dimitru-Valcea, E., "Termotechnica in Constructii", pp. 151-164, Editura Academiei Republicii Socialiste Romania (1970).

9 - Hassan, K., M.M. Hilal, and G.M. Zaki, "Convection from Pipe with Harmonic Internal Heat Generation", Int. J. Heat Mass Transfer, Vol. 17, pp. 1383-1390 (1974).

10 - Sohon, H., "Engineering Mathematics", pp. 143-152, D. Van Nostrand (1957).

11 - Doetsch, G., "Guide to the Applications of Laplace Transforms", pp. 68-71, Van Nostrand (1961).

METHOD FOR CALCULATION
OF THERMAL BEHAVIOUR
OF BUILDINGS

G. HAUSER

Institut für Bauphysik
815 Holzkirchen/Obb.
Postfach 1180
Federal Republic of Germany

A method for calculation of thermal behaviour of buildings of any
size, form, structure as well as any thermal charge from outside
and inside is being presented. All thermal processes within
buildings can be handled by calculation of room temperature or
energy consumption and handling of "regulating processes"
containing the transition from calculations of room temperature
to calculations of energy consumption and vice versa. After a brief
description of the mathematical method the application possibilities
of the method are explained. Confrontation of measured and
calculated indoor air temperature cycles of different buildings
serves for demonstration.

1. Introduction

The exact knowledge of thermal behaviour of whole buildings is
essential to find a satisfactory solution of many problems arising
in architecture, such as questions of comfort, energy saving and
statics. Therefore, a method has been developed to calculate in
advance thermal behaviour of even big buildings, taking into account
any possible problems to be solved. In the following - after a
brief description of the mathematical method, detailed presentation
in (1) - the application possibilities of the method are explained.
Results obtained by calculation of two completely different objects
are confronted with measured values, thus putting into relation
results obtained by the calculation method with conditions in
practice and showing the applicability of the method.

2. Mathematical Method

Calculation of the thermal behaviour of a whole building is
based on the following iterative process:

The building to be examined - as shown on Figure 1 - is divided
in single rooms in a way so that a neighbouring room in front
and at the back belongs to every room in one coordinate direction.
Thus every room has 6 neighbouring rooms at the most.
All rooms one beneath the other must have the same basal surface,
all rooms side by side the same side surface etc.
If this does not apply to the existing rooms, mathematical model
rooms are established by means of methodical walls - not really
existing, imagined walls without heat capacity and without
resistance of thermal transmittance - meeting the required
conditions.

Figure 1.
Schematic representation of
the method of coupling one
model room to the building.

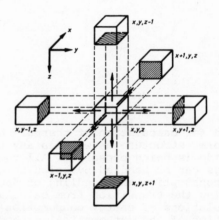

The heat balance of these mathematical model rooms comprises all
thermal processes in the model room. The heat flow through
one- and multi-layer inside and outside building elements
is calculated according to a specifically applied Crank-Nicolson-
Method (2) (3).
The solar radiation striking the building may either be computed
according to the geography, the selected day of the year and the
given haze of the atmosphere or may be put in at optionally
chosen intervals, using results of measurements. The radiation
energy reaching the rooms of a building through transparent
outdoor building elements strikes the indoor building elements
and/or the furniture and induces heat sources there.
The radiation energy can be distributed at discretion to the
different surfaces.

After calculation of the single rooms, these are reconnected to
the building by means of the room boundary surfaces, exchanging
transmission heat between the rooms, thus representing the
coupling elements of the system.

3. Application possibilities of the method

With the developed method any thermal processes within buildings
of any size, form, structure and usage can be handled.
The following basic problems have to be differentiated:

- Calculations of room temperature where the diurnal cycle of
  air temperature and surface temperatures of room boundary
  surfaces and/or the furniture is obtained using given data
  such as meteorological records and usage by the occupants.

- Calculations of energy consumption determinating the diurnal
  cycle of cooling or heating performance for every room of
  the building to guarantee the given specific diurnal cycle of
  room air temperature, considering known meteorological records
  and usage. It is also possible to air-condition the single
  rooms on different levels of temperature, every room can be
  provided with a chosen diurnal cycle of room air temperature.

- Regulating processes piloting transition from calculations
  of room temperature to calculations of energy consumption
  and vice versa. This exchange of set problems can be made
  as often as required in the course of one calculating operation
  and permits to achieve a practice orientated handling of many
  problems. If, for instance, in summer for several days
  heating would be required instead of air-condition to maintain
  room air temperature, there is the possibility to discontinue
  calculation of energy consumption for this time and to start
  calculation of room temperature.

The problems to be solved do not have to be the same or
synchronous in all rooms of the building but may concern only
single rooms or parts of the building.
In the northern part of a building e.g. room temperature can be
computed while in the southern part the cooling performance is
determined by calculation of energy consumption. With all these
problems, any processes such as thermal charge of the building
or habits of usage of the occupants can suffer optional changes
in time and duration. Periodicity of operations is not required.
Not only can be calculated the swing in of a building from
different starting conditions to midsummer level but also the
most varying weather cycles can be handled. The type of usage
too can change during a day as well as within several days,
regularily or erratically.
This applies for ventilation habits - brief or permanent ventilation-
as well as for the operation of blinds.
In that way it is possible to recognize the effects of daily or
several-day-changes, e.g. internal heat sources.
It is possible to find out whether heating is necessary on a cool
day during the transition seasons or to find out measures in
construction and operation to avoid the necessity of heating.

4. Confrontation of measured and calculated room air temperature cycles

Results of measured and with the method computed room air
temperature cycles of two completely different objects are
confronted to examine and to demonstrate the method.
Meteorological data and usage are considered as of the day
of measuring.

Object A: Office building (District Magistrate Office Rosenheim)

Figure 2.   North-eastern view
of the office building in Rosenheim where the first confrontation of calculated and in
Sept. 1975 measured room air
temperature cycles was made.

The District Magistrate Office shown on Fig. 2 is a two-haunch
six-storeyed office building of heavy construction. The office
rooms examined are 3,60 m wide, 5,20 m deep and 2.80 m high.
The windows consist of insulating clear-glass. During measuring
all windows and doors remain closed. The windows cover 40 %
of the surface of the main facades of the building. During the day of
measurement the outdoor air tem-
perature and global radiation
intensities shown on Fig. 3
were registered.

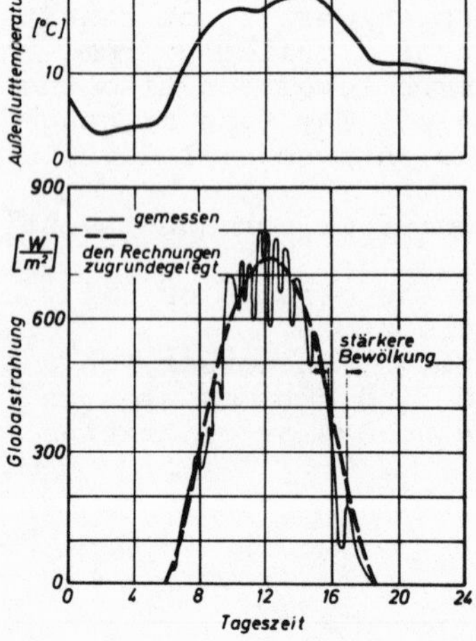

Figure 3.   Diurnal cycle of outdoor air
temperature and global radiation measured
on Sept. 14, 1975 in Rosenheim. The
measured radiation course is approximated
for calculations by a dotted line course.

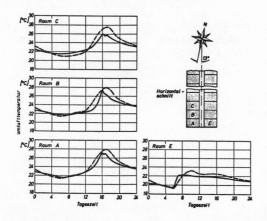

Figure 4. Comparison of measured (———) and calculated (-- ----) air temperature in different office rooms of the Rosenheim District Magistrate Office. The letter marking of the rooms can be taken from the horizontal cross-cut through the building (second storey) printed on the right side of this figure.

The confrontation of measured and calculated room air temperature cycles on figure 4 shows good agreement. Deviations of 2 K at the most, normally much smaller, show up between the peak temperature values shortly after 4 p.m.
An explanation herefore is found in the global radiation course on figure 3.
After 4 p.m. the actual radiation diminishes strongly due to a passing cloud and rises again later on to the value expected on the measuring day. This temporary diminution of radiation is not simulated by calculation as can be taken from the approximated global radiation course the calculation is based on, since the exact apply of existing and measured radiation course would have raised drastically the expenditure for data set up. The influence of the temporary heavier cloudiness is underlined by the measured temperature courses since these values rise again after passing of the clouds - especially noted in rooms A and C. The correct taking-in of the orientation influences by the developed method is proved by the east orientated room E where - contrary to the west rooms - peak temperature is reached before noon.

Object B: One-family-house

Figure 5. South-eastern view of the one-family-house in Elsendorf/Ndb, where the second confrontation of calculated and in July 1976 measured room air temperature cycles was made.

Photo: OKAL

The one-family-house used - Fig. 5 shows a view facing south-east - is a pre-fabricated house of wooden construction. It has a basal surface of 100 m², the floor plan is shown on fig. 6. It is provided with a cellar of same extent. The attic is not lined with masonry.

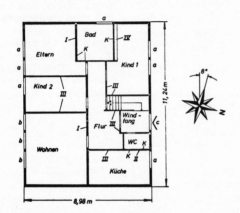

Figure 6. Floor plan of the measured one-family-house. Room clearance is 2,54 m. The attic is not lined with masonry. The roof has the form of a saddle roof with an inclination of 27O40′ .

The surface of the room doors is generally 1,75 m².

The house is made of pre-fabricated wooden elements with four different types of indoor walls (I to IV). The wall section marked with K is covered with tiles.

Surfaces of transparent outdoor building elements
(insulating clear-glass)
a: 1,46 m² rough measure : 0,95 m² glass surface
b: 2,45 m² rough measure : 1,73 m² glass surface
c: 2,57 m² rough measure : 1,34 m² glass surface

Figure 7. Measured outdoor air temperatures and global radiation intensities as of July 1 to July 9, 1976 in Elsendorf/Ndb.

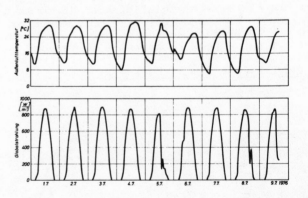

During the measuring period of 9 days the outdoor air temperatures and global radiation intensities shown on fig. 7 were noted. Within this period usage varies as follows:

Case I from July 1, 0 h to July 4, 7 p.m.

From 0 h to 12 p.m.

- external blinds are not drawn

- all windows and doors remain closed

Case II from July 4, 7 p.m. to July 7, 5 p.m.

From 0 h to 8 a.m. and from 6 p.m. to 12 p.m.

- external blinds are not drawn

- all windows are half opened and all room doors opened

From 8 a.m. to 6 p.m.

- all external blinds are drawn

- all windows and room doors remain closed

Case III from July 7, 5 p.m. to July 9, 4 p.m.

From 0 h to 8 a.m. and from 6 p.m. to 12 p.m.

- all external blinds are not drawn

- all windows and room doors remain closed

From 8 a.m. to 6 p.m.

- external blinds of all windows are drawn

- all windows and room doors remain closed.

The hourly exchange of air in the various cases was measured according to the method of diminution of Argon concentration and constitutes the base for the calculations.

Confrontation of measured and calculated room air temperature cycles is shown on fig. 8 for bedroom, living room and kitchen of the one-family-house. All other rooms do not show any new findings and are not mentioned therefore.
Comparison of measurement and calculation shows:

- In case I agreement is very good with the south orientated living room and bedroom, with the north orientated kitchen the calculated peak temperatures exceed the measured temperature by about 2 K. This may be explained by the position of the house being sited on a slope, thus weakening the diffuse radiation impinging on the kitchen window during the whole day and screening from direct solar radiation during early morning and late evening hours.

- Transition from case I to case II and case II itself are well covered by the method, except for the low fall of air temperature in the kitchen in the early morning hours of July 6. The reason for this extreme fall could be the fact that the exchange of air measured before the measuring period whereupon calculations have been based has been exceeded considerably during this time.

- Transition from case II to case III and case III itself are correctly imitated. However, in the bedroom the calculation shows permanently by 2 K too low air temperature.
  The reason herefore is obviously the blind not being as firmly closed as assumed in the calculations.

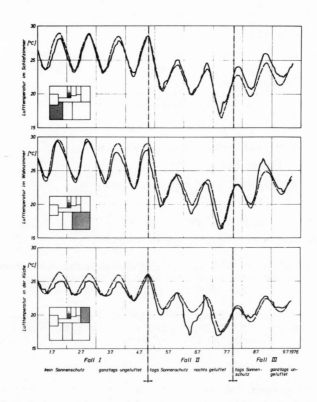

Figure 8. Comparison of measured (————) and calculated (- - - - -) air temperatures in different rooms of a one-family-house considering the meteorological data (fig. 7) and habits of usage prevailing on the measuring days
(Case I : All day no sunshade, no ventilation.
 Case II: During daytime sunshade, ventilation during the night.
 Case III: During daytime sunshade, no ventilation all day).

Finalizing it can be stated that the confrontation of measured and calculated room air temperature cycles for the separate rooms of the office building and of the one-family-house shows good agreement

REFERENCES

(1)   Hauser, G.: Rechnerische Vorherbestimmung des Wärmeverhaltens
       großer Bauten. Dissertation Universität Stuttgart (1977).

(2)   Brown, G.: A method of calculating heating and cooling loads
       with the aid of digital computers (swedish). VVS 34 (1963)
       No. 11, p. 401-410.

(3)   Rosenberg, D.U.: Method of the numerical solution of partial
       differential equations. American Elsevier Publishing Company,
       Inc., New York (1969).

ENERGY CONSERVATION IN NEW BUILDINGS

BO ADAMSON AND KURT KÄLLBLAD

Department of Building Science
Lund Institute of Technology
Fack 725 S-220 07
Lund 7, Sweden

ABSTRACT

Different methods for calculation of heat consumption in buildings are discussed in brief. Studies have been carried out of the way different factors
such as window size, facade orientation etc. influence energy consumption in
buildings. The risk of obtaining undesirably high indoor temperatures when
designing buildings with extremely low energy consumption is discussed.

CALCULATION OF ENERGY CONSUMPTION IN BUILDINGS

There are several methods for calculating energy consumption in buildings.
Some of these methods are

I. Computer programs which in different ways take into account most of the
different heat transfer phenomena in the building and calculate the energy
requirement hour by hour through the whole heating season.

II. Manual calculations which take into consideration insolation but neglect heat capacity in building materials. In this case monthly or weekly
average values of outdoor climate can be used.

The first method seems to be the most accurate of all available methods,
giving good agreement between measured and calculated values e.g. by Bradley
et al. Two other examples are shown in Figure 1.

Some comparisons between method I and II are given in Table 1 and as indicated one may use a manual calculation method with a fair degree of accuracy
by assuming normal losses as in older buildings. For low-energy buildings
the manual methods used are too inaccurate.

When selecting input data for calculations there are several uncertainties.
Metheorological stations are sometimes far away from the building site and
the local climate is influenced by e.g. heat islands over urbanised areas,
complex shadowing by other buildings, trees etc. Furthermore the building
may not be used as designed. For example it is unlikely that people will
open windows exactly as assumed in the calculations.

INFLUENCE OF DIFFERENT FACTORS ON ENERGY CONSUMPTION

Computer program

We have chosen the first method but, in order to reduce the computer costs,
we have simplified the mathematical model of the room, as in the following
example: If the room is surrounded by identical rooms so that no heat trans-

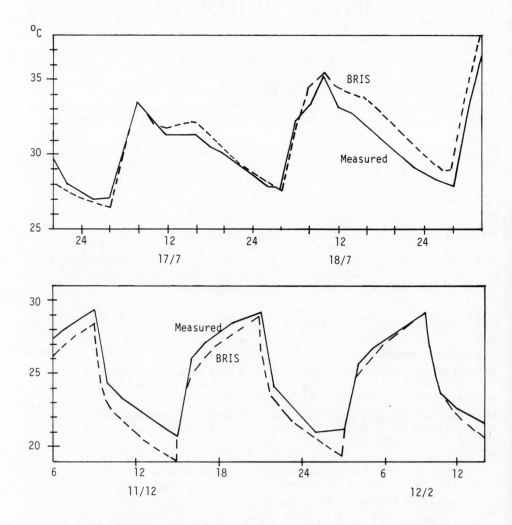

Figure 1   Comparison between measured room air temperature and computer
calculated temperature (BRIS program)
a) sunny hot summer days
b) cloudy winter days with periodically heated ceiling.

fer can take place between the rooms, then the influence of the walls on the heat transfer of the room can be described rather accurately by one single ordinary differential equation. Furthermore two inner walls can be treated as one if they can be assumed to remain at the same temperatures. This will reduce the number of differential equations and simplify the calculations e.g of heat transfer between inner surfaces.

The simplified model used in our calculations has been compared with more complex models by applying both to a number of specific cases and the agreement has been reasonably good. Furthermore, one has to keep in mind that absolute values are of less importance when energy requirements due to differences in design are to be examined.

Factors affecting annual heat consumption

The room examined is showm in Figure 2 and has, unless otherwise stated, 2-pane windows covering 20 per cent of the facade area. The room is assumed to be occupied by two persons monday to friday between 8 a.m. and 4 p.m. They contribute together with electrical appliances a total of 300 W.

A convective heating system is simulated. When using room thermostats the control is assumed to be "perfect" so when indoor temperature tends not to exceed the desired value (+22 ^{o}C) an exact amount of energy is supplied to maintain the temperature. When the temperature tends to rise above +22 ^{o}C all heat supply is assumed to be cut off. When using an outdoor thermostat the heat supply is equal to $C*(\vartheta_{in}-\vartheta_{out})$ if $\vartheta_{in}>\vartheta_{out}$, otherwise zero. ϑ_{in} and ϑ_{out} are indoor and outdoor temperatures and C is chosen so that indoor temperature will always be equal to or greater than +22 ^{o}C.

The ventilation system normally supplies 80 m^3/h of outdoor air. If the indoor temperature tends to rise above +22 ^{o}C the amount will be increased linearly up to three times the normal value at an indoor temperature of +24 ^{o}C. If the indoor temperature tends to increase above +25 ^{o}C cooling is added to limit the temperature to this value during working hours.

In Table 2-6 and Figure 3 the energy savings achieved by a number of different measures are presented. The influence of heat capacity is almost negligible but the influences of orientation and glazed area are quite large when using indoor thermostats. As no "perfect" control system exists, the differences in practice might be smaller.

As seen e.g. in Table 5 the energy saving due to one measure depends on other measures and this indicates the importance of proper calculations when discussing economy in energy conservation.

HIGH ROOM TEMPERATURES IN BUILDINGS WITH EXTREMELY LOW ENERGY CONSUMPTION

As previously mentioned there are a lot of measures that can be applied to a building in order to decrease the energy consumption for heating. Such measures are for example:

    a proper control of the heat supply
    better thermal insulation of walls and roof
    better insulation of windows
    windows facing south
    intermittent ventilation
    heat recovery from exhaust air

When several of these measures are applied, we will achieve a low energy consumption for heating. At the same time there is a risk of obtaining high ambient temperatures when there is a large heat input from insolation and a lot

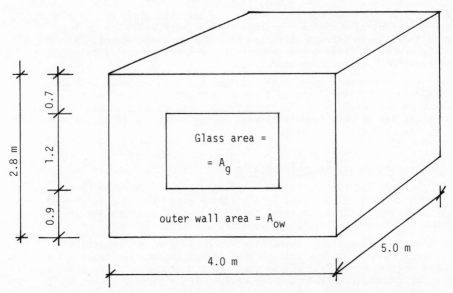

Figure 2   Room used as a basis for calculations. The room is assumed being surrounded by similar rooms with identical temperatures. Glass part $G = A_g/(A_g + A_{ow})$.

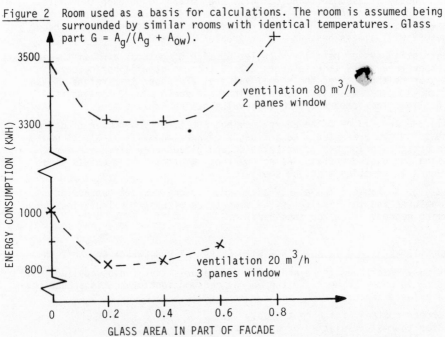

Figure 3   Influence of the glass area on the heat consumption when the facade is facing south.

Table 1 Calculated energy consumption for heating according to different calculation methods. Room according to Figure 2 and design alternative B in Table 2. Climate data for Stockholm Sept.1, 1945 - May 31, 1946. Heat supply control based on the indoor temperature of +22 °C. Free heat from people etc. 300 W monday - friday 8 a.m - 4 p.m.

Outer wall U-value	Window Number of panes	Venti-lation (m³/h)	Facade orien-tation	Energy consumption (kWh)		
				Computer calculation	Manual calculation	
				Δt=1h	Δt=1 week	Δt = 1 month
0.5 "	2 "	80 "	North South	4260 3390	4290 3370	4290 3340
0.2 "	3 "	20 "	North South	1144 674	1030 532	1002 465

Table 2 Design alternatives concerning heat capacity. Slabs and inner walls of room according to Figure 2. Light weight concrete with density = 500 kg/m³. Outer wall with U-value 0.37 W/°Cm².

Alterna-tive	Slabs	Inner walls
A	200 mm concrete	200 mm concrete
B	-"- -"-- " -	70 mm light weight concrete
C	200 mm light weight concrete	- " -

Table 3 The daily mean heat requirement $\bar{P}_h$ during an extremely cold day and annual heat consumption W_h during an ordinary heating season (Sept.1, 1945 - May 31, 1946). The daily mean cooling requirement $\bar{P}_c$ during a hot day and the cooling consumption during summer 1941. Design alternatives according to Table 2 and climate data for Stockholm. Glass part of facade = 0.20.

Facade Orientation	Design alternative	Heating to room temp. +22 °C		Cooling to room temp. +25°C	
		$\bar{P}_h$	W_h	$\bar{P}_c$	W_c
		W	kWh	W	kWh
North	A	1808	4140	-157	-14
"	B	1815	"	-190	-22
"	C	1824	"	-202	-33
South	A	1655	3300	-334	-43
"	B	1668	3320	-373	-56
"	C	1704	3410	-362	-82

Table 4 Energy savings by controlling heat supply by the indoor temperature (+22 °C) instead of the outdoor temperature. Room according to Figure 2 with 2.24 m² glass area. Design alternative B (Table 2). Climate data for Stockholm, Sept.1, 1945 - May 31, 1946. The over-all heat transfer coefficient of the outer wall = U_{ow} (W/°Cm²) and the absorptivity of the outer surface a_{ow} = 0.5. Between the indoor temperature +22 °C and +24 °C the ventilation varies linearly from 80 to 240 m³/h."Free heat" from people and electrical appliances = P (W) monday - friday from 8 a.m. to 4 p.m.

Heat supply controlled by the temperature	P (W)	Outer wall U-value (W/°Cm²)	Facade orientation	Energy consumption kWh	Energy saving kWh
outdoor	0.0	0.5	North	4932	213
indoor	"	"	"	4719	
outdoor	300	0.5	North	4932	673
indoor	"	"	"	4259	
outdoor	300	0.5	South	4932	1543
indoor	"	"	"	3389	
outdoor	300	0.2	South	4622	1497
indoor	"	"	"	3125	

Table 5 Influence of facade orientation on the heat consumption. Room according to Figur 2 with 2.24 m^2 glass area. Design alternative B (Table 2). Climate data for Stockholm Sept.1, 1945 - May 31, 1946. The over-all heat transfer coefficient of the outer wall U = 0.2 W/^{o}Cm^2 and the absorptivity of the outer surface a=0.5. Heat supply is controlled by the indoor temperature (min.+22 ^{o}C). Between the indoor temperature +22 ^{o}C and +24 ^{o}C the ventilation V varies linearly between V and 3∗V. "Free heat" from people and electrical appliances = 300 W monday - friday from 8 a.m. to 4 p.m.

Facade orien- tation	Number of panes in window	Ventilation m^3/h		Energy consump- tion	Energy saving
		monday- friday 8 a.m.- 4 p.m.	remaining hours	kWh	kWh
North South	2 "	80 "	80 "	3961 3125	836
North South	3 "	80 "	80 "	3723 2960	763
North South	3 "	80 "	20 "	1663 1029	634

Table 6 Energy saving by a third pane in windows. Room according to Figure 2 with 2.24 m^2 glass area. Design alternative B (Table 2). Climate data for Stockholm Sept.1, 1945 - May 31, 1946. The over-all heat transfer coefficient of the outer wall U = 0.2 W/^{o}Cm^2 and the absorptivity of the outer surface a=0.5. Heat supply is controlled by the indoor temperature (min.+22 ^{o}C). Between the indoor temperature +22 ^{o}C and +24 ^{o}C the ven- tilation V varies linearly between V and 3 ∗ V. "Free heat" from people and electrical appliances = 300 W monday - friday from 8 a.m. to 4 p.m.

Facade orien- tation	Number of panes in window	Ventilation m^3/h		Energy consump- tion	Energy saving
		monday- friday 8 a.m.- 4 p.m.	remaining hours	kWh	kWh
North "	2 3	80 "	80 "	3961 3723	238
South "	2 3	80 "	80 "	3125 2960	165
South "	2 3	80 "	20 "	1190 1029	161
South "	2 3	80[1] "	80[1] "	804 674	130

1) Heat recovery with efficiency of 75%.

Table 7 Distribution of room temperature above +22 °C during Sept.15, 1945 to May 15, 1946 for different design alternatives.

Design alternative	Glass area in part of facade G	Room temperature 15/9 - 15/5																		
		≥22 <23	≥23 <24	≥24 <25	≥25 <26	≥26 <27	≥27 <28	≥28 <29	≥29 <30	≥30 <31	≥31 <32	≥32 <33	≥33 <34	≥34 <35	≥35 <36	≥36 <37	≥37 <38	≥38 <39	≥39 <40	>40
Ia	0.2	5315	335	97	62	18	5	0	0	0	0	0	0	0	0	0	0	0	0	0
	0.3	4715	615	189	77	55	63	45	47	20	6	0	0	0	0	0	0	0	0	0
	0.4	4264	575	338	246	122	64	34	44	50	32	32	17	10	4	0	0	0	0	0
IIa	0.2	5205	422	128	50	18	7	2	0	0	0	0	0	0	0	0	0	0	0	0
	0.3	4691	537	235	160	90	43	38	16	12	5	5	0	0	0	0	0	0	0	0
	0.4	4267	515	295	220	144	142	77	55	38	32	19	11	8	4	5	0	0	0	0
IIIa	0.2	5113	398	172	93	35	13	4	4	0	0	0	0	0	0	0	0	0	0	0
	0.3	4589	504	272	149	111	82	56	30	18	11	3	4	3	0	0	0	0	0	0
	0.4	4314	336	315	203	187	120	98	75	61	42	27	19	15	7	6	3	2	2	0
Ib	0.2	5492	302	31	7	0	0	0	0	0	0	0	0	0	0	0	0	0	0	0
	0.4	4588	668	233	116	74	61	31	33	16	8	4	0	0	0	0	0	0	0	0
IIIb	0.2	5267	362	135	47	14	5	2	0	0	0	0	0	0	0	0	0	0	0	0
	0.4	4466	510	261	192	113	94	77	47	27	20	11	7	3	2	2	0	0	0	0

a) 15 m3/h + 30
b) 15 m3/h + 45

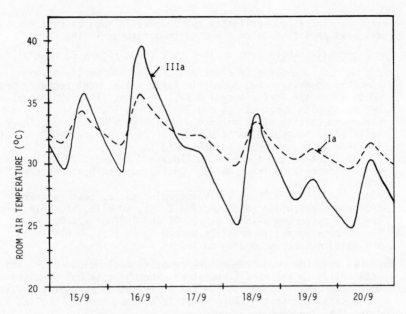

Figure 4   Room temperatures Sept.15 to Sept.20, 1945 in rooms according to
design alternatives Ia (concrete slab) and IIIa (light weight
concrete slab).

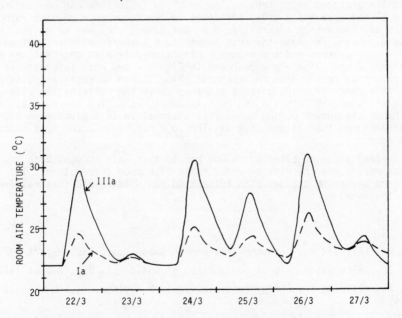

Figure 5   Room temperatures March 22 to March 27, 1946 in rooms according to
design alternatives Ia (concrete slab) and IIIa (light weight
concrete slab).

of free heat from electrical appliances and people. In such cases the heat capacity does have an influence on the high temperatures in the rooms.

Five design alternatives have been applied to a room, Figure 2, namely:

Ia: 200 mm concrete slabs in floor and ceiling, 70 mm light-weight concrete (density = 500 kg/m^3) in inner walls, heat recovery from ventilation air (efficiency = 0.67), 45 m^3/h, linearly increasing bypassing with 0% at +22 ^{o}C and 100% at +24 ^{o}C.

IIa: Floor and ceiling of 25 mm wood with 125 mm mineral wool, inner walls of 13 mm gypsum on both sides of 75 mm mineral wool, ventilation and heat recovery according to Ia.

IIIa: 200 mm light-weight concrete in floor and ceiling, 70 mm light-weight concrete in inner walls, ventilation and heat recovery according to Ia.

Ib: Floor, ceiling and inner walls according to Ia, heat recovery from ventilation air (efficiency = 0.75), 60 m^3/h, linearly increasing bypassing with 0% at +22 ^{o}C and 100% at +24 ^{o}C.

IIIb: Floor, ceiling and inner walls according to IIIa, ventilation and heat recovery according to Ib.

Heat is supplied when the room temperature tends to fall below +22 ^{o}C and is completely controlled by the room temperature. When heat is not supplied the heat exchanger is gradually bypassed so that at +24 ^{o}C no part of the ventilation air is passing the heat exchanger. It is assumed that no cooling system is used and that the room is unoccupied.

In Table 7 the room temperatures above +22 ^{o}C during the normal heating season, 15 Sept - 15 May, are shown for the five design alternatives. Most of the high temperatures arise from September. From Table 7 it can be concluded that the glazed area has a great influence on the occurrence of high room temperatures. The design alternative does not have a very marked effect. A concrete slab will decrease the room temperature compared with a light-weight slab by 1 or 2 degrees and the higher ventilation will also decrease the room temperature by approximately 2 degrees. In Figure 4 two room temperatures are shown, referring to six days in September 1945. It can be seen that alternative Ia with concrete slabs gives a smoother curve than alternative IIIa with light-weight slab. In Figure 5 the room temperatures during six days at the end of March are shown. During sunny days alternative IIIa gives much higher room temperatures than alternative Ia. Thus one has to take the heat capacity into account.

When selecting a design alternative one has to take into account both the heat consumption and the risk of undesirably high room temperatures. The window area has to be decided with respect to sun shading and internal free heat in the room.

REFERENCES

Adamson B., Källblad K., Computer programs in Sweden...., BKL, Report 1975:7.

Adamson B., Calculation of heat consumption of buildings, BKL, Report 1976:3.

Adamson B., Källblad K., Energy conservation in Buildings..., BKL, Report 1976:5.

Bradley A, et al, Dynamic Thermal Performance of an Experimental Masonry Building, U.S.Depart.of Commerce, July 1973.

BKL: Department of Building Science, LTH, Lund, Sweden.

INFLUENCE OF THE OUTDOOR TEMPERATURE VARIATION ON THE PERFORMANCE OF A BUILDING HEATING SYSTEM

CARLO CARRARA

Italsider, Gruppo Ricerca Operativa
Genova-Cornigliano, Italy

LUIGI FANTINI AND ANTONIO LORENZI

Università degli Studi di Genova
Genova, Italy

ABSTRACT

The joined influence of the outdoor temperature variations and structural characteristics of a building is investigated with reference to heating system performance.

The study is performed employing a simulation procedure. The outdoor temperature variations, considered in the calculations, are derived from a particular statistical analysis of real values referred to the zone considered (Genova-Italy).

The results of this investigation allow the evaluation ,in probabilistic form, of the optimum thermal design of buildings, in order to minimize the energy consumption and to get confortable indoor temperature.

NOMENCLATURE

a = thermal diffusivity of the building peripheral wall.

C_A = rooms' equivalent heat capacity;

E = daily energy consumption;

H = as defined by eq.6;

k = heat conductivity of the building peripheral wall;

L = thickness of building peripheral wall;

n = number of air change, per hour;

S = area of building peripheral walls;

t = peripheral wall temperature;

t_A = ambient temperature;

t_{Am} = minimum value of ambient temperature;

V = building's volume;

x = abscissas within building peripheral wall;

α' = heat transmission coefficient between air and inner surface of peripheral wall;

α'' = heat transmission coefficient between air and external surface of peripheral wall;

β = as defined by eq. 5;

θ = outdoor air temperature;

φ = heat power;

τ = time.

1. FOREWORD

Now more than ever in designing heating plants the choice of the design external temperature plays a fundamental role because it allows appreciable reductions of the operational and fuel costs, when properly selected.

The criteria for defining a design external temperature are numerous. As it is well known, irrespective of the structural characteristics of the building, one can define a design external temperature as an exclusive function of the meteorological conditions |1÷6| . Alternatively, one can define an external design temperature as a function of both the meteoroclimatic conditions of the zone considered and of the building characteristics |7÷11| .

The purpose of this paper is to present a numerical method which allows the evaluation of the power of a heating plant, by utilizing samples of temperature waves instead of a design external temperature, no matter how defined, keeping into account the building structural characteristics.

The parameters for making up the samples are obtained from a suitable statistical analysis of meteorological data about the region considered |12| .

2. DESCRIPTION OF THE SYSTEM

The present study has been developed with reference to a building having well defined geometrical characteristics and thermophysical parameters. With the aim of simplifying the mathematical model, which describes the heat transfer through the peripheral walls, these have been assumed opaque and homogeneous.

The study has been carried on assuming a building heated by a plant (e.g. using hot water) capable of supplying the ambients with a thermal power φ kcal/h , constant timewise. The heating system is controlled by means of an automatic regulator device having the purpose of limiting the temperature excursion of the ambient and avoiding energy wastes.

Namely, whenever the ambient temperature t_A, detected by a suitable sensor, is higher than the preset reference value (19°C in this case) the regulator takes over cutting off the heat supply to the ambient; and whenever it drops below the minimum preset value (18°C in this case) the regulator restarts the heat supply.

3. SYSTEM ANALYSIS

3a. Mathematical model of the building.

The analysis of the described operational behaviour of the system is developed by using a mathematical model which, by means of differential equations, expresses the heat balance of the building and the laws of heat transfer (in one dimensional unsteady state conditions) through the peripheral walls of the building.

These equations, together with the relative additional conditions, are presented in Table 1.

TABLE 1

$$\varphi = S \, \alpha' \, (t_A - t_{x=0}) + C_A \, \frac{\partial t_A}{\partial \tau} + 0,3 \, nV \, (t_A - \theta) \tag{1}$$

$$\varphi = 0 \rightarrow t_A > 19 \; °C$$

$$\frac{\partial t}{\partial \tau} = a \, \frac{\partial^2 t}{\partial x^2} \tag{2}$$

$$\alpha' \, (t_A - t_{x=0}) = - \, k \, (\frac{\partial t}{\partial x})_{x=0} \tag{3}$$

$$\alpha'' \, (t_{x=L} - \theta) = - \, k \, (\frac{\partial t}{\partial x})_{x=L} \tag{4}$$

The independent variables are time (τ) and abscissas (x) corresponding to the thickness (L) of the peripheral wall.

In particular, equation (1) describes the energy balance of the ambient under unsteady conditions. It includes: a term accounting for the leakage through the peripheral wall; another term representing the ambient thermal capacity, and, finally, a term accounting for the thermal requirement of the renewed air. Equation (1) is completed by two conditions representing the turning on of the heat source (having a power φ) when $t_A < 18°C$, and its cutting off when $t_A > 19°C$. Equation (2) refers to the peripheral wall of the building and describes the law of temperature variation of the same wall, in one dimensional unsteady state conditions. Equations (3) and (4) represent the boundary conditions of equation (2) and express the heat transfer between air and wall on the two faces of the same wall.

The external temperature θ , appearing in the aforementioned equations, is assumed to be variable with time and therefore can take on the values corresponding to the external conditions.

In this paper such external conditions have been specified having considered a great number of external temperature behaviours, obtained through a statistical analysis of the values of θ , measured in the zone. The method used for determining such a behaviours is briefly described in the following paragraph.

3b. Characterization of the external temperature.

The external temperature values, selected for the present paper, have been recorded in Genova (Italy) by the Servizio Meteorologico dell'Aeronautica as from November 1, 1972 to February 28,1973, with a frequency of two samples per hour. These data have been analysed in Fourier series according to the following expression:

$$\theta = \theta_0 + \sum_1^\infty \theta_n \cos \left(n \frac{2\pi}{T} \tau + \beta_n \right) \qquad\qquad n=1,2,\ldots\ldots \qquad (5)$$

where T is the period of the first harmonic, equal to 24 hours in this case.

The results are displayed in the form of hystograms: some of them are shown in figs. 1, 2 and 3. Figs. 1 and 2 represent the hystograms of θ_0 and θ_1 respectively, while fig.3 represents the hystogram of β_1.

Furthermore, plotting phase β_n versus amplitude θ_n values (as shown in figs. 4 and 5 referring to the first and second harmonic, respectively) it can be noticed that the amplitude and phase values may be correlated. Each of these diagrams shows the number of days for which θ_n and β_n are included in a specified interval $\Delta\theta_i$ and $\Delta\beta_i$. For instance, regarding fig.4 (fundamental harmonic) these intervals have been established equal to 1°C and $\pi/3$, respectively. Consequently, a generic value of θ_1 and β_1 can be calculated as:

$$\theta_1 = i \, \Delta\theta_1 \qquad\qquad i = 1,2,\ldots\ldots\ldots 6 \qquad\qquad\qquad (6)$$

$$\beta_1 = j \, \Delta\beta_1 \qquad\qquad j = 1,2, \ldots\ldots\ldots 6 \qquad\qquad\qquad (7)$$

Therefore, it is possibile to make up hystograms as shown in fig.6 (referring to fundamental harmonic) where the parameter H, determined as follow:

$$H = (i - 1) \, 6 + j \qquad\qquad\qquad\qquad\qquad (8)$$

is reported versus the relative number of days of the same H value. It may be notice that H univocally determines $\Delta\theta_i$ and $\Delta\beta_i$ intervals; in fact we have:

$$i = \text{integer of } |H/6| + 1 \qquad\qquad\qquad\qquad (9)$$

$$j = \text{rest} \quad \text{of } |H/6| \qquad\qquad\qquad\qquad\qquad (10)$$

Histograms of this kind have been assumed as reference, and the temperature waves,which are the input samples, are made up taking at random the amplitude and phase values of a preset number of harmonics (five in this case). For more details see reference |12| .

4.   RESULTS AND DISCUSSION

The calculation method presented in this paper can be used both for the optimal sizing of a heating plant and for the evaluation of energy consumption.

We report hereunder the results obtained in a building having well defined termophysical and geometrical characteristics, as specified in table 2. (A more

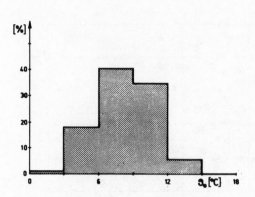

Fig.1 - Hystogram of θ_0

Fig.2 - Hystogram of θ_1

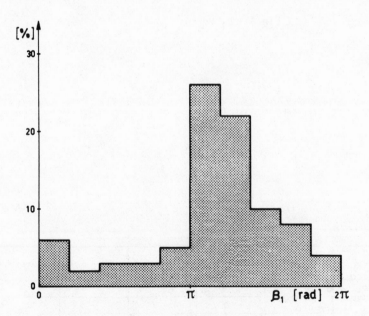

Fig.3 - Hystogram of β_1

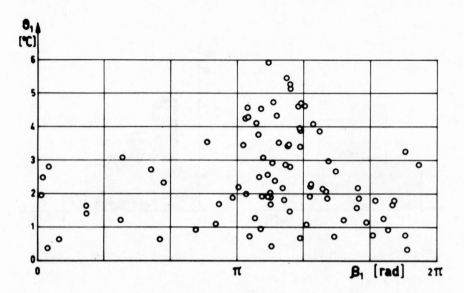

Fig.4 - θ_1 versus β_1

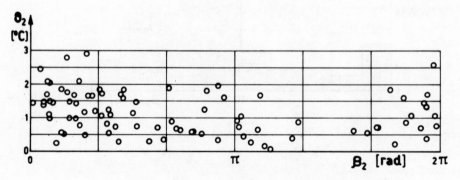

Fig.5 - θ_2 versus β_2

complete analysis is reported in |12| , where the results referring to several different buildings are shown; the differences mainly concern the weight of the peripheral walls).

TABLE 2

α' = 8 kcal/m^{2}h°C	a = 0,001 m^2/h
α'' = 20 kcal/m^{2}h°C	S = 3.500 m^2
n = 1	V = 10.000 m^3
L = 0,045 m	C_A= 34.800 kcal/°C
k = 0,1 kcal/mh°C	

The building is in Genova and it is therefore affected by the outdoor temperature wave as calculated through the above described procedure.

In fig.7, referring to the light wall building, results of the calculation performed, some examples, are shown. Each hystogram shows the probability that, inside the building, the temperature wave reaches minimum values corresponding to a series of installed heat power (80.000,110.000 and 180.000 kcal/h). From these diagrams it can be seen that the probability of unsatisfactory temperature conditions inside the building decrease with increasing installated thermal power.

With this procedure it is thus possible to determine the power required,related to a preset probability of minimum indoor temperature occurrence.

In order to evaluate how the installed power affects energy consumption, the hystograms reported in fig.8 show the probability of the occurrence of a daily energy consumption E, reported on the abscissae. The hystograms refer to powers of 80.000 and 180.000 kcal/h.

Comparing fig.7 with fig.8 it can be noticed, as already specified, that the probabity of unsatisfactory thermal conditions, inside the building, decreases with increasing thermal power. This implies, however, an increase in energy consumption.

The values obtained with the described procedure are clearly depending on the outdoor temperature variations and on the characteristics of the premise considered. Therefore, when considering different outdoor conditions and different building characteristics, it is advisable to plot hystograms analogous to those described above by using the calculation procedure described.

Through these hystograms it would therefore be possible to choose the most suitable heat power representing a compromise solution between a comfortable temperature inside the building and the need for the minimum installed power and energy consumption.

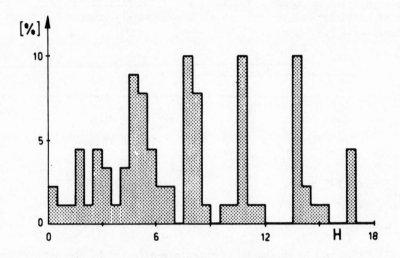

Fig.6 - Hystogram of H for first harmonic

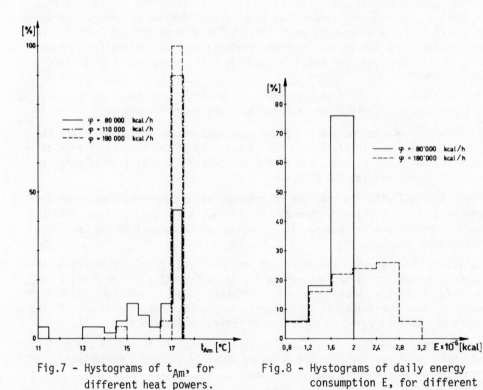

Fig.7 - Hystograms of t_{Am}, for different heat powers.

Fig.8 - Hystograms of daily energy consumption E, for different heat powers.

The calculation have been performed on the UNIVAC 1100/GR system of "CENTRO SIDERURGICO OSCAR SINIGAGLIA" Italsider, Genova (Italy). The programs have been written in FORTRAN 5 |12|.

REFERENCES

1. De Comelli G. Sul fabbisogno termico annuo di riscaldamento di quartieri urbani per le condizioni di clima italiane. La Termotecnica, 1967,n°5.

2. De Ponte et coll. Analisi del fabbisogno termico degli edifici con riscaldamento intermittente. Condizionamento dell'Aria, Riscaldamento, Refrigerazione, 1976, n°9.

3. Ullah M.B., Longworth A.L. A simpliefied multiple harmonic Fourier method for calculating periodic heat flow through building fabrics. The Building Services Engineer, June 1976, vol.44.

4. UNI-CTI 7357-74. Calcolo del fabbisogno termico per il riscaldamento degli edifici.

5. Rietschel H., Reiss W. Traité de chauffage et de climatisation. Dunod, Paris, 1973-74.

6. ASHRAE . Handbook of fundamentals (Chapter 33) 1972.

7. Piga M.G. Progettazione degli impianti di riscaldamento e scelta della temperatura esterna di progetto. Atti del XXX Congresso Nazionale A.T.I., Cagliari, Sept. 1975.

8. Ballantyne E.R., Airah M. Parametri climatici di progettazione e l'effetto del clima esterno sull'ambiente interno. Condizionamento dell'Aria, Riscaldamento, Refrigerazione, 1976, n°11.

9. Faggiani D., Magrini U., Reale G. Note sulla propagazione delle onde di temperatura negli edifici. Quaderno di Fisica Applicata n°4, a cura dell'Istituto di Fisica Tecnica dell'Università di Genova (Fac.Ingegneria)

10. Billington N.S. Isolamento termico e capacità termica degli edifici. Atti del VI Congresso Internazionale di Climatistica. Clima 2000-I-Milano, Marzo 1975.

11. Bondi P. et coll. Thermal performance of walls. Italian Studies and Researches. Consiglio Nazionale delle Ricerche -Roma, Marzo 1977.

12. Carrara C., Fantini L., Lorenzi A. A method for design heating building systems taking into account outdoor temperature variations. Pubblicazioni dell'Istituto di Fisica Tecnica e Impianti Termotecnici-Facoltà Ingegneria Università degli Studi di Genova - FTR.32, Aprile 1977.

COMPUTER PROGRAM "KLI"*

R. J. A. van der BRUGGEN AND J. T. H. LAMMERS

Eindhoven University of Technology
Eindhoven, The Netherlands

Abstract.

The heatbalance of a room is represented by a number of second order Fourier-
equations for the various layers of the walls and first order differential equations for the boundary conditions of the walls. The method of Crank Nicholson
is used for the numerical solution of the Fourier equation. The boundary conditions of the inside walls consist of two first order differential equations, one
for the inside of the room and the other for the outside of the room.
These two equations are used, by an iteration proces, for coupling neighbouring
rooms. The computerprogram is made in a conversational mode.

Introduction.

Within the group "Physical Engineering in Relation to Building design" of the
department of Architecture, Building and Planning of the Eindhoven University
of Technology has been developed a mathematical model for buildings, which can
be used to calculate the cooling and heating loads or, if required room-air temperatures.
The thermal environment in a building or a room is caused for an important part
by non-stationairy parameters, as outside-air temperature, solarradiation, etc.
As a result a number of problems cannot be solved accurately enough with analytical calculations; therefore a numerical solutionmethod has been developed.
Although there are many computerprograms to calculate cooling- and heating loads
this one is special because of:
1. For the numerical solution of the Fourier equation for the thermal conductance and the boundary equations   the discretisation-method of Crank Nicolson
   has been used.
2. The calculation can be done for a number of rooms at the same time. The rooms
   are coupled in the program by the heat exchange through the partition walls.
3. This program is written in a conversational mode so that the required input
   data can be entered as answers to the questions the computer asks.
This has two important advantages:
- In the first place it is not necessary for the program-user to be a specialist
  in the field.
- The conversational mode gives the user the possibility to enter in a simple
  way changes in the input data to compare different solutions. In that way it
  is possible to examine the influence of windowsize, composition of walls etc.

* A computing method to calculate both the maximum heating and cooling load under
  extreme conditions and the total energy demand by heating and cooling during a
  reference year for buildings.

The components of the heatbalance of a room.

The thermal environment in a building is caused by a number of external and internal factors acting upon that building. These factors are:
- Outside air-temperature.
- Solar-radiation absorbed by and/or transmitted through the facade.
- Wind verlocity and direction.
- Internal heat gain from occupants, lighting etc.
- Installed devices to control the indoor air-temperature at a certain level.
The next picture shows the places where and how in a room the heat transfer takes place.

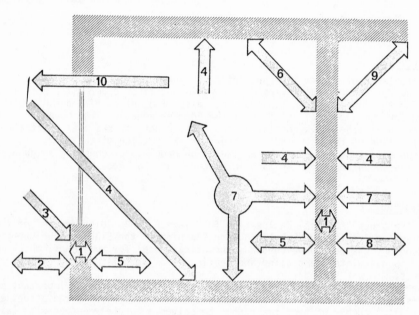

Figure 1.

The heat flows in figure 1 represent:
1. Heat conduction through the walls.
2. Heat exchange between the external walls and external environment by radiation and convection.
3. Absorption of solar radiation by the non-transparent part of the external walls.
4. Solar transmission through the windows.
5. Convective heat exchange between the walls and the indoor air.
6. Radiative heat exchange between the walls in the room.
7. Heat transfer by radiation and convection as a result of internal heat sources.
8. Convective heat exchange between the innerwalls and adjacent rooms.
9. Radiative heat exchange between the innerwalls and the walls in adjacent rooms.
10. Heat exchange between in- and outdoor air by infiltration and/or ventilation.

Heat transfer equations.

1. The non-stationary heat conduction in walls is described by the Fourier equation:

$$\frac{\delta T}{\delta t} = a_j \cdot \left(\frac{\delta^2 T}{\delta x^2}\right), \text{ where } a_j = \frac{\lambda_j}{\rho_j \cdot c_j} \tag{1}$$

$T(x,t)$ = temperature at place x and time t ($^\circ$C).

a_j = thermal diffusivity in layer j ($m^2 . s^{-1}$).

λ_j = thermal conductivity ($W.m^{-1}.K^{-1}$).

$\rho_j . c_j$ = volumetric heat capacity ($J.m^{-3}$).

2. Heat exchange between the walls and windows and the external environment:

$$-\lambda_1 \cdot \left(\frac{\delta T}{\delta x}\right)_{x=0} = a.Qz(t) + \alpha e (Te(t) - T(0,t)) \tag{2}$$

λ_1 = thermal conductivity in the first layer.

a = absorption factor.

$Qz(t)$ = incident solar radiation at time t ($W.m^{-2}$).

αe = external heat transfer coefficient for convection and radiation ($W.m^{-2}.K^{-1}$).

$Te(t)$ = outside air temperature at time t.

$T(0,t)$ = external wall surface temperature.

3. The heat exchange in the room can be subdivided in three ways of heat-transfer:

   a. Heat exchange between the walls and the indoor air.
   b. Heat exchange between the walls.
   c. Heat transfer by sun radiation transmitted through the windows and by internal sources.

The inside boundary condition for the Fourier equation is then described by:

$$-\lambda_j \cdot \left(\frac{\delta T}{\delta x}\right)_{x=d} = \alpha c(t).(T(d,t) - Ti(t)) + \alpha s . \sum_g F_{i,g}.(T(d_i,t)-T(d_g,t))-$$

$$-\{ (\sum_g A_g.Qzd(t) + (1 - fr).Qi(t))/Atot\}. \tag{3}$$

$\alpha c(t)$ = convective heat transfer coefficient at time t.
$T(d,t)$ = internal wall surface temperature.

$Ti(t)$ = room air temperature.
αs = radiative heat transfer coefficient.
$F_{i,g}$ = geometric factor between wall i and wall g.

$T(d_g,t)$ = internal surface temperature of wall g.

A_g = area of wall g (m^2).

$Qzd(t)$ = transmitted sun radiation at time t ($W.m^{-2}$).
fr = convective fraction of the internal sources.
$Qi(t)$ = internal heat sources at time t. (W).
Atot = total area of all the walls.

4. The heat exchange between the innerwalls and the adjacent rooms is described by:

$$-\lambda_1 \cdot \left(\frac{\delta T}{\delta x} \right)_{x=0} = \alpha c(t) \cdot (T(0,t) - Ti'(t)) + \alpha s \cdot \sum_z F_{i,z} \cdot (T(0,t) - T_z(d_z,t)) -$$

$$-\{\sum_z A_z \cdot Qzd_z(t) + (1 - fr') \cdot Qi'(t))/Atot'\}. \qquad (4).$$

$Ti'(t)$ = air temperature in the adjacent room.
$F_{i,z}$ = geometric factor between wall i and wall z in the adjacent room.
$T_z(d_z,t)$ = internal surface temperature of wall z in the adjacent room.

5. The heat exchange between the room and the external environment by ventilation and infiltration together with the convective heat transfer from the walls and the convective part of the internal heat sources gives the next equation:

$$\rho \cdot c \cdot V \cdot \frac{\delta Ti}{\delta t} = \sum_i \alpha c_i(t) \cdot A_i \cdot (T_i(d,t) - Ti(t) + fr \cdot Qi(t) +$$

$$+ \rho \cdot c \cdot V \cdot (Te(t) - Ti(t)) \cdot v/3600. \qquad (5).$$

ρ = specific mass of air $(kg.m^{-3})$.
c = specific heat of air $(J.kg^{-1}.K^{-1})$.
V = room volume (m^3)
v = ventilation rate (h^{-1}).

When in a room the temperature at a certain level has to be maintained, the equation changes into:

$$Q(t) = \rho \cdot c \cdot V \cdot \frac{\delta Tg}{\delta t} - \Sigma \alpha c_i(t) - A_i(T_i(d,t) - Tg(t)) - fr \cdot Qi(t) -$$

$$- \rho \cdot c \cdot V \cdot (Te(t) - Tg(t)) \cdot v/3600. \qquad (6).$$

$Q(t)$ = cooling or heating load at time t (W).
$Tg(t)$ = required room air temperature.

Discretisation of the Fourier equations.

For the discretisation of the Fourier equations the method of Crank Nicolson is used (ref. 1).

1. The heat conduction equation (1) is then given by:

$$-T(x-1,t) + 2 \{ \frac{h_j^2}{k.a_j} + 1 \} \cdot T(x,t) - T(x+1,t) = T(x-1,t-1) +$$

$$+ 2\{\frac{h_j^2}{k.a_j} - 1\} \cdot T(x,t-1) + T(x+1,t-1). \qquad (7).$$

h_j = semi-infinite displacement in layer j (m).
k = time step (s).

The discretisation of the boundary conditions is solved by using the next Taylor expansions around x=0 (or x=d):

$$T(1,t) = T(0,t) + h_1 \left(\frac{\delta T}{\delta x} \right)_{x=0} + \frac{h_1^2}{2} \left(\frac{\delta^2 T}{\delta x^2} \right)_{x=0} + \cdots\cdots$$

$$T(2,t) = T(0,t) + 2h_1 \left(\frac{\delta T}{\delta x} \right)_{x=0} + 2h^2{}_1 \left(\frac{\delta^2 T}{\delta x^2} \right)_{x=0} + \cdots\cdots \quad (8).$$

After elimination of the second-order derivatives:

$$\left(\frac{\delta T}{\delta x} \right)_{x=0} = (-3T(0,t) + 4T(1,t) - T(2,t))/2h_1 \qquad (9).$$

2. The heat exchange between the walls and windows and the external environment (2) using (9) is then given by:

$$(\frac{3\lambda_1}{2h_1} + \alpha e).T(0,t) - \frac{2\lambda_1}{h_1}. \ T(1,t) + \frac{\lambda_1}{2h_1}. \ T(2,t) =$$

a. $Qz(t) + \alpha e.Te(t)$. $\qquad (10).$

3. The inside boundary condition (3):

$$(\frac{3\lambda_i}{2h_i} + \alpha c(t) + \alpha s). \ T(d,t) - \frac{2\lambda_i}{h_i} \ . \ T(d-1,t) + \frac{\lambda_i}{2h_i} \ . \ T(d-2,t) -$$

$$- \alpha c(t). \ Ti(t) - \alpha s \sum_g F_{i,g}. \ T(d_g,t) = (\sum_g A_g.Qzd(t) + (1-fr)Qi(t))/Atot$$

$$\qquad (11).$$

4. The external boundary condition for the innerwalls (4):

$$(\frac{3\lambda_1}{2h_1} + \alpha c(t) + \alpha s). \ T(0,t) - \frac{2\lambda_1}{h_1} \ . \ T(1,t) + \frac{\lambda_1}{2h_1} \ . \ T(2,t) =$$

$$\alpha c(t). \ Ti'(t) + \alpha s \sum_z F_{i,z}. \ T(d_z,t) +$$

$$(\sum_z A_z. \ Qzd_z(t) + (1-fr'). \ Qi'(t))/Atot'. \qquad (12).$$

5. The heat balance equation (5) is described by:

$$\{\rho.c.V(1 + k.v/3600) + k \sum_i A_i.\alpha c(t)\}.Ti(t) \ - k\sum_i A_i.\alpha c(t). \ T_i(d,t) =$$

$$\rho.c.V. \ (Te(t-1) + k.v/3600. \ Te(t)) + k.fr. \ Qi(t). \qquad (13).$$

When the temperature in the room has to be maintained at a certain level the equation changes into:

$$k.Q(t) + k \sum_i \alpha c(t). \ A_i T_i(d,t) = \{ \rho.c.V. \ (1+v.k/3600)+$$

$$+ k.\sum_i A_i. \ \alpha c(t) \}.Tg(t) - \rho.c.V.\{Ti(t-1) + v.k/3600. \ Te(t)\} -$$

$$- k.fr. \ Qi(t). \qquad (14).$$

Method of solution.

For each wall in a room the discrete equations are, with exception of the balan-
ce-equation, placed in a matrix equation.
In each room can be a maximum of ten walls subdivided in six walls and maximum
four windows in the vertical panes.
The last equation of each tridiagonal matrix is placed, together with the balance equation in a new matrix. This matrix is solved by using the Crout method
(ref. 2). After solving this matrix equation the following quantities are known:

1. The indoor air temperature or the cooling or heating load.
2. The inside wall surface temperatures of all the walls in the room.

With these temperatures the original matrix equations of the walls are solved
and then the temperatures at each step in the walls are known.
The first equation of the matrices of the innerwalls (12) contains in the right
part two terms, that are unknown at time t.
These terms are: $\alpha e(t)$. $T_i'(t)$ and $\alpha s\sum_z F_{i,z}$. $T_z(d_z,t)$ where $T_i'(t)$ and $T_z(d_z,t)$
are the unknown temperatures in the adjacent room at time t.
An iterative calculation method is used to solve this problem.

The computerprogram can be used for:
1. Design calculations.
   The cooling- and or heating load or the resulting room-air temperatures in the
   various rooms of a building are calculated, using extreme outdoor conditions.
2. Energy- cost calculations.
   During a longer period the total energy-demand of a building in relation to
   cooling- and or heating can be determined. The outdoor conditions, necessary
   for the energy-demand calculations are provided by a reference-year wich will
   be discussed in the next part.

A reference year for energy-cost calculations in buildings.

The reference year is intended to be used for absolute and comparative energy
cost calculations. For that reason the determination of the reference year does
originate with the energy consumption of buildings.
It was calculated how energy has to be supplied in a month to a certain room for
maintaining a desired room air-temperature. Therefore the heatloss by the hour,
by transmission and ventilation was determined. The heat gain by the sun radiation and internal heat sources was subtracted.

The 10 comparable months were ranked in order of the agreement of the monthly
energy consumption with the average for these months.
To form a good notion of the behaviour of the monthly energy consumption the
following situations were researched:
For a good and a bad insulated room:
all combinations of the orientations: North, East, South or West, with a high
or low number of air changes and with or without internal heat sources.

The calculation of the monthly energy consumption has the following limitations:
1. The heat capacity of the construction is not taken into account,
2. The used room had one exterior wall, including a window,
3. During occupation hours (8-18 hour) a desired air temperature was 15^oC,
4. The used meteorological data referred only to the center of Holland (De Bilt).
   The period is 1961 - 1970.

To check the limitations, the following calculations were done:
1. For one situation the energy consumption of the months January and December

has been calculated with the computerprogram, mentioned before, that does take into account the heat capacity of the construction.
2. In some situations the time of occupation was 24 hours and the desired air temperature was 22°C. This was also done for the month of January for one situation, by using the program, mentioned in 1.
From this study the following conclusions are to be drawn:
It is not possible to construct one reference year for the cooling and the heating; two different reference years are necessary. These composed years are valid, without regard to the construction or orientation of the building. This is in agreement with the results of H. Saito and Y. Matsuo (9).

The composition of these two reference years is shown in the next table.

Month	Heating	Cooling
January	1961	1966
February	1965	1962
March	1965	1963
April	1964	1964
May	1965	1969
June	1961	1969
July	1964	1968
August	1961	1968
September	1967	1962
October	1970	1970
November	1967	1961
December	1961	1966

Composition of the reference years selected from the period 1961 - 1970.

The verification of the computermodel.

The verification of the computermodel (10) has been checked in a number of practical cases, out of which two will be described. The first is a well insulated house with a relatively high heatcapacity made of concrete and having double pane windows, and the second one is an officeroom in the main building of the Eindhoven University of Technology, which is badly insulated and has a relatively low heatcapacity.
The physical parameters measured were:
1. solarradiation, total global as well as the diffuse radiation.
2. windspeed and direction.
3. ventilationrate.
4. wall surface temperatures.
5. airtemperatures.
6. airvelocities inside, along the heattransmitting wall-surfaces.
The parameters were measured, for both cases, during two weeks continuously, and the data were primarily treated by a data-acquisition system (11), consisting of:
1. multiplexer (100 channels).
2. A.D. converter.
3. calculator (16 K bytes semi-conductor memory).
4. digital cassette-recorder (250 K bytes storage on one cassette-tape).
After concluding the measurements, the data were read into the Burroughs 6700/ 7700 computer system of the University.

The well insulated house (12).

The livingroom, in this house, is used for the verification. The vertical facades consist of concrete (thicknes 0.26 m) with glasswool (thickness 0.05 m) and

Figure 2.   The well insulated house in a block of three. Measurements were done in the one in the middle.

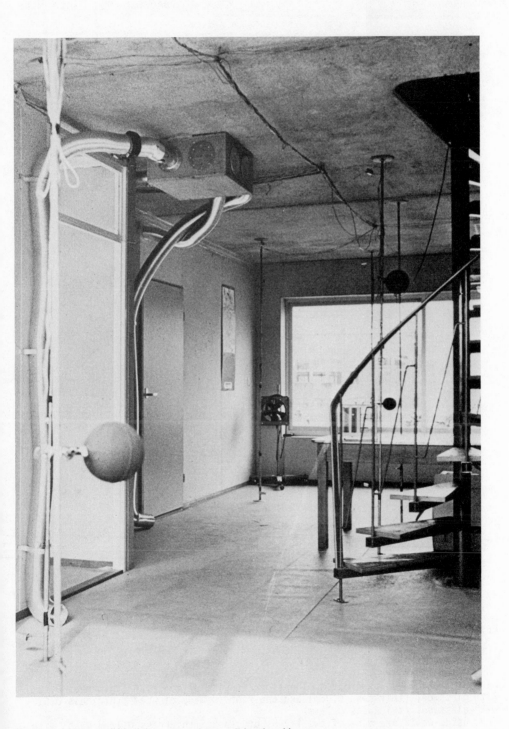

Figure 3. Picture of the living room of the well insulated house.

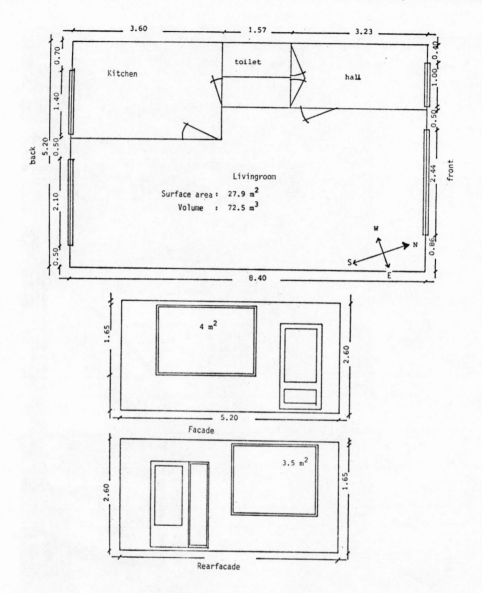

figure 4.

Layout of the groundfloor of the well insulated house.

windows with double pane glass. The floor consist of concrete (thickness 0.13m) with glasswool (thickness 0.05m) and the roof is made of wood (thickness 0.013m) with glasswool (thickness 0.12 m).
The inside walls are made of two layers of gypsum (thickness 0.12 m each) with a small airspace in between. The ceiling of the livingroom is made of concrete (thickness 0.13 m).

The measurements and calculations were done without heating or cooling and the results are given in figure 5.

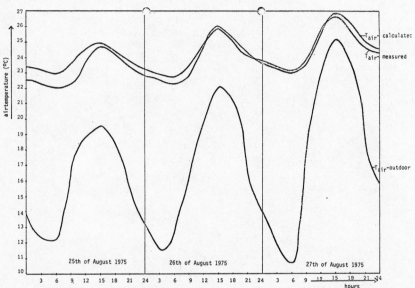

Figure 5: airtemperatures measured and calculated in a well insulated house.

The officeroom.

The officeroom has two facades consisting of wired glass at the outside (thickness 0.006 m), cork (thickness 0.025 m), steel plates at the inside and double pane windows. The inside walls consist of brick (thickness 0.1 m). The floor is made of concrete (thickness 0.45 m) and the ceiling is made of wood (thickness 0.02 m) with an aluminum cover.
The measurements and calculations were done without heating or cooling and the results are given in figure 9.

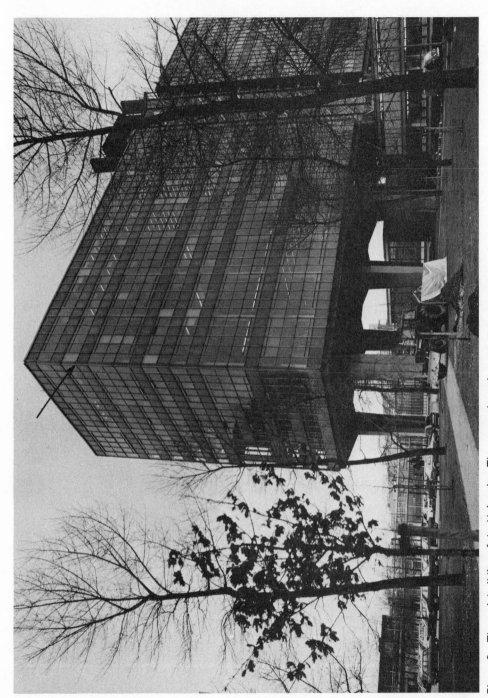

Figure 6. The mainbuilding of the University. The room where the measurements were done is situated at the ninth floor and is shown in the picture by an arrow.

Figure7. Picture of the inside of the officeroom.

461

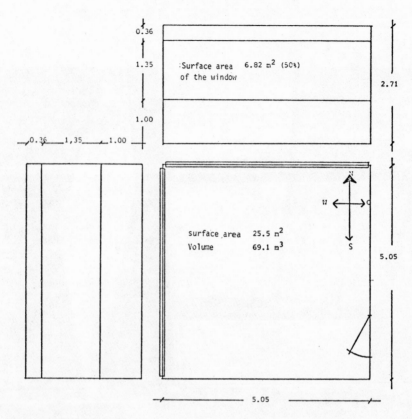

figure 8.

Layout of the officeroom.

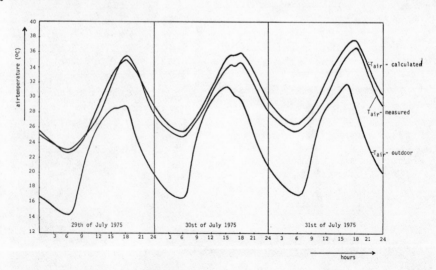

figure 9.

Airtemperatures measured and calculated in a room of the mainbuilding of the
Eindhoven University of Technology.

Conclusions.

As shown in figures 5 and 9 the calculated and measured temperatures match very well. This gives confidence in the validity of the computermodel. More recent results show that in buildings which were heated and/or cooled the calculation of the energy consumption over a period of a few weeks has an accuracy of 6%, when compared with the actual consumption.

References:

1. Prof. dr. G.W. Veltkamp, lecturebook "Numerical Methods", Technische Hogeschool Eindhoven.

2. R.C. Information 5, Computercentre, Technische Hogeschool Eindhoven.

3. Imura, Natural-convection heat transfer from a plate with arbitrary inclination. International Journal of Heat and Mass transfer, Vol. 15, pp.755-767.

4. Ir. S.W.T.M. Oegema, The digital calculation of solarradiation, Report 332-III, Technisch Physische Dienst.-Toegepast Natuurwetenschappelijk Onderzoek.-Technische Hogeschool Delft.

5. Ir. S.W.T.M. Oegema, The digital calculation of reflection, absorption and transmission of solarradiation for a number of windowsystems. Report 332-IV, Technisch Physische Dienst,-Toegepast Natuurwetenschappelijk Onderzoek, Technische Hogeschool Delft.

6. Gröber, Erk, Grigull, Wärmeübertragung.

7. Ir. S.W.H. de Haan, Calculation of solarradiation on a plane with an arbitrary slope by using the solarradiation on a horizontal plane. Report 032-VI, Technisch Physische Dienst.-Toegepast Natuurwetenschappelijk Onderzoek.-Technische Hogeschool Delft.

8. Hans Lund, The "Reference Year", paper A3 of the second symposium on the use of computers for environmental engineering related to buildings, Paris (France) June 13-15 1974.

9. Heizo Saito and Yo Matsuo, Standard Weather Data for Shase Computor Program of Annual Energy Requirements and Example Results of Hourly load for Ten Years in Tokyo, Paper A4 of the same symposium as 1.

10. D.S.M. intermezzo 1975/8, p. 14-21, Information Centre, P.O. Box 64, Heerlen, The Netherlands,

11. A.J. Lammers, Dataacquisition-programm for a compucorp 425 calculator, Internal report, Technische Hogeschool Eindhoven.

12. Ir. R.J.A. v.d. Bruggen, Ir. J.T.H. Lammers, Verification of Computerprogram "KLI", to be published in July 1977 in Klimaatbeheersing.

APPLICATION OF THE DIGITAL COMPUTER TO EVALUATE ENERGY CONSERVATION OPTIONS IN BUILDING AIR CONDITIONING BASED ON TEMPERATURE DEVIATION FROM DESIGN

A. IQBAL, J. K. NEWLIN, AND P. R. FISH

Atkins Research and Development
Epsom, Surrey KT18 5BW, England

SYNOPSIS

The use of computers in building airconditioning heat transfer calculations
is now well established, and we have come a long way from the time when the
computer was used simply to crunch numbers of total building heat gains.

Today, the use of computers is opening up many new and interesting avenues
for practising engineers. With the added incentive to conserve energy,
studies are now being carried out which would have been inconceivable without
a computer.

This paper is based on one such set of studies, where the impact of allowing
the temperature to swing has been quantified and the many ways it can be used
to conserve energy in buildings described.

The authors have been developing computer programs over the last two years,
with the aim of getting a better understanding of building heat transfer and
consequently the energy behaviour due to the interaction of weather, building
characteristics, plant characteristics and control systems. This work has
culminated in a Fortran package, the starting point of which is a conventional
cooling load calculating program. The loads output by this program form the
input to the family of programs one of which is described here.

1. INTRODUCTION

The first generation of airconditioning computer programs simple calculated
the cooling loads on which engineers have traditionally based their equipment
selection.

The present energy awareness has necessitated a fresher and a much deeper
analysis of the thermal behaviour of buildings. The use of computers is no
longer limited to plant selection or just plant selection on the basis of
meeting the worst conditions. The emphasis is clearly on energy analysis, and
design of buildings and equipment based on minimum energy consumption.

The second generation of programs can take these analyses a step further and
closer to reality.

2. THE PRESENT WORK

This paper describes a system for energy analyses and outlines in detail, the analyses possible by studying the deviation of actual temperature from the design condition.

The complete system is shown in Fig. (i). The starting point of this is a conventional cooling load calculating program. This calculates the hourly cooling and heating loads for a constant design internal temperature and writes the results onto a magnetic disc file. This in essence, defines the building behaviour in a given configuration.

The second phase programs are essentially post processors which then operate on the results file created by the central cooling load program.

The post processor that has been of considerable interest in the present energy climate is KOSWING which evaluates the energy behaviour when temperature is allowed to deviate from its design point and forms the subject of this paper.

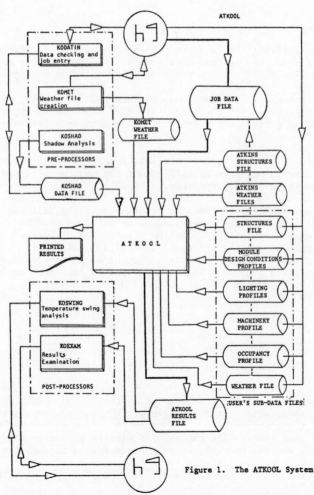

Figure 1. The ATKOOL System

3.   THEORY OF KOSWING

Koswing is part of a larger system called ATKOOL based on the thermal response factor technique developed originally in Canada (1).   The overall system concept uses response factors to calculate 1) instantaneous heat gains through fabric conduction given inside/outside sol-air temperature difference 2) cooling loads on inside air given instantaneous heat gains and 3) most important of all, the inside air temperature deviation from the design value given cooling-load/heat extraction rate differences.   Each application of the thermal response factor is called a "convolution".

3.1 The Convolution Principle

The ATKOOL load calculation procedure makes extensive use of a convolution principle to account for the thermal storage effect of the building structure. In this it differs from older techniques which use such concepts as equivalent temperature differences and storage load factors.

A time-dependent variable, for example, air temperature, T, is usually expressed as $T(t)$ to mean that T is a function of time t.   Another way of expressing the time-dependent variable is by a set of numbers such as T_1, T_2, T_3 ... T_m to mean that the set consists of a series of numerical values, which are air temperatures at t = 1, 2, ...   This set is called the time series and expressed by $[T]$.

When one time series $[A]$ is influenced by another time series $[B]$, the relation between these two series may be expressed in a linear form as:

$$A_t = \sum_{j=0}^{m} X_j * B_{t-j} \quad \text{for } t = 0, 1, 2 ... \tag{1}$$

In this example, the value of A at time t is expressed as a linear function of all the time values of B at t = t, t-1, t-2, ... t-m with X_0, X_1 ... X_m being time-independent coefficients.   The above equation is called the convolution and X_0, X_1 ... X_m are called the "filter coefficients" in the mathematics of "time series analysis".   They are called the response factors when referring to wall or roof heat conduction, and weighting factors (2) when referring to the hourly load calculation in the ATKOOL load calculation procedure.   In the above expression, the time series $[A]$ is said to be calculated "by convolving" the time series $[B]$ with the response factors $[X]$.

The convolution scheme is employed in three different places in the ATKOOL calculation procedure.   Firstly, the transient heat conduction through exterior walls and roofs is calculated by convolving the outside and inside surface temperatures with wall response factors.   Secondly, the space cooling load is calculated by convolving the instantaneous heat gain with its weighting factors.   Thirdly, the most important in the context of this paper, the calculation of inside air temperature (dry-bulb or environmental) deviation from the design value given cooling load/heat extraction rate differences.

The value of m in the convolution equation depends upon the degree that the time parameter B_{t-m}, or B at m hours previous to time t, would influence the value of A_t.   If the response of B_{t-m} upon A_t is insignificant, X_m is nearly zero and the values of B beyond $(t-m)^{th}$ hour are of no importance.   If the time lag effect does not exist for the relation between two time series, $[A]$ and $[B]$, the value of m will be zero or the response factors X_j will all be zero except for the first term X_0.   In this case, thermal load is calculated on a steady state basis.

3.2 Cooling Load and Actual Heat Extraction Rate

Cooling load and heat extraction rate are two separate parameters which are often confused. Cooling load is the rate at which heat <u>should</u> be extracted from a building to maintain a constant internal temperature (i.e. the design temperature). Heat extraction rate is the heat that is actually extracted and this may or may not match the cooling load. The inside temperature for any hour depends on the design temperature and cooling load vs the heat extraction rate for the present hour and all other hours. It also depends on the thermal storage capacity of the building.

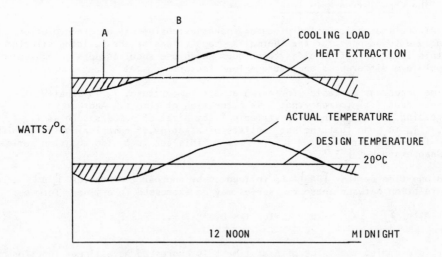

Figure 2 Heat Extraction and Actual Temperatures

Consider a building for which the cooling load has been calculated to be as shown by curve B in Figure 2. Suppose that the heat extraction rate is constant for all hours of the day as shown by line A in Figure 2. As the figure shows, the inside temperature will vary throughout the day. For those times of day when the heat extraction rate is higher than the cooling load, the actual temperature will fall below the design temperature and conversely the temperature will swing above the design temperature when the extraction rate falls short of the cooling load.

The amount of variation of space temperature depends upon the amount of heat supplied or extracted by the heating/cooling system, and the admittance of the building. In this case, the admittance of the building is implied in the weighting factors, and the transfer function is expressed in the form (3):

$$HE_t = CL_t + \sum_{j=0}^{v} x_j * \delta_{t-j} - \sum_{j=1}^{w} y_j * (HE_{t-j} - CL_{t-j}) \qquad (2)$$

Where δ is the deviation of actual air temperature from the reference point.

HE : Heat extraction rate
CL : Cooling load
x_j
y_j : Weighting factors implying admittance

The heat extraction rate given by equation (2) must match the characteristic of the heating or cooling system.

A simple proportional control has a characteristic of the form:

$$HE_t = C + D . \delta_t \qquad (3)$$

Where C : heat extraction rate of the system of the unit operating at the design condition

D : change in rate of heat extraction by one degree rise/fall in space temperature

Equation (2) and (3) give:

$$\delta_t = \frac{CL_t - C + \sum_{j=0}^{v} X_j . \delta_{t-j} - \sum_{j=1}^{w} y_j * (HE_{t-j} - CL_{t-j})}{D - X_o} \qquad (4)$$

Heat extraction rate can be set at any level and as a special case HE=0 corresponds to plant shut off. It is therefore possible to evaluate the swing of temperatures experienced during high gain periods in non-air conditioned buildings. This forms a very useful tool for architects to establish whether a new building would require air conditioning. In many cases the temperature swings can be brought within acceptable limits, by changing building design parameters such as the use of shading devices, tinted glazing, or variable ventilation rates.

It is possible to calculate the resultant temperatures with any given heat extraction rate; it can conveniently model a control device which could govern the extraction rate to be proportional to the resultant temperature. The user can choose the minimum and maximum plant capacities and the bandwidth thereby determining the proportions of the control response curve (Figure 3). The curve can be shifted left or right by the choice of set point temperature which may vary from hour to hour. The set point choice is analogous to a thermostat setting; when the control senses a room temperature above the set point it calls for more heat extraction (up to the limit of maximum capacity). The maximum and minimum plant capacities may be both positive (to represent cooling plant) both negative (to represent heating plant) or positive and negative respectively (to represent both cooling and heating plant). The minimum may also be set to zero to correspond to plant shut off.

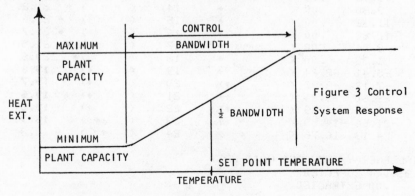

Figure 3 Control System Response

4.   APPLICATIONS OF KOSWING

Having explained the theory of KOSWING, let us examine the results of its application to one sample hypothetical building as an example of its power.

4.1 Energy Savings with Temperature Swing

First compare the energy requirements for maintaining the design temperature precisely with those of allowing inside temperature to vary either side of the design value. Figure 4 shows the energy supplied in each hour (as negative heat extraction) and the resulting temperatures varying between 18.2°C and 21.7°C. The cost of maintaining exactly 20°C can be obtained by summing the cooling load and is 318,000 Wh in heating and 54,000 Wh in cooling. Note that in letting the temperature swing, no cooling is required at all. (This is not intended to be the lowest cost plan for heating this building on this day).

Figure 4 Energy Savings with Temperature Swing

```
ON/    COOLING
OFF     LOAD    --HEAT EXTRACTION--              -- TEMPERATURE --
       (THOUSAND WATTS)                                   (C)
                    MIN           MAX                   20.0
                  -50.00          .00            16.4       21.7
       ON   -25.92 -18.42 (      ♦    )     1    (    ♦    )     18.7
       ON   -27.83 -20.11 (      ♦    )     2    (    ♦    )     18.6
       ON   -29.39 -21.54 (      ♦    )     3    (    ♦    )     18.4
       ON   -30.82 -22.86 (     ♦     )     4    (    ♦    )     18.4
       ON   -32.15 -24.09 (     ♦     )     5    (   ♦     )     18.3
       ON   -33.17 -25.08 (     ♦     )     6    (   ♦     )     18.2
       ON   -29.46 -22.89 (      ♦    )     7    (    ♦    )     18.4
       ON   -24.92 -19.92 (      ♦    )     8    (    ♦    )     18.6
       ON    -5.83  -7.22 (          ♦)     9    (       ♦)     19.5
       ON    -1.45  -3.79 (          ♦)    10    (       ♦)     19.7
       ON     1.36  -1.46 (           ♦    11    (        ♦     19.9
       ON     4.29   .00  (           ♦    12    (        ♦     20.2
       ON     6.96   .00  (           ♦    13    (        )♦    20.7
       ON     9.80   .00  (           ♦    14    (        )  ♦  21.2
       ON    11.31   .00  (           ♦    15    (        )   ♦ 21.6
       ON    10.88   .00  (           ♦    16    (        )   ♦ 21.7
       ON     8.04   .00  (           ♦    17    (        )  ♦  21.5
       ON     1.38   .00  (           ♦    18    (        )♦    20.6
       ON    -5.41  -2.24 (           ♦    19    (       ♦)     19.8
       ON    -7.12  -3.87 (          ♦)    20    (       ♦)     19.7
       ON    -8.60  -5.24 (          ♦)    21    (       ♦)     19.6
       ON    -9.85  -6.39 (          ♦)    22    (       ♦)     19.5
       ON   -21.66 -14.59 (        ♦  )    23    (     ♦   )    18.9
       ON   -24.13 -16.76 (        ♦  )    24    (     ♦   )    18.8

       TOTAL ENERGY(THOUSAND WATT-HRS)
          236.44 SUPPLIED
            .00 EXTRACTED
```

```
ON/   COOLING
OFF   LOAD   --HEAT EXTRACTION--              -- TEMPERATURE --
      (THOUSAND WATTS)                              (C)
                    MIN       MAX                        20.0
                  -40.00       .00           15.0      21.3
 ON   -25.92  -9.07 (        ♦ )      1     ( ♦)     :    16.5
 ON   -27.83 -11.34 (       ♦  )      2     ( ♦)     :    16.4
 ON   -29.39 -13.22 (      ♦   )      3     ( ♦)     :    16.3
 ON   -30.82 -14.93 (     ♦    )      4     ( ♦)     :    16.3
 ON   -32.15 -16.48 (    ♦     )      5     ( ♦)     :    16.2
 ON   -33.17 -17.74 (   ♦     )       6     ( ♦)     :    16.1
 ON   -29.46 -15.43 (     ♦    )      7     ( ♦)     :    16.2
 ON   -24.82 -27.49 ( ♦        )      8          (♦ )     18.6
 ON    -5.88 -12.17 (       ♦  )      9          ( ♦)     19.4
 ON    -1.45  -7.62 (         ♦ )    10          ( ♦)     19.6
 ON     1.36  -4.50 (          ♦)    11          (  ♦     19.8
 ON     4.29  -1.47 (           ♦    12          (  ♦     19.9
 ON     6.96   .00 (            ♦    13          (  ♦     20.2
 ON     9.80   .00 (            ♦    14          ( )♦     20.8
 ON    11.31   .00 (            ♦    15          ( ) ♦  21.2
 ON    10.88   .00 (            ♦    16          ( ) ♦  21.3
 ON     8.04   .00 (            ♦    17          ( ) ♦  21.1
 ON     1.38   .00 (            ♦    18          ( ♦     20.2
 ON    -5.41   .00 (            ♦    19     ( )     ♦:    19.2
 ON    -7.12   .00 (            ♦    20     ( )    ♦ :    18.8
 ON    -8.60   .00 (            ♦    21     ( )  ♦  :    18.5
 ON    -9.85   .00 (            ♦    22     ( )  ♦  :    18.2
 ON   -21.66  -3.62 (          ♦)    23     ( ♦     :    16.8
 ON   -24.13  -6.72 (         ♦ )    24     ( ♦     :    16.7

TOTAL ENERGY(THOUSAND WATT-HRS)       Figure 5 Energy Savings with
   161.80 SUPPLIED                    Night Set Back
      .00 EXTRACTED
```

```
ON/   COOLING
OFF   LOAD   --HEAT EXTRACTION--              -- TEMPERATURE --
      (THOUSAND WATTS)                              (C)
                    MIN       MAX           20.0
                  -50.00       .00          20.0      22.0
 ON   -27.24 -26.42 (     ♦    )      1     ♦           20.0
 ON   -28.38 -27.67 (    ♦     )      2     ♦           20.0
 ON   -29.49 -28.87 (    ♦     )      3     ♦           20.0
 ON   -30.37 -29.83 (    ♦     )      4     ♦           20.0
 ON   -31.36 -30.89 (    ♦     )      5     ♦           20.0
 ON   -32.04 -31.63 (    ♦     )      6     ♦           20.0
 ON   -32.73 -32.37 (   ♦      )      7     ♦           20.0
 ON   -33.17 -32.86 (  ♦       )      8     ♦           20.0
 ON   -10.45 -10.19 (        ♦ )      9     ♦           20.0
 ON    -2.85  -2.62 (         ♦)     10     ♦           20.0
 ON     6.35   .00 (            ♦    11     )      ♦    21.0
 ON     9.15   .00 (            ♦    12     )        ♦  21.5
 ON    11.39   .00 (            ♦    13     )         ♦ 22.0
 ON     6.13   .00 (            ♦    14     )        ♦  21.3
 ON     4.93   .00 (            ♦    15     )       ♦   21.2
 ON     1.99   .00 (            ♦    16     )     ♦     20.8
 ON     -.71   .00 (            ♦    17     ) ♦         20.4
 ON    -1.56   .00 (            ♦    18     )♦          20.2
 ON    -9.46  -7.13 (          ♦)    19     ♦           20.0
 ON   -10.13  -8.26 (         ♦ )    20     ♦           20.0
 ON   -10.50  -8.96 (         ♦ )    21     ♦           20.0
 ON   -11.08  -9.79 (         ♦ )    22     ♦           20.0
 ON   -23.75 -22.65 (     ♦    )     23     ♦           20.0
 ON   -25.56 -24.62 (     ♦    )     24     ♦           20.0

TOTAL ENERGY(THOUSAND WATT-HRS)       Figure 6 Mild Weather Case - Full
   334.77 SUPPLIED                    Time Heating
      .00 EXTRACTED
```

4.2   Energy Savings with Night Set-Back

Next, for the same building and weather conditions, we can examine the energy
savings obtained through setting the thermostat back to 16.5° between the hours
of 19 and 7.   Figure 5 shows the inside temperature falling slowly from hour
19 onward until heating is first called for in hour 23.   A low level of heating
results until hour 8 when the set point is restored to normal and a heavy dose
of heating is required in the first hour to bring the temperature up.   The
savings over the scheme shown in Figure 4 is a further 74,000 Wh per day.

4.3   Mild Weather Case

Next we can measure the energy savings and comfort increase obtained by
turning the heating plant off at night.   In this case the cooling plant is
also assumed to be off all day.   Figure 6 shows the plant operation pattern
and resulting temperatures (rising to 22°C) when the heating plant is on full
time.   Here the temperature has been assumed to be controlled precisely and
the total daily energy requirement is 335,000 Wh.

Figure 7 shows the same parameters when the plant only operates from 9a.m. to
5p.m.   Here we see the building interior cooling considerably before the plant
starts heating at 9a.m.   Heat supplied is greater in every hour because the
cold interior structure is absorbing the extra.   It is this same cold
structure which results in the midday temperature rising to 1°C above the
design value and the overall daily energy consumption is reduced to 88,000 Wh.

4.4   Effects of Weather Severity on Start Up Times

Here we have made ATKOOL runs for several different weather severities, where
the severity is expressed as the probability that outside air temperatures will
be warmer than those stated to the ATKOOL program.   Using KOSWING the optimum
start-up time has been found for each weather severity as the latest possible
first hour of plant operation which will give 20°C inside air temperature by
9a.m. (for a fixed plant capacity of 50,000 W).   Figure 8 shows the inside
temperature and energy consumption profiles for 90% severity weather when the
plant is started at hour 3 (the latest possible hour in this case).   Note that
severity of N per cent in month M is defined as that temperature in each hour
which has a probability of being exceeded at that hour in month M and that
cloud cover which has a probability of N per cent of exceeding the cloud cover
for a day in month M.

ON COOLING
OFF LOAD --HEAT EXTRACTION-- -- TEMPERATURE --
(THOUSAND WATTS) (C)

ON OFF	LOAD	MIN -50.00	MAX .00			hr		11.6 ·	20.0 20.1		temp
OFF	-27.24	.00	(		)	1		♦		)	14.2
OFF	-28.38	.00	(		)	2		♦		)	13.8
OFF	-29.49	.00	(		)	3		♦		)	13.3
OFF	-30.37	.00	(		)	4		♦		)	13.0
OFF	-31.36	.00	(		)	5		♦		)	12.6
OFF	-32.04	.00	(		)	6		♦		)	12.2
OFF	-32.73	.00	(		)	7		♦		)	11.9
OFF	-33.17	.00	(		)	8		♦		)	11.6
ON	-10.45	-35.47	(	♦	)	9			♦		20.0
ON	-2.85	-22.27	(	♦	)	10			♦		20.0
ON	6.35	-9.21	(	♦)		11			♦		20.0
ON	9.15	-3.64	(	♦)		12			♦		20.0
ON	11.39	.00	(	♦		13			♦		20.1
ON	6.13	-2.89	(	♦)		14			♦		20.0
ON	4.93	-2.83	(	♦)		15			♦		20.0
ON	1.99	-4.71	(	♦)		16			♦		20.0
ON	-.71	-6.53	(	♦)		17			♦		20.0
OFF	-1.56	.00	(		)	18			♦)		19.0
OFF	-9.46	.00	(		)	19		♦		)	17.8
OFF	-10.13	..00	(		)	20		♦		)	17.6
OFF	-10.50	.00	(		)	21		♦		)	17.4
OFF	-11.08	.00	(		)	22		♦		)	17.2
OFF	-23.75	.00	(		)	23		♦		)	15.3
OFF	-25.56	.00	(		)	24		♦		)	14.7

TOTAL ENERGY(THOUSAND WATT-HRS)
87.55 SUPPLIED
.00 EXTRACTED

Figure 7 Mild Weather Case - Night Shut Off

ON/ COOLING
OFF LOAD --HEAT EXTRACTION-- -- TEMPERATURE --
(THOUSAND WATTS) (C)

ON/ OFF	LOAD	MIN -50.00	MAX .00			hr		8.9	20.0 20.0		temp
OFF	-48.17	.00	(		)	1		♦		)	9.6
OFF	-49.33	.00	(		)	2		♦		)	8.9
ON	-50.86	-50.00	♦		)	3			♦	)	15.7
ON	-51.77	-50.00	♦		)	4			♦	)	16.0
ON	-52.79	-50.00	♦		)	5			♦	)	16.2
ON	-53.50	-50.00	♦		)	6			♦	)	16.3
ON	-54.61	-50.00	♦		)	7			♦	)	16.4
ON	-54.68	-50.00	♦		)	8			♦	)	16.5
ON	-31.98	-50.00	♦		)	9			♦		20.0
ON	-23.60	-38.28	(♦		)	10			♦		20.0
ON	-14.00	-26.15	(	♦	)	11			♦		20.0
ON	-10.77	-21.00	(	♦	)	12			♦		20.0
ON	-8.13	-16.85	(	♦	)	13			♦		20.0
ON	-12.96	-20.46	(	♦	)	14			♦		20.0
ON	-14.13	-20.62	(	♦	)	15			♦		20.0
ON	-17.03	-22.67	(	♦	)	16			♦		20.0
ON	-19.70	-24.60	(	♦	)	17			♦		20.0
OFF	-20.91	.00	(		)	18			♦	)	16.3
OFF	-28.79	.00	(		)	19		♦		)	14.8
OFF	-29.82	.00	(		)	20		♦		)	14.2
OFF	-30.61	.00	(		)	21		♦		)	13.7
OFF	-31.59	.00	(		)	22		♦		)	13.2
OFF	-44.28	.00	(		)	23		♦		)	11.1
OFF	-46.50	.00	(		)	24		♦		)	10.3

TOTAL ENERGY(THOUSAND WATT-HRS)
540.63 SUPPLIED
.00 EXTRACTED

Figure 8 Optional Start Up - 90 Per Cent Severe

```
ON/   COOLING
OFF   LOAD    --HEAT EXTRACTION--              -- TEMPERATURE --
      (THOUSAND WATTS)                                (C)
                   MIN        MAX                    20.0
                  -50.00      .00          9.0       20.0
OFF  -39.39   .00 (          )       1      ◆        ) 11.4
OFF  -40.54   .00 (          )       2     ◆         ) 10.8
OFF  -41.88   .00 (          )       3    ◆          ) 10.3
OFF  -42.78   .00 (          )       4    ◆          )  9.8
OFF  -43.79   .00 (          )       5   ◆           )  9.4
OFF  -44.48   .C0 (          )       6  ◆            )  9.0
ON   -45.45 -50.00 ◆         )       7             ◆ ) 15.9
ON   -45.64 -50.00 ◆         )       8              ◆) 16.5
ON   -22.94 -48.17 ◆         )       9              ◆  20.0
ON   -14.90 -35.12 (   ◆     )      10              ◆  20.0
ON    -5.48 -22.09 (      ◆  )      11              ◆  20.0
ON    -2.41 -16.33 (       ◆ )      12              ◆  20.0
ON     .07 -11.76 (        ◆ )      13              ◆  20.0
ON    -4.95 -15.09 (      ◆  )      14              ◆  20.0
ON    -6.14 -14.90 (      ◆  )      15              ◆  20.0
ON    -9.05 -16.65 (      ◆  )      16              ◆  20.0
ON   -11.73 -18.35 (     ◆   )      17              ◆  20.0
OFF  -12.78   .00 (          )      18             ◆ ) 17.3
OFF  -20.67   .00 (          )      19           ◆   ) 15.9
OFF  -21.55   .00 (          )      20           ◆   ) 15.5
OFF  -22.19   .00 (          )      21          ◆    ) 15.1
OFF  -22.99   .00 (          )      22         ◆     ) 14.8
OFF  -35.67   .00 (          )      23      ◆        ) 12.7
OFF  -37.71   .00 (          )      24      ◆        ) 12.0
```

TOTAL ENERGY(THOUSAND WATT-HRS)
  298.46 SUPPLIED
    .00 EXTRACTED

Figure 9 Optimal Start Up - 70 Per Cent Severe

```
ON/   COOLING
OFF   LOAD    --HEAT EXTRACTION--              -- TEMPERATURE --
      (THOUSAND WATTS)                                (C)
                   MIN        MAX                    20.0
                  -50.00      .00         10.2       20.0
OFF  -33.31   .00 (          )       1      ◆        ) 12.8
OFF  -34.46   .00 (          )       2     ◆         ) 12.3
OFF  -35.68   .00 (          )       3     ◆         ) 11.8
OFF  -36.57   .00 (          )       4    ◆          ) 11.4
OFF  -37.57   .00 (          )       5    ◆          ) 11.0
OFF  -38.26   .00 (          )       6   ◆           ) 10.6
OFF  -39.09   .00 (          )       7   ◆           ) 10.2
ON   -39.41 -50.00 ◆         )       8             ◆  ) 17.3
ON   -16.70 -41.32 ( ◆       )       9              ◆  20.0
ON    -8.87 -28.41 (   ◆     )      10              ◆  20.0
ON     .44 -15.48 (        ◆ )      11              ◆  20.0
ON    3.37  -9.89 (        ◆ )      12              ◆  20.0
ON    5.73  -5.48 (        ◆)       13              ◆  20.0
ON     .59  -9.00 (        ◆ )      14              ◆  20.0
ON    -.61  -8.87 (        ◆ )      15              ◆  20.0
ON   -3.53 -10.69 (        ◆ )      16              ◆  20.0
ON   -6.22 -12.44 (        ◆ )      17              ◆  20.0
OFF  -7.17   .00 (          )       18             ◆ ) 18.1
OFF  -15.07   .00 (          )      19           ◆   ) 16.9
OFF  -15.84   .00 (          )      20           ◆   ) 16.5
OFF  -16.35   .00 (          )      21          ◆    ) 16.3
OFF  -17.04   .00 (          )      22          ◆    ) 16.0
OFF  -29.71   .00 (          )      23       ◆       ) 14.0
OFF  -31.64   .00 (          )      24      ◆        ) 13.4
```

TOTAL ENERGY(THOUSAND WATT-HRS)
  191.59 SUPPLIED
    .00 EXTRACTED

Figure 10 Optimal Start Up - 50 Per Cent Severe

Figure 9 illustrates the results of optimal start-up on a day when the weather
is of 70% severity. Here start up was delayed until the 7th hour and energy
demand was reduced from 540,000 to 300,000 Wh. (Note how the temperature
jumps from 16.5 to 20°C in hour 9 because the heating demand(negative cooling
load) was reduced dramatically at that hour by incidental gains of occupancy).

Finally, Figure 10 illustrates the results for 50% severity (average) weather
where the optimal start-up was at the 8th hour. Note the further reduction in
energy required and, as in Figure 6, the cool building structure continuing to
draw heat from the air even in hours when there is a positive cooling load
(calculated on constant inside air temperature). Further reductions in
weather severity result in optimal start-up in the 9th hour with the building
coming up to temperature within that hour.

5.   CONCLUSION

The results presented are not to be taken as generalisations which apply to all
buildings, weather patterns, occupancy profiles and loads. They are merely an
example of the power of the ATKOOL package as a tool which is capable of
handling all the variations of individual problems and allows an engineer to
make his own generalisations based on experience and experiment.

References

1.   D.G. Stephenson and G.P. Mitalas - National Research Council of Canada:-
      Technique presented in a paper "Cooling Load Calculations by Thermal
      Response Factors" at ASHRAE semi-annual conference. Detroit-Michigan
      1967, Paper No. 2018.

2.   G.P. Mitalas and D.G. Stephenson - "Room Response Factors" ibid Paper No.
      2019.

3.   "Procedures for Determining Heating and Cooling Loads for Computering
      Energy Calculations" - compiled by The Task Group on energy requirements
      for heating and cooling. ASHRAE 1975.

THERMAL DYNAMICS
OF A BUILDING REPRESENTED
BY RESPONSE FUNCTIONS

R. S. SOELEMAN

TNO Research Institute for Environmental Hygiene
Postbus 214
Delft WIJK 8, The Netherlands

ABSTRACT

Computing methods based on the use of the system modeling program itself
might have the objection to be expensive for energy consumption analyses
as a consequence of computing time.
The introduction of the transfer function (in this case the impuls response
function) concept could be the remedy to this end. In this contributing
paper a concised description is given regarding the seasonal heat consumption of a dwelling.
Computed figures are the result of a computer program executed by means
of the response functions of the simulated dwelling system.
With simple statistical methods a comparison has been made between calculated
values and actual readings.

1. INTRODUCTION

To characterize building performance as it affects energy consumption
during the heating season to control indoortemperature, the response
function concept has been adopted.
Heat loss calculations by means of specific response functions of separate
parts of the building structure accept any wall and roof construction and
unsteady, nonperiodic outdoorconditions.
Calculated energy consumption figures of different types of dwellings
provide better appreciation of building performance, if the calculation
method can directly accept "real" weather data.

Although numerous studies have been carried out to improve accuracy of
computing methods, relatively little has been done to characterize thermal
performance of dwellings by an explicit expression.
It was expected that our study could be a contribution to this end.

If one wants to predict the energy demand for heating a dwelling during
the winterseason, the heat balance concept is the crucial scheme.
Each dwelling has its specific heat balance and the mutual relationship
between heat input and heat loss will be dependent on the type of dwelling,
on the life behaviour of the occupants and the weather.

It is therefore of great importance to have a sound comparative base as
to the building performance itself.
Conservation of energy for domestic heating could mean improved insulation
of outer walls and roofs, while some attention has to be paid to decrease
air infiltration rate.

The problem to face is to gain knowledge in a quantitative sense about the
thermal dynamics of different types of dwellings.
Heat loss will be affected by wall and roof construction, thermal insulating
properties of window panes, airinfiltration cracks and ventilation ducts.
Weather conditions are continuously changing on an unpredictable way.
All these items are essential and have to be treated in the computerprogram to study energy consumption of dwellings.

2. BASIC CONSIDERATIONS

2.1. Room modeling system

To meet the requirements for a presentation of building performance
related to "real" weather conditions, a computerprogram is developed
to compute impulseresponses of room modeling systems.
The modeling system includes the computation of:

(a) energy transfer through the structure by means of an integrating
    version of the finite difference method

(b) energy exchange inside the room by convective and radiative modes
    of heat transfer to the room air and between the enclosing
    surfaces respectively

(c) energy exchange between indoor and outdoor air by infiltration/
    ventilation

(d) energy supply to the room air to keep indoortemperature under
    controlled conditions

2.2. Weather data input

Air temperature data are readily available for various locations on an
hourly basis.
Predicting the intensity of the solar radiation incident upon a surface
is a problem of the relative position of the sun and the surface.
Orientation and inclination define the position of the surface; latitude, time of day, and the sun's declination define the position of
the sun.
This information is used to calculate the angle of altitude of the sun,
from which the solar intensity normal to the radiation is derived.
The angle of incidence is calculated and used to find the intensity
of solar radiation on the surface.
Weather station readings of daily total global radiation, daily sunshine hours and cloud cover are used to calculate the distribution
of direct and diffuse radiation.

3. CALCULATION OF ENERGY DEMAND

Energy demand for heating the room air is computed as the solution of a
convolution integral of two time dependent functions, the impulse response
function of the room system and the weather input function respectively.
This computing method enables the user to handle arbitrary input functions,
not necessarily periodic.

To verify the utility of the computing method, which was based on the use
of response functions of the "dwelling" instead of the room modeling
system itself, the "energy"-computerprogram had been run with "real"
weather data of a complete heating season.
In connection with our study the occupant of the apartment under investigation was requested to make gasmeter readings twice a day.
This was necessary because of the applied setback of the thermostat to
15º C for the night periods.

Building information data would never be as accurate and complete as
desired. For recently built houses or buildings under construction, however,
complete floor plans and detailed information concerning structural
composition of the building fabric are usually available.

Precalculations of "geometric factors" needed in connection with longwave
radiant interaction of the enclosing surfaces of the room, and of essential
response functions of the room modeling system had to be executed in
advance of the energy-program.
Finally the energy-program had been run on the basis of half-hourly values
of weather input-data.

4. ANALYSIS OF ENERGY CONSUMPTION FIGURES

Because of a lack of empirical data with respect to air infiltration characteristics of the dwelling, the energy-program had been executed with a
constant ventilation rate as well as with an assumed relationship between
ventilation rate and daily mean of windvelocity.

The computed results just as the gasmeter readings were compiled statistically and brought into comparison.
Table 1 shows the weekly figures.
These figures, the predicted as well as the real values, were correlated
with outside temperature (weekly mean values).
As an effort to express quantitatively the validity of the room modeling
system, that means the response functions of the system, statistical
measures have been derived.
Heat consumption figures were correlated not only with outside temperature,
but also with insolation and with windvelocity.
Results of statistical analysis are tabulated in tables 2(a), 2(b) and 2(c).

TABLE 1. FIGURES OF WEEKLY HEAT CONSUMPTION (MJ/week).

WEEK NR	ACTUAL READINGS	CALCULATED I*	CALCULATED II*	WEEK NR	ACTUAL READINGS	CALCULATED I*	CALCULATED II*
1	2543	2159	1828	14	2837	2979	2559
2	2904	2653	2145	15	1974	2236	2144
3	2484	2538	2408	16	2463	2658	2671
4	3247	3540	3379	17	2775	2338	2878
5	2751	2895	2719	18	2704	2663	2925
6	3456	3582	2882	19	2763	2523	2617
7	3084	3058	2922	20	2205	1754	1835
8	2867	2746	2866	21	1622	1022	1386
9	2161	3053	3212	22	855	684	778
10	2624	2776	2567	23	784	686	816
11	2712	2595	2282	24	1309	1417	1368
12	2303	2653	2651	25	1774	1610	1563
13	2258	2509	2280				

Seasonal sum (MJ)	59459	59327	57681
Seasonal mean (MJ/week)	2378	2373	2307

* I = ventilation rate dependent on windvelocity
 II = ventilation rate constant

Table 2(a). Correlation of heat consumption (Y) with outside
         temperature (X)

		ACTUAL READINGS	CALCULATED I	CALCULATED II	
correlation coefficient	r	−0.658	−0.765	−0.900	
mean heat consumpt.	$\overline{Y}$	2378	2373	2307	(MJ)
mean outside temp.	$\overline{X}$	5.9	5.9	5.9	(°C)
regression coefficient	b_y	−184	−245	−258	
regression coefficient	b_x	−0.002	−0.002	−0.003	
stand. error of estimate	S_y	502	492	297	(MJ)
stand. error of estimate	S_x	1.8	1.5	1.0	(°C)
number of weeks	n	25	25	25	

Table 2(b). Correlation of heat consumption (Y) with global radiation (X)

		ACTUAL READINGS	CALCULATED I	CALCULATED II	
correlation coefficient	r	−0.871	−0.894	−0.848	
mean heat consumpt.	$\overline{Y}$	2378	2373	2307	(MJ)
mean global radiation	$\overline{X}$	5.87	5.87	5.87	(MJ/m^2)
regression coefficient	b_y	−113	−132	−112	
regression coefficient	b_x	−0.006	−0.006	−0.006	
stand. error of estimate	S_y	327	342	363	(MJ)
stand. error of estimate	S_x	2.5	2.3	2.7	(MJ/m^2)
number of weeks	n	25	25	25	

Table 2(c). Correlation of heat consumption (Y) with windvelocity (X)

		ACTUAL READINGS	CALCULATED I	CALCULATED II	
correlation coefficient	r	0.452	0.484	0.154	
mean heat consumpt.	$\overline{Y}$	2378	2373	2307	(MJ)
mean windvelocity	$\overline{X}$	7.7	7.7	7.7	(m/s
regression coefficient	b_y	148	181	52	
regression coefficient	b_x	0.001	0.001	0.0004	
stand. error of estimate	S_y	594	668	676	(MJ)
stand. error of estimate	S_x	1.8	1.7	2.0	(m/s)
number of weeks	n	25	25	25	

5. CONCLUSIONS

Although the results of comparison seem to be encouraging, we are conscious of their incompleteness.
Comparisons were made as to the weekly figures, which as a consequence will cancel out daily fluctuations.
However, dealing with energy consumption figures of an entire heating season, the question arises whether we should go that far to compute hourly quantities of consumption.
From the results of table 1 we have learned that the seasonal consumption quantity can be predicted quite well in spite of deviations in the weekly figures.
As far as it concernes efforts to reduce computing time, we have got the experience that the application of response functions in the calculation is suitable.

REFERENCES

1. Mitalas, G.P., and Stephenson, D.G.
   Room thermal response factors.
   ASHRAE Transactions 73, I (1967)

2. Mitalas, G.P.
   Calculation of transient heat flow through walls and roofs.
   ASHRAE Transactions 74, II (1968)

3. Yamasaki, N.
   Puls transfer function and its application related to buildings.
   Proc. Symp. on Use of Computers for Environmental Engng. related
   to buildings. N.B.S. (U.S). Building Sci. Ser. 39 (Sept. 1971)

4. Matsuo, Y., and Takeda, H.
   The calculation method of heat load and its examples by Response
   Factormethod.
   Transactions of the SHASE, Vol. 9 (1971)

5. Choudhury, N.K.D., and Warsi, Z.U.A.
   Weighting function and transient thermal response of buildings.
   Part I and Part II, Int. J. Heat and Mass Transfer (1964)

6. Brown, G.
   Verfahren zur Berechnung des Kühl- und Wärmebedarfs mittels
   Digitalrechner.
   VVS 34 (1963), Nr. 11

7. Heynert, P. and Bauerfeld, W.L.
   Die Behandlung instationärer Wärmeleitprobleme in Räumen mit
   Hilfe des Methode der digitalen Simulation.
   Ges. Ing. 93 (1972)

CALCULATION OF ENERGY CONSUMPTION OF A BUILDING BY COMPUTER SIMULATION

A. H. C. van PAASSEN AND E. N. 't HOOFT

Delft University of Technology
Department of Mechanical Engineering
Mekelweg 2, Delft, The Netherlands

Abstract

A computer simulation is made of the thermal behaviour of an existing
office building and its air conditioning system, a so called
"variable volume system".

The yearly energy consumption and the influence on energy consumption of the
various properties of the building and its air conditioning system are calculated and discussed.

To reduce computer time much attention is paid to the construction of compact
mathematical models of the various components of the system without loss of
accuracy. For example the behaviour of the building is calculated by means of
the method of room response factors.
To obtain a further reduction of computer time a new approach of this calculation
is suggested in this paper.
A very simple calculation model is used to simulate the water cooler and the
heat recovery system (twin coil system), which is derived from a complex model
describing the physical behaviour in detail.

Introduction

Designing an optimal air conditioning system with respect to capital and
energy costs requires a flexible computer program, making it possible to
simulate any airconditioning system.
However it showed to be very difficult to develop such a program. Therefore,
as a first start, a computer simulation of an existing office building with a
specific airconditioning system, a socalled variable volume system, was set up.

Only improvements of this building and this air conditioning system are
considered with respect to energy consumption.
The description of the computer simulation is restricted to the mathematical
models of the most important parts such as the thermal behaviour of the
building and characteristics of the air cooler and heatrecovery system.

The office building and its air conditioning system.

The office building is of a traditional style with considerable heat capacity,
in which 100 individual rooms are available for 150 employees.
The outside plan dimensions of the building are: 42 m long, 20 m high and
15 m wide. The exposed wall is well insulated and much attention is paid to
its construction to avoid air leakages. Moreover only 30 % of the facade is
taken up by windows consisting of double pane (heat absorbing on the outside
and normal pane on the inside) with on the inside Venetian blinds as sun-
protection. The orientations of the facades are NW and SE (angles between the
normal vector of the facade and the south are respectivily -115 and + 65°,
taken positive towards east).
The internal load, caused by the computer and other devices, is 25 kW,which
is 30 % of the mean value of the yearly consumption of electricity.
The indoor climate is controlled by a combined water/air system. The water
is used as heating- and the air as cooling medium. With the water flow,heat
is transferred into the room by radiators under the window, while with the
air flow the required fresh air and cooling load of the room are controlled by
means of a socalled variable volume system (VAV). The supply air temperature is
more or less constant between 14 and 15°C depending on the outdoor temperature.
The radiators are controlled by means of an outdoor thermostat and are
transferring such a surplus of heat into the room that the variable volume box,
which is controlled by the room thermostat, always can supply the required
minimum amount of fresh air.
If cooling is required the indoor thermostat controls the VAV box, by changing
the supply of cooled air.
The exhaust air is used to ventilate the lightings and next its sensible heat
is used to heat the fresh air by means of a twin-coil heat recovery system.
The heat is supplied by two atmospheric gas boilers delivering water of 135°C,
that is used to heat the radiators and to supply steam for the humidifiers.
Cooled water of 6°C is delivered by two waterchillers driven by reciprocating
compressors. These chillers are of different size, 164 and 193 KW and are

controlled in three steps. This cooled water is used for the air cooler in
the central air handling installation of the VAV system. The condensor is
cooled by river water.

The air handling system has three functions:

1) Supplying cooled air for the VAV boxes. Its temperature is 15°C at an outdoor temperature of -10°C and decreases lineair to 14°C for higher outdoor
temperatures.

For outdoor temperatures higher then 15°C the supply air temperature is kept
constant at 14°C.

2) The second function is controlling the humidity of the supply air. For outdoor temperature between -10°C and 15°C the relative humidity is controlled
by the steam humidifier between respectively 50 and 90 %, resulting in room
air humidities of respectively 30 and 50 %. At higher outdoor temperatures
the humidifier is switched off and no active humidity control is available. At
absolute humididies of the oudoor air higher then 11 g/kg the air cooler will
dehumidify the air keeping the maximum relative humidity in the building below
65 %.

3) The third function is supplying the required amount of air by controlling
the fans for the supply and return air flow. The supply air flow is controlled
by a throttling air damper. However, in the simulation it is assumed that both
fans are controlled by the inletvanes. The required air flow is controlled by
keeping the pressure, measured halfway the airducts, at constant level. This
air flow is at fully opened VAV boxes 34000 m^{3}/h.

Mathematical models used for simulation.

In this part a more detailed description is given of the mathematical models
used for the characteristics of the building, the air cooler and the recuperative
twin coil heat recovery system. The other components such as fans, chillers,
boilers will be discussed briefly.

Heating and cooling load calculation of the building.

The load is calculated by a computer program based on the method of room thermal
response factors as outlined by Stephenson and Mitalas [1] .   In short
the method will be described. The dynamic thermal behaviour of the walls is
expressed by response factors. They predict the rate of heat conduction, q_n,
in the time interval, n, into the surface as a sum of responses caused by
series of indivudual temperature pulses occurring at the boundary surface during the

preceeding time intervals. This time series of triangular pulses, can be
considered as a good approximation of the continuous time function of the wall
temperature. See Figure 1a.

The response factors can be obtained by calculating the variation of the rate
of heat q(t), which is caused by the mentioned pulswise change of the surface
temperature on both sides of the wall. The discrete values of this variation,
occurring on the discrete moments,are the response factors. In figure 1b this
is shown schematically.

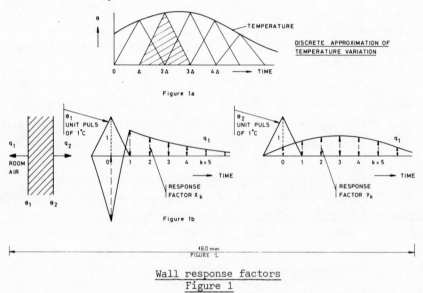

Figure 1a

Figure 1b

Wall response factors
Figure 1

The factors x_k are the result of a pulswise temperature change of the wall
surface on the room side and y_k is the result of the temperature change of the
wall surface on the other side. The rate of heat conduction into the wall
surface can now be written as:

$$q_n = \sum_{k=0}^{\infty} (x_k \, \theta_{1,n-k} - y_k \, \theta_{2,n-k})$$

In a simular way as described for the wall,a response of cooling or heating
load of a room can be calculated, which is caused by a triangular pulswise
variation of the indoor- or outdoor air temperature or by a simular variation
of heat flux caused by lighting or sunshine.

Again the values of these responses at discrete moments are the room response
factors (R_k) for respectivily the room air temperature, the lighting, the
sunshine and the outdoor temperature. Furtheron these variables will be called
the excitation variables.

The cooling or heating load, q_c, can now be given by following function, which connect the excitation variables to the cooling load by the room response factors.

$$q_{c,n} = \sum_{k=0}^{\infty} \left\{ R_{v,k} \, q_{v,n-k} + R_{s,k} \, q_{s,n-k} + R_{u,k} \, (\theta_{u,n-k} - L) + R_{i,k}(\theta_{i,n} - L) \right\}$$

....(1)

labels v,s,u,i,k indicate respectivily lighting, sunshine, outdoor
temperature, indoor temperature, time interval.

L = reference temperature

For the calculation of the room air temperature the preceeding equation can be rewritten explicit in $\theta_{i,n}$.

Ofcourse, the room response factors are only valid for the linearised thermal system for which they are calculated.

Consequently the room response factors are dependent of nonlineair effects such as the position of the Venetian blinds and windvelocity. For the energy calculation the room cooling load response factors are calculated for 2 positions of the blinds, namely open and closed, and for various regions of windvelocities. During the hour by hour calculation the proper response factors are choosen each hour for all the excitation variables.

In equation (1) the number of response factors has to be limited to a minimum value to reduce computer time. To study the effect of this limitation the response of the indoor temperature and cooling load of a heavy room on a daily variation of the sunshine is calculated. The maximum values of the responses calculated with different numbers of response factors are compared with the exact values in figure 2. It shows, that at least 100 response factors have to be used to calculate cooling loads as well as indoor temperatures.

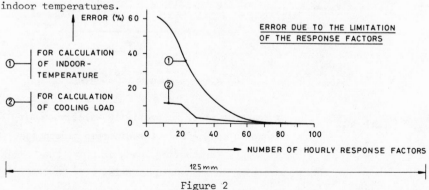

Figure 2

A faster cooling or heating load calculation model

Although the computer time of the method, described before, is reasonable, a further reduction in computer time is necessary in order to make optimising calculations economical.   This is realised by using one or more representative wall temperatures as an intermediate step of cooling load calculations. A representative wall is a combination of walls with the same structure and without mutual heat exchange by radiation. As an approximation, only one wall can be considered, which exchanges heat with the room air by convection and with the window surface by radiation.

Calculation of the room response factors eliminates the wall temperature. This temperature can be considered as the integrated thermal history of the room. By eliminating the wall temperature the integrator is eliminated too. Therefore the complete detailed history must be described by the response factors. This can only be done by using a large number of room response factors for each of the excitation functions. If one or more wall temperatures are used as an intermediate step of calculation the history can be described more compactly by just a small number of wall response factorsFor example, for a normal inside wall the 20-th wall response factor is almost zero, while for the room response factors this is the case for numbers higher than 100. Consequently it is advantaguous to develop a calculation method, which uses one or more representative walls.

By solving analytically the heat balances, describing the thermal behaviour of the room, by elimination of all variables except the representative wall temperature(s) and room air temperature the following two equations can be derived:

$$\theta_{i,k} = \frac{1}{c_2} \sum_{j=1}^{\infty} (x_{ij} - y_{ij}) \cdot \theta_{i,k-j} - \frac{c_3}{c_2} \cdot \theta_{1,k} - \frac{c_4}{c_2} \cdot q_{z,k} - \frac{cv_4}{c_2} \cdot q_{v,k} + \frac{c_5}{c_2}$$

$$\cdot \theta_{u,k} \qquad \ldots \ldots (2)$$

$$q_k = c_6 \cdot \theta_{i,k} + c_7 \theta_{i,k} + c_8 \cdot q_{z,k} + cv_8 \cdot \eta_{v,k} + c_9 \cdot \theta_{u,k} - v_c \cdot q_{v,k} \ldots \ldots (3)$$

with $\theta_{i,k}$, $\theta_{1,k}$, $\theta_{u,k}$ = temperature of respectivily wall, room air and outside air in the discrete moment k.

q_k, $q_{z,k}$, $q_{v,k}$ = cooling- or heating load, in coming sunshine and lighting

The coefficients c_i are complex functions of parameters such as heat transfer coefficients, and surface area's. Changing position of the Venetian blinds or wind velocity causes different coefficients. However the wall response factors (x_{ij}, y_{ij}) remain constant in contrast with room response factors used in the cooling load calculation.

Air cooler and twin coil heat recovery system.

In cases the air cooler is dry and not working at its maximum capacity the heat transferred from the cooled water to the air is defined by its temperature control system and can be calculated by a simple heat balance .Wet coolers, however, dehumidify the air too. The amount of dehumidification is free then and must be calculated in order to know the heat transferred to the air. The model as described by Green [2] is used for this purpose, because the known input- and unknown output variables of the cooler are related by lineair equations sothat no iteration is necessary in case of a counter flow heat exchanger. The coefficient of these equations are complex functions of the cooler parameters and the air- and cooling water flow in the coils.

These coefficients are calculated by a separate computer program.

For a serpentine cross flow heat exchanger with 8 rowes of tubes, arranged in line in the direction of the air flow, the mathematical model was checked by experiments; 23 measurements at dry cooler conditions and 29 at wet cooler conditions. A good agreement was obtained by matching following relation for the local heat transfer of the fins (without fin efficiency)

$$Nu = c_1 \ (P\acute{e} \ \frac{Deq}{D} \ ^{0,4}) \text{ with } Deq = \frac{4AA.D}{A_{tot}}$$

c_1 = 2,7 - dry cooler

$\quad$ = 1,5 - wet cooler

AA = front area of cooler minus that of
$\qquad\quad$ coils and fins

D = depth of cooler

A_{tot} = area of fins + coils

The heat transfer coefficient on the water side was calculated by the relation given by Hausen [3] . Using these relations the agreement is good for most of the measurements. A histogram for the difference between the calculated and measured values of the cooled air temperature showes that the mean value is zero and the standard deviation $0,3^{\circ}C$.

For the absolute humidity the mean error is 0,2 and the standard deviation is 0,2 g/kg.

It was found that as soon as condensation starts the sensible heat transfer
coefficient on the air side drops and remains constant, independently of rate
of dehumidification. This in contrast with other investigations [4] .
Calculations with this model showed following phenomenon, which is very useful
for the system simulation: the air is always cooled to almost saturation before
the condensation proces starts. This means that the effect of the cooler
on the air can be represented by a predictable curve in the h-x diagram of humid
air. Consequently the heat transferred to the air can be calculated easily
without losing much accuracy.

The twin coil heat recovery system as used in this air conditioning system,
consists of two heat exchangers which are coupled on the waterside. Each heat-
exchanger can be described by the same lineair equations as derived for the
air cooler. By eliminating the in- and outlet water temperature only one lineair
equation is left, which relates the unknown outlet air temperature $T_{Au,1}$ to
the known variables such as the inlet temperature and absolute humidities of
both heat exchangers.

$$T_{Au,1} = W_1 T_{Ai,1} + W_2 X_{i,2} + W_3 T_{Ai,2} + W_4 X_{i,2} + W_5$$

The coefficients W_i can be calculated by the same computer program as for the
air cooler. For a dry heat exchanger the coefficients are more or less
independent of the inlet conditions except for the airflow. In a VAV system the
airflow is changing and consequently these coefficients have to be expressed
as functions of the airflow. By these functions and preceeding equation the
behaviour of the heat recovery system can now be described for all load conditions.
When condensation occurs in one or both heat exchangers the situation is more
complex.
In that case the coefficients are also dependent on the unknown water temperature and consequently an iteration proces will be necessary.

Other components of the air conditioning system are simulated by polynomials
derived from firms data, according to the procedures of ASHRAE [5] . For
example the fan power P, in case inlet vane control is used, is simulated as a
function of the load L by following polynomial:

$$P_i/P_{max} = 0,4149 - 0,1124.L + 0,2424 L^2 + 0,4798 L^3$$

The water chiller cooled by river water, is simulated by a function, relating
the electricity consumption with only the cooling load. In future a more detailed
model will be developed to calculate the effects of alternatives such as air

cooled condensors or different control systems.

Simulation of the air conditioning system.

The mathematical models of the components of the VAV system are combined
such that the information flow agrees with that of the real system.
The simulation was strictly limited to the basic features of the VAV system.
An attempt will be made to develop a fully flexible computer simulation which
can simulate any air conditioning system just by changing its input data.
With the computer simulation of the VAV system various alternatives of this
system can be investigated by calculating its behaviour hour by hour.
As outdoor climate the hourly meteorological data of the period 1961-1970,
recorded by the Dutch Weather bureau K.N.M.I., are used. From this 10 years a
socalled building oriented reference year is selected [6] .

Effects of various parameters and control procedures on energy consumption

The existing office building is occupied day and night, consequently the air
conditioning system has to be in operation during 24 hours. All calculations
are therefore carried out for continuous operation, except one calculation to
show the effect of the more frequently applied intermittent operation.
Ten variants of the existing system are considered. Each of them differs from
the existing system by one or more improvements or changes. Furtheron   the
existing system is considered as variant 1.
A comparison is made of all variants in diagrams 1 and 2. Diagram 1 gives a
visual impression of the electricity consumed by water chillers, fans and
lighting. Diagram 2 shows the total consumption of electricity and natural
gas and the energy cost. The latter is calculated by weighing both energy
items by following prices: 1 kwh electricity costs 0.16 and 1 m^3 natural gas
costs 0.23 Dfl. In the diagram 2 the mentioned 100 % corresponds to 85.500 Dfl.
Now the variants and their effects on energy consumption will be discussed
successively.
Variant 2. The applied static outdoor thermostat is replaced by a dynamic one,
whose behaviour is described by a simple first order differential equation.
The coefficients of the equation are approximated by means of the step responses
of the cooling load on the various excitation functions.
A reduction in natural gas of 6.4 % can be obtained using this thermostat.
Variant 3. A sort of computerised ideal outdoor thermostat is used, which leads
to a further reduction of 2.4 %.

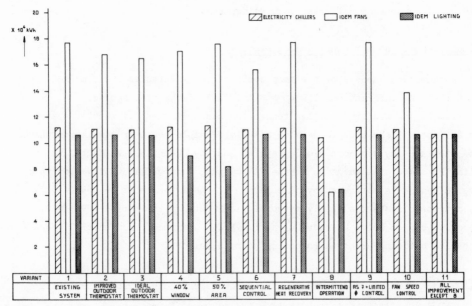

Comparison of the yearly consumption of electricity
of chillers, fans and lighting for various variants

diagram 1.

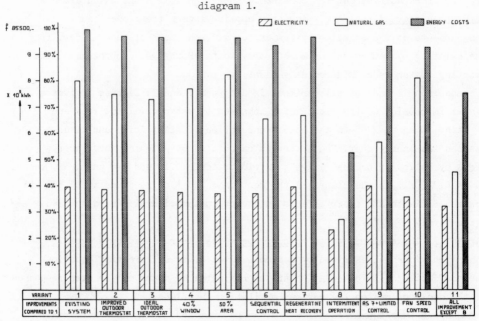

Comparison of yearly energy consumption and costs
for various variants.

diagram 2.

Variant 4 and 5. The window area is increased from 30 % to respectively 40 %
and 50 % of the total fascade area. Compared with variant 3 the consumption
of natural gas is increased by respectively 5.6 and 8.9 %, while the
electricity is decreased by respectively 2.7 and 3.0 %. Both opposite effects
result in a weak optimum with respect to energy cost at a window area of 40 %.
The small effect of window area is not general. For example, when normal
glass is used instead of heat absorbing glass an increase in window area causes
a strong increase in energy cost.

Variant 6. Sequential control of radiators and VAV box by the same room thermostat is applied. Now the required surplus of heat delivered by the radiators
is not always necessary. The consumption of natural gas is reduced by 18 %
compared to variant 1.

Variant 7. The twin coil heat recovery is replaced by a regenerative, sensible
and latent heat transferring system, reducing natural gas consumption by 17 %.

Variant 8. Intermittant operation of the AC system is applied:
only between 7.00 and 19.00 o'clock the room air temperature is controlled at
22°C. In order to realise 22°C at 7 o'clock the radiators or AC system are
taken in operation by means of a switching criterium, dependent of indoor- and
outdoor temperature. The consumption of natural gas and electricity are decreased now by respectively 66 and 44 %.

Variant 9. When the absolute humidity of outdoor air is higher than 6 gr/kg the
steam humidifiers are switched off. Also the improvement of variant 7 is applied.
This reduces the consumption of natural gas by 29 %.

Variant 10. Replacing the inlet vane control by fan speed control reduces
the total consumption of electricity by 10 %.

Variant 11. An optimal system is designed combining the reduced humidity
control, sequential indoor temperature control, the regenerative heat recovery
and the fan speed control system. This optimal system decreases the consumption
of natural gas and electricity by respectively 43 and 19 %. Weighed by the
energy prices this means a reduction of 25 % in cost.

Conclusions
- Simple and accurate mathematical models of components of an air conditioning
  system can be derived through which the hour by hour calculation of the
  yearly energy consumption can be carried out economically.
- A reduction of 25 % in energy cost can be obtained for the system considered
  in this paper by optimising the air conditioning system.
- The optimal window area showed to be 40 % of the fascade area. However, this

optimum is very smooth.

This statement is only valid for the system considered in this paper.

References

1   D.G. Stephenson and G.D. Mitalas; Cooling load calculation by thermal
     Response Factor Method.
     ASHRAE Transactions no 1018.

2   Green, Chandra, Shebar, S. Water cooling and dehumidifying coil.
     ASHRAE nr. 2141, 1970

3   Hansen, H. Neue Gleichungen für Wärmeübertragung bei freier oder
     erzwungener Strömung Alg. Wärmetechnik 9. 75/19, 1959.

4   Ziel, G. Die Auslegung von Gas-Dampf-Gemisch-Kühlern
     Dissertation, Technischen Hochschule Aachen

5   Proposed procedures for simulating the performance of components and
     systems for energy calculations, second edition, ASHRAE

6   Hoogen, A.J.J. v.d. Reference year for the Netherlands,
     Verwarming en Ventilatie, februari 1976, no. 2.

SEASONAL OPERATING PERFORMANCE OF GAS-FIRED HYDRONIC HEATING SYSTEMS WITH CERTAIN ENERGY-SAVING FEATURES

JOSEPH CHI[*]

Center for Building Technology
National Bureau of Standards
Washington, D.C., USA 20234

ABSTRACT

DEPAB (DEsign and Performance Analysis of Boilers) is an NBS computer program for simulation of fossil-fuel-fired boilers for residential heating systems. It is based upon an analytical model which accounts for cyclic (on-and-off) operation of a boiler fuel burner and water circulating pump. This paper illustrates the use of DEPAB for evaluating quantitatively the effectiveness of several selected energy-saving features for gas-fired hydronic heating systems. Sufficient information is also provided to demonstrate the important factors of the simulation program DEPAB.

NOMENCLATURE

c	specific heat at constant pressure
C	heat capacity
$\dot{C}$	heat capacity rate
D_F	draft factor defined as the ratio of gas flow rate at off-cycle to that at on-cycle while both are reduced to the same temperature
G	heat conductance
HHV	higher heating value of fuel
K	mass flow ratio of air to fuel
L	height or length
M	kg moisture formed per kg of fuel fired
$\dot{m}$	mass flow rate
$\dot{Q}'$	heat distribution rate per unit pipeline length
t	time
T	temperature

Subscripts

a	combustion products or draft air
b	heat exchanger wall
c	water in boiler
d	boiler jacket

---

[*]The author is also known as S.W. Chi.

e   water in pipeline

f   pipeline wall

fp   fuel

o   outdoor

p   pipeline

pp   combustion products

r   boiler room

INTRODUCTION

Widespread interest in evaluating and ultimately reducing seasonal and annual energy consumption in heating and/or cooling of buildings has led to the development of complex computer programs at the U.S. National Bureau of Standards and several other laboratories [1-6]. DEPAB (DEsign and Performance Analysis of Boilers) is a computer program for simulation of fossil-fuel-fired boilers. An extension of DEPAB is being made to include features for commercial and industrial boilers; we will restrict this paper to a description of boilers for residential heating systems. This simulation program for residential boilers is based upon an analytical model which accounts for cyclic (on-and-off) operation of a boiler fuel burner and water circulating pump. Transmission of heat at on-cycle uses theories of radiative and convective heat transfer; and transmission of heat at off-cycle uses theories of turbulent and conductive heat transfer.

While reference 2 documents details of the mathematical formulation and computer algorithm of the simulation program (DEPAB) for design and performance analysis of boilers, this paper is concerned mainly with an application of DEPAB in evaluating quantitatively the effectiveness of several selected energy-saving features for the gas-fired residential hydronic heating system. However, sufficient information is also provided to demonstrate the important factors and organization of DEPAB. A boiler physical model and its governing equations, together with the program organization, will be described below, followed by a comparison of the simulation results with the available experimental data, in order to establish confidence in DEPAB. Finally, seasonal performance of a gas-fired hydronic system incorporating design and operational changes is calculated using DEPAB, and results are discussed and compared with the system which is common for many existing homes.

BOILER SIMULATION MODEL

The analytical model illustrated in Figure 1 is used as the basis for this study of the gas-fired hydronic boiler systems. A burner supplies the required fuel. Combustion products (or draft air when the burner is off) flow through the gas side of the heat exchanger. A stack draws combustion products (or draft air) to its outlet. When the heat demand is less than the steady-state full-load output, the boiler will operate in a cyclic manner. Cyclic operation of the boiler can conveniently be divided into four successive time periods, namely: 1) burner on and pump off, 2) burner on and pump on, 3) burner off and pump on, and 4) burner off and pump off.

A room thermostat calling for heat starts off period 1. Heat is transferred from the combustion products to the circulating water through a heat exchanger. When the circulating water leaving the boiler reaches a certain set-point temperature, the pump comes on and period 2 begins. During this period heat

a COMBUSTION PRODUCTS OR DRAFT AIR
b HEAT-EXCHANGE WALL
c WATER IN HEAT-EXCHANGE TUBES
d BOILER JACKET WALL
e WATER IN PIPELINE
f PIPELINE WALL
r BOILER-ROOM

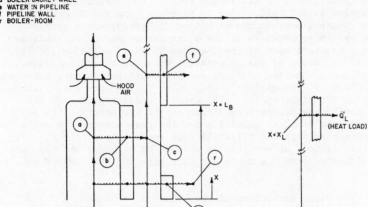

Figure 1   Schematic of a Boiler Model

is delivered continuously to the desired locations. When heat demand for the
residence is satisfied, the room thermostat signals a control to shut off the
burner; period 3 begins. Finally, the pump is turned off, and the circulating
water flow rate is drastically reduced. During those periods when the burner
is off, heat is transferred from the water and heat-exchanger wall to draft air
passing through the boiler and up the stack. The above-described cycles are
repeated, when the residence heat loss is less than the boiler steady-state
output.

It was established by this author [2] that the governing equations and
the boundary and initial conditions for the system described above can be
written as follows:

$$G_{ab}(T_b - T_a) = C_a \frac{\partial T_a}{\partial t} + \dot{C}_a L_b \frac{\partial T_a}{\partial x}$$

$$G_{ab}(T_a - T_b) + G_{bc}(T_c - T_b) = C_b \frac{\partial T_b}{\partial x}$$

$$G_{bc}(T_b - T_c) + G_{cd}(T_d - T_c) = C_c \frac{\partial T_c}{\partial t} + \dot{C}_c L_b \frac{\partial T_c}{\partial x}$$

$$G_{cd}(T_c - T_d) + C_{dr}(T_r - T_d) = C_d \frac{\partial T_d}{\partial t}$$

$$G_{ef}(T_f - T_e) = C_e \frac{\partial T_e}{\partial t} + \dot{C}_e L_p \frac{\partial T_e}{\partial x} + L_p \dot{Q}''(x)$$

$$G_{ef}(T_f - T_f) + G_{fr}(T_r - T_f) = C_f \frac{\partial T_f}{\partial t} \tag{1}$$

$$@t = 0, \quad T_a = T_b = T_c = T_d = T_e = T_f = T_r \tag{2}$$

$$@x = 0, \quad T_a = 298 + \frac{HHV - (2.44 \times 10^6)M - (C_{fp} + Kc_a)(298 - T_r)}{C_{pp}(1 + K)} \tag{3}$$

In the above system of equations, equation 1 set represents heat balance for each differential element of the fluids and walls (i.e., the temperatures, T_a through T_f, are temperatures of gas in the boiler, heat-exchanger wall, water in boiler, boiler jacket wall, water in pipeline, and pipeline wall, respectively). In general, performance of boiler at "steady-state cycles" is of interest; so system temperature at the start can be set at an arbitrary level and equation 2 has been used as the required initial condition for the system. With cyclic variation of gas temperature entering the boiler (i.e., alternately that of the combustion products and the draft air passing through the boiler heat exchanger), successive "steady-state cycles" are established after several initial transient cycles. The cyclic characteristics of the system are controlled by the boundary condition of the gas entering the boiler. Equation 3 represents the boundary condition for the gas entering the boiler. It was derived from a consideration of equilibrium combustion of the fuel [2]. The value of air fuel ratio K at on-cycle is assumed to be known and its value at off-cycle is calculated from the draft air flow rate and fuel supply rate to the burner pilot.

In addition, the cyclic variations of the values of conductance G and heat capacity rate $\dot{C}$ for the equation 1 set, in response to the on-and-off operation of the fuel burner and water pump, are required to be accounted for. Details of these variations are discussed in reference 2; but a brief description of several important factors is given below. During the steady-state full-load operation, the heat transfer rate, gas and water flow rates and their inlet and exit temperatures can all be measured in a straightforward manner; hence gas- and water-side conductances and heat capacity rates at on-cycle can be calculated from these measured values. During the off-cycle, the water flow rate and the water-side conductance are calculated by the theory of free convection. A consideration of flue-gas hydrostatic pressure difference from the ambient pressure and the flow resistance in turbulent flow yields an expression for the draft air flow rate at off-cycle in terms of the flue-gas flow rate at on-cycle, i.e.:

$$\frac{\dot{m}_{a,off}}{\dot{m}_{a,on}} = D_F \left(\frac{T_{a,off} - T_o}{T_{a,on} - T_o}\right)^{0.56} \left(\frac{T_{a,on}}{T_{a,off}}\right)^{1.19} \tag{4}$$

where D_F is a draft constant whose value is on the order of unity for natural draft burners and on the order of 0.4 for forced draft burners, and other symbols are defined in nomenclature. With off-cycle draft air flow rate calculated by equation 4, gas-side conductance at off-cycle can be found to be related to the convective component of the conductance at the on-cycle, using the theory of turbulent heat transfer, and the gas-side heat capacity rate is simply the product of the gas specific heat and its flow rate.

The computer program DEPAB has been developed to solve the equation 1 set with coefficients and initial and boundary conditions outlined above. A numerical method based upon explicit difference scheme [7] is used to solve the equation 1 set. After a network of nodel points has been constructed inside a boiler model (see Figure 2), a complete time history of system temperature at all nodes can be evaluated. Time history of flue gas flow rate can be evaluated using equation 4. Also the stack gas flow rate can be calculated by the same equation, with T_a interpreted as the stack gas temperature which can be determined from adiabatic mixing of relief air and flue gas leaving the boiler. From the time history of the mass flow rates and temperatures, performance of the boiler at difference outdoor temperature and cyclic conditions can be evaluted in a straightforward manner.

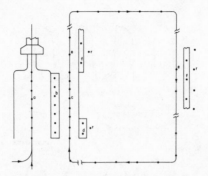

Figure 2   Schematic of Layout of Network of Nodel Points

COMPARISON WITH EXPERIMENTS

A setup for boiler experiments is available in the NBS Mechanical Systems laboratories. Data on boiler efficiencies at part-load have been reported by Kelly and Didion [8]. Figure 3 shows a comparison between the predicted and measured part-load efficiency for a boiler with input rate of 88 kW. The computer was run under the same conditions as the experiments. Excellent

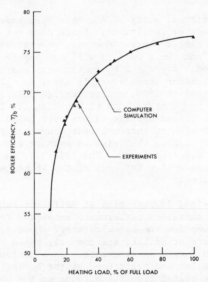

Figure 3   Predicted and Measured Part-Load
Performance of a Boiler

agreement between the theoretical prediction and experiments can be observed. Computer simulation has predicted accurately not only the part-load efficiency but also the time history of system temperatures, as can be seen in Figure 4 where temperatures of exhaust gas, supply water and return water are plotted versus time.

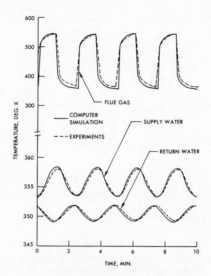

Figure 4   Predicted and Measured Cyclic Temperature
Variation of a Boiler at Half Load

ENERGY SAVING MEASURES

Hundreds of computer runs, using DEPAB, have been made for a variety of
design, operating, and weather conditions. Space does not allow a presentation
of complete results in a single paper. However, several sample runs will be
described in some detail and their results will be compared from the viewpoint
of annual fuel consumption. Average U.S. continental weather data will be
used: i.e., the design outdoor temperature is equal to -12.5C, degree-C days
per heating season are equal to 2800, and percentage distributions of the
residence heat load in 6 temperature bins with median at -10, -5, 0, 10, 15C
are as shown in Figure 5. These average values were based upon the design
indoor temperature of 20C with an internal heat source equivalent to a ΔT of
2.5C; and were obtained by taking the weighted average of the weather data [9]
for the 48 states on the U.S. continent. The 1970 U.S. census data on
residence heating fuel consumption for each of the 48 states [10] have been
used as the weighting factor.

With residence heat-load distribution at different temperatures (see
Figure 5) and the boiler part-load performance generated from DEPAB runs,
annual heating fuel consumption can be calculated. However, it should be
noted that laboratory tests of boilers are commonly based upon either the
exhaust loss or input and output measurements. In a residential heating
system, air enters the boiler and draft hood at the boiler room temperature.
This air has in fact been heated from the outdoor to the boiler room temperature.
Useful heat available for heating a home is therefore less than the laboratory-
measured efficiency. The heat spent in heating the boiler air (including
combustion, draft and relief air) from the outdoor to the boiler room temperature
must be accounted for. This heat should not, however, be charged completely
against the boiler, because the presence of a stack reduced the boiler room
pressure which tends to decrease the normal building exfiltration. Janssen
and Bonne [11] indicated that the average value for the boiler infiltration
is 0.7 (i.e., on the average, 70% of boiler exfiltration is chargeable against
the boiler). Field investigations are being undertaken by several research
groups to determine the infiltration parameter under different conditions.

In this study, the infiltration value 0.7 is used throughout.

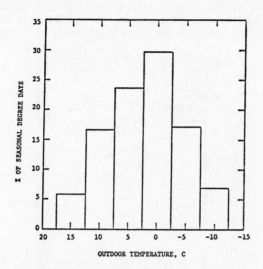

OUTDOOR TEMPERATURE, C

Figure 5   Average Distribution of Degree Days for U.S.
Continent. (Note: Total Degree - C Days Per
Year = 2800)

Based upon the above-described conditions, DEPAB was first run for a
boiler under prescribed conditions and then with certain changes in design
and/or operating variables from these prescribed conditions. Specifications
for each of the eight DEPAB runs to be discussed are as follows:

Run 1: A gas-fired hydronic boiler with an atmospheric burner is
used as the reference system. This reference system burns natural gas
with 50% excess air. Its full-load input rate is 24.9 kW and the fuel
supply rate to the pilot flame is 0.8% of the full-load input rate. The
boiler has a steady-state efficiency (excluding infiltration loss) of 80%.
The number of cycles per hour at which the burner operates varies parabolically
with respect to the heat demand; and at the heat demand being 50% the boiler
full-load output the burner operates at 3 cycles per hour. The water
circulating pump turns on at supply-water temperature equal to 65C and off at
85C. At the outdoor design temperature -12.5C, the residence heat loss is
equal to 58.8% of the boiler full-load output (i.e., the boiler is 70%
oversized).

Run 2: Conditions for this run are the same as those for Run 1 with the
exception that the pilot ignition is assumed to have been replaced by an
intermittent electronic ignition.

Run 3: The changes in this run from Run 1 include the use of an
intermittent ignition instead of the pilot ignition and the use of a power
burner instead of the atmospheric burner. It is assumed that the power burner
reduces the off-cycle draft factor from 1 to 0.4.

Run 4: In this run, an intermittent ignition device and an automatic
stack damper is assumed to have been added to the reference system. The
automatic stack damper is assumed to reduce the off-cycle draft factor from
1 to 0.15.

Run 5 to 8: These runs are made under the same conditions as those for Runs 1 to 4, respectively, with the exception that the fuel input rate has been reduced from 24.8 kW (i.e., 70% oversize) to 17.5 kW (i.e., 20% oversize) and air for combustion has been reduced to maintain same air-to-fuel ratio as in reference case.

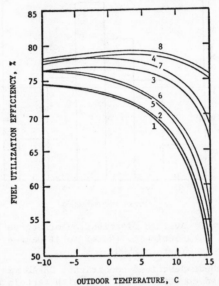

OUTDOOR TEMPERATURE, C

Figure 6   Fuel Utilization Efficiency Versus Outdoor Temperature for a Boiler Under Different Conditions (Viz: 1, Reference System; 2, Intermittent Ignition; 3, Intermittent Ignition and Power Burner; 4, Intermittent Ignition and Automatic Stack Damper; 5-8, Same as 1-4 Except that Firing Rate and Combustion Air are Reduced by 30%)

The resultant fuel utilization efficiencies of a boiler (including infiltration losses) obtained from the above-described DEPAB runs are plotted in Figure 6 versus outdoor temperature. Combination of Figures 5 and 6 yields annual performances. Table 1 lists for comparison the calculated annual fuel consumption, efficiency and operating cost for the system under eight different conditions. It can be seen in this table that changes in boiler design and operating conditions can have a considerable effect on the annual performance of the boiler heating systems.

SUMMARY AND CONCLUSIONS

A computer simulation program --DEPAB-- for design and performance analysis of boilers has been developed. It accounts for cyclic (on-and-off) operation of the boiler fuel burners and water circulating pump for the residential heating boiler. Formulation of the boiler model DEPAB is based upon rigorous heat transfer analyses. Results of computer runs are compared with data from actual tests under the same conditions; excellent agreements have been observed.

For these reasons, sufficient confidence has been established in DEPAB for its use in evaluating the seasonal performance and operating costs of the gas-fired heating systems. Sample runs of DEPAB for boiler systems with energy-saving features, using average U.S. weather data, are described in some detail. Results of these sample runs are summarized in Table 1. It can be seen in this table that seasonal savings in energy consumption can often be considerable with judicious choices of boiler design and operating variables.

Table 1.   Energy Usage Comparison for Various Gas-Fired Hydronic
           Heating Systems.

Run	System Parameters	Residence Heat Loss 10^9J/Yr	Gas Usage 10^9J/Yr	Seasonal Efficiency %	Gas* Bill $/Yr	Energy Savings %
1	Reference	84.8	125.8	67.4	315	---
2	Intermittent Ignition	84.8	122.1	69.5	305	2.9
3	Intermittent Ignition, Power Burner	84.8	112.8	75.2	282	10.4
4	Intermittent Ignition, Automatic Stack Damper	84.8	108.8	78.0	273	13.5
5	30% Firing[†] Reduction	84.8	121.1	70.1	303	3.8
6	30% Firing[†] Reduction Intermittent Ignition	84.8	117.5	72.2	294	6.6
7	30% Firing[†] Reduction, Intermittent Ignition, Power Burner	84.8	110.9	76.5	277	11.9
8	30% Firing[†] Reduction Intermittent Ignition, Automatic Stack Damper	84.8	108.0	78.6	270	14.2

*Gas Rate is assumed to be at $2.5 per 10^9J.

[†]Air for combustion also reduced to maintain same air-to-fuel ratio as in
reference case.

REFERENCES
1.   Bonne, U., and Johnson, A.E., Thermal Efficiency in Non-Modulating
     Combustion Systems, First NBS/ASHRAE HVAC Equipment Conference,
     Purdue University, Lafayette, IND., USA, October 1974.

2.   Chi, J., Computer Simulation of Fossil-Fuel-Fired Hydronic Boilers,
     Second NBS/ASHRAE HVAC Equipment Conference, Purdue University,
     Lafayette, IND., USA, April 1976.

3.   Gable, G.K., and Koenig, K., Seasonal Performance of Gas Heating
     Systems with Certain Energy-Saving Features, Preprint for ASHRAE
     Transactions, Vol. 83, Part 1, 1977.

4.   Chi, J., DEPAF - A Computer Model for Design and Performance Analysis
     of Furnaces, Preprint of an ASME Paper for ASME/AIChE National Heat
     Transfer Conference, Salt Lake City, Utah, USA, August 1977.

5.   Larsen, B.T., Digital Simulation of Energy Consumption in Residential
     Buildings, Norwegian Building Research Institute Report, Norway, 1976.

6.   Inoue, U., and Lee, H., Simulation of Refrigeration System for Energy
     Conservation and the Results Verified by Actual Measurements, J. of
     Japanese Soc. Heating, Air-Conditioning and Sanitary Engineering, Vol.50,
     pp. 93-108, June 1976.

7.   Dusinberre, G.M., Calculation of Transient Temperatures in Pipes and Heat
     Exchangers by Numerical Methods, Trans. ASME, Vol. 76, pp.421-426, 1954.

8.   Kelly, G.E., and Didion, D.A., Energy Conservation Potential of Modular
     Gas-Fired Boiler Systems, National Bureau of Standards, Building Science
     Series 79, 1975.

DYNAMIC MODELING OF GLASSHOUSE CLIMATE, APPLIED TO GLASSHOUSE CONTROL

G. P. A. BOT, J. J. VAN DIXHOORN, AND A. J. UDINK TEN CATE

Department of Physics and Meteorology
Agricultural University
Wageningen, The Netherlands

ABSTRACT

A systems approach to glasshouse modeling and control is outlined by presenting two types of dynamical models.
The first type of modeling uses a simple black box model. This is updated by on-line estimation and is incorporated in an adaptive computer control system. Results of field trials are included. The second model is more elaborate and based on the heat and water vapour balances in the glasshouse, which are presented in bond graph notation. The application of these models in a hierarchical control system configuration is discussed.

NOMENCLATURE

c_p	specific heat at constant pressure $(J\ kg^{-1}K^{-1})$	u	controller output signal
		v	velocity (ms^{-1})
e	vapour pressure (Nm^{-2})	α	heattransfercoefficient $(Js^{-1}m^{-2}K^{-1})$
k	masstransfercoefficient (ms^{-1})		
K	gain	Δ	difference
l	characteristic length (m)	ε	error signal
m	molar weight $(kmol\ kg^{-1})$	ψ	flow $((J\ or\ kg)s^{-1})$
R	gasconstant $(J\ kmol^{-1}K^{-1})$	ϕ	angle or position
s	(Laplace)differential-operator	ρ	density $(kg\ m^{-3})$
T	temperature (K)	τ	time constant (s)

Subscripts

f	feed	p	pipe
g	glasshouse	r	return
h	heat	v	volume
in	in	vent	ventilation
out	out	w	water vapour

Dimensionless Numbers

Gr	Grashof	Number	Pr	Prandtl Number
Nu	Nusselt	Number	Re	Reynolds Number

INTRODUCTION

Glasshouse cultivation is an important part of Dutch agricultural activity. It is a very intensive way of growing fine vegetables, fruits and flowers in every season. This is accomplished by manipulating the climate in the glasshouse roughly by heating when the temperature is too low and ventilating when it is too high. The main factors determining the cost of glasshouse cultivation are energy, investment and labour. At present the energy cost is only one third of the total cost, because cheap natural gas is used.
However, energy becoming more expensive and scarce, ways to reduce energy cost are to be investigated. Direct methods to prevent losses are applied already on a large scale:
- thermal screens, closed at night and open during daytime
- exhaust gas condensors, using the condensation heat of the water vapour in the exhaust gases of the natural gas burners; this low temperature heat is used for soil heating.
Another direct approach to reduce energy cost is the use of plant species that give a high production at low temperatures. Search in this direction is already successful.
A different approach - often called the <u>systems approach</u> is to consider plant production and its associated energy cost a production system that should be controlled to yield an "optimal" performance. Roughly stated three hierarchical levels of control can be distinguished.
I) The control of heating and ventilation to ensure a desired glasshouse climate, independent of the disturbances caused by the weather. Good control on this level will result in energy savings due to the prevention of overheating and of simultaneous heating and ventilation. The usually applied automatic analogue controllers do work on this level.

II) The control of short term conditions to ensure optimal plant growth. Whereas the glasshouse climate itself is controlled on the previous level, the desired values of the climate factors are determined and controlled on this level. The airtemperature, for instance, is often automatically increased in accordance with the radiation intensity. Condensation on leaves and flowers should be avoided to prevent diseases and damage. The underlying ideas on this level are partly physiological and partly empirical.

III) The ultimate control level is that of the long term plant development. The short term plant situation should be related to plant development and production. Some empirical rules and much of the growers experience is decisive here. This field is a main challenge for horticulturists.

A basic tool of the systems approach is the mathematical model. In the discussed subject models of the micro-climate will be needed on all levels, models of plant growth are needed on level two and three and of plant development on the third level. The ultimate object being improved control, one approach is to incorporate models in a learning control strategy. This implies that the models can be more simple than the conceptual or explanatory models usually applied in research. If the model contains a few unknown parameters an on-line parameter estimation technique can be used for updating. As will be shown even a simple black box model, not containing physical but only input-output relations, can give a significant improvement over conventional control.

This paper reports our systems oriented approach to glasshouse climate control. It is performed in cooperation with plant physiological and horticultural groups outside our laboratory. Furthermore close cooperation is established with the

Glasshouse Crops Research and Experimental Station at Naaldwijk, The Netherlands, where an extensive digital computer control and data logging system is in operation.

An important motivation for our work is that in this country computer control is not restricted to experimental stations like Naaldwijk. At present over a hundred installations are in use or ordered by commercial growers. These installations are generally minicomputer or microprocessor based. Equipment cost is about $ 20.000,- and is at the break-even point with a sophisticated analogue controller for about six glasshouses.

The computer provides the grower with attractive data logging and averaging facilities. Algorithms used however, are generally a digital version of the conventional analogue controller. A control philosophy fully exploiting computer capabilities is still lacking. The systems approach, as followed by our group, making use in a stepwise way of simple and increasingly more elaborate models and modern control methods, could lead in this direction.

Our first step, not reported here, has been the construction of a black box simulation model of a glasshouse, its heating system and   associated control system. The model was validated with experimental data from the Naaldwijk installation. It was used to improve the conventional control algorithms. The results were used to improve the Naaldwijk computer control and showed a considerable improvement. It became clear, however, that no adjustments for the standard control algorithms could be found that gave satisfactorily control performance under the strongly varying environmental conditions.

So our next step, briefly reported in this paper and more extensively elsewhere 1 has been to develop an adaptive (=selftuning) control algorithm for glasshouse heating. This algorithm continously tunes the control algorithm depending on the actual situation. To track the actual situation a simple black box model with one adjustable parameter is incorporated in the control system. The adaptive algorithm is presently in operation in all computer control loops at Naaldwijk. Research on this level will be continued on ventilation control.

Another parallel step, also briefly reported in this paper and elsewhere 2 has been the development of a not too complicated simulation model of the glasshouse climate. In contrast to the previously mentioned approach this model is based on the heat and mass balances and on the physical transport phenomena and includes a model of plant behaviour in this respect. The bond graph notation is used to obtain a clear representation and more convenient interactive simulation. The model will be used for accurate simulation and adaptive control.

GLASSHOUSE CLIMATE

Glasshouses are built to improve the environmental factors for plant production. The factors of interest are radiation, temperature, relative humidity, carbon dioxide concentration and wind velocity. First of all these factors are influenced by the reduction of the turbulent exchange. This effect mainly effects the temperature rise in the glasshouse.3 .

Another effect is the so called "greenhouse effect": the glass is transparant for the solar short wave radiation but not for the thermal radiation emitted by the soil and plants. This effect causes ten to twenty percent of the temperature rise.

Compared with other buildings a glasshouse is an open system in which the climate is influenced directly by the outdoor climate. But compared with the climate in the open, the glasshouse climate is that of a closed system in which the heat and vapour exchange inside directly influences the climate. So both a translation from outdoor to indoor climate and a description of plant behaviour

is needed to determine the environmental factors near the plant. These factors
in turn determine plant behaviour so only a description of the total system
will be succesful.

To understand the heat and mass transfer between plants and atmosphere a short
description of this aspect of plant behaviour is essential. The main plant processes involved here are photosynthesis and transpiration.
    Photosynthesis is determined as main factor by short wave radiation, car-
bondioxide concentration and temperature. Only a small, almost negligeable amount
of the short wave radiation is directly absorbed in the photsynthesis proces.
The main part is absorbed by the leaf, leading to a temperature rise of the
leaf. To prevent a too large temperature rise it is cooled by transpiration,
the water being transported from the roots to the inner parts of the leaf and
then evaporated to the atmosphere.
    For the absorption of carbondioxide and the evaporation of water vapour,
the leaf has small holes or stomata in the surface. For photosynthesis these
holes have to be wide to support carbondioxide. In this situation, however, water
is evaporating producing a possible water shortage in the leaf. To control these
phenomena the aperture of the stomata is regulated by internal plant processes,
dependent on photosynthesis, transpiration and water content of the leaf 4.5
which in turn are dependent on the local environmental factors.

ADAPTIVE CONTROL

The control problem

    To introduce the adaptive controller 1 , the control problem has to be
defined.
It is recalled that the glasshouse climate is influenced by the external factors
showed in fig. 1, where also the heating system control loop is given.

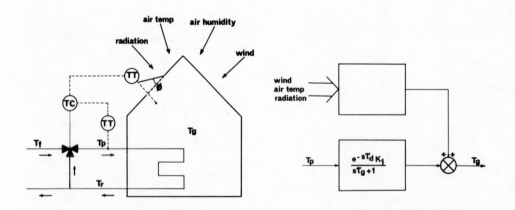

fig. 1. Heating system control   fig. 2. A dynamic model

    The objective of the heating system is to regulate the glasshouse tempera-
ture T_g. This temperature is influenced by the external factors and can be mani-
pulated by ventilation which is caused by the position (with angle ϕ) of the
control windows and by regulating the heating-pipe temperature T_p. The control
problem can thus be regarded as a multi-input process. The inputs are the control
variables ϕ and T_p, the output is the controlled variable T_g. The characteristics

of this process are highly influenced by the external factors. In order to prevent heat losses it is desirable to manipulate T_p, since manipulating of ϕ causes opening of the windows and subsequent heat loss. Therefore, the control of the windows is separated from the heating system control. Manipulation of ϕ acts as a disturbance for the heating system control. Usually the setpoint for the window control is set higher than the heating setpoint so that interaction of both control loops is decreased to an acceptable level - from control theoretical point of view - and heat loss is prevented.

In the heating control loop, the heating-pipe temperature is controlled by a three-way valve that mixes the return water with temperature T_r with the feed-water from the main boiler with temperature T_f (fig.1). The response of T_p on a change of the valve position is relatively fast if T_p has to increase, but slow if T_p has to decrease. The cooling response of the valve is rather slow. Since the temperature fall is dominated by T_r and T_r decreases slowly because of the large heat content of the pipe-water, the position of the mixing valve is in fact the proper control variable but is not selected because of the asymmetric relation between the valve position and T_p. This allows the selection of a simple control model, that produces specific control problems when large transients of T_p are required.

A simple model

In order to design a control loop it is necessary to construct a dynamic model of the heating system. When the partial differential equations governing the heat (and vapour) flows from the pipes into the glasshouse are lumped into an approximate simple linear first order transfer function with a time delay, a dynamic model results as shown in fig. 2, with input T_p and output T_g. In this model the external influences are included following Roots 6 . Experiments were performed in the Naaldwijk glasshouse that consists of 24 identically, individually controlled compartments of 56 m^2 each. Under different conditions typical results for the parameters of the simple model of fig. 2 were: time delay $\tau_d \simeq 6$ minutes, the time constant of the first order model $\tau \simeq 30$ minutes and associated gain $K_1 \simeq 0.25\text{-}0.5$; τ_d and τ being fairly constant. The external factors cause a disturbance signal, of which the significant part is relatively slowly time-varying. It is therefore assumed in this paper that the offset caused by the external influences and the dynamic gain K_1 can be lumped together producing a time variant gain K_g. This gain K_g is not easily determined because the dynamic model of fig. 2 assumes known offsets on T_p and T_g. Usually a model is defined by linearization around a nominal operation point. In the glasshouse this point is subject to large variations so that this approach cannot be used. Therefore, zero offsets were assumed producing a gain K_g related to a static plus dynamic model. In the Naaldwijk glasshouse typical values of $K_g \simeq 0.2\text{-}1.0$.

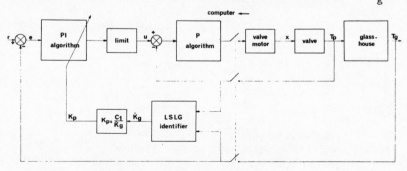

fig. 3. Heating system control loop

The adaptive control loop (fig.3)

In the adaptive controller of fig.3 the adaptation compensates for variations in K_g, which means that K_g has to be computed from input-output observations. This is performed by a recently reported "least-squares like gradient" identification technique 7 , that is physically similar to the well-known recursive least-squares technique 8 .

In the identification procedure, the simple model of fig.2 is discretized and K_g is estimated. The estimate $\hat{K}_g$ of K_g is used to adjust a PI (proportional plus integral) algorithm that is a discrete version of the continuous PI controller with inputsignal e and output u:

$$u = K_p(e + \frac{1}{\tau_i} \int_0^t e(\tau)d\tau \tag{1}$$

The time constant τ_i = 25 min. The gain K_p of the PI controller is varied proportional to the inverse of the gain $\hat{K}_g$, thus keeping the product $K_p \hat{K}_g$ constant. The signal u acts as setpoint for the pipe temperature control (fig.3). In the controller, the signal u is limited between the minimum and maximum T_p. The values of T_p min./max. are time-varying and are based on horticultural requirements as well as on practical control considerations.

Field results

The adaptive control algorithm was programmed in the computer at Naaldwijk. Fig.4 gives some results on February 8, 1977.
The weather conditions on that day were: sunny, mean outside air temperature $\simeq 7^\circ$C, mean wind velocity 5 m/s.

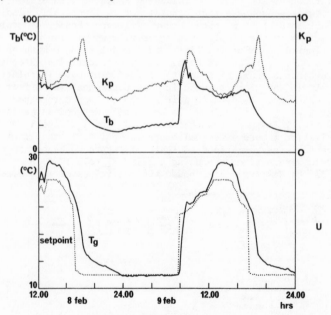

fig. 4. Results of a field test at Feb. 8, 1977.

Shown are the responses of the adaptive gain K_p, T_p, T_g and the setpoint of T_i; the setpoint is varied according to the amount of light. From the results it can be concluded that the controller follows the setpoint satisfactorily. An advantage of the adaptive control is that the value of the product $K_p \bar{K}_g$ was selected in January 1977 and since then (in May) there has been no reason to tune the controller for warmer weather conditions.

The responses stress the interesting features of the adaptive controller: after an initial tuning the controller is continuously and automatically adapted to varying weather conditions, leading to a control loop that is insensitive to external influences. It is remarkable that this succesful adaptive control is based on an almost too simple model of the glasshouse heating dynamics. On the other hand; the choice of this particular simple model and the detailed configuration are based on quite some engineering knowledge which makes the simple model rather the product of the design procedure than the starting point.

CLIMATE MODEL

In the climate model the heat and water vapour flows are considered in a compartimented glasshouse.
The compartments in the model are the glass-cover, the inside air, the plants (or canopy), and some layers in the soil.
This is only a rough compartimentation, but measurements justify this approach for a model which only includes total heat and mass exchange of the canopy.
The main incoming energy flow is short wave solar radiation. Sometimes a glasshouse is regarded as a large radiation collector, converting solar energy into plant production. The incoming solar energy is partly absorbed, reflected and transmitted by the glass-roof and walls. The transmitted radiation is again partly absorbed, reflected and transmitted by the canopy and then by the soil. From the known optical properties of the cover, the canopy and the soil, the total absorbed amount of incoming solar energy is determined. In the model the optical properties were assumed to be constant, though the reflection, which is dependent on the direction of the incoming radiation is changing during a day and during a year. For the glasshouse under consideration the absorbed percentages were 40% for the cover, 27% for the canopy and 16% for the soil, 12% was reflected directly by the cover and the remaining 5% was reflected by the canopy and the soil and transmitted to the atmosphere.
The incoming thermal or terrestrial radiation is described in literature by empirical formulas because of the complexity of the mechanism. Wartena 9 reviewed various methods and compared them with measurement data. The results turned out to be so inaccurate that it is not justified to include terrestial radiation in the model. Under certain meteorological conditions however, e.g. in clear nights without wind, or even on clear days, this will be a shortcoming.
On the contrary, thermal radiation exchange in the glasshouse between cover, canopy and soil is of interest. Because of the relatively small temperature differences, the Stephan Boltzmann formula is linearised.
As stated in the discussion on the "greenhouse effect" the reduction of the turbulent exchange is of main importance. The related transport process is ventilation, transporting both heat and water vapour to and from the air compartment.
The incoming heat flow by ventilation $\psi_{h,\text{vent. in}}$ is given by:

$$\psi_{h,\text{vent. in}} = \psi_v \, \rho \, C_p \, T_{out} \qquad (2)$$

The incoming water vapour flow $\psi_{w,\text{vent. in}}$ is also determined by ψ_v according to:

$$\psi_{w,\text{vent. in}} = \psi_v \, (m \,/\, R \, T_{out}) \, e_{out} \qquad (3)$$

The same formulas with of course T_{in} and e_{in} instead of T_{out} and e_{out} are valid
for the outflowing heat and water vapour flow.
The amount of exchanged air ψ_v is dependent (see fig.1) on the position of
the window, the outside wind velocity and direction, and to a lesser degree on
the temperature difference between in- and outside air.
Satisfying relations are not reported in literature. We started model experiments
in a windtunnel to find the relation between air exchange, window position and
wind velocity. Full scale experiments should validate the results.
In the present model ψ_v is based on forced ventilation data or estimations in
a natural ventilated glasshouse.
     In the model the mechanism of convection is treated in the usual way by
using dimensionless relations. For the convective exchange between cover and
outside turbulent airflow, the heat transfer coefficient is assumed to be 10 :

$$Nu = 0,03 \ Re^{0,8} \qquad\qquad\qquad (4.a)$$

resulting in $\alpha = 5,5 \ v^{0,8} \ l^{-0,2}$ $\qquad\qquad\qquad (4.b)$

with air properties at about $10^{o}C$. Determining a more accurate relation is not
so interesting, since the largest convective resistance exists between cover
and inside air. Inside the glasshouse a combination of circulation and free
convection is present along the glass, due to the temperature difference between
glass and inside air. The natural convective heat transfer coefficient between
a vertical wall and inside air is for the turbulent region given by:

$$Nu = 0,1 \ Gr^{1/3} \ Pr^{1/3} \qquad\qquad\qquad (5.a)$$

resulting in:
$$\alpha = 0,125 \ (\Delta T)^{1/3} \qquad\qquad\qquad (5.b)$$

The forced convective heat transfer is still given by (4).
For the low inside air velocity of about $0,1 - 0,25 m/s$ the forced and free
convective heat transfer have about the same magnitude. The result is a heat
transfer coefficient of about $5 \ W/m^2 K$. For the almost horizontal roof (an angle
of 26° between roof and horizon is applied for a common type of glasshouse),
the natural convection will be somewhat smaller, the total result however will
be similar. So the inside heat transfer coefficient is fixed to a value of
$5 \ W/m^2 K$. More accurate figures can be expected when the inside circulation
as a function of ventilation and natural convection along the heating pipes is
determined. In the present research programme on ventilation this subject is
pursued. For the convective heat transfer between inside air and soil, natural
convection will be the most important mechanism; the circulation is decreased
in the canopy. Therefore, a heat transfer coefficient of $3,5 \ W/m^2 K$ is selected.
Along the leaves forced convection will dominate.
A great variety of relations is found in literature:

$$\alpha = (4 \ \grave{a} \ 13) \ v^{0,5} \ l^{-0,5} \qquad\qquad\qquad (6)$$

This might be motivated because the relations with the low coefficient have to
be applied on both sides of the leaf but in the relations with the high coefficients both sides of the leaf are already incorporated. This is however not
always stated clearly. As a compromise $\alpha = 5 \ (v/l)^{0,5}$ is used, applied on both
sides of the leaf. With low windspeed of about $0,1 \ m/s$ and a leaf width of
$5 \ cm$, α will be $7 \ W/m^2 K$.

The convective mass transfer coefficient k is related to the heat transfer coefficient α according to

$$k \underset{\sim}{=} \alpha / \rho \, C_p \tag{7}$$

So in the water vapour model the convective mass transport is determined from the convective heat transport in the temperature model. For the plant compartment we have taken into account the stomatal resistance according to section 2.

In the soil, heat transfer by conduction is considered between the soil compartments. Due to the assumed constant water content of the soil (in glasshouses the soil is kept moisty) the heat conductivity is assumed constant at 2 W/mK.

BOND GRAPH REPRESENTATION AND INTERACTIVE SIMULATION

The relatively novel bond graph notation 12,13 is chosen for the representation of the model for the following reasons:
-    it represents the physical structure in an easily recognisable, compact way
-    computational problems can be detected and remedied in the bondgraph
-    it is readily converted in a simulation programme, suited for interactive
     simulation by using the block oriented simulation language THTSIM 14 .
Moreover it can also easy be translated into a set of differential equations, that can be solved by programming in CSMP.

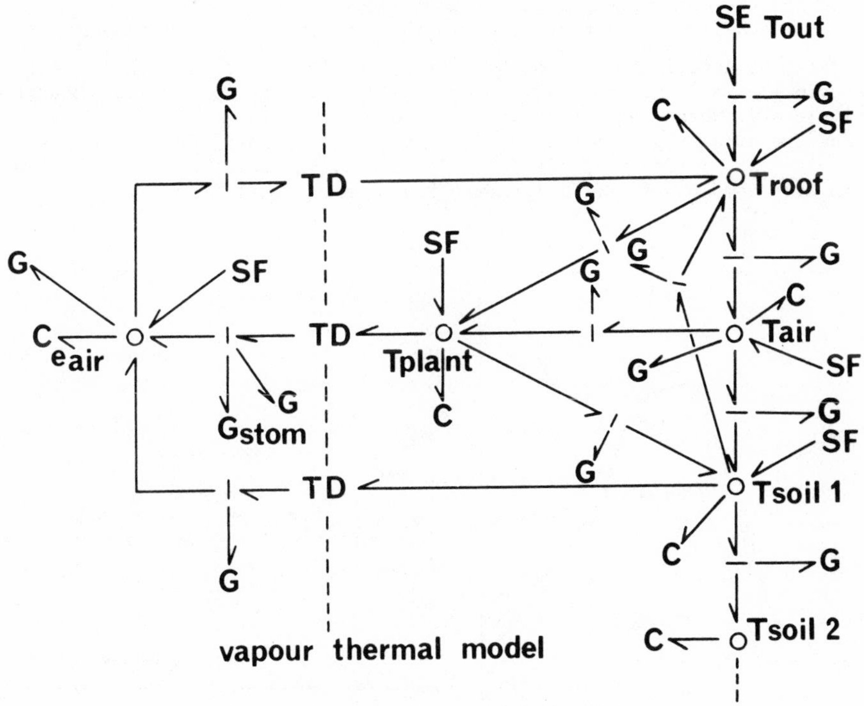

fig. 5. Bond graph of simplified glasshouse

Fig. 5 shows the bond graph of the simplified glasshouse model. The nodes or "0-junctions" represent the distinct temperatures or vapour pressures in the model. The line elements or "power bonds" correspond with energy flow, the half arrow indicating the prescribed positive direction. A bond is actually a shorthand notation for the interaction between components and involves two signals.

On the top of the right side or thermal part of the model the direct and diffuse radiation is represented by a time dependent source of heat flow (SF), acting on the heat capacity (C) of the roof. From this "T roof" the heat flows in several directions via "1-junctions" to which heat transfer components "G" are connected. The G's represent the linear or non-linear heat transfer by conduction (in the soil), by convection and by radiation. On the bottom right side the heat flow is seen to enter a regular conductance - capacitance network, representing the compartimented soil.

The left side of the bond graph is the vapour model. There the mass balance of the vapour in the inside air is represented by the vapour capacity C, with its associated state variable e_{air}. An incoming flow source SF and an outgoing G represent the vapour which comes in and goes out by ventilation.

The coupling between the vapour and thermal parts of the model involves evaporation or condensation at the soil surface (lower coupler) and at the plant surface (middle coupler). At the glassroof (top coupler) only condensation takes place. The couplers are represented by transducers (TD). Due to the interactive nature of the energy exchange each TD represents two relations: one between the vapour pressure and its associated temperature and one between the vapour mass flow and the corresponding latent heat flow.

The stomatal behaviour is incorporated in the non-linear G_{stom} component.

A more detailed description of the bond graph model and its simulation is given in a previous paper 2 . The THTSIM programme accepts the bond graph structure in a simple way. When the constant or time dependent parameters of the components and the simulation control data (timing, plot outputs) are specified the numerical or plotted response can be obtained. The user keeps all the time in touch with components and a structure having a physical meaning.

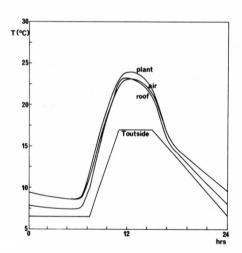

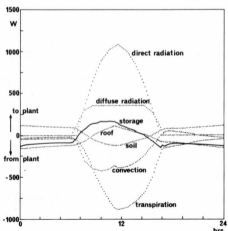

fig. 6. Some temperatures over 24 h.   fig. 7. Terms of the energy balance of
        in an unheated glasshouse   the plant compartment

Simulation results are shown in figs. 6, 7 and 8. The temperatures of the outside air, the cover-, the air-, the plant- and the upper soil-layer compartment are given in fig. 6. The glasstemperature has a value between that of the inside air and outside air at night, by day it has nearly the same value as the inside airtemperature though the outside heat transfer coefficient is much higher than the inside one. This is caused by the high rate of radiation absorption. By day, the plant temperature is a few degrees higher than the air temperature, at night they are nearly the same. The adjustment of the plant temperature is given in fig. 7. In the daytime, incoming radiation is mainly compensated by transpiration. The thermal radiation to and from the soil and the glass nearly compensate each other. Also for varying weather condition simulations can be made.

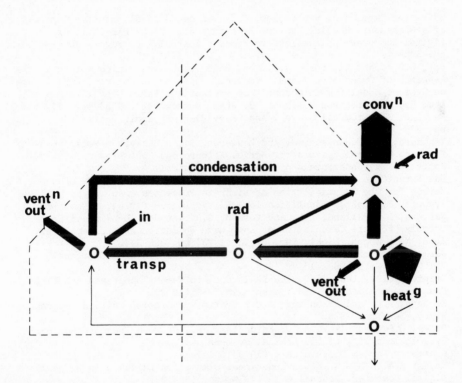

Fig. 8. Temperatures and heat flows during a cold rainshower

Fig. 8 shows the heat flows and temperatures at one moment during a cold rainshower. It is represented in a formalised bond graph way (Sankey diagram), constructed from the printer outputs.
This type of diagram gives a good insight in the heat and vapour flows in the system under varying conditions.

CONCLUSIONS

Optimal plant growth is formulated as a hierarchical control problem with three control levels. At the first, basic, level the systems approach was succesfully applied by using a simple approximation for the glasshouse dynamics in an adaptive heating control system. A physical climate model is promising

for application in a control algorithm for both heating and ventilation.
This model is kept simple because it is applied in a closed control loop and
it is not yet possible to quantify accurately all involved transport processes.
By sensitivity analysis, the sensitive parts of the model can be located. For
control in the second level, not only plant exchange but also plant processes
have to be included in the model. Presently, plant behaviour can be described
as a function of the environmental parameters and used in the simulation of the
microclimate 15 . At the third level, a lot of horticultural knowledge is
available, but only in static form. In the systems approach dynamic optimization
has to be performed. A first attempt is published recently 16 .

REFERENCES

1.   Udink ten Cate, A.J. and Vooren, J.J. van de, Digital Adaptive Control
     of a Glasshouse Heating System. Preprints 5th. IFAL/IFIP Int.conf. on
     Digital Computer Applications to Process Control. The Hague, June 1977,
     pp. 505-512.
2.   Bot, G.P.A. and Dixhoorn, J.J. van, Bond Graphs and Minicomputers in the
     Research of Greenhouse Climate Control. Paper presented to EPPO/IOBC Conf.
     on Systems Modelling in modern Crop Protection. Paris Oct. 1976.
3.   Lee, R., The Greenhouse Effect. J. Appl. Meteor. 12 (1973) pp. 556/557.
4.   Raschke, K., Stomatal Action. Ann. Rev. Plant Physiol. 26 (1975)
     pp. 309-340.
5.   Takakura, T., Goudriaan, J. and Louwerse, W., A Behaviour Model to
     Simulate Stomatal Resistance. Agric. Meteorology 15 (1975) pp. 393-404.
6.   Roots, W.K., Fundamentals of Temperature Control. Ac.P., New York (1969).
7.   Udink ten Cate, A.J. and Verbruggen, H.B., Least-squares Like Gradient
     Method for the Identification of Discrete Processes. rep. to be published.
8.   Eykhoff, P., System Identification. Wiley, London (1974).
9.   Wartena, L., Palland, C.L. and Vossen, G.H. van de, Checking of Some
     Formulae for the Calculation of Long Wave Radiation from Clear Skies.
     Arch. Met. Geoph. Biokl. Ser. B. 21 (1973) pp. 335-348.
10. Wärmeatles, V.D.I. ed. 2. Berechnungblätter für den Wärmeübergang.
     VDI Verlag GmbH. Düsseldorf 1974.
11. Stigter, C.J., Leaf Diffusion Resistance to Water Vapour and its Direct
     Measurement. 1. Meded. LH, Wageningen 72-3 (1972).
12. Karnopp, D.C. and Rosenberg, R.C., System Dynamics: A Unified Approach,
     Wiley, New York 1975.
13. Oster, G.F., Perelson, A.S., Katchalsky, A., Network Thermodynamics:
     dynamic modelling of biochemical systems.
     Quarterly Review Biophysics 6 (1973) pp. 1-34.
14. Kraan, R.A., THTSIM: A Conversational Simulation Language on a Small
     Digital Computer. Journal A 15 (1974) pp. 186-190.
15. Goudriaan, J., Crop Meteorology: A Simulation Study. Pudoc, Wageningen.
     diss. LH, Wageningen 1977.
16. Challa, H., An Analysis of the Diurnal Course of Growth, Carbondioxide
     Exchange and Carbohydrate Reserve Content of Cucumber.
     Agric. Res. Rep. 861. diss. LH, Wageningen 1977.

TEMPERATURE REGIME IN HOTHOUSES WITH SOLAR ENERGY ACCUMULATION AND CLOSED HYDROLOGIC CYCLE

R. BAIRAMOV AND L. E. RYBAKOVA

(Authors' address can be obtained from V. P. Motulevich, National Committee
for Heat and Mass Transfer, Academy of Sciences of the USSR, Moskva V-71,
Lenjinsky prospekt 14, USSR)

The expenses required for the heating of hothouses (including the heating system, fuel, service staff) are rather high,
and constitute about 50-60% of the total operation expenses. In
this connection it should be noted that a problem of saving fuel
and most efficient use of the fuel resources is very topical at
the present time.

At the conditions of Middle Asia, and the south of Turkmenistan, in particular, the quantity of the solar heat incident
on the unit area of the soil is greater than 10^6 kcal/m^2 a year.
In Turkmenistan there happen during the year about 200 clear
days and about 140 days with 4-6 shining hours when the solar installations can operate. Such favourable climatic conditions make it possible to use the solar energy for a partial or complete
heating of hothouses, hotbeds, lemonaries, which may considerably decrease the cost price of agricultural products cultivated
during autumn-winter and spring seasons.

In order to maintain the necessary temperature regime in
the nighttime, there are used, at the present time, the heat-
accumulating arrangements in the solar hothouses. A heat accumulator represents a system of batteries characterized by high
coefficients of specific heat. The battery surface interacts with
the heat transfer agent (air medium of the hothouse),, which
results in the heat transfer process. In the daytime the heated
air of the hothouse gives up a part of its energy to the accumulators. After the sunset the air temperature in the hothouse decreases and becomes less than that of the accumulator. The heat
accumulated in the daytime will warm the air in the hothouse.
The accumulator is charged in the daytime due to the convective
heat exchange, and during the nighthours it is dicharged. As a
result, the temperature regime of the air medium in solar hothouses is somewhat improved.

The accumulators should be simple in design and operation,
and should occupy the least part of the hothouse useful area.
In order to develop an optimal design of the accumulator chamber
we have undertaken an experiment in two identical solar hothouses having differently designed ground accumulators. A comparison of temperature and humidity regimes of the two hothouses
has been performed at equal environmental conditions.

The experimental solar hothouses with the ground heat accumulator of 100 m^2 area represent a lean-to building with a
longitudinal east-west axis. The transparent roofing is south
oriented and consists of two parts, i.e. the main part of the

517

cover glass is positioned at 45° to the horizon, and the subsi-
diary one, at 20°. The vitrification is unary. The nothern part
of the hothouse is heat insulated. In the back wall of the hothouse a ground heat accumulator is mounted, which consists of
the troughs, whose length is equal to that of the hothouse with
the height of 25 cm, and width, 100 cm. The troughs are located
on five shelves, distanced 40 cm, and filled with the moist
soil (Fig. 1).

The difference in design of the ground accumulators is the
following: in the first case the accumulator is isolated from
the hothouse by a protection wall with airguides, while another
experimental hothouse has no such a protection, i.e. the accumulator is open for solar beams.

The measurements showed that the air temperature in the
hothouse begins to heighten from about 8 o'clock a.m. and reaches
its maximum at 2 or 3 o'clock p.m. As a result of heating of the
entire building of a hothouse, soil and plants, there occurs
charging of accumulators due to the convective heat exchange,
At 8 o'clock p.m. after the sunset, when the air temperature in
the hothouse is less than the temperature of the soil and
ground in accumulators, the fluxes of warm air propagate from
the accumulator, hence maintaining the temperature inside the
hothouse higher than that of the external air.

By comparing temperature regimes of two experimental hothouses it became possible to ascertain that the most efficient
is the solar hothouse with the ground accumulator having no
protection wall, and besides it has the improved economic-engi-
neering properties.

In order to work out the service conditions and to determine
the terms of growing of agricultural products, we have studied
temperature and humidity regimes of solar hothouses during several seasons of operation. For instance, within the testing period
from March 1975 to May 1976 the hothouse with the ground heat
accumulator has not been heated at all. From the analysis of
weekly data obtained with the help of thermographs installed at
different places of a hothouse, it follows that at minimum temperatures of the outside air, from -12°C to -17°C, the air temperature in the hothouse was not lower than +5°C and 3°C.

Table I

Month	Days	Temperature of the outside air	Temperature of air in the hot-house
December 1975	9	− 9.0	+ 3.5
January 1976	14	− 3.0	+ 10.0
	27	− 7.0	+ 10.0
	28	− 7.0	+ 11.0
	29	− 7.0	+ 11.0
February 1976	4	− 7.0	+ 8.0
	15	−8.0	+ 3.0
	19	− 12.0	+ 5.0
	26	− 17.0	+ 3.0
March 1976	3	− 10.0	+ 7.0
	13	− 8.0	+ 5.0
	27	− 4.0	+ 6.0
April 1976	3	− 3.0	+ 9.0

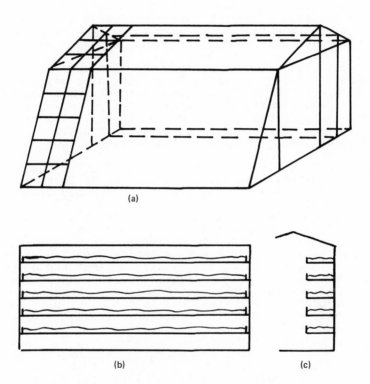

(a)

(b)   (c)

Fig. 1. A scheme of the solar hothouse with the ground heat accu-
         mulator (a – general view, b,c, – cross-section of the
         accumulator).

Minimal temperatures were mainly observed in the nighttime
during 2 or 3 hours. The experiments showed that the accumulator
is not discharged completely. It stores an amount of energy,
which is sufficient to maintain the positive temperature in the
hothouse during subsequent hours of cloudy days.

It should be noted that in a solar hothouse the temperature
heightens in cluody days as well. It is a result of the scattered
radiation. For example, the temperature of the outside air in
cloudy days was invariably equal to -3.5°C, but the air temperature in the hothouse varied from +8.0°C to + 10.0°C.

At the conditions of prolonged cloudy weather and negative
temperature of the outside air, beginning from the second day,
the nighttime temperature of air in the hothouse may be lower
than 0°C.

In spring time, beginning from the middle of March, when
the temperature of the outside air in the solar days is rather
high (+ 18.0 - +20.0°C) the air temperature in hothouses reaches 140.0°C and higher, which results in the overheating. In
such cases it is necessary to air the hothouse, and beginning
from April, the casement windows are left open for the night.
The same situation is being observed in autumn up to the middle
of October.

Let us consider the humidity regime of solar hothouses
with the ground heat accumulator during twenty four hours typical of the spring time, at the condition of sunny weather in
the previous two days.

Relative humidity of air was measured with the August psychro-
meter over the length and width of the hothouse, 2.5 m above the
soil surface, and over the height of the hothouse.

Relative humidity of the air beginning from 7 o'clock p.m.
to 8 o'clock in the morning was mainly constant and amounted
to 90-96%. The flux of solar radiation increasing, there occurs
the heating[+]of the hothouse, the temperature difference of dry
and wet thermometers heightens, and at 1 o'clock p.m. the relative humidity of the air in the solar hothouse becomes less
stable and varies from 60 to 80%. Relative humidity is higher
in places with vegetation than in barren areas. The humidity
regime in hothouses seems to be affected also by the transpiration of plants. Maximum humidity of the air is observed near
the surface of the soil, i.e. at 7 o'clock a.m. when the air temperature in the hothouse is the smallest, and at 3 o'clock p.m.
when the soil has already accumulated the heat during the daytime. The least relative humidity of air is observed in the upper layers near the glassy roofing.

The psychrometers installed at the height of 2.5 m from the
soil show that at noon the relative humidity of air drops in the
direction from the accumulator to the glassy roofing from 80 to
70%.

It has been found that in the daytime when the active biological processes occur, the air humidity in the solar hothouse
does not correspond to the state of saturation.

The closed unheated film-covered hothouses situated in the
middle regions of this country, are characterized by high relative humidity. In the daytime it ranges from 85 to 90%, and at
night, from 94 to 100%, which may be due to the lower air temperature in hothouses than in the solar hothouses at daytime.
The geographical conditions of Turkmenistan are characterized
by high intensity of the solar radiation, which defines the dif-

ferenc in the value of relative humidity.

The conditions of heat- and mass-exchange in the studied solar hothouse have a specific qualitative picture. When the flux of solar radiation is directed to the surface of soil and leaves of the plants, the intensity of the convective and radiative heat exchange determines the process of evaporation from the soil and plants, and at the condition $t_{air} > t_{soil} = t$ the quantity of heat spent for evaporation is equal to that obtained by the soil from its surface. At the condition $t_{air} > t_{soil} < t$ the quantity of heat required for evaporation will decrease. At the same time, a part of the heat delivered to the soil increases. Hence, the temperature of the soil, and of the accumulating ground of the hothouse will depend on the hydrothermal state of the medium and the velocity of air motion in the hothouse.

In the nighttime the heated air masses from the soil surface, accumulator and the leaves are moving upward, reaching the glassy roofing, and cool down there. As a result, the relative air humidity in the upper layers increases. A further cooling of the air with high relative humidity leads to the vapor condensation, even resulting in "dripping". Such a phenomenon has been observed from 24.30 to 1.30 o'clock in the morning.

A particular feature of the studied solar hothouse is the fact that a design of the ground accumulator without a protecting wall makes it possible to use the accumulating ground on the shelves as the sowing area for the undersized plants. Thus, the total sowing area of the hothouse increases by 30-40%. On the shelves one can grow such plants as strawberry, lettuce, garden-cress, parsley, fennel, sorrel and so on.

As a result of investigating temperature regimes of hothouses with the ground heat accumulator we have established the terms of growing the agricultural cultures. By cultivating vegetables in hothouses, one can save up to 70% of fuel, and if the hothouse is used as a lemonary, then about 100% of fuel will be saved. The developing of deserts is a complicated problem, which is closely connected with the provision of these regions with drinking water for people, ponds for cattle, and watering of plants.

The cultivation of agricultural plants by using water-fresheners or other methods (such as delivering water with transport, gathering of precipitations) can't be profitable because of the high cost of such methods. A way out of this situation is in the construction of hothouses with a closed hydrologic cycle.

The solar hothouses with the closed hydrologic cycle, that might be used in remote deserted regions and sea coasts, represent the constructions with the vapor circulation by the system soil-plant-soil. Such hothouses differ from the solar hothouses by high airtightness, the presence of a tank for distilled water and the cooling system. The water evaporating from the soil surface, tank and the plant leaves, enriches the hothouse air with water vapors. After the condensation on the surface of a transparent protection wall or in the cooling system, the water is used again for watering of the plants.

In order to study temperature regimes with the closed hydrologic cycle we have elaborated and tested a number of operating models of hothouses at natural conditions. The temperature regime defined by the dry and wet thermometers in the hothouse and outside, for clear days, is shown in Fig. 2.

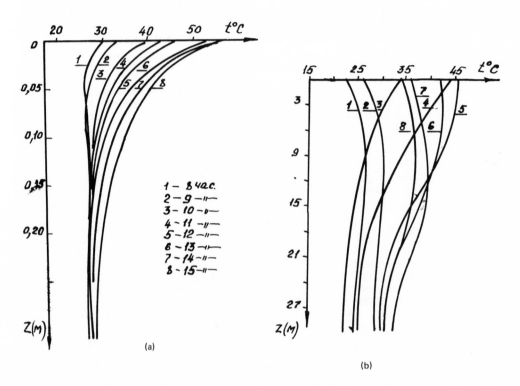

Fig. 2.   Variation of the soil temperature (a) and that of the distilled water (b) from the layer depth in the hothouse
with closed hydrologic cycle.

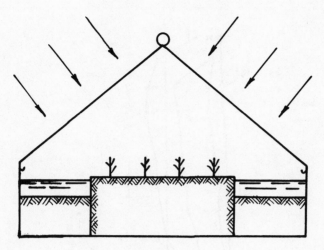

Fig. 3.   A scheme of the hothouse with closed hydrologic cycle.

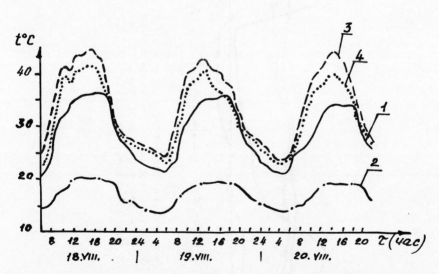

Fig. 4.   Diurnal temperature behavior of the external air (1,2)
and the air in the hothouse model with a closed hydrologic cycle (3,4) measured with the dry and wet thermometers,
respectively.

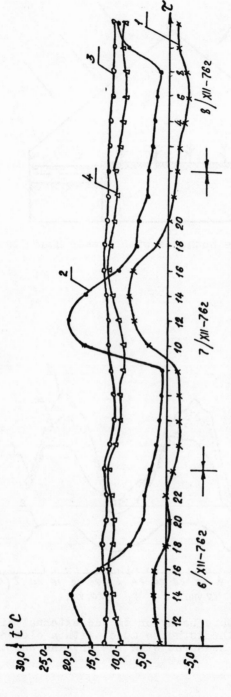

Fig. 5. Temperature regime of the air in the hothouse with closed hydrologic cycle (1 - temperature of the external air, 2,3,4 - temperature of the air, soil, distilled water in the hothouse, respectively).

After the sunrise the temperature in the hothouse sharply increases. When the temperature reaches 40°C the cooling system switches on. During cooling at noon the mean temperature of the hothouse air is equal to the mean diurnal value.

Along with the air temperature regime, of great importance for growing of plants is the temperature regime of soil and distilled water (Fig. 3).

In the daytime with the introduction of solar energy the surface of the soil is heated, and hence, the heat is accumulated in the soil, and at night, otherwise, the accumulated heat goes from the soil to the hothouse. Thus, in order to study theoretically the temperature distribution in the soil and air medium of solar hothouses with the closed hydrologic cycle, one can use the same approaches as for the solar hothouses with the accumulated solar energy. Using these considerations as the base, it is assumed that the temperature of the soil surface and of the presoil air layer changes sinusoidally in time with the same periods and amplitude. It is easy to allow for the additional surface of the heat exchanger in calculations, In the system of equations there are added the components of heat fluxes and the equation of heat balance for the heat exchange surface is included.

It is more complicated to allow for the influence of the tank with distilled water on the heat and humidity regimes of the hothouse. The authors of /1/ investigated a nonstationary operation of hothouse distillers. They have drawn a conclusion that the temperature field of the distilled water in the trough may be described by a solution of the differential equation for the solid-state heat conductivity. In this case a concept of the equivalent heat conductivity coefficient λ_{eq} is introduced, i.e. the coefficient of the solid state heat conductivity transmitting the heat flux of the same density as that transmitting the liquid layer under the action of the same temperature difference at the conditions of the convective heat transfer. Since the temperature pressure for the given water layer changes negligibly during twenty-four hours, and the temperature oscillations attenuate with depth, one can assume than λ_{eq} = const.

In this case the temperature distribution by the water thickness in the hothouse tank is described by the same conductivity equation as for the soil. It can be qualitatively confirmed by a comparison of curves represented in Fig. 3.

On the basis of testing results we have elaborated a project and realized a construction of the hothouse with the closed hydrologic cycle with 100 m² area. The hothouse represents a two-slope builsing with the north-south longitudinal axis. Inside the hothouse two tanks with distilled water are located along the whole length with the sowing area between them. The condensed water from the glass goes to the gutters, and in order to regulate the heat regime in the hothouse, to draw aside the heat, the water was poured off over the external glassy roofing (Fig.4). The temperature regime of the air in the hothouse with closed hydrological cycle is represented in Fig. 5.

The experiments showed that in the closed volume of the hothouse even when it is hot time, one can achieve the temperature regime necessary for growing of the plants.It is also possible to obtain the extra 3 l/m² of the distilled water, which is extremely useful especially for the regions suffering from the lack of the distilled water.

REFERENCES
1. Bairamov, R.,Toiliev,K.Izv.ANTSSR,ser.Fiz. i GN,1968,N 4.

THE ECONOMICS OF HEAT RECOVERY SYSTEMS FOR ANIMAL SHELTERS

B. S. LARKIN

Division of Mechanical Engineering
National Research Council
Ottawa, Canada

J. E. TURNBULL

Engineering Research Service
Research Branch, Agriculture Canada
Ottawa, Canada

ABSTRACT

In severe climates, livestock are kept in a building over the winter. Animals generate moisture and the building must be ventilated to control moisture. The heat required to heat the incoming air can be extracted from the exhaust air by a heat exchanger. The energy savings that result have been calculated for laying chickens housed in cages or growing-finishing pigs. The heat exchanger cannot always satisfy design requirements for the building during extremely cold weather but most farmers would accept short periods of substandard ventilation. In one of the cases considered the load factor of the system was low and the rate of return on investment was only average. In three cases with a higher load factor the savings were large.

INTRODUCTION

The two components of winter heat loss from an animal shelter are the conduction losses (through the structure) and the ventilation losses (heat required to bring the incoming ventilation air up to the room temperature). With well-insulated construction to minimize the conduction component, the ventilation component is the larger of the two; it depends on the ventilation rate and the outside temperature. The livestock generate considerable quantities of heat, sufficient to maintain the desired room temperature with adequate ventilation when outside temperatures are moderate. But, if the outside air temperature is too low, then supplementary heat is required. Otherwise, even the minimum acceptable ventilation rate will cause the room temperature to fall.

A conventional supplementary heat system uses electric heat or burns oil or natural gas. One alternative is a heat recovery system, using a heat exchanger to extract heat from the exhaust air to preheat the inlet air. Theoretically, it is simple to design and construct a heat exchanger that, when clean, will recover the required heat. In practice, heat recovery systems have not been used because heat exchanger performance falls off rapidly due to fouling in typical farm conditions.

An experimental installation in a poultry house has shown that a system consisting of filters and a thermosiphon heat exchanger is practical [1,2]. The filters remove most of the fouling and can be cleaned quickly and easily. The heat exchanger too must be cleaned, at longer intervals, and is designed for easy, convenient cleaning. This paper discusses the economics of such a system for

Contribution No. 650, from Engineering Research Service, Research Branch, Agriculture Canada, Ottawa, K1A 0C6. Prepared for the Seminar on Heat and Mass Transfer in Buildings, Dubrovnik, Yugoslavia, Aug. 1977.

typical farm applications. Although it is based on the thermosiphon heat exchanger, many of the points raised are relevant for other types of heat
exchangers.

CONVENTIONAL SUPPLEMENTARY HEAT SYSTEM

The usual ventilation system in northern U.S.A. and Canada consists of
several fans controlled by thermostats set in temperature steps so that the number of fans operating and hence the ventilation rate decrease with outside
temperature. The smallest fan is sized to give the minimum ventilation rate that
will control relative humidity during the coldest weather anticipated.

If supplementary heat is provided it is switched on and off by thermostat
to maintain the desired room temperature while the smallest fan is running.
Figure 1 shows how the deficit in the room heat balance increases as the outside
ambient decreases, for a typical set of conditions; this is the heat that should
be supplied by the supplementary heating system.

HEAT RECOVERY SYSTEM

If a heat recovery system is used it replaces the smallest fan and the supplementary heating system. One possible scheme for a 500-pig grower-finisher
unit is shown in Figure 5. Fresh air is ducted into the inlet side of the heat
exchanger, through the inlet fan and into an insulated duct in the attic from
which it is distributed throughout the building. Exhaust air enters the heat
exchanger compartment through filters, goes through the exhaust fan, the exhaust
side of the heat exchanger, then to outdoors.

Ideally, a thermostat-controlled supplementary heat system automatically
supplies just enough heat to maintain room temperature. A heat recovery system
behaves differently; the heat recovered depends upon the inlet conditions of the
two flows. Figure 1 shows the variation in heat recovery with outside temperature for a typical set of conditions. During mild winter weather the heat
exchanger recovers more heat than is required to maintain the room temperature.
In this case the room temperature will rise, and the next larger fan in the
ventilation system will switch on and off to control the temperature in the
building.

FREEZING

Another characteristic requires explanation. If the exhaust flow is cooled
too far then the saturated exhaust will deposit ice in the heat exchanger,
beginning at the coldest part of the exhaust section. If this is allowed to
continue the heat exchanger will block up entirely. Freezing imposes a natural
limit on heat recovery, as shown in Figure 1. One satisfactory method of control
is to install at the outlet from the exhaust side of the heat exchanger a
thermostat which switches the inlet fan to half speed just before freezing
starts. The cooling effect on the exhaust flow is then reduced. Any ice which
has formed is melted and, when the final exhaust air has warmed sufficiently,
the thermostat returns the inlet fan to full speed and normal operation resumes.
During very cold weather the inlet fan cycles between full and half speed.

The heat recovery may be limited either by the effectiveness of the heat
exchanger or by the freezing limit. In Figure 1 the extended dashed part of the
'heat recovery line' cannot be achieved in practice because of the freezing
limit.

At the point where either the freezing line or the heat recovery line
(whichever is lower) intersects the heat deficit line the heat balance is in
equilibrium. If the outside temperature falls lower, the heat recovery is insufficient to maintain design conditions in the building. A room thermostat
then cycles the exhaust fan between full speed and half speed, reducing the

average ventilation rate and allowing the humidity to rise above the design
level.

FOULING

In the system tested the main effect of fouling is to increase the flow
resistance of the filters. Secondary effects are to increase the flow resistance
and decrease the heat transfer coefficients in the heat exchanger. These effects
all tend to decrease the heat recovery and some loss of performance is inevitable
between cleanings. The cleaning routine must be a satisfactory compromise between the work involved and the heat recovery required.

In the system tested, the fouling in the filters over a period of two days
reduced the exhaust flow by 15%. As freezing is normally the controlling limit
on heat recovery, the possible heat recovery is reduced by 15%. The filters can
be cleaned in a few minutes with a vacuum cleaner. Over a period of 30 days,
fouling in the heat exchanger reduced heat recovery by 6%. The heat exchanger
is cleaned by removing the top cover of the exhaust section and washing the tube
banks with a water jet, taking at most 30 minutes. With this cleaning routine,
just before cleaning filters and heat exchanger after 30 days, the heat recovery
was about 79% of the "clean" heat recovery. However, this situation is not as
unfortunate as it may appear at first glance.

During most of the winter the outside air temperature is in the range where
79% of the "clean" heat recovery is sufficient to maintain the design minimum
ventilation rate. Whenever the weather is cold enough to require the maximum
heat recovery, the system could be cleaned more frequently (e.g. every day).
Similarly, during milder winter weather the intervals between cleanings can be
extended.

ENERGY REQUIREMENTS

The desirability of a heat recovery system depends on the severity of the
winter, the heat and moisture production of the livestock, the shelter construction and insulation, the cost of energy, etc. The procedure used here is to
select typical cases, define the basis for comparison and calculate the useful
heat recovery for these particular sets of conditions.

Conventional supplementary heating is normally designed to supply enough
energy to maintain the minimum ventilation rate down to the design outside
temperature. Usually the design temperature is slightly above the expected
minimum temperature. Occasional short periods of extreme cold result in substandard ventilation; this is accepted for the sake of reducing capital costs.
In this study -34°C has been taken as the design outside temperature.

Theoretically, a heat recovery system could also be designed to correspond
exactly to design conditions by variation of the heat exchanger parameters.
For convenience, the authors used the basic design of the experimental system
described in previous papers [1,2], varying only the number of banks of finned
tubing to give a close approximation to design requirements.

Heat recovery system specifications

        Filter - face velocity 30 m/min (100 ft/min)
               - filter mesh (approx.) 1.8 mm (0.07 in)

        Thermosiphon heat exchanger
               - effectiveness 0.40
               - face velocity through tube banks 2 m/s (400 ft/min)
               - pressure drop per tube bank 23 Pa (0.09 in water gage)
               - fin/tube surface ratio 20.3/1

```
        - tubes, outside diam. 25.4 mm (1.00 in)
                 inside diam. 22.9 mm (0.90 in)
                 spacing in fin banks 76.2 mm (3.0 in)
        - fins (pressed onto tubes), thickness 0.38 mm (0.015 in)
                                     depth 50.8 mm (2 in)
                                     spacing 4 mm (0.16 in)
```

CLIMATE

The location selected for this analysis was Saskatoon, Saskatchewan, representative of the coldest climates in which commercial Canadian livestock operations are common (see Table 1). Temperature records were searched to find, for each winter month, an example with average temperature close to the 30-year average temperature for that month. In this way an "average" winter was assembled and the number of hours at each temperature was counted for use in the analysis.

TABLE 1 - SASKATOON WINTER TEMPERATURES

	Average of daily maximums (°C)	Average of daily minimums (°C)
October	+11	- 1
November	- 1	-11
December	- 9	-19
January	-13	-24
February	- 9	-21
March	- 3	-14
April	+ 9	- 3

LAYING CHICKENS IN CAGES

Building specifications (CPS Plan 5212)[a]

- exterior length and width, 46.8 × 10.2 m (156×34 ft)
- floor to ceiling height, 2.95 m (9.8 ft)
- wall and ceiling insulation, 150 mm glass fiber (R-20)
- 10,000 hens capacity in triple-deck cages

Chicken population

- 10,000 white leghorns, average weight 1.8 kg (4 lb)
- heat and moisture production per chicken, from Ota et al [3]

	Day		Night	
	16°C	21°C	16°C	21°C
Room temperature				
Sensible heat, (W)	7.8	7.6	6.2	5.6
Vaporized moisture (g/h)	4.9	5.1	3.4	4.0

Case 1   Room temperature 16°C
                  Relative humidity 75%
                  Four thermosiphon tube banks

[a]Canada Plan Service, published by Canada Dept. of Agriculture, Ottawa, K1A 0C6 in cooperation with all provinces of Canada.

The Canadian Farm Building Code [4] recommends a continuous minimum ventilation rate of 0.23 ℓ/s (0.5 ft³/min) per laying hen. However, using removal of vaporized moisture [3] as the basis for ventilation design, the daytime minimum rate was calculated to be 0.14 ℓ/s (0.3 ft³/min) per chicken. This was the figure used in this analysis.

For 'day operation', figure 1a is the heat balance diagram. The 'heat deficit' line shows the supplementary heating rate calculated to maintain the design conditions defined above. It is obtained by subtracting the sensible heat produced by the chickens from the sum of the building heat loss and the heat required to bring the incoming air up to room temperature. This is explained in detail in [5]. The heat recovery line shows the heat recovered by the heat exchanger. The freezing limit shows the restriction on heat recovery due to freezing in the last tube bank of the heat exchanger.

A conventional supplementary heat system supplies just the heat required to maintain the minimum ventilation rate. This corresponds to the heat deficit line until the design outside temperature of -34°C is reached. If the outside temperature falls further the supplementary heat system runs continuously at full power.

A heat recovery system will recover heat as shown by the lower of the heat recovery line and the freezing limit. Until one of these lines crosses the heat deficit line, more heat is recovered than is necessary to achieve design conditions within the building. This results in more ventilation and relative humidity less than 75%, but no commercial value was assessed to this improvement. Over the range of outside temperature above the equilibrium point, the heat recovery system is given credit only for the heat required to maintain the minimum ventilation rate. At lower temperatures, when the heat recovery is insufficient to make up the heat deficit, the heat recovery system is given credit for all the heat recovered.

For the 'night operation', figure 1b shows the heat balance conditions. At night chickens give off less heat and moisture than during the day. Based only on humidity control, the minimum night ventilation rate should be 0.1 ℓ/s (0.21 ft³/min) per bird. In practice this variation between day and night is usually ignored. Similarly, in this analysis it will be assumed that the minimum ventilation rate and supplementary heat system designed for daytime conditions will be used unaltered at night.

As the chickens are generating less heat than in the daytime the supplementary heat system can only maintain the ventilation rate of 0.14 ℓ/s (0.3ft³/ min) down to an outside temperature of -27°C. If the temperature drops further the ventilation rate will fall. The chickens generate less moisture than in the daytime so that the relative humidity will not rise above 75% until the outside temperature is about -40°C. But, humidity is not the only criterion by which ventilation should be judged. The ventilation at low temperatures may be substandard from the point of view of ammonia and odors. As before the heat recovery system is given credit only for heat required to maintain the minimum ventilation rate until the equilibrium point is reached, after which it gets credit for all heat recovered. See Table 2 for the results of this analysis.

TABLE 2 - ANNUAL ENERGY SAVING - 10,000 HENS @ SASKATOON

	Case 1	Case 2
Design conditions	16°C, 75% RH	21°C, 0.14 ℓ/s-bird
Heat recovery	8,400 kWh	24,780 kWh
Energy consumption of heat exchanger fans	1,200 kWh	2,500 kWh
Capacity of equivalent supplementary heat system	19 kW	30 kW

Case 2   Room temperature 70°F
                    Design minimum ventilation rate 0.14 ℓ/s (0.3 ft³/min)
                       per bird
                    Five thermosiphon tube banks

The higher room temperature is of interest because it can increase feed
conversion efficiency. If the minimum ventilation rate is calculated to give a
relative humidity of 75% at design conditions, the ventilation rate is found to
be 0.12 ℓ/s (0.25 ft³/min) per hen, compared with 0.23 ℓ/s recommended in the
Canadian Farm Building Code. It is probably unwise to reduce the ventilation
rate too far from the point of view of odors and ammonia; 0.14 ℓ/s has been
selected for this analysis, giving the heat balance diagrams in Figure 2. At
night, if the outside temperature falls below -22°C the heat recovery is not
sufficient to maintain 75% R.H. See Table 2 for details of the energy saving.

GROWING-FINISHING PIGS

It is assumed that the cleaning work load will be the same as in the chicken
house. This has not been justified experimentally. The system will probably foul
up less quickly than in a chicken house but the fouling may be more difficult to
remove, and these opposed effects were assumed to cancel each other.

Building specifications (CPS Plan 3028)[a]

- exterior length and width, 10.8 × 31.8 m (36×104 ft)
- floor to ceiling height 2.6 m (8.5 ft)
- wall insulation 90 mm (R-12) glass fiber
- ceiling insulation 150 mm (R-20) glass fiber
- foundation perimeter insulation 50 mm (R-8) polystyrene board

Pig population

- 500 growing and finishing pigs weight range 22-90 kg (54 kg ave.)
- heat and moisture production per pig, from Bond et al [6]

Room temperature	16°C	21°C
Sensible heat (W)	88	70
Vaporized moisture (g/h)	84	100

Case 3   Room temperature 16°C
                    Relative humidity 75%
                    Six thermosiphon tube banks

The winter minimum design flow rate is 2.5 ℓ/s-pig (5.3 ft³/min-pig);
separate data are not available for waking and sleeping pigs so that only an
average calculation is possible.
Figure 3 is the heat balance diagram. Pigs are considerably less self sufficient than chickens as far as heat is concerned. The equilibrium point is at
-25°C outside temperature. At lower temperatures the relative humidity in the
building will increase as shown in Figure 3. See Table 3 for details of energy
savings.

Case 4   Room temperature 70°F
                    Relative humidity 75%
                    Seven tube banks

---

[a]Canada Plan Service, published by Canada Dept. of Agriculture, Ottawa, K1A 0C6
in cooperation with all provinces of Canada.

TABLE 3 - ANNUAL ENERGY SAVING - 500 GROWING/FINISHING PIGS @ SASKATOON

	Case 3	Case 4
Design conditions	16°C, 75% RH	21°C, 75% RH
Heat recovery	30,480 kWh	48,970 kWh
Energy consumption of heat exchanger fans	3,180 kWh	5,290 kWh
Capacity of equivalent supplementary heat system	22.8 kW	31.4 kW

As in the case of chickens a higher room temperature is of interest because it can give better feed conversion efficiency. The minimum ventilation rate is 2.0 ℓ/s (4.3 ft^3/min) per pig.

Figure 4 is the heat balance diagram. The equilibrium point is at an outside temperature of -24°C. The increase in humidity at lower temperatures is shown in Figure 4. See Table 3 for details of energy saving.

COST COMPARISON

Capital costs and running costs vary considerably around the world, depending on the energy sources available. Also, any comparisons are very sensitive to future increases in energy costs. Thus an analysis can only be very general and must be interpreted in the light of local conditions. The calculations are simple so that it is easy to substitute local costs in the illustrations.

Electrical energy is widely used and is the most convenient for comparing heating-ventilating costs. Installation of electric heating was estimated to cost $100/kW, and electric energy was charged at $0.02/kWh with a demand charge of $2/kW for each of the four winter months.

The cost of the heat exchanger has been estimated by comparison with a heat exchanger made by Q-Dot Corp., Dallas, Texas. This is basically similar to the experimental NRC unit but with some detail differences.

Labor for cleaning the heat recovery system was charged at $5.00/h.

TABLE 4 - COST COMPARISON - ELECTRIC SUPPLEMENTARY HEAT SYSTEM
vs. HEAT RECOVERY SYSTEM

	Laying chickens		Growing/finishing pigs	
	Case 1	Case 2	Case 3	Case 4
(1) Room temperature	16°C	21°C	16°C	21°C
(2) Design standard	75% RH	0.14 ℓ/s	75% RH	75% RH
CAPITAL COSTS ($)				
(3) Electric supplementary heat system	1900	3000	2280	3140
(4) Step 1 fan in conventional system	200	200	200	200
(5) Cost of heat recovery system	4200	4500	4240	3650
(6) Additional cost of heat recovery system = (5)-(4+3)	$2100	1300	1760	310

TABLE 4 (Continued)

	Laying chickens		Growing/finishing pigs	
	Case 1	Case 2	Case 3	Case 4
(7) Energy cost for electric heat	168	495	610	980
(8) Demand charge	144	240	184	250
(9) Energy cost for heat exchanger fans	24	50	64	106
(10) Cost of labour for cleaning heat recovery system @ $5/hr	100	150	150	150
(11) Operating cost saving due to heat recovery system = (7)+(8)-(9+10)	$188	535	580	974

DISCUSSION

The results show the importance of the load factor of the heat recovery system. The less self-sufficient in heat production the livestock are the more attractive a heat recovery system becomes. Factors which increase the economic advantage of a heat recovery system are

a) High room temperature. Apart from considerations of feed efficiency, some livestock require a high room temperature.

b) High standard of ventilation. If, in fact, the usual flow rate of 0.23 ℓ/s (0.5 ft^3/min) is necessary, rather than the figure of 0.14 ℓ/s (0.3 ft^3/min) used here, the economics of the heat recovery system would be much enhanced.

c) Low standard building insulation. Heat recovery systems may be attractive for old buildings with inadequate insulation.

d) Low winter temperatures. Saskatoon has a very cold winter. In less severe climates a heat recovery system will be less attractive.

e) The cost of energy. There are some areas where electricity already costs significantly more than $0.02/kWh as assumed here. The cost of energy will probably continue to increase; a heat recovery system could be viewed as protection against future high prices.

High room temperature alone does not make a heat recovery system desirable. Farrowing barns for sows and their litters require high temperatures but very little ventilation. Brooding chickens need only minimum ventilation at the time when the building must be kept at a high temperature. In neither case is a heat recovery system attractive. Similarly a requirement for high ventilation does not call for heat recovery if the building temperature is low.

In areas where the power supply is unreliable a heat recovery system has the advantage that the standby generator required is an order of magnitude smaller than would be required for an electric heating system.

The analysis reported here is conservative in some respects. No credit has been taken for the heating effect of the two heat exchanger fans. All of the electrical energy supplied to the inlet fan is useful heat input to the building and about 40% of the energy supplied to the exhaust fan is recovered. This amounts to about 1½ kW while the fans are running. The power input to the exhaust fan also raises the freezing limit slightly.

It was assumed the freezing limit occurs when the temperature of the last tube bank, as calculated in the heat exchanger computer program, reaches 0°C. Experimental evidence appears to show that ice blockage does not occur until this temperature is lower than 0°C. It is not clear if this discrepancy is due to an approximation in the computer program or whether the assumption of 0°C as the critical temperature is incorrect. In either case there is reason to believe that the freezing limits given here are pessimistic.

One important consideration is the durability of the heat recovery system. The filters, fans and thermostat controls should present no special problems. The thermosiphon heat exchanger is comparatively novel. It is a very simple device; research and development have shown that it can be made to operate satisfactorily for long periods. One possible problem is corrosion due to contamination such as ammonia in the exhaust air. The cost estimates for the heat exchanger are based on all-aluminum construction. In the experimental work there was no significant corrosion of the aluminum fins after three winters so that aluminum should be a satisfactory material.

The design of the heat recovery system has not been optimized. Changes in face velocity, fin pitch, etc., may improve the economics of the system. Use of a larger filter area would reduce the cleaning work load.

No reduction in performance due to fouling has been assumed. This is equivalent to assuming that the system will be cleaned as frequently as necessary to maintain a close approximation to "clean" performance.

CONCLUSIONS

Chickens, case 1. Room temperature 16°C, relative humidity 75%.

The heat recovery system can recover sufficient heat during the day to fulfill design requirements. At night, the daytime ventilation rate cannot be maintained when outside temperatures are below -27°C but the relative humidity does not exceed 75%. The saving in operating costs (with the assumptions used here) does not justify the extra capital cost of a heat recovery system.

Chickens, case 2. Room temperature 21°C, minimum ventilation 0.14 ℓ/s.

During the day a heat recovery system can recover sufficient heat to fulfill design requirements. At night the design ventilation rate cannot be maintained when outside temperatures are less than -22°C, although the relative humidity does not exceed 75%. The extra capital cost of a heat recovery system is justified by the saving in annual operating cost.

Pigs, case 3.   Room temperature 16°C, relative humidity 75%
and case 4.   Room temperature 21°C, relative humidity 75%

A heat recovery system cannot recover sufficient heat to maintain the design standard of ventilation during the colder parts of the winter (below -25°C). In an average Saskatoon winter there are about 200 hours below -29°C, giving a significant increase in humidity. A dual system could be used - a heat recovery system to recover most of the heat required and a conventional supplementary heat system which will operate only when the heat recovery is insufficient. The annual energy input of the conventional system would be small. Probably most farmers would accept a reduced standard of ventilation some of the time rather than go to the complication and expense of a dual system. The large saving in energy cost makes a heat recovery system very attractive.

REFERENCES

1. Larkin, B.S., Turnbull, J.E., and Gowe, R.S., 1975. Thermosiphon Heat Exchanger for Use in Animal Shelters. Can. Agric. Eng. 17:2, pp. 85-89 December.

2. Larkin, B.S., and Turnbull, J.E., 1976. Effect of Poultry Dust on Performance
    of a Thermosiphon Heat Recovery System. Can. Agric. Eng. (in press).

3. Ota, H., and McNally, E.H., 1961. Poultry Respiration Calometric Studies of
    Laying Hens. ARS 43-43, Agricultural Research Service, U.S. Department of
    Agriculture.

4. Standing Committee on Farm Building Standards, 1975. Canadian Farm Building
    Code. NRCC No. 13992, National Research Council, Ottawa, K1A OR6.

5. Turnbull, J.E., and Bird, N.A., 1976. Conefinement Swine Housing, Pub. 1451,
    Agriculture Canada, Ottawa, K1A OC6.

6. Bond, T.E., Kelly, C.F., and Heitman, H.Jr., 1959. Hot House Air Conditioning
    and Ventilation Data. Trans. Amer. Soc. Agr. Eng., Vol. 2, No. 1, pp. 1-4.

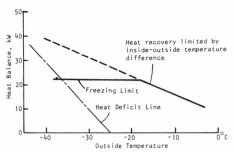

Fig. 1a   Heat Balance Diagram, Case 1 Chickens Daytime

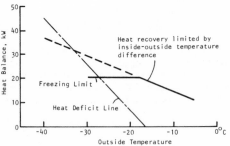

Fig. 1b   Heat Balance Diagram, Case 1 Chickens Nighttime

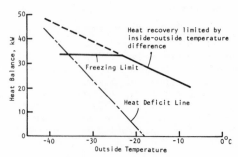

Fig. 2a   Heat Balance Diagram, Case 2 Chickens Daytime

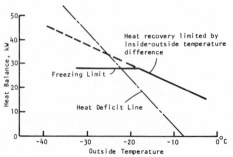

Fig. 2b   Heat Balance Diagram, Case 2 Chickens Nighttime

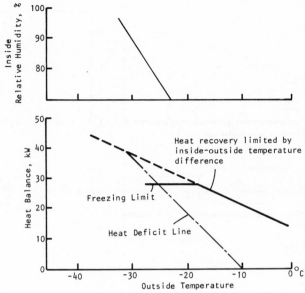

Fig. 3   Heat Balance Diagram, Case 3 Pigs

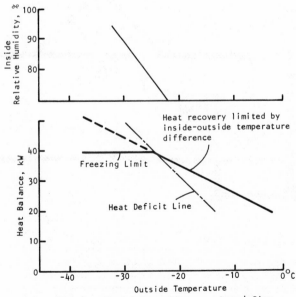

Fig. 4   Heat Balance Diagram, Case 4 Pigs

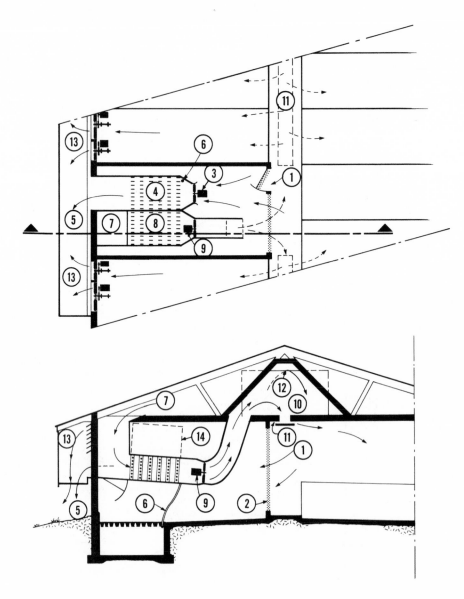

Fig.5 Proposed arrangement of thermosiphon heat recovery system in a swine finishing barn (500 hogs, CPS Plan 3428)

1. warm room air to dust filter ②
2. dust filter 2 x 2m, includes access door
3. exhaust fan to heat exchanger ④, 1200 1/s @ 13 Pa static
4. heat exchanger cooling section, 6 tube banks 0.9 x 0.7m
5. cooled exhaust to outdoors
6. condensate drain at lower corner, hose to manure trench
7. cold air from ventilated attic to ⑧

8. heat exchanger warming section, 6 tube banks **0.9** x 0.7m
9. intake fan, to ⑩
10. insulated attic duct, to ⑪
11. adjustable baffled air inlet slot
12. summer air inlet doors 2.4 x 1.2 into ⑩ , both gables
13. warm weather exhaust fans
14. access covers hinge up for washing ④ and ⑧

538